PHYSICS OF p–n JUNCTIONS AND SEMICONDUCTOR DEVICES

FIZIKA ELEKTRONNO-DYROCHNYKH PEREKHODOV I POLUPROVODNIKOVYKH PRIBOROV

ФИЗИКА ЭЛЕКТРОННО-ДЫРОЧНЫХ ПЕРЕХОДОВ И ПОЛУПРОВОДНИКОВЫХ ПРИБОРОВ

PHYSICS OF p-n JUNCTIONS AND SEMICONDUCTOR DEVICES

Edited by

S. M. Ryvkin and Yu. V. Shmartsev

A. F. Ioffe Physicotechnical Institute
Academy of Sciences of the USSR
Leningrad, USSR

Translated from Russian by
Albin Tybulewicz
Editor, *Soviet Physics - Semiconductors*

⊂⁄♭ Springer Science+Business Media, LLC • 1971

The original Russian text, published by Nauka Press in Leningrad in 1969
for the A. F. Ioffe Physicotechnical Institute of the Academy of Sciences of
the USSR, has been corrected by the editors for the present edition. The
English translation is published under an agreement with Mezhdunarodnaya
Kniga, the Soviet book export agency.

**Физика электронно-дырочных переходов
и полупроводниковых приборов**

Library of Congress Catalog Card Number 72-128510

ISBN 978-1-4757-1234-6   ISBN 978-1-4757-1232-2 (eBook)
DOI 10.1007/978-1-4757-1232-2
© 1971 Springer Science+Business Media New York
Originally published by Consultants Bureau, New York in 1971.

CONTENTS

CONTENTS   vii

DIFFERENTIAL RESISTANCE AND DIFFUSION CAPACITANCE OF JUNCTIONS IN p−n−n+ STRUCTURES AT HIGH INJECTION LEVELS

D. A. Aronov and Ya. P. Kotov

Physicotechnical Institute
Academy of Science of the Uzbek SSR, Tashkent

An analysis is made of the current and frequency dependences of the impedance of p−n and n−n+ junctions in a p−n−n+ structure (an expression for this impedance has been derived earlier by the present authors [1]). The analysis is carried out for high injection currents through the structure, when it is necessary to take into account the drift component of the current in the base and the injection leakage through the contacts. The impedance of the junctions is represented by an equivalent circuit in the form of a differential resistance and diffusion capacitance, connected in parallel. An investigation is made of the influence of the leakage on the frequency characteristics of the junctions in the case of low and high rates of leakage through the contacts, compared with the carrier combination velocity in the base. Expressions are deduced for the low-frequency differential resistance and diffusion capacitance whose values can be used, in some cases, to determine the length of the base and the lifetime of nonequilibrium carriers in the structure. It is shown that strong deviations of the frequency dependence of the junction impedance from the suggested law are possible at high frequencies. Thus, for example, at low rates of leakage in short diodes at moderate frequencies, the junction capacitances are frequency-independent and the resistances are inversely proportional to the square of the frequency. It is also shown that the differential resistance of the junctions is inversely proportional to the current for any ratio of the leakage rate and the carrier recombination velocity in the base. The junction capacitance is proportional to the current only in some cases: for example, at high leakage rates, the junction capacitance is proportional to the square root of the current. The results obtained can be used in an analysis of the reactive and active (resistive) components of the impedance of the bulk of a diode.

The present paper gives an analysis of the expressions derived in [1] for the impedance of a junction in a p−n−n+ structure. The aim is to determine the dependences of the active (resistive) and reactive components of the impedance on the forward-current density, the alternating-signal frequency, and the base length for various values of the ratio of the recombination velocity in the base and the corresponding velocity in the low-resistivity regions.

Representing Eqs. (13) and (14) from [1] in the form

$$Z_{p-n}^{-1} = R_{p-n}^{-1} + i\omega C_{p-n}, \quad Z_{n-n+}^{-1} = R_{n-n+}^{-1} + i\omega C_{n-n+}, \tag{1}$$

we obtain the following expressions for the differential resistance and diffusion capacitance of the junctions (the meanings of the symbols are the same as in [1]):

$$R_{p-n}^{-1} = \frac{e^2(b+1)p_0(0)}{kT} \frac{1}{\left(bV_p^* + \frac{D_1}{L_p}\right)^2 + \left(\frac{D_2}{L_p}\right)^2} \left\{ \frac{D_2}{L_p}\left[(V_p^* + V_n^*)\frac{D_4}{L_p} + \omega D\right] + \left(bV_p^* + \frac{D_1}{L_p}\right)\left[V_p^* V_n^* + \frac{D_3}{L_p}(V_p^* + V_n^*)\right] + \left(\frac{D}{L_p}\right)^2 \right\},$$

(2)

$$C_{p-n} = \frac{e^2(b+1)p_0(0)}{kT\omega} \frac{1}{\left(bV_p^* + \frac{D_1}{L_p}\right)^2 + \left(\frac{D_2}{L_p}\right)^2} \left\{ \left(bV_p^* + \frac{D_1}{L_p}\right)\left[(V_p^* + V_n^*)\frac{D_4}{L_p} + \omega D\right] - \frac{D_2}{L_p}\left[V_p^* V_n^* + \frac{D_3}{L_p}(V_p^* + V_n^*) + \left(\frac{D}{L_p}\right)^2\right] \right\},$$

(3)

where

$$D_1 = 2D \frac{\varphi \, \mathrm{sh}\,\varphi \,(b\,\mathrm{ch}\,\varphi + \cos\Psi) + \Psi \sin\Psi\,(\mathrm{ch}\,\varphi + b\cos\Psi)}{\mathrm{ch}\,2\varphi - \cos 2\Psi},$$

(4a)

$$D_2 = 2D \frac{\Psi \, \mathrm{sh}\,\varphi \,(b\,\mathrm{ch}\,\varphi + \cos\Psi) - \varphi \sin\Psi\,(\mathrm{ch}\,\varphi + b\cos\Psi)}{\mathrm{ch}\,2\varphi - \cos 2\Psi},$$

(4b)

$$D_3 = D \frac{\varphi \, \mathrm{sh}\,2\varphi + \Psi \sin 2\Psi}{\mathrm{ch}\,2\varphi - \cos 2\Psi}, \quad D_4 = D \frac{\Psi \, \mathrm{sh}\,2\varphi - \varphi \sin 2\Psi}{\mathrm{ch}\,2\varphi - \cos 2\Psi},$$

(4c)

$$\varphi = \bar{\varphi}\frac{W}{L_p} = \frac{W}{L_p}\sqrt{\frac{\sqrt{1+\omega^2\tau_p^2}+1}{2}}, \quad \Psi = \bar{\Psi}\frac{W}{L_p} = \frac{W}{L_p}\sqrt{\frac{\sqrt{1+\omega^2\tau_p^2}-1}{2}}.$$

(5)

Expressions analogous to Eqs. (2)-(4) can be derived also for the n−n⁺ junction [1].

The formulas (2) and (3) differ from those usually employed (see, for example, [2, 3]) in their allowance for the drift component of the current and for the carrier recombination in the low-resistivity p- and n⁺- type regions (allowance for these factors sometimes results in different dependences of the relevant quantities on the current and frequency).

The expressions (2) and (3) are fairly cumbersome and, therefore, we shall consider some of the more interesting limiting cases under dc conditions.

1. The range of currents in which the carrier recombination in the base is dominant corresponds to the conditions

$$V_p^*, \; V_n^* \ll \frac{D}{L_p}\,\mathrm{th}\,\frac{W}{L_p},$$

(6)

which allows us to drop the terms with V_p^* and V_n^* in Eqs. (2) and (3). This case corresponds to the Hall region $[j_0 \propto \exp(eV/2kT)]$ in the current–voltage characteristic [4, 5]. The carrier densities at the boundaries, given by

$$p_0(0) = p_0(W)\frac{b\,\mathrm{ch}\,\frac{W}{L_p}+1}{\mathrm{ch}\,\frac{W}{L_p}+b} = \frac{j_0 L_p}{2beD_p}\frac{b\,\mathrm{ch}\,\frac{W}{L_p}+1}{\mathrm{sh}\,\frac{W}{L_p}},$$

(7)

are proportional to the current and it follows from Eqs. (2) and (3) that the differential conductance and the diffusion capacitance of the junctions increase linearly with increasing current density. The frequency dependences are given by the formulas

$$R_{p-n} = \frac{L_p}{4beu_p p_0(0)} \frac{b^2(\mathrm{ch}\,2\varphi + \cos 2\Psi) + 4b\,\mathrm{ch}\,\varphi\cos\Psi + 2}{\varphi\,\mathrm{sh}\,\varphi\,(b\,\mathrm{ch}\,\varphi + \cos\Psi) - \Psi \sin\Psi\,(\mathrm{ch}\,\varphi + b\cos\Psi)},$$

(8)

$$C_{p-n} = \frac{4beu_p p_0(0)}{\omega L_p} \frac{\Psi\,\mathrm{sh}\,\varphi\,(b\,\mathrm{ch}\,\varphi + \cos\Psi) + \varphi \sin\Psi\,(\mathrm{ch}\,\varphi + b\cos\Psi)}{b^2(\mathrm{ch}\,2\varphi + \cos 2\Psi) + 4b\,\mathrm{ch}\,\varphi\cos\Psi + 2}.$$

(9)

Since at low frequencies $\Psi \ll 1$ (even in the case of long diodes, the ratio W/L_p does not exceed 5-15 [6]), we find that Eqs. (8) and (9) yield

$$R_{p-n}^0 = R_{n-n^+}^0 \simeq \frac{kT}{ej_0}, \tag{8a}$$

$$C_{p-n}^0 = \frac{ej_0\tau_p}{2kT}\left[1 + \frac{W}{L_p}\frac{b + \mathrm{ch}\dfrac{W}{L_p}}{\mathrm{sh}\dfrac{W}{L_p}\left(b\,\mathrm{ch}\dfrac{W}{L_p}+1\right)}\right] \simeq \frac{ej_0\tau_p}{2kT}\begin{cases} 2 \text{ when } W \ll L_p, \\ 1 \text{ when } W \geqslant 3L_p. \end{cases} \tag{9a}$$

According to Eq. (8a), the differential (low-frequency) resistance at a given temperature is governed solely by the value of the current and is independent of the base length and carrier lifetime. The diffusion capacitance is proportional to τ_p for short and fairly long diodes. It follows that the carrier lifetime can be determined by measuring C_{diff} in structures of this type.

At high frequencies, when $\mathrm{ch}\,\varphi \gg b$ (W/L_p not too small), we obtain from Eqs. (8) and (9)

$$\frac{1}{R_{p-n}^\infty} = \omega C_{p-n}^\infty = \frac{ej_0}{bkT}\sqrt{\frac{\omega\tau_p}{2}}\,\frac{b\,\mathrm{ch}\left(\dfrac{W}{L_p}\right)+1}{\mathrm{sh}\left(\dfrac{W}{L_p}\right)} \simeq \frac{ej_0\sqrt{\omega}}{kT}\begin{cases} \dfrac{\tau_p}{W}\sqrt{\dfrac{b+1}{b}D_p} & \text{when } W \ll L_p, \\[2mm] \sqrt{\dfrac{\tau_p}{2}} & \text{when } W \geqslant 3L_p, \end{cases} \tag{8b}$$

i.e., as in the case of a p–n junction with an ohmic rear contact and an infinite base [2], the resistance and diffusion capacitance of the junctions decrease with increasing frequency, in accordance with the law $\omega^{-1/2}$. In this case, $R_{p-n}^{-1} \propto \tau_p/W$ and $C_{p-n} \propto \tau_p/W$ if the diode is short, but $R_{p-n}^{-1} \propto \tau_p^{1/2}$ and $C_{p-n} \propto \tau_p^{1/2}$ (the conductance and diffusion capacitance are independent of the base dimensions), if the diode is long. In the case of long diodes, the high-frequency junction impedance (for a given current j_0) is governed, in contrast to [2], not only by the diffusion of carriers into the n-type region but also by the carrier lifetime.

We note that for short-base diodes at moderately high frequencies ($\varphi \simeq \Psi < 1$), the expression for C_{diff} is again given by Eq. (9a). The value of R_{diff} for $W \ll L_p$, $\omega\tau_p \gg 1$, and $\varphi \simeq \Psi \ll 1$ is

$$R_{p-n} \simeq \frac{kT}{ej_0}\left(1 + \frac{2b-1}{12b}\frac{\omega^2\tau_p W^2}{D_p}\right)^{-1}, \tag{8c}$$

which may differ considerably from the low-frequency value given by Eq. (8a). In a relatively narrow range of frequencies, R_{diff} is inversely proportional to the carrier lifetime, the square of the frequency, and the base length [if the second term in the parentheses of Eq. (8c) is larger than the first term]. Moreover, the expression for the differential resistance includes a dependence on the ratio of the electron and hole mobilities and on the carrier diffusion coefficient.

2. In the range of currents in which the recombination velocity (relative to the velocity in the base) is high in one region (for example, V_p^*) and low in another region (V_n^*), the following conditions are satisfied*

$$bV_p^* \gg \frac{D_1}{L_p}, \quad \frac{D_3}{L_p}V_p^* \gg \left(\frac{D}{L_p}\right)^2, \quad \frac{D_4}{L_p}V_p^* \gg \omega D, \quad V_n^* \ll \frac{D_1}{L_p}. \tag{10}$$

---

*Using Eqs. (4) and (5), we can show that the third inequality in Eq. (10) is always more stringent than the second; it is also more stringent than the first inequality at low ($\omega\tau_p < 1$) and high ($\omega\tau_p > 1$) frequencies. In the last case, when $\mathrm{ch}\,\varphi \gg b$, the conditions in Eq. (10) reduce to two inequalities:

$$V_p^* \gg \left(\frac{D}{L_p}\right)\sqrt{2\omega\tau_p}, \quad V_n^* \ll \left(\frac{D}{L_p}\right)\sqrt{\frac{1}{2}\omega\tau_p}.$$

In this range of currents, we have

$$\left.\begin{aligned}
p_0(0) &= \frac{j_0 L_p}{2eD_p}\,\mathrm{th}\,\frac{W}{L_p} \propto j_0, \\[2mm]
p_0(W) &= \sqrt{\frac{j_0 n_n\left(b + \mathrm{ch}\,\dfrac{W}{L_p}\right)}{(b+1)\,e\vartheta_p^*\,\mathrm{ch}\left(\dfrac{W}{L_p}\right)}} \propto j_0^{1/2}
\end{aligned}\right\} \tag{11}$$

and $j_0 \propto \exp(eV/ckT)$ or $j_0 \propto V^4$, depending on whether the nonequilibrium carrier density in the base varies monotonically or has a minimum [5].

In this case, the expressions for the junction resistances and capacitance are of the form:

$$R_{p-n} = \frac{L_p}{2eu_p p_0(0)}\,\frac{\varkappa\mathrm{ch}\,2\varphi - \cos 2\Psi}{\varphi\,\mathrm{sh}\,2\varphi + \Psi\sin 2\Psi}, \tag{12}$$

$$C_{p-n} = \frac{2eu_p p_0(0)}{\omega L_p}\,\frac{\Psi\,\mathrm{sh}\,2\varphi - \varphi\sin 2\Psi}{\mathrm{ch}\,2\varphi - \cos 2\Psi}, \tag{13}$$

$$R_{n-n^+} = \frac{kT n_n}{2e^2(b+1)\vartheta_p^* p_0^2(W)}\,\frac{\mathrm{ch}\,2\varphi + \cos 2\Psi + 4b\,\mathrm{ch}\,\varphi\cos\Psi + 2b^2}{\mathrm{ch}\,2\varphi + \cos 2\Psi + 2b\,\mathrm{ch}\,\varphi\cos\Psi}, \tag{14}$$

$$C_{n-n^+} = \frac{4e^2 b(b+1)\vartheta_p^* p_0^2(W)}{kT n_n \omega}\,\frac{1}{\mathrm{ch}\,2\varphi + \cos 2\Psi + 4b\,\mathrm{ch}\,\varphi\cos\Psi + 2b^2} \times$$
$$\times\left\{\mathrm{sh}\,\varphi\sin\Psi + \frac{D}{bV_p^* L_p}[\Psi\,\mathrm{sh}\,\varphi(\mathrm{ch}\,\varphi + b\cos\Psi) + \varphi\sin\Psi(b\,\mathrm{ch}\,\varphi + \cos\Psi)]\right\}. \tag{15}$$

According to Eqs. (11)–(14), the conductances of both junctions and the capacitance of the p–n junction increase linearly with the current. However, the capacitance of the n–n$^+$ junction is proportional to the current only at low and moderately high frequencies; at very high frequencies, when the second term in the braces of Eq. (15) is dominant, we have $C_{n-n^+} \propto j_0^{1/2}$.

At low frequencies and — in the case of short diodes — at high frequencies ($\varphi \simeq \Psi \ll 1$), we obtain the following expressions from Eqs. (12), (13), and (15) using Eq. (11):

$$R_{p-n}^0 = 2R_{n-n^+}^0 = \frac{kT}{ej_0}, \tag{12a}$$

$$C_{p-n}^0 = \frac{ej_0\tau_p}{2kT}\left(1 - \frac{2W}{L_p}\,\mathrm{sh}^{-1}\frac{2W}{L_p}\right), \tag{13a}$$

$$C_{n-n^+}^0 = \frac{bej_0\tau_p}{kT}\,\frac{W}{L_p}\,\frac{\mathrm{th}\left(\dfrac{W}{L_p}\right)}{b + \mathrm{ch}\left(\dfrac{W}{L_p}\right)}. \tag{15a}$$

It follows from Eqs. (13a) and (15a) that, when $W \ll L_p$, we have

$$C_{p-n}^0 = \frac{b+1}{3b}\,C_{n-n^+}^0 = \frac{(b+1)ej_0}{6bkT}\,\frac{W^2}{D_p}. \tag{16}$$

We can easily see that the diffusion capacitance of a p–n junction for $W \geq 3L_p$ is independent of the base length and proportional to the carrier lifetime, whereas when $W \leq 0.1L_p$, this capacitance is independent of the lifetime and proportional to the square of the base length. Consequently, measurements of C_{p-n}^0 can be used to determine the carrier lifetime if the base of the structure is long, and the length of the base if it is short. This can be done also by measuring $C_{n-n^+}^0$; however, in practice, the lifetime τ_p cannot be determined from $C_{n-n^+}^0$. Comparison of Eqs. (13a) and (14a) shows that the greatest contribution to the diffusion capacitance of short diodes is made by the n–n$^+$ junction, but in the case of diodes with sufficiently long bases ($W/L_p \geq 4$ for germanium and silicon, and $W/L_p \geq 6$ for gallium arsenide and silicon carbide), the p–n junction plays the dominant role. Physically, this is due to the fact that when $W < L_p$ the diffusion capacitance is determined by the more highly mobile injected carriers, but when

$W/L_p \geq 4$ the diffusion capacitance is governed by the ratio of the electron and hole mobilities and by the ratio of the leakage rates through the contacts.

At high frequencies ($\omega \tau_p \gg 1$), when $\mathrm{ch}\, 2\varphi \simeq \mathrm{sh}\, 2\varphi \gg 1$, we obtain the following expression for the p–n junction from Eqs. (11)-(13):

$$\frac{1}{R_{p-n}^{\infty}} = \omega C_{p-n}^{\infty} = \frac{ej_0}{kT}\sqrt{\frac{\omega \tau_p}{2}}\,\mathrm{th}\,\frac{W}{L_p} \simeq \frac{ej_0}{kT}\sqrt{\frac{\omega}{2}} \begin{cases} \dfrac{W}{\sqrt{\dfrac{2b}{b+1}D_p}} & \text{when } W \ll L_p, \\[2mm] \sqrt{\tau_p} & \text{when } W \geq 3L_p. \end{cases} \tag{12b}$$

If we also assume that $\mathrm{ch}\,\varphi \gg b$, we obtain from Eqs. (11), (14), and (15)

$$R_{n-n+}^{\infty} = \frac{kT}{2ej_0}\frac{\mathrm{ch}\left(\dfrac{W}{L_p}\right)}{b + \mathrm{ch}\left(\dfrac{W}{L_p}\right)}, \tag{14b}$$

$$C_{n-n+}^{\infty} = \frac{bej_0}{kT}\left[\sqrt{\frac{eD_p n_n\left(b + \mathrm{ch}\dfrac{W}{L_p}\right)}{bj_0\vartheta_p^* \omega\,\mathrm{ch}\left(\dfrac{W}{L_p}\right)}} + \frac{4\left(b + \mathrm{ch}\dfrac{W}{L_p}\right)}{\omega\,\mathrm{ch}\left(\dfrac{W}{L_p}\right)}e^{-\varphi}\sin\varphi\right]. \tag{15b}$$

It is evident from Eq. (12b) that the resistance and capacitance of the p–n junction have the frequency dependence $\bar{\omega}^{-1/2}$, like ideal junctions [see Eq. (8b)].

However, in this case, $R_{p-n}^{-1} \propto W$ and is independent of τ_p in the case of short bases but $R_{p-n}^{-1} \propto \tau^{1/2}$ and is independent of W for long bases, whereas, according to Eq. (8b), $R_{p-n}^{-1} \propto \tau_p/W$ for $W \ll L_p$ and is independent of W and τ_p when $W \geq 3L_p$. It follows from Eqs. (14b) and (15b) that the resistance of the n–n$^+$ junction is independent of the frequency but the capacitance is proportional to $\omega^{-1/2}$ when the second term in the square brackets of Eq. (15b) is much smaller than the first. However, if these two terms are comparable, the frequency dependence of C_{n-n+} may deviate strongly from the $\omega^{-1/2}$ law.

These results can be easily generalized to the case when the recombination velocity in the p-type region is high and that in the n$^+$-type region is low. In this case, the roles of the junctions are interchanged: the formulas derived for the p–n junction now describe the n–n$^+$ junction (b is replaced by 1/b, and conversely).

3. The range of currents in which carrier recombination in the low-resistivity regions predominates over recombination in the base satisfies the conditions under which we can drop those terms from Eqs. (2) and (3) which do not include the quantities V_p^* and V_n^*. This case corresponds to the $j_0 \propto V^2$ region in the current–voltage characteristic, and [5]

$$p_0(W) = \gamma p_0(0) = \sqrt{\frac{j_0 n_n}{(b+1)e\vartheta_p^*}}, \tag{17}$$

where

$$\gamma = \frac{V_n^*}{bV_p^*} = \left(\frac{\vartheta_n^* n_n}{b\vartheta_p^* p_n}\right)^{1/2}. \tag{18}$$

Retaining the largest terms ($\sim V_p^{*2}$, V_n^{*2}, $V_p^* V_n^*$) in Eqs. (2) and (3), we obtain the following expressions for the differential resistances of both junctions and the diffusion capacitance of the p–n junction:

$$R_{p-n} = R_{n-n+} = \frac{kT}{2ej_0}, \tag{19}$$

$$C_{p-n} = \frac{e}{kT} \sqrt{\frac{2ej_0 D_p p_n}{\vartheta_n^* \tau_p}} \frac{1}{\omega\,(\mathrm{ch}\,2\varphi - \cos 2\Psi)}\,[\Psi\,\mathrm{sh}\,2\varphi - \varphi \sin 2\Psi + 2\gamma\,(\varphi\,\mathrm{ch}\,\varphi \sin \Psi - \Psi\,\mathrm{sh}\,\varphi \cos \Psi)]. \tag{20}$$

The expression for C_{n-n^+} is obtained from Eq. (20) by replacing D_p, p_n, ϑ_n^*, and γ with bD_p, n_n, ϑ_p^*, and $1/\gamma$, respectively.

It follows from Eq. (19) that, in this range of currents, the differential resistances of both junctions are equal, inversely proportional to the current, and independent of the frequency and base length. However, the capacitances of the junctions are, according to Eq. (20), proportional to the square root of the current; the frequency dependence of C_{diff} is governed, to a considerable extent, by the ratio of the carrier leakage rates through the contacts in the absence of current flow. According to Eq. (20), the low-frequency capacitance is

$$C_{p-n}^0 = \frac{ej_0}{kT} \sqrt{\frac{eD_p p_n \tau_p}{2j_0 \vartheta_n^*}}\,\mathrm{sh}^{-2}\frac{W}{L_p}\left[\frac{1}{2}\,\mathrm{sh}\frac{2W}{L_p} - \frac{W}{L_p} + \gamma\left(\frac{W}{L_p}\,\mathrm{ch}\frac{W}{L_p} - \mathrm{sh}\frac{W}{L_p}\right)\right] \simeq \frac{ej_0}{2kT}\sqrt{\frac{(b+1)ep_n}{bj_c\vartheta_n^*}}\begin{cases}\dfrac{W}{3}\,(\gamma+2)\ \text{when } W \ll L_p, \\[2ex] L_p\left[1 + 2\gamma\left(\dfrac{W}{L_p}-1\right)e^{-\frac{W}{L_p}}\right] \\ \text{when } W \geqslant 3L_p,\end{cases} \tag{20a}$$

which shows that the junction capacitance of short diodes is proportional to the base length and independent of the carrier lifetime. Consequently, we can determine the base length (but not the carrier lifetime) by measuring C_{diff}^0.

At high frequencies (ch $\varphi \gg b$), we obtain from Eq. (20)

$$C_{p-n}^\infty = \frac{ej_0}{kT}\sqrt{\frac{eD_p p_n}{j_0 \vartheta_n^* \omega}}\,(1 + 2\sqrt{2}\,\gamma e^{-\varphi}\sin\varphi), \tag{20b}$$

i.e., when γ differs slightly from unity, the junction capacitances are proportional to $\omega^{-1/2}$ but independent of τ_p; at high or low values of γ, the frequency dependence of C_{p-n}^∞ (or $C_{n-n^+}^\infty$) may deviate considerably from the quadratic law.

Conclusions

Using the expression for the impedance given in [1], an analysis is made of the current and frequency dependences of the differential resistances and the diffusion capacitances of the junctions in a p-n-n$^+$ structure at high injection levels. Allowance is made for the fact that an increase in the current alters the relationships between the leakage rates of carriers through the contacts and the recombination velocities of the carriers in the bulk of the base.

An analysis of some limiting cases yields the following conclusions.

1. The differential resistances of the junctions are always inversely proportional to the current but the direct proportionality between the diffusion capacitance and the current is observed only at a low leakage rate through a corresponding contact. When the rate of leakage through one of the contacts is high and low through the other, the capacitance of the junction with the high leakage rate is proportional to the current at low frequencies and to the square root of the current at fairly high frequencies. However, if carrier recombination in the low-resistivity regions is the dominant process, the dependence of the diffusion capacitance on the current is quadratic at all frequencies.

2. Measurements of the low-frequency diffusion capacitance can, in some cases (but not for all currents), be used to determine the nonequilibrium carrier lifetime corresponding to a high injection level, as well as the base length.

At high frequencies, the differential resistances and the diffusion capacitances of the junctions are not always proportional to the square root of the frequency; in some cases, considerable deviations from this relationship may be observed.

Literature Cited

1. D. A. Aronov and Ya. P. Kotov, Radiotekhn. i Élektron., 12:1479 (1967).
2. W. Shockley, Electrons and Holes in Semiconductors, Van Nostrand, New York (1950).
3. S. P. Sinitsa, Radiotekhn. i Élektron., 7:1427 (1962).
4. R. N. Hall, Proc. IRE, 40:1512 (1952).
5. A. Yu. Leiderman and P. M. Karageorgii-Alkalaev, Radiotekhn. i Élektron., 10:720 (1965).
6. V. I. Stafeev, Zh. Tekhn. Fiz., 28:1631 (1958).

STATIC CURRENT—VOLTAGE CHARACTERISTIC OF A p—n—p—n STRUCTURE IN THE ON STATE

A. A. Lebedev

Physicotechnical Institute
Academy of Sciences of the USSR, Leningrad

An analysis is made of the influence of the injection efficiency of the emitter p—n junctions in a p—n—p—n structure on the voltage drops across the p—n junctions and across the bulk of the wide and lightly doped n-type base in which a high injection level has been established.

A calculation of the static current—voltage characteristic of a p—n—p—n structure is given in [1] for the case of a high injection level in the wide lightly doped base. In this calculation, the injection efficiencies of the emitter p—n junctions are assumed to be unity. However, this condition is not always satisfied even in the case of simple p—n—p—n structures, and it is certainly inapplicable to many-collector structures [2, 3], which are combinations of two or more p—n—p—n structures. The present paper reports the results of a calculation of the current—voltage characteristic of a p—n—p—n structure in which the injection efficiencies of the emitter p—n junctions are assumed to be arbitrary. In other respects, the calculation is carried out using the standard approximations in the theory of transistors, as has been done in [1].

Under steady-state conditions in the narrow p-type base (Fig. 1), where the injection level is low, electrons obey the equation

$$L_1^2 \frac{d^2 n_1}{dx^2} - n_1 = 0. \tag{1}$$

In the wide n-type base, where the injection level is high, the holes obey the equation

$$L_{2h}^2 \frac{d^2 n_2}{dx^2} - n_2 = 0. \tag{2}$$

Here, $n_1(x)$ is the nonequilibrium electron density in the p-type base; $n_2(x)$ is the nonequilibrium electron density in the n-type base; $L_1 = \sqrt{D_1 \tau_1}$ is the diffusion length of the electrons;

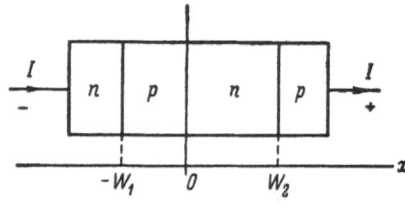

Fig. 1. Schematic representation of a p—n—p—n structure.

D_1 is the diffusion coefficient of the electrons; $L_{2h} = \sqrt{\frac{2b}{b+1} D_2 \tau_2}$ is the diffusion length of the holes at a high injection level; D_2 is the diffusion coefficient of the holes; $b = \mu_1/\mu_2$ is the ratio of the electron to the hole mobility. The boundary conditions, which follow from the quasineutrality of the base regions, are of the form

$$\frac{dn_1}{dx}\bigg|_{x=-W_1} = -\frac{\gamma_1 I}{qD_1}, \tag{3}$$

$$\frac{dn_2}{dx}\bigg|_{x=W_2} = \frac{I[\gamma_2(b+1)-1]}{2qD_2 b}, \tag{4}$$

$$\frac{2b}{b+1} D_2 \frac{dn_2}{dx}\bigg|_{x=0} - D_1 \frac{dn_1}{dx}\bigg|_{x=0} = \frac{Ib}{q(b+1)}, \tag{5}$$

$$\frac{n_1}{n_{01}}\bigg|_{x=0} = \frac{n_2}{n_{02}}\bigg|_{x=0}, \tag{6}$$

where q is the electronic charge; γ_1 and γ_2 are the injection efficiencies of the emitter p−n junctions j_1 and j_3, respectively; W_1 and W_2 are the widths of the two base regions; n_{01}, n_{02} are the equilibrium densities of the minority electrons and holes.

The distribution of the nonequilibrium electron density $n_1(x)$ in the narrow heavily doped p-type base is described by the following law (for a given value of the current I):

$$n_1(x) = A_1 \operatorname{sh} \frac{x}{L_1} + B_1 \operatorname{ch} \frac{x}{L_1}. \tag{7}$$

The coefficients A_1 and B_1 are, respectively,

$$A_1 = -\frac{IL_1 b}{qD_1(b+1)M_{sh2}} \left\{ a_1\gamma_1 s_{2h} \operatorname{th} \frac{W_2}{L_{2h}} + \left[1 - a_{2h}\frac{\gamma_2(b+1)-1}{b}\right] s_1 \operatorname{th} \frac{W_1}{L_1} \right\},$$

$$B_1 = \frac{In_{01}b}{q(b+1)M_{sh2}} \left(\frac{b+1}{b}\gamma_1 a_1 + \frac{\gamma_2(b+1)-1}{b} a_{2h} - 1 \right),$$

where

$$M_{sh2} = s_1 \operatorname{th} \frac{W_1}{L_1} + \frac{b}{b+1} s_{2h} \operatorname{th} \frac{W_2}{L_{2h}},$$

$$s_1 = \frac{n_{01}D_1}{L_1}, \quad s_{2h} = \frac{n_{02}2D_2}{L_{2h}}, \quad a_1 = \operatorname{sech}\frac{W_1}{L_1}, \quad a_{2h} = \operatorname{sech}\frac{W_2}{L_{2h}}.$$

The distribution of the nonequilibrium holes $n_2(x)$ in the wide lightly doped n-type base, in which the injection level is high, is described by the following law:

$$n_2(x) = A_2 \operatorname{sh} \frac{x}{L_{2h}} + B_2 \operatorname{ch} \frac{x}{L_{2h}}, \tag{8}$$

$$A_2 = \frac{IL_{2h}}{2qD_2 M_{sh2}} \left[\frac{\gamma_2(b+1)-1}{b} a_{2h}s_1 \operatorname{th} \frac{W_1}{L_1} + \left(1 - \frac{b+1}{b}a_1\gamma_1\right)\frac{b}{b+1} s_{2h} \operatorname{th} \frac{W_2}{L_{2h}} \right],$$

$$B_2 = B_1 \frac{n_{02}}{n_{01}}.$$

Using Eqs. (7) and (8), we can easily find expressions for the voltage drops across the p−n junctions and across the wide base. We shall give these expressions for the case when $n_{01} \ll n_{02}$. In this case, the voltage drop across the emitter p−n junction j_1 is

$$U_1 = \frac{kT}{q} \ln \frac{n_1(-W_1)}{n_{01}} = \frac{kT}{q} \ln \left[\frac{I\gamma_1 \operatorname{th}\frac{W_1}{L_1}}{qs_1} + 1 \right], \tag{9}$$

where k is the Boltzmann constant and T is the absolute temperature. The voltage drop U_3 across the emitter p−n junction j_3 is

$$U_3 = \frac{kT}{q} \ln \left\{ \frac{I}{qs_{2h} \, \text{sh} \, \frac{W_2}{L_{2h}}} \left[\frac{b+1}{b} a_1\gamma_1 - 1 + \frac{\gamma_2(b+1)-1}{b} \, \text{ch} \, \frac{W_2}{L_{2h}} \right] + 1 \right\}. \tag{10}$$

The voltage drop across the collector p–n junction j_2 is given by

$$U_2 = \frac{kT}{q} \ln \left[\frac{I \, \text{cth} \, \frac{W_2}{L_{2h}}}{qs_{2h}} \left(\frac{b+1}{b} \gamma_1 a_1 + \frac{\gamma_2(b+1)-1}{b} a_{2h} - 1 \right) + 1 \right]. \tag{11}$$

It is evident from the last expression that, in order to observe voltage inversion at the collector p–n junction, we must satisfy the condition

$$\frac{b+1}{b} a_1\gamma_1 + \frac{\gamma_2(b+1)-1}{b} a_{2h} > 1. \tag{12}$$

If $\gamma_1 = \gamma_2 = 1$ the above condition transforms into a relationship derived in [4]. The second necessary condition for the turn on of a p–n–p–n structure, which [together with Eq. 12] is also a sufficient condition, is satisfied when the minimum value of the nonequilibrium hole density in the n-type base is higher than the density of the majority equilibrium carriers in this base.

The voltage drop across the bulk of the n-type base, U_T, is given by expressions which depend on the value of the injection efficiency γ_2 of the emitter p–n junction. In order to obtain simple expressions, we must make the following additional assumptions: $a_1\gamma_1 \simeq 1$, $W_2/L_{2h} > 1$.

When $\gamma_2 > 1/(b+1)$, we obtain

$$U_T = \frac{kT}{q} \varkappa \exp \left(\frac{W_2}{2L_{2h}} \right), \tag{13}$$

where

$$\varkappa = \frac{2b}{(b+1)[\gamma_2(b+1)-1]^{\frac{1}{2}}} \, \text{arctg} \, \frac{\left(e^{\frac{W_2}{L_{2h}}} - 1 \right) [\gamma_2(b+1)-1]^{\frac{1}{2}}}{\gamma_2(b+1) \, e^{\frac{W_2}{2L_{2h}}}}.$$

When $\gamma_2 \leq 1/(b+1)$, the voltage drop U_T can still be described by Eq. (13) but now the coefficient $\varkappa$ is

$$\varkappa = \frac{b}{(b+1)[1-\gamma_2(b+1)]^{\frac{1}{2}}} \ln \frac{\gamma_2(b+1) + 2\sqrt{1-\gamma_2(b+1)} \, \text{sh} \, \frac{W_2}{2L_{2h}}}{\gamma_2(b+1) - 2\sqrt{1-\gamma_2(b+1)} \, \text{sh} \, \frac{W_2}{2L_{2h}}}.$$

If $\gamma_2 = 1/(b+1)$, the voltage drop across the wide base is

$$U_T \simeq \frac{kT}{q} \frac{2b}{b+1} e^{\frac{W_2}{L_{2h}}}. \tag{14}$$

It follows from Eqs. (13) and (14) that, in order to obtain low voltage drops across the wide base, it is desirable to have $\gamma_2 > 1/(b+1)$. If the actual type of conduction in the base regions is opposite to that assumed in our calculations, we must replace b with 1/b in all the expressions given so far.

Literature Cited

1. V. A. Kuz'min, Radiotekhn. i Elektron., 8:171 (1963).
2. I. V. Grekhov and V. B. Shuman, present collection, p. 224.
3. A. A. Lebedev, present collection, p. 321.
4. A. A. Lebedev and A. I. Uvarov, Fiz. Tekh. Poluprov., 1:211 (1967).

DISTRIBUTION OF THE INTENSITY OF RECOMBINATION RADIATION AND OF THE VOLTAGE DROP IN DIFFUSED p—n JUNCTIONS IN GALLIUM ARSENIDE

V. G. Voronin, A. V. Petukhov, I. V. Ryzhikov, and V. F. Titova

A description is given of a method which can be used to determine which regions in a crystal with a p–n junction emit recombination radiation. The method is based on the use of combined electrical and optical probes. It is shown that this method can only be applied to crystals which absorb their intrinsic recombination radiation. The method can be used not only to determine that region in which the light emission is concentrated, but also to estimate its electrical and optical properties (absorption coefficient, diffusion length, and carrier lifetime). The method was applied to diffused p–n junctions in gallium arsenide, which emitted infrared radiation. It was established that radiation was emitted mainly by that part of the p-type layer which adjoined the space-charge region of the p–n junction. The absorption coefficient of the recombination radiation in this region was 700 cm^{-1}, the diffusion length of the electrons was 4-6 μ, and the electron lifetime in the p-type region was $1.5 \cdot 10^{-8}$-$3.5 \cdot 10^{-8}$ sec.

Most investigators attribute the electroluminescence of p–n junctions in GaAs to the recombination of electrons at the Zn levels in the p-type region [1, 2]. We investigated the distribution of the electric potential and the spatially identical distribution of the intensity of the recombination radiation in order to determine the region where recombination took place and to estimate the electrical parameters of the n- and p-type regions adjoining the space-charge layer of the p–n junction.

Samples and Experimental Method

Our investigation was carried out on p–n junctions in GaAs produced by the diffusion of Zn. The diffusion annealing temperature was 850-900°C and the annealing duration was 6-7 h. The measurements were carried out using oblique sections cut from either the p- or n-type side. The apparatus used in the simultaneous measurement of the electric potential and the recombination radiation intensity is shown in Fig. 1. It was based on a microphotometer MF-4. The stage on which the section was placed could be moved along two mutually perpendicular directions x and y, and could be rotated through an angle φ. Micrometer screens were used for this purpose. Since the measurements were carried out using sections cut at an angle of 3°,

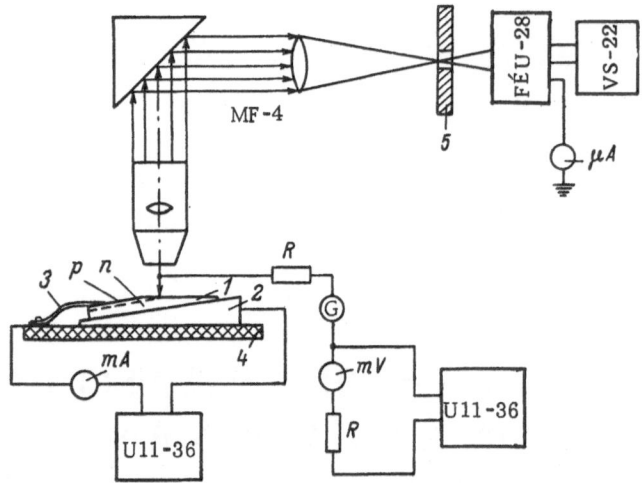

Fig. 1. Apparatus used for the simultaneous measurement of the electric potential and the intensity of recombination radiation. 1) Oblique section through a p—n junction; 2) tilted stage which also acted as the lower contact; 3) upper contact; 4) insulating plate; 5) slit in front of the detector.

the stage was tilted through 3° so that the face on which the measurements were carried out remained horizontal. The voltage drop along an oblique section with a diffused p-type layer was investigated using a probe which was moved along the surface of the section. An optical system, which focused the emitted radiation, was placed above the stage. An FÉU-28 photomultiplier was used as a detector. Before the measurements were made, the apparatus was adjusted so that the p—n junction line was parallel to the detector slit. Then the radiation reached the optical system from that point on the surface of a crystal on which a metal probe was placed (to measure the potential).

The first stage of the adjustment consisted of the rotation of the section about the vertical axis until the photoresponse of the detector reached its maximum value. In the second stage, we exploited the principle of the reversibility of the rays in an optical system: the detector was replaced by an incandescent lamp and a light spot was projected through the slit and the optical system onto the surface of the crystal.

The electric probe was placed in the illuminated area of the spot. The metal wire used as the probe was polished electromechanically until a point of 10–20 μ diameter was obtained. The slit in front of the detector was 10–30 μ wide.

Experimental Results and Discussion

Typical distributions of the electric potential and the intensity of recombination radiation are shown in Figs. 2 and 3. It is evident from these figures that the p-type region absorbed intrinsic recombination radiation more strongly than the n-type region. In the case of a crystal ground on the n-type side, the radiation was observed on the right and left of the space-charge plane of the p—n junction, but in the case of a crystal ground on the p-type side, the radiation was concentrated in the p-type layer adjoining the space-charge region of the p—n junction. Since the intrinsic recombination radiation was strongly absorbed in the p-type region, these observations indicated that the recombination radiation was generated in the p-type layer and that the contribution of the n-type layer or the space-charge region of the p—n junction could be neglected.

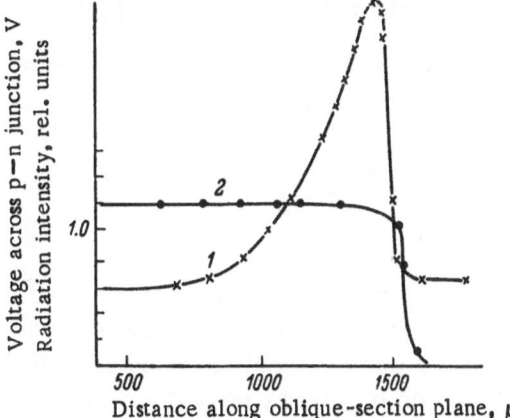

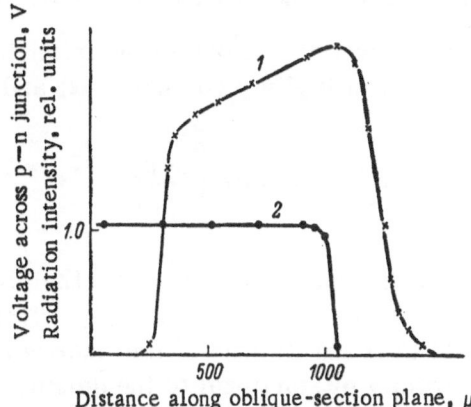

Fig. 2. Distribution of the potential and of the radiation intensity for a crystal with a p–n junction ground on the p-type side: 1) radiation intensity; 2) potential; section No. 3; i_d = 150 mA; 30 μ slit.

Fig. 3. Distribution of the potential and of the radiation intensity for a crystal with a p–n junction ground on the n-type side; 1) radiation intensity; 2) potential; section No. 22; i_d = 200 mA; 10 μ slit.

The optical system used in our measurements recorded the radiation which emerged at an angle of 18° with respect to the ground surface of the crystal. Taking into account the difference between the refractive indices of GaAs and air, we found that the divergence of the radiation in the crystal did not exceed 5°. In view of the narrowness of the radiation beam, we simplified our calculations by assuming that the recombination radiation was emitted from a layer of GaAs in the shape of a rectangular parallelepiped whose base was defined by the geometrical dimensions of the microphotometer slit and the "height" H (Fig. 4). The error due to this assumption did not exceed 7%.

We shall now calculate the intensity of the radiation [dΦ(y)] emerging from a layer of infinitesimal thickness dy in a parallelepiped lying at a distance ~y from the p–n junction plane:

$$d\Phi(y) = \frac{\Delta n(0)}{\tau_{rad}} e^{-\frac{y'}{L_n}} e^{-(H-y)x} dy, \tag{1}$$

where $\Delta n(0)$ is the difference between the nonequilibrium and equilibrium electron densities in the p-type layer near the p–n junction; τ_{rad} is the radiative electron lifetime in the p-type layer; L_n is the diffusion length of the electrons in the p-type layer; H is the distance from the

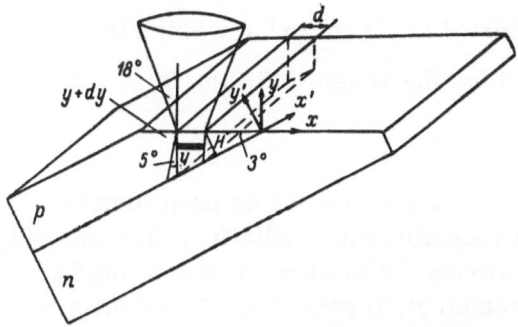

Fig. 4. Geometrical construction used to calculate the intensity of radiation recorded at a given point.

p–n junction plane to the ground surface of the crystal; $\varkappa$ is the absorption coefficient.

After integration, we obtain the intensity of the radiation $\Phi(H)$ emerging from a layer of thickness H [because y' = y cos (90° − α) and α = 3°, we can assume that y' ≈ y]:

$$\Phi(H) = \int_0^H \frac{\Delta n(0)}{\tau_{rad}} e^{-\frac{y}{L_n}} e^{-(H-y)\varkappa} dy = \frac{\Delta n(0)}{\left(\varkappa - \frac{1}{L_n}\right)\tau_{rad}} \left(e^{-\frac{H}{L_n}} - e^{-H\varkappa}\right). \tag{2}$$

For distances H, small compared with the diffusion length, we can easily show, by expanding the function $\Phi(H)$ as a series, that $\Phi(H) = (\Delta n(0)/\tau_{rad})$ H, i.e., the radiation intensity is directly proportional to H and its rate of increase with H is inversely proportional to the radiative lifetime and directly proportional to the density of nonequilibrium electrons near the p–n junction. For H/L > 1, we find that Eq. (2) yields

$$\Phi(H) = -\frac{\Delta n(0)}{\tau_{rad}\left(\varkappa - \frac{1}{L_n}\right)} e^{-H\varkappa}, \tag{3}$$

i.e, $\Phi(H)$ decreases exponentially with increasing H and the argument of the exponential function is proportional to the absorption coefficient.

Thus, the function $\Phi(H)$ has a maximum. Investigating this maximum of $\Phi(H)$, we find that

$$H_{max} = \frac{1}{\frac{1}{L_n} - \varkappa} \ln \frac{1}{L_n \varkappa}. \tag{4}$$

The position of the maximum of the radiation intensity thus depends on the absorption coefficient and the diffusion length of electrons in the p-type layer. The expression for the function $\Phi(H)$ in the case H/L > 1 can be used to determine the absorption coefficient $\varkappa$ of the p- and n-type regions:

$$\varkappa = \frac{1}{H_2 - H_1} \ln \frac{I_1(H_1)}{I_2(H_2)}, \tag{5}$$

where H_1 and H_2 are the thicknesses of the layers through which the radiation has to pass; $I_1(H_1)$ and $I_2(H_2)$ are the recorded radiation intensities corresponding to two different thicknesses.

It is evident from Fig. 2 that $\varkappa_p \approx 700$ cm^{-1}, which is in satisfactory agreement with the results of the direct determination of this quantity. The experimentally measured values of H_{max} lie within the range 7–9 μ. Knowing the absorption coefficient of the crystal, we can use Eq. (4) as the basis of the graphical determination of the diffusion length of electrons in the p-type layer, which is found to be 4–6 μ. Knowing the diffusion length, we can employ the formula $L_n = \sqrt{\frac{kT}{q}\mu\tau_t}$ and the results reported in [3] to find τ_t, which is the total electron lifetime in the p-type layer. This lifetime is within the range $(1.5-3.5)\cdot10^{-8}$ sec.

Conclusions

A description is given of a method which can be used to determine which regions of a crystal with a p–n junction emit recombination radiation. The method is based on the use of combined electrical and optical probes. It is shown that this method can be employed only if a crystal absorbs its intrinsic recombination radiation. The method can be used not only to determine that region in which the light emission is concentrated but also to estimate its electrical and optical properties (absorption coefficient, diffusion length, and carrier lifetime).

The method was applied to diffused p—n junctions in gallium arsenide, which emitted recombination radiation. It was established that the radiation was emitted mainly by that part of the p-type layer which adjoined the space-charge region of the p—n junction, which was in agreement with the results reported in [1, 2]. The absorption coefficient of recombination radiation in this region was 700 cm^{-1}, which was in good agreement with the results given in [4]; the diffusion length of electrons was 4-6 μ and the lifetime of electrons in the p-type layer was (1.5-3.5)$\cdot 10^{-8}$ sec.

Literature Cited

1. G. Lucovsky, Appl. Phys. Letters, 5:37 (1964).
2. W. N. Carr and J. R. Biard, J. Appl. Phys., 35:2777 (1964).
3. V. P. Sushkov, Fiz. Tverd. Tela, 8:3431 (1966).
4. D. E. Hill, J. Appl. Phys., 36:3405 (1965).

LOW-ENERGY RECOMBINATION RADIATION
OF p–n JUNCTIONS IN GaAs

V. M. Lomako, V. D. Tkachev, and D. S. Domanevskii
V. I. Lenin Belorussian State University, Minsk

The results are reported of an investigation of low-energy (0.3-0.7 eV) recombination radiation emitted by p–n junctions in GaAs. The diodes were produced by the diffusion of zinc in gallium arsenide with an electron density 10^{15}-10^{16} cm^{-3}. At 80°K, bands were observed at 0.36, 0.58, 0.59, 0.61, and 0.65 eV. It was found that the intensities of the bands varied strongly from sample to sample even when a batch was prepared from a single cut. This indicated a nonuniform radial distribution of the compensating impurities in the original single crystal. The 0.65 eV band was due to the presence of oxygen. The 0.36 and 0.59 eV bands were observed at room temperature. No shifts of the maxima or changes in the band profiles were found when the samples were heated to room temperature.

The present paper reports the results of an investigation of low-energy recombination radiation generated by the injection of nonequilibrium carriers through a p–n junction in GaAs. Junctions were produced in n-type GaAs, grown by the directional crystallization method and having an equilibrium carrier density $n_0 = 10^{15}$-10^{16} cm^{-3} and a carrier mobility $\mu = 4500$ cm$^2 \cdot$ V$^{-1} \cdot$ sec^{-1} at 300°K. The junctions were produced by the diffusion of zinc at 900-920°C.

An analysis of the recombination radiation emitted by the p–n junctions showed that, in addition to the known bands at 1.47, 1.28, and 1.0 eV (80°K), there were several long-wavelength bands in the 0.3-0.7 eV range (Fig. 1).

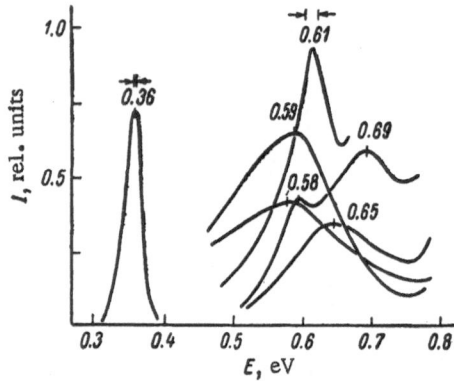

Fig. 1. Low-energy radiation bands of p–n junctions in GaAs. The intensities of the bands are given in relative units.

The strongest long-wavelength band was that at 0.36 eV (its width at half the intensity was 0.025-0.03 eV). In some samples, the intensity of this band reached that of the 1.0 eV band and decreased rapidly with increasing temperature. However, the 0.36 eV band was sometimes observed also at room temperature, at which its profile and energy position were the same as at lower temperatures.

In addition to the 0.36 eV band, we always observed one or two radiation bands (near the middle of the forbidden energy band) adjoining the wing of the 1.0 eV radiation band. The intensities of these bands were three or four orders of magnitude lower than the intensity of the principal band at 1.47 eV.

The recombination radiation spectra of these p—n junctions were characterized in this spectral region by considerable variations of the intensity and band profile from sample to sample in a batch prepared from the same cut of a single crystal. This indicated a nonuniform radial distribution of the impurities in the original single crystal. The observed bands were identified by the positions of their maxima and half-widths. It was found that in the 0.5-0.7 eV region there were bands at 0.58 (0.27) eV, 0.59 (0.21) eV, 0.61 (0.1) eV, and 0.65 (0.17) eV; the figures in parentheses are the widths of the bands, in eV, at half intensity. The positions of these bands varied from sample to sample within 0.03 eV. The 0.59 eV band was also observed at room temperature with the same profile and energy position as at lower temperatures.

It has been reported in [1] that the photoluminescence spectra of GaAs grown in an oxygen atmosphere include several luminescence bands located near the middle of the forbidden energy band. We produced some p—n junctions in GaAs grown in an oxygen atmosphere (10 mm Hg). The recombination radiation spectra of these diodes showed that the 0.65 eV band was dominant, which indicated the participation of oxygen in the formation of the recombination radiation centers.

It was interesting to note a shift of the 0.65 eV band to 0.67-0.69 eV in the presence of a radiation band near 0.6 eV. This shift could be explained by the interaction of oxygen with other lattice defects, resulting in a change of the structure of the centers responsible for the 0.65 eV band and a shift of the energy levels of these centers. The different values of the energy levels of oxygen compensating impurities in GaAs [2] could be explained in the same way.

Diodes made of undoped GaAs with a high electron density (10^{16}-10^{17} cm^{-3} at 300°K) did not usually have the low-energy radiation bands: some samples did exhibit such bands but their intensities were very low (close to the sensitivity threshold of our apparatus). The low-energy recombination radiation was not observed for p—n junctions with still higher electron densities ($n_0 > 10^{17}$ cm^{-3}).

The results reported indicate that as the quality of single crystals is improved and shallow-level impurities are removed, secondary recombination centers (corresponding to deep levels of compensating impurities) become increasingly important. The main role in the formation of these deep compensating levels is played by oxygen, which is one of the principal impurities in GaAs (oxygen is introduced accidentally during the preparation of this material).

Literature Cited

1. W. J. Turner, G. D. Pettit, and N. G. Ainslie, J. Appl. Phys., 34:3274 (1963).
2. C. Hilsum and A. C. Rose-Innes, Semiconducting III-V Compounds, Pergamon Press, Oxford (1961).

ELECTROLUMINESCENCE AND CATHODOLUMINESCENCE OF p–n JUNCTIONS IN GaAs

V. M. Lomako, D. S. Domanevskii, and V. D. Tkachev

V. I. Lenin Belorussian State University, Minsk

An investigation was made of the spectral distribution of the impurity (extrinsic) recombination radiation of diffused p–n junctions made of n-type GaAs crystals with a range of defect concentrations. The impurity luminescence region included bands whose maxima were located at 0.99-1.06, 1.14-1.18, 1.21-1.29, 1.36, and 1.396 eV. The 1.396 eV band was ~kT wide at 80°K and it was attributed to radiative recombination in the space-charge region of the p–n junction. A comparison of the luminescence spectra obtained by injection through the p–n junction and by electron bombardment of different regions of the junction showed that the 1.0 and 1.28 eV bands originated from the n-type region in the p–n junction. The luminescence spectrum of this region differed from the spectrum of the n-type region far from the junction. This gave rise to a difference between the luminescence spectrum of the p–n junctions and the cathodo-luminescence spectrum of GaAs subjected to heat treatment.

The luminescence spectra of diffused p–n junctions prepared from n-type GaAs by the diffusion of Zn, Cd, or Be consists of three bands (at 77°K): 1.47, 1.28, and 1.0 eV. The principal band is due to the radiative recombination of carriers via the shallow acceptor level of the diffusant. The nature of the other two impurity luminescence bands has not yet been determined.

It has been reported in [1] that the 1.0 eV band is emitted by the n-type region of the p–n junction and that it is due to the capture of free electrons by a level located 1.0 eV below the bottom of the conduction band. It has also been reported [2-4] that additional doping with copper increases the intensity of the 1.0 and 1.28 eV bands. It is suggested in [4] that these bands are due to recombination in the n-type layer rather than in the space-charge region. The appearance of these bands is attributed to recombination through the first and second acceptor levels of copper. However, the first acceptor level of copper is located 0.15 eV above the valence band [5] and, therefore, it should correspond to a radiative recombination band with a maximum at 1.36 eV and not at 1.28 eV.

An investigation of the influence of heat treatment on the photoluminescence of n-type GaAs has demonstrated [6] that heating to 700°C shifts the observed bands from 0.96 and 1.25 eV to 1.01 and 1.28-1.30 eV, respectively. Moreover, heating to a higher temperature (800-900°C) produces a band at 1.35 eV with phonon replicas. It is concluded in [6] that the 1.35 eV band is associated with the first acceptor level of copper, whose concentration increases during heat treatment. Moreover, a study of the cathodoluminescence of GaAs [7] has indicated the presence of luminescence bands at 1.35-1.36, 1.21, 0.96, and 0.83 eV. The floating-zone treatment has been found to reduce considerably the intensities of the 1.35 and 0.96 eV bands without altering the 1.21 eV band, which is evidently due to structure defects.

It must be mentioned that the recombination radiation spectra of p–n junctions exhibit a wide band at 1.39 eV [8], which is due to a manganese acceptor level at $E_v + 0.1$ eV.

The purpose of our investigation was a further study of the influence of heat treatment and the degree of doping of the original single crystals on the spectral distribution of the radiative recombination of diffused p–n junctions in GaAs.

Experimental Method

We investigated undoped n-type GaAs (electron density $n_0 = 1 \cdot 10^{15} - 5 \cdot 10^{17}$ cm^{-3} and electron mobility $\mu = 2000$-4500 cm$^2 \cdot$ V$^{-1} \cdot$ sec^{-1} at 300°K), in which p–n junctions were produced by the diffusion of zinc. We also investigated p–n junctions produced in GaAs doped with Te ($2 \cdot 10^{18}$ cm^{-3}) and GaAs ($5 \cdot 10^{15}$ cm^{-3}) grown in an oxygen atmosphere at a pressure of 10 torr.

The diffusion treatment was carried out in a two-zone furnace. The temperature of the diffusant source was 620-700°C and the temperature of the sample was 830-920°C. The diffusion of Zn in Te-doped GaAs was carried out at an equilibrium pressure of As vapor.

Ohmic contacts were provided by firing vacuum-electrolytically deposited Pd. In some cases alloyed contacts were used. The samples were prepared by cleaving, and the p–n junction area was 0.5-2 mm^2.

Before the measurements, a sample was placed in a pressure contact holder, which was soldered to a heat conductor of a cryostat. The holder had a heater for raising the temperature, which was measured with a thermocouple and kept constant (by an automatic controller) to within ±1 deg C.

The radiation emerging from the edge of a p–n junction was focused by mirrors onto the entry slit of an IKS-12 monochromator with glass ($\lambda = 0.5$-2.5 μ) or LiF ($\lambda = 2.5$-4 μ) optical components. The radiation was detected by an FÉU-28 photomultiplier and by cooled PbS or Ge:Au photoresistors. The detector signal was passed on to a pre-amplifier, a U2-6 amplifier, an SD-1 synchronous detector, and an automatic recorder.

Experimental Results and Discussion

An investigation of the recombination radiation spectra of p–n junctions indicated that the spectral distribution and the intensity of the radiation depended strongly on the number of defects in the original crystals and on the diffusion conditions.

The luminescence spectra (all the results were obtained at 80°K) usually consisted of several bands. The main band at 1.47 eV was well known: it was attributed to recombination through shallow acceptor levels in the p-type region. We found that in relatively pure GaAs the maximum of this band, located at 1.48 eV, did not shift when the excitation level was altered. This suggested the radiative capture of electrons from the conduction band by the acceptor level of zinc at $E_v + 0.034$ eV.* The impurity region of the spectrum always had two wide bands, whose intensities and energy positions varied from sample to sample and were governed primarily by the degree of doping of the original single crystal. For example, the samples pre-

---

*Similar behavior of the 1.48 eV band, observed when the excitation level was increased, indicated the absence of overlap of the shallow donor and acceptor levels with the edges of the corresponding bands.

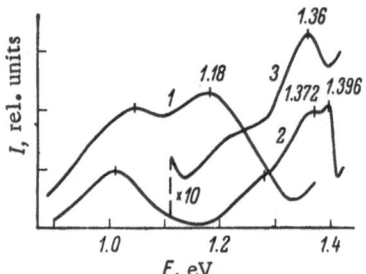

Fig. 1. Recombination radiation spectra of p—n junctions in the 1.0-1.4 eV range.

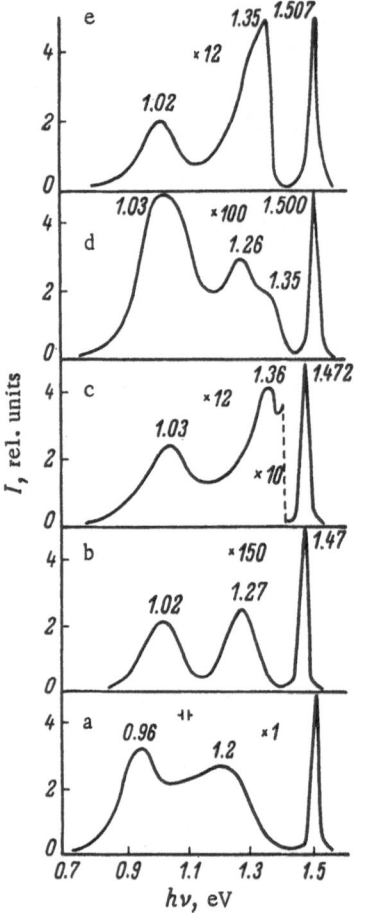

Fig. 2. Recombination radiation spectra: a) original single crystal; b) p—n junction diode; c), e) p- and n-type regions, respectively; d) junction region. T = 80°K. The factors given in each of the figures represent an increase in the intensity of a given spectrum compared with the spectrum of the original single crystal.

pared from n-type GaAs, with carrier densities of $1 \cdot 10^{16}$ and $2 \cdot 10^{18}$ cm^{-3} at 300°K, had a band at 1.26-1.29 and 1.21-1.26 eV, respectively. The energy position of the second impurity luminescence band ranged from 0.99 to 1.06 eV and was not related to the position of the 1.21-1.29 eV band. Some samples also had a band at 1.14-1.18 eV (curve 1 in Fig. 1). Attempts to identify all these impurity luminescence bands with the currently known energy levels of accidental impurities or other defects in GaAs were not successful. Moreover, it was not clear why the energy positions of these bands were different for samples prepared from different single crystals.

In addition to these typical recombination radiation bands, the spectra of diodes prepared by the diffusion of zinc at 900°C in n-type GaAs ($n_0 = 10^{15}$–10^{16} cm^{-3}) also had a characteristic band at 1.396 eV, observed at low excitation levels. This band exhibited a steep drop (~kT at 80° K) on the high-energy side and it had a phonon replica at 1.372 eV (curve 2 in Fig. 1). A study of the spectra and of the dependences of the band intensities on the current showed that the 1.396 eV band was observed as long as there was an energy barrier at the p—n junction. As the forward bias was increased, bands appeared at 1.47, 1.02, and 1.27 eV in succession and the intensity of the 1.396 eV band tended to saturation when the current was increased. The spectrum had its usual form (Fig. 2b) when the forward bias across the p—n junction was considerable and the thermal injection of carriers became important. A band with a similar profile was observed by us at 1.35-1.36 eV in an earlier investigation of the cathodoluminescence [7]. We therefore concluded that the 1.396 eV band observed in diodes was due to recombination through the same centers, but its maximum was shifted as a result of the heat treatment during the preparation of the diodes. To check this hypothesis, we compared the luminescence spectra of the diodes with the recombination radiation spectra obtained by the successive grinding away of regions of the p—n junction. The recombination radiation spectra obtained in this manner are shown in Fig. 2. The spectrum shown in Fig. 2a was obtained by excitation with a 50 keV electron beam of the original single crystal of GaAs from which the investigated diodes were prepared. This spectrum consisted of three relatively weak bands with maxima at 0.96, 1.20, and 1.507 eV. Figure 2b gives a typical radiation spectrum of a diode when carriers are injected through the p—n junction. The next spectrum (Fig. 2c) is that of a diode without a contact to the p-type region and it represents the luminescence spectrum of the p-type region excited by electron bombardment. We can see that the 0.96 eV band shifts, as usual, in the direction of higher energies (up to 1.03 eV) and becomes 12 times as strong as in the original single crystal; the 1.20 eV band disappears altogether but we have, instead, a band at 1.36 eV (Fig. 2c). However, the dominant component of the luminescence of the p-type region is a band at 1.472 eV. Gradual grinding of the p-type region uncovers the transition (junction) layer, whose luminescence spectrum consists of very strong bands at 1.03, 1.26, 1.35, and 1.5 eV (Fig. 2d). Continued grinding uncovers the n-type region: it is now found that the luminescence intensity decreases and the spectrum consists of bands at 1.02, 1.35, and 1.507 eV (Fig. 2e). A control heating cycle of the original crystal, carried out under the conditions employed during the diffusion of zinc, shows that its luminescence spectrum is completely identical with the luminescence spectrum of the n-type region of the p−n junction.

The 1.36 eV band corresponds to the energy position of the first acceptor level of copper, which is captured from the quartz ampoule during the heat treatment [6] or which may be present in the original single crystal and become activated by the heat treatment [9]. However, a redistribution of the cathodoluminescence band intensities, observed when the p−n junction is ground away, indicates a complex impurity distribution profile. The relative reduction in the intensity of the 1.36 eV band, observed in the transition (junction) region, may be explained by the expulsion — in the diffusion process — of the mobile interstitial copper atoms from the p−n junction region by an internal electric field [10]. This reduces the concentration of the E_v + 0.15 eV centers. The reduction in the concentration of the copper in the p−n junction region can explain a general increase in the luminescence intensity in the transition region. These results, taken together with the low value of the diffusion length of the carriers, explain why the luminescence spectra of the diodes are dominated by the 1.28 eV band but not by the 1.36 eV band, which is usually observed in heat-treated GaAs and in the diode base far from the p−n junction.

It is worth noting the difference between the positions of the luminescence maximum at 1.35-1.36 eV under electron bombardment and injection conditions (in the latter case, this band is at 1.396 eV). The shift of this band can be explained by a change in the structure of the corresponding centers in the internal field of the p−n junction during the diffusion process; this change shifts the maximum from 1.35 to 1.396 eV. Consequently, the 1.396 eV band should originate from the space-charge region of the p−n junction, which is confirmed by the dependence of the luminescence intensity on the current.

The high temperature of the formation of these centers (∼900°C) is evidently due to the participation of arsenic vacancies, at which copper atoms may be located, giving rise to singly ionized $\{Cu_{As}\}$ or $\{V_{As}Cu_{Ga}V_{As}\}$ centers [11]. However, it is possible that the centers responsible for the radiative recombination of the 1.396 eV band consist not of copper but of other impurity atoms, giving rise to a level at 0.118 eV from the edge of one of the energy bands. This hypothesis is supported by the observation that a weak 1.36 eV band (curve 3 in Fig. 1) is observed in the luminescence spectra of some diodes when the intensity of the 1.28 eV band is low compared with that of the 1.0 eV band.

Since Mn gives rise to a band with a maximum at the same energy as the 1.396 eV band, we prepared control p−n junctions by the double diffusion of Mn and Zn. An investigation of the dependences of the luminescence spectra of these junctions on the current and temperature showed that the 1.396 eV band differed from the luminescence band of manganese not only in its profile but also in the different dependences on the current and temperature. Consequently, this band could not be due to manganese impurities, which gave rise to a luminescence band with a 0.11 eV half-width and a maximum at 1.39 eV. We must mention also that the 1.396 eV band was observed for p−n junctions prepared from GaAs grown in an oxygen atmosphere. A characteristic feature of these junctions was their integrated luminescence intensity, which was several times greater than the luminescence intensity of "oxygen-free" samples. Moreover, such junctions had two or three low-energy luminescence bands in the 0.3-0.7 eV range [12].

Conclusions

The following conclusions can be drawn from our investigation of the recombination radiation spectra of p–n junctions in GaAs.

1. The low-energy luminescence bands with maxima near 1.0 and 1.28 eV are due to radiative recombination at structure defects, as reported in [13].

2. At high concentrations of shallow donors ($\sim 10^{18}$ cm^{-3}), an interaction between these donors and residual defects shifts the 1.26-1.29 eV band to 1.21-1.26 eV.

3. The difference between the electroluminescence spectrum of a p–n junction in GaAs and the cathodoluminescence spectrum of an original single crystal subjected to high-temperature treatment is due to the expulsion of copper from the transition region by the electric field of the p–n junction itself.

4. In the 1.0-1.5 eV region there are luminescence bands at 1.14-1.18, 1.36, and 1.396 eV with a phonon replica at 1.372 eV. The 1.396 eV band originates from the space-charge region of the p–n junction and is characterized by a narrow profile, which may indicate the localization of electrons and holes before their recombination.

Literature Cited

1. M. I. Nathan, Solid State Electron., 6:425 (1963).
2. T. L. Larsen, Appl. Phys. Letters, 3:113 (1963).
3. T. N. Morgan, M. Pilkuhn, and H. Rupprecht, Phys. Rev., 138:A1551 (1965).
4. M. F. Millea and L. W. Aukerman, J. Appl. Phys., 37:1788 (1966).
5. F. D. Rosi, D. Meyerhofer, and R. B. Jensen, J. Appl. Phys., 31:1105 (1960).
6. Zh. I. Alferov, D. Z. Garbuzov, and E. P. Morozov, Fiz. Tverd. Tela, 8:3236 (1966).
7. D. S. Domanevskii and V. D. Tkachev, Fiz. Tekh. Poluprov., 1:377 (1967).
8. K. Weiser and R. S. Levitt, J. Appl. Phys., 35:2431 (1964).
9. B. I. Boltaks and F. S. Shishiyanu, Fiz. Tverd. Tela, 7:1021 (1965).
10. G. B. Larrabee and J. F. Osborn, J. Electrochem. Soc., 113:564 (1966).
11. C. S. Fuller, H. W. Allison, and K. B. Wolfstirn, J. Phys. Chem. Solids, 25:1329 (1964); Y. Furukawa and C. D. Thurmond, J. Phys. Chem. Solids, 26:1535 (1965).
12. V. M. Lomako, V. D. Tkachev, and D. S. Domanevskii, present issue, p. 16.
13. J. Blanc, R. H. Bube, and L. R. Weisberg, J. Phys. Chem. Solids, 25:225 (1964).

FILM DEVICES PREPARED BY
THE ION BOMBARDMENT METHOD

G. A. Kachurin, A. E. Gorodetskii,
V. M. Zelevinskaya, and L. S. Smirnov

Institute of Semiconductor Physics
Siberian Branch, Academy of Sciences of the USSR, Novosibirsk

It is demonstrated that, in principle, it is possible to prepare photodiodes and photoresistors by the ion bombardment of thin ($\sim 1\ \mu$) polycrystalline films of cadmium telluride. Possible applications of this method are considered.

Ion bombardment of semiconductors has become a widely used method in device fabrication. Although this method is particularly suited to the manufacture of microminiature devices, no work has been published on the use of the ion bombardment method in the preparation of thin-film elements. The purpose of our investigation was to demonstrate, in principle, that film devices could be prepared by the ion bombardment technique.

Photoresistors

Photoresistors were prepared from polycrystalline films of CdTe about $1\ \mu$ thick, in which the incident light was practically completely absorbed. After the films had been sensitized by annealing in air at 350–400°C for 1 h, metal contacts were evaporated onto them. All the measurements were carried out on samples with aluminum contacts. Introduction of In atoms into CdTe by ion bombardment lowered the resistivity of this material and raised its photosensitivity [1].

The main increase (by a factor of almost 20) in the photocurrent, ΔI_{ph}, occurred for a dose of ~ 1 mC (Fig. 1). Annealing in vacuum after irradiation increased somewhat the sensitivity of the photoresistors.

The resistors prepared in this way had a high sensitivity in the visible and infrared regions (a maximum was found at $\lambda = 0.83\ \mu$). The dark resistance of these devices was $\sim 10^8$–$10^9\ \Omega$ and the resistance decreased by a factor of ~ 20–70 when they were illuminated with 200 lux. The threshold sensitivity at room temperature was not less than 10^{-7} lm. The time constant deduced from the photoconductivity decay was $\sim 2 \cdot 10^{-4}$–$8 \cdot 10^{-5}$ sec. No saturation of the photocurrent was observed right up to $5 \cdot 10^4$ lux.

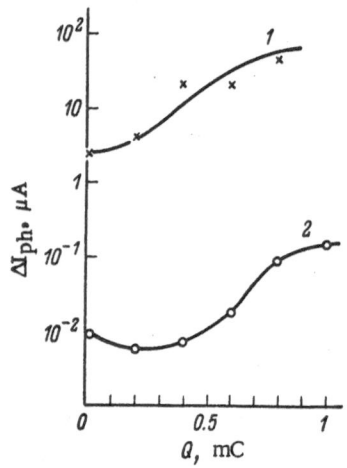

Fig. 1. Dependence of the photocurrent on the radiation dose (area of the sample: ~2 cm^2): 1) annealed sample (400°C for 1 h); 2) unannealed sample.

We tried to prepare photoresistors with a field applied transversely across the thickness but the results were not easily reproducible because of variations in the thickness of the film and frequent short circuits between the upper and lower electrodes. In the best case, we obtained a sensitivity of 0.5 mA · lm^{-1} · V^{-1}.

All the photoresistors exhibited an increase in the sensitivity and a reduction in the resistance under electric load conditions. We did not investigate the ageing process.

Photodiodes

Since it is difficult to prepare a planar diode ~1 thick, our results may be of some interest, although the quality of the obtained p–n junctions was far from satisfactory.

There have been many attempts to produce p–n junctions in thin polycrystalline films of $A^{II}B^{VI}$ compounds [2-4]. The main difficulty is to dope a very thin layer (0.1-0.01 μ) and to produce a sufficiently high impurity concentration. It is reported that rectification has to be produced by a heterojunction or by a metal–semiconductor contact.

When impurities are introduced by ion bombardment, their distribution with depth and concentration can be tailored to a specification quite accurately simply by varying the energy of the ions and the radiation dose.

To obtain a good-quality photodiode, it is necessary to ensure that it have a low series resistance. The preparation of low-resistivity p-type CdTe is still difficult. We used films which were annealed in air at 400°C and had a good sensitivity and weak hole conduction. The thickness of these films was about 1 μ. Thin layers of Ni, Te, Cu, Ag, or Au were used as the lower contact. The best results were obtained using nickel contacts. The n-type region was generated by bombardment with In ions. An aluminum grid-like contact was deposited on top of the film.

The maximum photo-emf was ~0.4 V, produced by illumination with light whose spectral distribution was close to that of solar radiation. The rectification coefficient in darkness was 50 for an applied voltage of 5 V. Because of the high series resistance, the forward and reverse currents were extremely low (Fig. 2a).

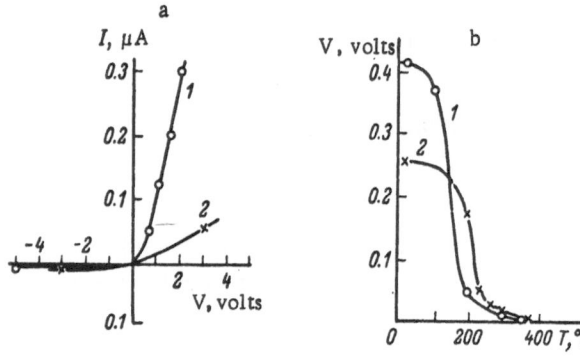

Fig. 2. Influence of annealing on the properties of photodiodes (~2 cm^2 area): a) dark current-voltage characteristics before (1) and after (2) annealing at 350°C; b) changes in the photo-emf due to annealing of lightly (1) and heavily (2) doped samples, prepared using radiation doses of 0.5 mC (30 keV) and 9 mC (10 keV), respectively.

These properties were obtained for an ion energy of 30 keV and doses of 0.5-1 mC. We assumed that a heterojunction could not be produced by such doses. These properties could not be attributed either to a metal—semiconductor contact because no rectificiation was observed before ion bombardment. The presence of a p—n junction was supported also by a reduction in the photo-emf with increasing annealing temperature (Fig. 2b), which was probably associated with diffusion [5].

One of the ways of lowering the series resistance of a photodiode is to produce a metal-saturated surface layer in a semiconductor; this can be done quite easily by ion bombardment. Such a layer ensures that a highly conducting contact is obtained, it absorbs practically no light (its thickness is ~0.01-0.1 μ), and it stabilizes the surface. A p—n junction is probably formed because of a "tail" of the impurity distribution with depth. Such a layer was produced in our p-type CdTe samples by bombardment with low-energy (10 keV) In ions in doses of 5-10 mC.

Illumination with light close to solar radiation produced a short-circuit photocurrent of ~25 μA. Low-temperature annealing (up to 150°C) improved the properties of the photodiodes but further increase in the temperature reduced the photo-emf.

The high series resistance of these photodiodes limits their possible practical applications. Nevertheless, the ion bombardment method seems a very promising technique in the fabrication of multilayer film structures and microelectronic elements.

Literature Cited

1. G. A. Kachurin, A. E. Gorodetskii, Yu. V. Loburets, and L. S. Smirnov, Fiz. Tverd. Tela, 9:494 (1967).
2. D. A. Cusano, IRE Trans. Electron Devices, ED9:504 (1962).
3. H. I. Moss, RCA Rev., 22:29 (1961).
4. A. P. Landsman and R. N. Tykvenko, Radiotekhn. i Élektron., 12:503 (1967).
5. A. E. Gorodetskii, G. A. Kachurin, and L. S. Smirnov, Abstracts of Papers presented at Second All-Union Conf. on Diffusion in Semiconductors, Gor'kii, 1966.

KINETICS OF THE ESTABLISHMENT OF
THE CURRENT IN POLYCRYSTALLINE FILMS OF PbS

V. G. Butkevich and I. A. Drozd

A study was made of polycrystalline films prepared by sublimation in vacuum. Measurements were carried out of the dependence of the film resistance on the frequency (50 cps–10 Mc) of the applied voltage using various field intensities. It was found that at a low field intensity of 10 V/cm the resistance was practically independent of the frequency but in a field of 200 V/cm the resistance fell appreciably when the frequency was increased. An investigation was made of the processes during the establishment of the current in films subjected to voltage pulses. It was found that the application of a square voltage pulse to a film produced initially a current corresponding to the value of the resistance at low field intensities but the equilibrium value of the current in the sample, established after some time, was governed by the static current–voltage characteristic. The time for the establishment of the equilibrium value of the current depended on the field intensity: it decreased with increasing field. The experimental results were in agreement with the assumption of the presence of potential barriers which had voltage-dependent intrinsic resistance and capacitance. Simple estimates showed that nonlinear properties of the films appeared only after the voltage across such an elementary barrier exceeded 0.01 V. Above this voltage, the barriers had rectifying properties and their resistance and capacitance depended on the applied voltage.

Introduction

There are two different approaches to the explanation of the mechanism of conduction and photoconduction in polycrystalline films of PbS. In one case, a photoconducting film is regarded as a single crystal with a spectrum of energy levels in the forbidden band, whose parameters determine the electrical and photoelectric properties of the film. This approach is most evident in the work of Giroux [1]. Other workers [2, 3] assume that the polycrystalline structure of a film is manifested by differences between the physical properties of crystallites and of the layers between them, which give rise to potential barriers exerting an influence on the flow of the current and the photoconductivity of the film. Chasmar [4] explains the frequency dependence of the resistance of PbS films by the shunting of the high-resistivity layers between the crystallites by their capacitances. This explanation has been criticized by others [5, 6]. It has been suggested that a film is shunted not by the capacitances of the barriers but by the distributed capacitance of the film as a whole. However, these critics have ignored the nonlinearity of the current–voltage characteristics of PbS films. They have not stated the field intensities used in their measurements. We investigated the dependence of the resistance of polycrystalline PbS films on the frequency of the applied voltage. We took into account the fact that the capacitance

26

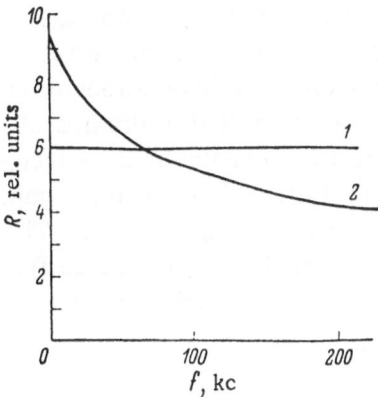

Fig. 1. Frequency dependence of the resistance of a PbS film. E (V/cm): 1) 10; 2) 200.

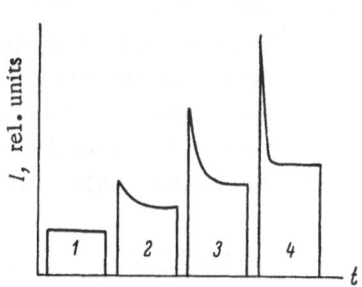

Fig. 2. Dependence of the changes in the current pulse shape on the amplitude of the voltage pulse. E (V/cm): 1) 50; 2) 200; 3) 500; 4) 1000.

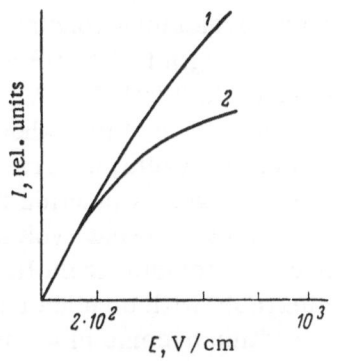

Fig. 3. Current–voltage characteristics of a PbS film, recorded using voltage pulses of different durations (μsec): 1) 2; 2) 300.

and resistance of p−n junctions in such films could depend on the applied voltage. We also investigated the kinetics of the establishment of the current in a film subjected to voltage pulses of various durations and amplitudes.

Results of Measurements

We investigated photoconducting films of PbS prepared by sublimation in vacuum, followed by high-temperature activation. Figure 1 shows the dependences of the resistance of one of the films on the frequency for different applied fields. In a relatively weak field of 10 V/cm (curve 1 in Fig. 1), the resistance of a film was practically constant up to 10 Mc. In a field of 200 V/cm, the resistance clearly depended on the frequency. These dependences were obtained using a conventional circuit with a load resistance whose value was much lower than the film resistance. The voltage was supplied by an audiofrequency oscillator or a signal generator; it was measured with a high-frequency voltmeter. Bearing in mind that the frequency dependence of the film resistance should also be observed in films subjected to voltage pulses, we investigated the processes of establishment of the current under pulse conditions. This method made it possible to carry out a study in a wide range of field intensities, which could not be done using static voltages because of the possible heating of the films. Figure 2 shows the changes in the shape of current pulses observed when the amplitude of the 300 μsec voltage pulses was increased. These curves were obtained for a film whose resistance was $2 \cdot 10^3$ Ω. The distance between the electrodes was 0.5 mm and the width of the film was 20 mm. The interelectrode capacitance (measured with a PIMEL meter) was 2.4 pF. Taking into account the stray capacitances, the intrinsic relaxation time of the equivalent circuit should be less than $5 \cdot 10^{-8}$ sec. The pass band of the oscillograph used in these measurements was 0.1 Mc.

When a sample was replaced by an equivalent resistance, with a 50 pF capacitance in parallel, the shape of the current pulse was identical with the shape of the voltage pulse. The possible role of the contacts was determined using samples with a large distance between the electrodes and illuminating the contact regions with strong (up to 1000 lux) radiation. Since these samples had a high photosensitivity, the shape of the current pulse should have been affected by the motion of the light beam along the sample from one contact to the other if illumination of the contact regions had been the dominant factor. However, no such change in the shape of the current pulse was observed. Figure 3 shows the current—voltage characteristics of one of the samples, recorded using voltage pulses of 2 μsec (curve 1) and 300 μsec (curve 2) duration. The pulse duration of 300 μsec was sufficiently long to observe the decay of the current pulse from its initial peak to a steady-state value, which was assumed to be the equilibrium value.

Discussion of Results

The frequency dependence of the film resistance, observed in high fields, cannot be explained by the distributed capacitance of the film, as suggested by Lax and Sachs [6]. The distributed capacitance considered in [6] is independent of the applied voltage and is present in any film. However, the barrier capacitance is governed by the properties of crystallites and of the layers between them, and it depends on the applied voltage. The general properties of p—n junctions can be used to explain qualitatively the results obtained, assuming that potential barriers are present in a film. It is known [7] that films used as photoresistors consist of crystallites 0.1 μ in size, i.e., the number of such crystallites per unit length is 10^4-10^5 cm^{-1}. At field intensities up to 100 V/cm, the voltage across a single barrier is 0.001-0.01 V. On the other hand, measurements of the temperature dependence of the carrier mobility show [8] that the barrier height is 0.06-0.08 eV. Thus, at field intensities up to 100 V/cm, we are dealing with very low (lower than the barrier height) voltages. At these voltages, a junction represented by a barrier does not exhibit appreciable rectification and the current—voltage characteristic remains linear. A deviation of the current—voltage characteristic from linearity occurs in fields in which a voltage applied to a junction is comparable with or higher than the barrier height, i.e., when junctions connected in the reverse direction become blocked. Since the establishment of the current in a junction takes a finite time, the film resistance naturally depends on the frequency and an initial current peak is observed when voltage pulses are used. This time is governed by the rate of dispersal of the space charge at a junction, and depends on the applied voltage. This is illustrated in Fig. 2. The relaxation time of the initial current peak, defined as the time taken by the current to decrease to 0.63 of its maximum value, decreases from 100 μsec in a field E = 200 V/cm to 10 μsec in a field E = 1000 V/cm.

The authors are planning to carry out a more detailed study of the role of barriers in photoconductivity processes by investigating the influence of illumination and temperature on the kinetics of establishment of the current in lead sulfide films.

Literature Cited

1. G. Giroux, Can. J. Phys., 41:1840 (1963).
2. J. C. Slater, Phys. Rev., 103:1631 (1956).
3. D. P. Snowden and A. M. Portis, Phys. Rev., 120:1983 (1960).
4. R. P. Chasmar, Nature, 161:281 (1948).
5. V. V. Balakov and V. A. Smeshkova, Zh. Tekh. Fiz., 24:989 (1954).
6. M. Lax and R. Sachs, Phys. Rev., 107:650 (1957).
7. R. H. Harada, J. Chem. Phys., 24:447 (1956).
8. R. L. Petritz, Phys. Rev., 104:1508 (1956).

INVESTIGATION OF THE TIME CONSTANTS
OF AN INDIUM ARSENIDE LASER

V. B. Buber, V. V. Nikitin, and K. P. Fedoseev

P. N. Lebedev Physics Institute
Academy of Sciences of the USSR, Moscow

An investigation was made of the time constants of an InAs p−n junction laser. The carrier recombination lifetime was measured in the active region of the diode. The dependence of the delay time of a light pulse, relative to an injection current pulse, was determined. Near the laser threshold, the delay time was ~33 nsec, but when the threshold current was exceeded fivefold the delay decreased to ~6 nsec.

The present paper reports an investigation of the time constants of an indium arsenide injection laser and a determination of the carrier recombination time in the active region of the p−n junction in the laser.

Similar investigations have already been carried out for GaAs injection lasers [1-4].

The delay time was measured by oscillographic display of coherent emission and exciting current pulses.

An indium arsenide injection laser was placed in a Dewar containing liquid nitrogen. It was excited with square current pulses whose rise time was 5 nsec. The amplitude of the injection current pulse could be varied continuously from 10 to 70 A. An indium antimonide photodiode, operating at liquid-nitrogen temperature and having a response time of 10^{-8} sec, was used as the detector. The laser radiation was focused by an optical system onto the sensitive area of the photodetector. The photodetector signal was applied to one input of a double-beam oscillograph. A voltage taken from a calibrated induction-free resistor connected in series with the laser diode was applied to the second input. The laser and injection current pulses were displayed on the oscillograph screen and photographed. Figure 1 shows the oscillograms of these pulses. The delay time was measured with a comparator; it was defined as the time interval between the leading edge of the current pulse and the beginning of the light pulse.

The response time of the recording system was determined using radiation of a fast-response GaAs semiconductor laser ($\tau \sim 10^{-9}$ sec). These measurements showed that the response time was ~$2.0 \cdot 10^{-8}$ sec.

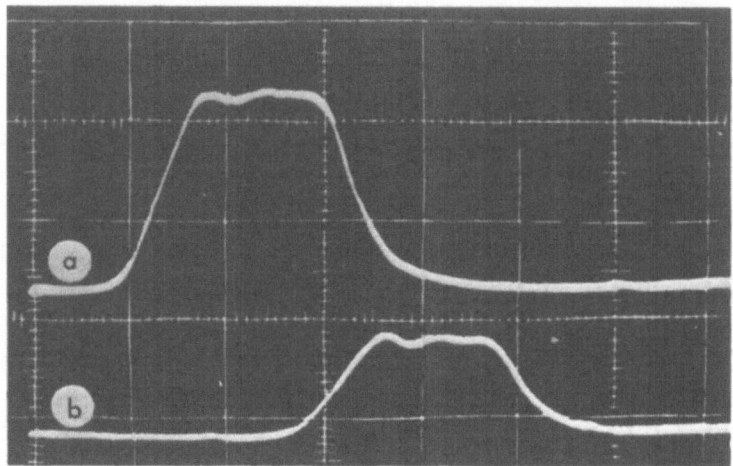

Fig. 1. Oscillograms of the injection current (a) and laser
radiation (b) pulses.

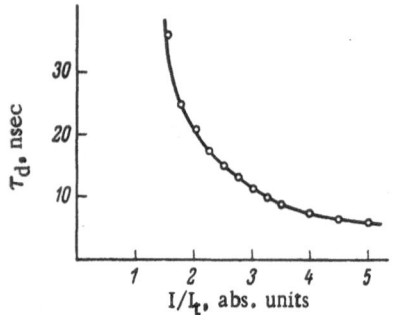

Fig. 2. Dependence of the delay
time of a light pulse on the injection current in InAs diodes.

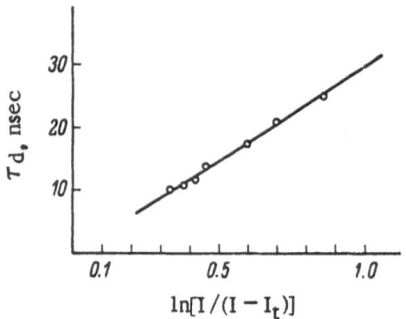

Fig. 3. Dependence of the delay
time on the quantity $\ln[I/(I - I_t)]$.

The dependence of the delay time of the InAs laser on the excitation level is shown in
Fig. 2. It is evident from this figure that the delay time of the InAs injection laser was approximately 33 nsec near the threshold current ($I \approx 1.51 I_t$), but it decreased fairly rapidly with
increasing excitation level so that it was only 6 nsec at an injection current which was five
times higher than the threshold value.

It can be shown [2] that the delay time of coherent emission by a semiconductor laser,
measured relative to the excitation pulse, depends on the injection current in the following way:

$$\tau_d = \tau_{sp} \ln\left(\frac{I}{I - I_t}\right),$$

where τ_{sp} is the spontaneous lifetime of the carriers in the active region of the p−n junction; I_t
is the threshold value of the injection current; I is the injection current.

Plotting this dependence on a semilogarithmic scale, we can determine the carrier lifetime for spontaneous recombination involving transitions between acceptors and the conduction
band.

The dependence $\tau_d = f\{\ln[I/(I-I_t)]\}$, shown in Fig. 3, yields a spontaneous carrier lifetime of $3 \cdot 10^{-8}$ sec. Estimates made using Dumke's quantum-mechanical theory [5] show that this value of the spontaneous lifetime corresponds to a concentration of 10^{17} cm^{-3} uncompensated Zn acceptors in the active region of the p−n junction. This is in satisfactory agreement with an estimate of the acceptor concentration obtained from the conditions during the diffusion of Zn into InAs.

Thus, when the recombination lifetime is 30 nsec the delay time of an InAs injection laser varies from 33 nsec near the threshold to 6 nsec for an injection current five times higher than the threshold value.

Since the recombination lifetime is inversely proportional to the concentration of uncompensated Zn acceptors near a p−n junction [5], we may expect the delay time of an InAs semiconductor laser to decrease with increasing acceptor concentration.

The authors are grateful to the optics specialist E. I. Chigidin for the preparation of the samples, and to A. S. Semenov and V. A. Yakovlev for their valuable advice.

Literature Cited

1. B. S. Goldstein and R. M. Weigand, Proc. IEEE, 53:195 (1965).
2. K. Konnerth and C. Lanza, Appl. Phys. Letters, 4:120 (1964).
3. Yu. A. Drozhbin, V. V. Nikitin, A. S. Semenov, B. M. Stepanov, A. M. Tolmachev, and V. A. Yakovlev, Izmeritel'naya Tekhnika, No. 11, p. 92 (1966).
4. N. G. Basov, Yu. A. Drozhbin, Yu. P. Zakharov, V. V. Nikitin, A. S. Semenov, B. M. Stepanov, A. M. Tolmachev, and V. A. Yakovlev, Fiz. Tverd. Tela, 8:2816 (1966).
5. W. P. Dumke, Phys. Rev., 132:1998 (1963).

TRANSIENT PROCESSES IN SEMICONDUCTOR INJECTION LASERS WITH STRONG OPTICAL COUPLING

V. A. Grekhnev, V. D. Kurnosov, A. A. Pleshkov,
O. N. Prozorov, L. A. Rivlin, A. T. Semenov,
V. V. Tsvetkov, and V. S. Shil'dyaev

A time-scanning image converter with a resolution of $5 \cdot 10^{-12}$ sec was used to observe the interaction between gallium arsenide injection lasers sharing a common junction. A strong optical coupling was maintained through those regions of the junction which were shared by the resonators of different lasers. The absolute value of the absorption coefficient of these regions decreased under the saturating effect of the radiation field of any of the lasers. Finally, this resulted in the switching on or off of one of the lasers by the optical signals of the other lasers. The characteristic times of the transient (switching) processes were of the order of 10^{-10} sec and the durations of the emitted light pulses were of the same order.

1. The operation of semiconductor lasers with isolated injection regions [1] has been investigated before [2-6, etc.], and it has been established that the coupling between different regions, via the common radiation field, plays a dominant role.

The present paper describes the operation of GaAs lasers consisting of several Fabry—Perot resonators linked by a common p—n junction, some regions of which were shared by different resonators. Such regions maintained a strong optical coupling between the lasers, which was manifested by a decrease in the absolute value of the absorption coefficients of these regions under the saturating action of the photon field in each of the resonators. When the absorption coefficient of the interaction region was positive (weak injection), radiation from one of the coupled lasers reduced the losses in the other laser. When the absorption coefficient of the interaction region was negative (strong injection), the gain in one of the lasers decreased under the influence of the other interacting laser.

2. The two effects just described produced basically a change in the self-excitation threshold of a laser under the influence of an external optical signal at a fixed value of the injection current. In particular, the injection level could be set so that a shift of the threshold would switch a laser on or off.

Fig. 1. Semiconductor lasers coupled by the common parts of resonators (a photomicrograph with linear dimensions of each region amounting to about 200 μ).

The transient processes associated with these phenomena had characteristic times of less than 10^{-9} sec. Following the method described in [7], we used a time-scanning image converter with a time resolution of $5 \cdot 10^{-12}$ sec. Our experiments were carried out in an evacuated liquid nitrogen cryostat using gallium arsenide samples with diffused p−n junctions and resonators made by cleaving. Some parts of the injecting region in the p-type material were isolated from one another by photoetched grooves which did not intersect the p−n junction. The resistances due to the presence of these grooves were 1-5 Ω. Figure 1 shows a photomicrograph of one such sample, consisting of nine isolated regions, each of about 200 μ linear size. Different combinations of injecting and noninjecting regions in such samples could be used to produce various coupled laser arrangements.

The absolute values of the injection currents in each of the regions did not exceed a few amperes.

3. The simplest configuration investigated (Fig. 2a), which may be called a triode (because it has three contacts), consists of two crossed resonators (cleaved faces are shown by double lines) with isolated injecting regions 1 and 3 on the p-type side, a common noninjecting (interaction) region 2, and a common base contact on the n-type side of the sample.

The absorption of light, which produces electrons and holes in the noninjecting region 2, can be saturated by the radiation field of either of the lasers sharing the interaction region 2. Therefore, the conditions for the self-excitation and maintenance of the laser emission con-

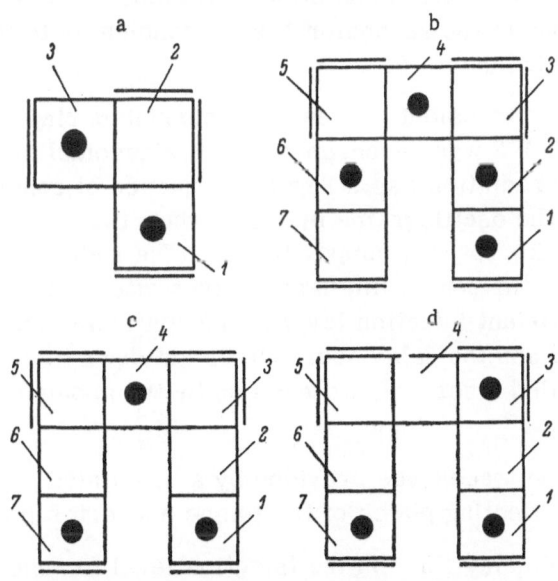

Fig. 2. Investigated laser configurations: a) triode (logic repetition function); b) triode chain (neuristor); c) tetrode (logic multiplication or addition function); d) tetrode (inhibition function).

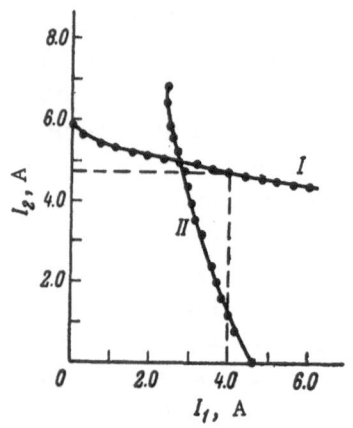

Fig. 3. Dependence of the threshold current density in one of the coupled lasers in a triode configuration (Fig. 2a) on the injection current density in the other laser.

ditions, governed by the amplification of light in regions 1 and 2, respectively, as well as by the absorption of light in region 2, are mutually related, as can be seen from the static threshold characteristics which give the dependences of the threshold current of one of the lasers on the current in the other laser (curves I and II in Fig. 3). Before the point of intersection of these falling characteristics, an increase in the transmission (reduction in the absorption coefficient) in region 2 takes place under the action of the spontaneous radiation from the neighboring injecting region. Beyond the point of intersection, this increase in transmission is due to the coherent photons emitted by the controlling laser. Some samples showed practically no further fall of the threshold current beyond the point of intersection. This was usually due to the nonuniformity of the distribution of radiation in the controlling laser across its p−n junction. Consequently, an increase in the transmission occurred only in a limited part of region 2, which did not affect appreciably the self-excitation threshold.

The coupling between the lasers was not due to spreading of the injection current through the dividing grooves because the reduction in the threshold was observed both under the action of the spontaneous radiation from the neighboring injecting region when it was totally isolated by a groove intersecting the p−n junction, as well as under the action of the coherent radiation from the controlling laser, in which case the injecting region and the interaction region 2 were separated by an additional buffer region. This was true also of all the other experiments described in the present paper.

When injection in region 3 was maintained at the level shown in Fig. 3 by a horizontal dashed line, the excitation of the laser consisting of regions 2 and 3 could occur only in the presence of coherent photons in the resonator 1-2 perpendicular to this laser. In other words, laser 2-3 was switched on by laser 1-2.

In order to observe the transient processes occurring during such switching, the radiation of the controlling laser 1-2 was synchronized by a sinusoidal current of 0.5 Gc frequency. Thus, the controlling laser radiation was a regular sequence of coherent light pulses [5] of about 10^{-10} sec duration. The oscillograms in Fig. 4 show two such triggering light pulses, separated by an interval of $2 \cdot 10^{-9}$ sec (upper trace). The radiation pulses of the controlled laser (lower trace) followed the controlling laser pulses after a delay time of less than 10^{-10} sec (Fig. 4a). When the constant injection level in the controlled laser was increased, the duration of its pulses increased and the delay decreased right down to a negative value (Fig. 4b). This meant that the controlled laser was switched on by the spontaneous radiation of the controlling laser.

Time calibration of the traces was provided by a modulated sinusoidal voltage of 0.5 Gc applied to the transverse deflecting plates of the image converter (middle trace).

The dependence of the delay time on the injection level for one of the samples is shown in Fig. 5.

4. The optical connection of two lasers in series was obtained in a three-link chain (Fig. 2b). The sinusoidal synchronizing signal was applied to region 1; the injection level in regions 2, 4, and 6 was constant, and regions 3 and 5 were the interaction regions where there is no injection. The excitation was propagated from laser 1-2-3 to laser 5-6-7 through laser 3-4-5 with an average delay time of about 10^{-10} sec per link.

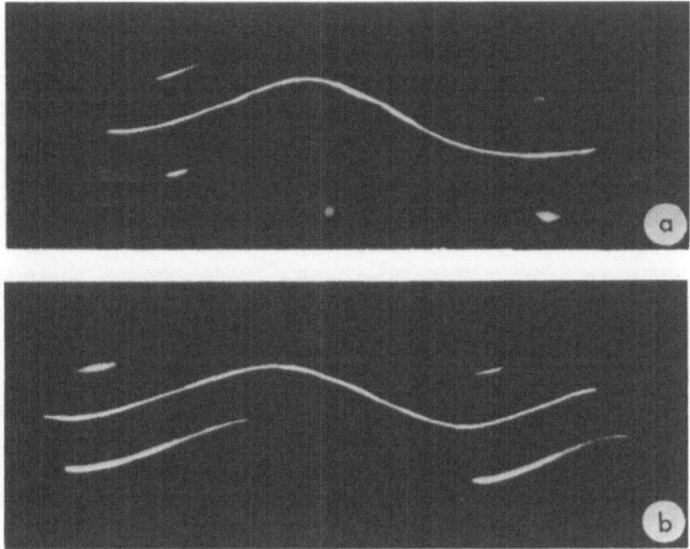

Fig. 4. Oscillograms of the interaction of coupled lasers in a triode configuration (Fig. 2a): a) the triggering pulses of coherent light, repeated at intervals of $2 \cdot 10^{-9}$ sec (upper trace) and the light pulses of the controlled laser, following after a delay time of less than 10^{-10} sec (lower trace); b) the pulses of a laser triggered by spontaneous radiation (lower trace) leading the pulses of the controlling laser (upper trace). The middle trace is a time calibration sinusoid with a frequency of 0.5 Gc.

5. In the tetrode configuration shown in Fig. 2c, we observed switching of laser 3-4-5 by the simultaneous action of lasers 1-2-3 and 5-6-7. Sinusoidal synchronization was applied to regions 1 and 7 and the injection in region 4 was maintained at a constant level. The oscillograms in Fig. 6 show switching-on by the simultaneous action of the two controlling lasers (Fig. 6a) and by a relative shift of the controlling light pulses by $1.5 \cdot 10^{-10}$ sec (Fig. 6b); the switching action stopped when the shift reached $2.5 \cdot 10^{-10}$ sec (Fig. 6c).

When the injection level in region 4 was increased, or when the relative length of this region was made greater, we observed the switching-on of laser 3-4-5 by either of the two controlling lasers and by both of them.

6. The coupling of lasers through an interaction region subjected to strong injection was observed in the tetrode configuration shown in Fig. 2d. Sinusoidal synchronization was applied to regions 1 and 7, a constant injection current was applied to an interaction region 3, and no injection took place in the interaction region 5.

Laser 3-4-5 was switched on by laser 5-6-7 when the gain in region 3, maintained by the injection current, was sufficiently high (Fig. 7a).

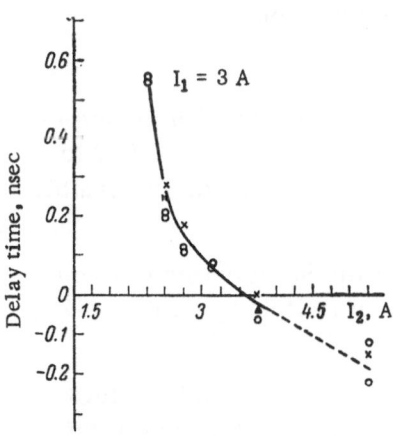

Fig. 5. Dependence of the delay time on the level of the constant component of the injection current for the triode configuration shown in Fig. 2a.

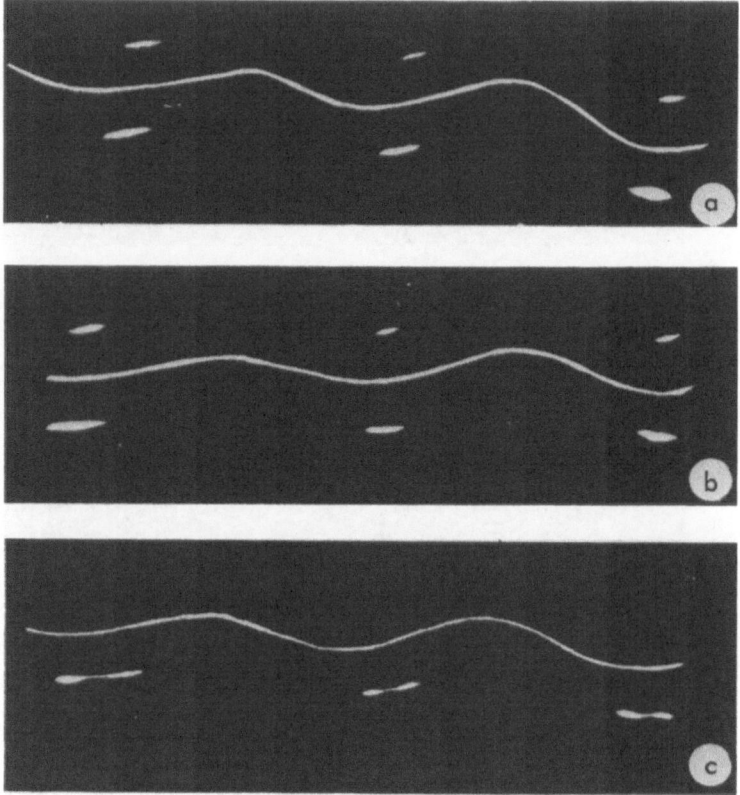

Fig. 6. Oscillograms of the interaction of coupled lasers
in a tetrode configuration (Fig. 2c): a) two switching-on
pulses, separated by a time interval $1.5 \cdot 10^{-10}$ sec, trigger the controlled laser; switching-on pulses separated
by a time interval of $2.5 \cdot 10^{-10}$ sec do not trigger the controlled laser. The upper trace represents the pulses emitted
by the controlled laser. The lower trace shows the triggering pulses, and the middle trace is a time calibration
sinusoid of 0.5 Gc frequency.

When laser 1-2-3 produced a short light pulse, synchronized by sinusoidal injection in
region 1, we observed a reduction in the gain in the interaction region 3 which could fall to a
level at which the subsequent switching-on of laser 3-4-5 by laser 5-6-7 became impossible
(Fig. 7b).

Reestablishment of the initial gain in the interaction region 3 by the injection current
required a certain time after which laser 5-6-7 was again capable of switching-on the controlled laser (Fig. 7c).

It was thus impossible to switch-on the controlled laser during a time which, following
neurophysiological terminology, could be called the refractory period. The duration of this
period decreased with increasing injection level in the interaction region.

7. Semiconductor laser configurations with strong optical coupling of the types considered here and in other combinations can be used to control effectively the emission from
a laser by the coherent radiation from another laser, and the switching time is less than 10^{-9}
sec. This opens up some promising applications in fast switching techniques. In this con-

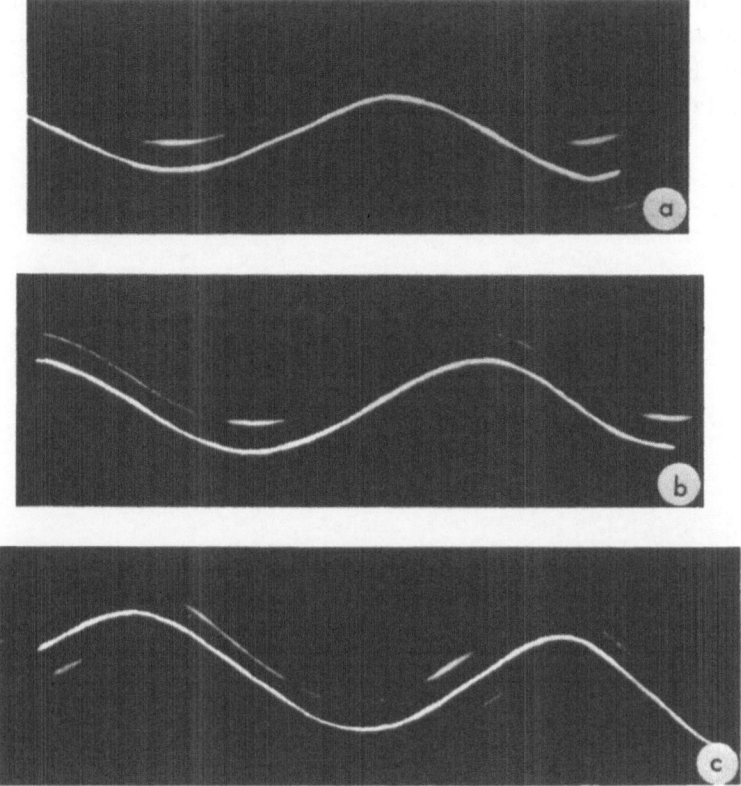

Fig. 7. Oscillograms of the interaction of coupled lasers in a tetrode configuration (Fig. 2d): a) switching-on pulse from laser 5-6-7 (upper trace) triggers laser 3-4-5 (lower trace); b) switching-on pulse from laser 5-6-7 is delayed by a time interval of $5 \cdot 10^{-10}$ sec after blocking pulse from laser 1-2-3 (upper trace) and the controlled laser is not switched-on; c) switching-on pulse from laser 5-6-7 follows a blocking pulse from laser 1-2-3 after an interval of $6 \cdot 10^{-10}$ sec (upper trace) and triggers the controlled laser 3-4-5 (lower trace).

nection, it is worth mentioning that the schemes described earlier in the paper carry out essentially elementary logic operations. Thus, for example, a triode scheme executes the repetition function (in neuristor terminology [8] it is called the T-junction); the three-link chain is a segment of a neuristor chain [8]; a tetrode scheme can be used for logical multiplication (coincidence circuit) or logic addition, or as an inhibition element (R-junction [8]).

Literature Cited

1. G. J. Lasher, Solid State Electron., 7:707 (1964).
2. M. I. Nathan, J. C. Marinace, R. F. Rutz, A. E. Michel, and G. J. Lasher, J. Appl. Phys., 36:473 (1965).
3. P. G. Eliseev, A. A. Novikov, and V. B. Fedorov, Zh. Éksp. Teor. Fiz. Pis. Red., 2:58 (1965).
4. N. G. Basov, Yu. P. Zakharov, V. V. Nikitin, and A. A. Sheronov, Fiz. Tverd. Tela, 7:3128 (1965).
5. V. I. Magalyas, A. A. Pleshkov, L. A. Rivlin, A. T. Semenov, and V. V. Tsvetkov, Zh. Éksp. Teor. Fiz. Pis. Red., 6:550 (1967).

6.   L. A. Rivlin and V. S. Shil'dyaev, Zh. Éksp. Teor. Fiz. Pis. Red., 6:659 (1967).

7.   V. D. Kurnosov, V. I. Magalyas, A. A. Pleshkov, L. A. Rivlin, V. G. Trukhan, and V. V. Tsvetkov, Zh. Éksp. Teor. Fiz. Pis. Red., 4:449 (1966).

8.   H. D. Crane, Proc. IRE, 50:2048 (1962).

POSSIBILITY OF PAIR CORRELATION OF ELECTRONS AND HOLES IN A SANDWICH CONSISTING OF n- and p- TYPE SEMICONDUCTOR FILMS

V. G. Kogan and B. A. Tavger

N. I. Lobachevskii Gor'kii State University

A theoretical analysis was made of the energy spectrum for a system comprised of n- and p-type semiconductor films, separated by a dielectric film sufficiently thick to prevent tunneling. This system is of interest due to the possible formation of bound states. The various current flows possible with different circuit configurations lead to some new results. Considering the average distance between electrons to be of the same order as the film thickness, spectrum excitation, interaction of the holes and electrons across the dielectric, and critical-temperature effects are discussed.

The present paper gives a theoretical analysis of the energy spectrum of a system consisting of n- and p-type semiconductor films separated by a thin dielectric layer. It is assumed that the dielectric layer is sufficiently thick to prevent the tunneling of electrons and holes from one film into the other and, therefore, n- and p-type films can be regarded as separated in space. The system is interesting because there is a Coulomb attraction between the electrons in film a and the holes in film b and such attraction may, in principle, give rise to bound states.

The problem is formally similar to that considered by some workers [1, 2] who studied the possible instability of the semimetallic state due to the formation of electron−hole pairs. The conditions necessary for the appearance of such pairs in a p−n junction between two bulk regions are considered in [3]. Since the motion of a neutral electron−hole pair does not give rise to an electric current, the formation of such pairs in a semimetal represents the transition to the dielectric state.

The spatial separation of the electrons and holes in our system introduces a completely new aspect: an electron−hole pair moving as a unit need not represent zero current if the films a and b are connected in series (Fig. 1). In this case, the motion of a pair, for example, to the left, gives rise to a current flowing to the left in the p-type film and to the right in the n-type film. When the films are connected in parallel (Fig. 2), the electron−hole pairs no longer produce a current. Their existence is then simply manifested by the special features of the current−voltage characteristic of the sandwich, which are absent from the characteristic of a

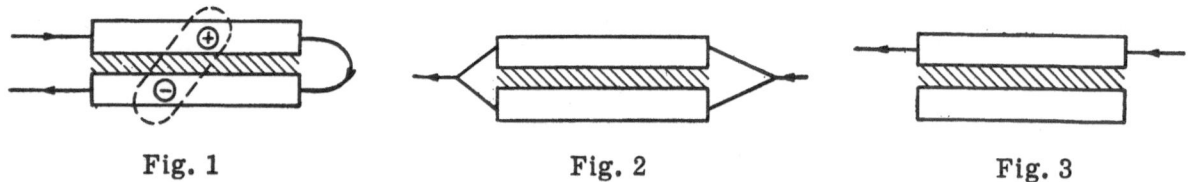

Fig. 1   Fig. 2   Fig. 3

single film. The flow of a current through one of the films in the sandwich (Fig. 3) should also give rise to some special effects. As in the preceding case, the appearance of a pair correlation may reduce the electrical conductivity, which can affect the temperature dependence of the current.

Excitation Spectrum

The state of an electron or a hole in a thin film can be represented by the projections p_x, p_y of the crystal momentum along directions parallel to the film and by the integer n, which specifies the state in transverse finite motion. The separation between the energy levels with different values of n depends on the film thickness L and on the transverse effective mass: for $L \sim 10^{-6}$ cm this separation may reach a value corresponding to 0.001°K. Quantization of the electron energy, due to the finite thickness of the film, can occur under some conditions [5] and has been observed experimentally in Bi films [4].

When the carrier density N is increased and the number of occupied levels in transverse motion becomes correspondingly larger, the Debye screening of the Coulomb interaction becomes stronger. Therefore, we shall consider the case when only the states with n = 1 are occupied in both films. We can easily show that, in this case, we must have $N \propto L^{-3}$ [5]; then, the average distance between the electrons is of the order of the film thickness, and the screening radius of the Coulomb interaction is at least of the same order of magnitude.

We shall assume that both semiconductor films are degenerate. The Fermi momenta for longitudinal motion p_0 depend on the carrier density and are assumed to be equal. The Hamiltonian of the system is

$$H = \sum_p (\xi_p^a a_p^+ a_p + \xi_p^b b_p^+ b_p) - \sum_{p'+q'=p+q} \gamma_{p'q', pq} a_{p'}^+ b_{q'}^+ b_q a_p, \tag{1}$$

where $2m_{a, b}\xi_p^{a, b} = \hbar^2 (p^2 - p_0^2)$; a_p, b_p are the annihilation operators of, respectively, an electron of momentum p in film a and a hole of momentum p in film b; $-\gamma_{p'q',pq}$ is the matrix element for the scattering of an electron of momentum p and a hole of momentum q into states with p',q'.

The spectrum of excitations can be found by diagonalizing Eq. (1) using a linear transformation for the operators, such as the Bogolyubov transformation for fermions

$$\left.\begin{aligned} a_p &= u_p a_p - v_p b_{-p}^{\pm}, \\ \beta_p &= u_p b_p + v_p a_{-p}^{\pm}. \end{aligned}\right\} \tag{2}$$

We can easily show that, if $u_p^2 + v_p^2 = 1$, the new operators satisfy the anticommutation relationships

$$\{a_p, a_q^+\} = \{\beta_p, \beta_q^+\} = \delta_{pq}, \quad \{a, \beta\} = 0,$$

since all anticommutators obey $\{a, b\} = 0$.

Substituting Eq. (2) in Eq. (1), we obtain the following expression for the energy:

$$E = \sum_p v_p^2 (\xi_p^a + \xi_p^b) + u_p^2 (\xi_p^a n_p^\alpha + \xi_p^b n_p^\beta) - v_p^2 (\xi_p^b n_p^\alpha + \xi_p^a n_p^\beta) - \sum_{pq} \gamma (p-q) u_p v_p (1 - n_p^\alpha - n_p^\beta) u_q v_q (1 - n_q^\alpha - n_q^\beta), \tag{3}$$

$$\gamma (p-q) = \gamma_{p, -p; q, -q'}$$

which can be minimized with respect to the parameter u_p to give

$$u_p^2 = \frac{1}{2}\left(1 + \frac{\xi_p}{\sqrt{\Delta_p^2 + \xi_p^2}}\right), \qquad v_p^2 = \frac{1}{2}\left(1 - \frac{\xi_p}{\sqrt{\Delta_p^2 + \xi_p^2}}\right), \tag{4}$$

where

$$2\xi_p = \xi_p^a + \xi_p^b \text{ and } \Delta_p = \sum_q \gamma\,(p-q)\,u_q v_q\,(1 - n_q^\alpha - n_q^\beta). \tag{5}$$

The equation for Δ_p is obtained, as usual, by substituting Eq. (4) into Eq. (5):

$$\Delta_p = \frac{1}{2}\sum_q \gamma\,(p-q)\,\Delta_q\,\frac{1 - n_q^\alpha - n_q^\beta}{\sqrt{\Delta_q^2 + \xi_q^2}}. \tag{6}$$

For $\xi_p^a = \xi_p^b$, this equation is identical with Cooper's equation. The spectrum of α and β excitations is of the form

$$\left.\begin{array}{l} \varepsilon_p^\alpha = \dfrac{\delta E}{\delta n_p^\alpha} = \dfrac{1}{2}\,(\xi_p^a - \xi_p^b) + \sqrt{\Delta_p^2 + \xi_p^2}, \\[2mm] \varepsilon_p^\beta = \dfrac{\delta E}{\delta n_p^\beta} = \dfrac{1}{2}\,(\xi_p^b - \xi_p^a) + \sqrt{\Delta_p^2 + \xi_p^2}. \end{array}\right\} \tag{7}$$

(The same spectrum of excitations is obtained in [1] for an electron−hole system.) When $p = p_0$, we find that $\xi_{p_0}^a = \xi_{p_0}^b = \xi_{p_0} = 0$ and $\varepsilon^\alpha = \varepsilon^\beta = \Delta_{p_0}$, so that the quantity Δ_{p_0} represents, as usual, an energy gap at the Fermi level. If, in addition to this relationship for the Fermi momenta, we also have $m_a = m_b$, we find that $\varepsilon_p^\alpha = \varepsilon_p^\beta = \sqrt{\Delta_p^2 + \xi_p^2}$.

The nature of the two types of excitation described by Eq. (7) becomes clear when we consider momenta differing greatly from p_0, so that Δ is unimportant:

$$\left.\begin{array}{l} \varepsilon^\alpha = \dfrac{1}{2}\,(\xi^a - \xi^b) + \dfrac{1}{2}\,|\xi^a + \xi^b| = \left\{\begin{array}{l} \xi^a, \;\; p > p_0, \\ -\xi^b, \;\; p < p_0, \end{array}\right. \\[4mm] \varepsilon^\beta = \dfrac{1}{2}\,(\xi^b - \xi^a) + \dfrac{1}{2}\,|\xi^a + \xi^b| = \left\{\begin{array}{l} \xi^b, \;\; p > p_0, \\ -\xi^a, \;\; p < p_0, \end{array}\right. \end{array}\right\} \tag{8}$$

i.e., when $p > p_0$ an excitation is an electron in film a (n-type film) and a hole in a film b (p-type film); when $p < p_0$, a hole appears in film a and an electron in film b.

The energy of the ground state (T = 0) of the system E_{0s} is obtained from Eq. (3) by assuming that $n^\alpha = n^\beta = 0$; in order to obtain the energy of the ground state of a normal system we must also assume that $\Delta = 0$. We can easily show that $E_{0n} - E_{0s} \propto \Delta^2$, so that the effects considered may take place if Eq. (6) has real solutions.

Interaction of a and b Particles

This interaction is described by an equation which can be represented as follows:

$$\tag{9}$$

where γ^{ab} denotes the matrix element of the Coulomb interaction between two particles, one of which is in film a and the other in film b (this element takes into account the interactions with other a and b particles); γ^{bb} has a similar meaning. The quantities denoted by V are the matrix elements of the Coulomb interaction without allowance for the correlation with other particles.

Equation (9) must be supplemented by an equation for γ^{bb} :

so that we obtain a closed system of equations. Because of spatial isolation these equations do not include loops consisting of the Green's functions of a and b particles. (The lines G_0^a and G_0^b cannot begin at the same point in space: in other words, transitions from a to b are forbidden. This assumption is justified because a dielectric layer $\sim 10^{-6}$ cm thick is sufficient to make the tunnel effect negligible.)

Writing Eq. (9) in the momentum representation, we must bear in mind that the Green's function $G_0(\mathbf{r_1}, \mathbf{r_2}; \varepsilon) = \sum_\nu G_{0\nu}(\varepsilon) \chi_\nu^*(\mathbf{r_1}) \chi_\nu(\mathbf{r_2})$ $[\chi_\nu(\mathbf{r})$ is the wave function of an electron in state $\nu]$ for a film is of the form

$$G_0(r_1 z_1,\ r_2 z_2;\ \varepsilon) = \sum_n \int G_{0n}(p,\ \varepsilon) \chi_{np}^*(r_1 z_1) \chi_{np}(r_2 z_2) \frac{S dp}{(2\pi\hbar)^2} .$$

Here, r is the radius vector of a particle, directed parallel to the film; z is the transverse coordinate; p is the two-dimensional crystal momentum corresponding to the longitudinal motion; S is the area of the film.

We shall assume that the motion of an electron along a film is free and, therefore,

$$\chi_{np}(r,\ z) = \frac{1}{\sqrt{SL}} \varphi_n(z) e^{ipr}, \tag{10}$$

where $\frac{1}{\sqrt{L}} \varphi_n(z)$ gives the state of an electron in transverse motion. We can now show easily that the matrix element is $V_{p_1 p_2,\ p_1' p_2'} = V(k)$ $(k = p_1' - p_1 = p_2 - p_2'$ is the longitudinal momentum being transferred) and we find from Eq. (9) that:

$$\gamma_{mn,\ ij}^{ab}(k) = V_{mn,\ ij}^{ab}(k) + \sum_{st} V_{ms,\ it}^{aa}(k) \Pi_{st}^{a}(k) \gamma_{tn,\ sj}^{ab}(k) + \sum_{st} V_{ms,\ it}^{ab}(k) \Pi_{st}^{b}(k) \gamma_{tn,\ sj}^{bb}(k) \tag{11}$$

(the subscripts represent the transverse states).

The polarization operator, which occurs in Eq. (11), is given by

$$\Pi_{st}^{a,\ b}(k) = -i \int G_{0s}^{a,\ b}(p,\ \varepsilon) G_{0t}^{a,\ b}(p + k,\ \varepsilon)\, d\varepsilon\, \frac{S dp}{(2\pi)^2} .$$

Integration at T = 0 without allowance for the delay effects and assuming that $kL \ll 1$ (only the low values of k are important in the Coulomb interaction) gives

$$\Pi_{s,\ t}^{a,\ b} = -\frac{m_{a,\ b}^* S}{2\pi\hbar^2} \left(1 + \frac{2k}{p_0} \ln \frac{4p_0}{k}\right), \tag{12}$$

if the smaller of the two numbers s and t represents an occupied level; otherwise, $\Pi = 0$ (we can now omit the subscripts in the polarization operator).

A considerable simplification of the system of equations (11) is obtained because the matrix elements of the transitions in which the transverse state is altered are considerably smaller than the matrix elements for the transitions in which each particle remains at its own level. In fact, for eigenfunctions of the type given by Eq. (10), we easily obtain the matrix elements

$$V_{ij,\,ij}^{aa} = \frac{2e^2}{SL_a^2\varepsilon k} \iint\limits_{0}^{L_a} e^{-k|z_2-z_1|}\varphi_i^2(z_1)\,\varphi_j^2(z_2)\,dz_1 dz_2,$$

$$V_{ij,\,ij}^{ab} = -\frac{2e^2}{SL_aL_bk\varepsilon}\,e^{-kd} \int\limits_{0}^{L_a}\int\limits_{0}^{L_b} e^{-k(z_2+z_1)}\varphi_i^2(z_1)\,\varphi_j^2(z_2)\,dz_1 dz_2 \qquad (13)$$

(d is the thickness of the dielectric layer). If the transverse state changes, then, for example,

$$V_{ij,\,il}^{aa} = \frac{2e^2}{SL_a^2 k\varepsilon} \int\limits_{0}^{L_a} e^{-k|z_2-z_i|}\varphi_i^2(z_1)\,\varphi_j(z_2)\,\varphi_l(z_2)\,dz_1 dz_2,$$

and expanding the exponential function in the integrand (for $kL \ll 1$) we obtain $V_{ij,il} \ll V_{ij,ij}$.

Now, instead of Eq. (11), we have the following system (the second pair of indices is identical with the first):

$$\gamma_{il}^{ab} = V_{il}^{ab} + \sum_{s}^{n_a} V_{is}^{aa}\Pi^a\gamma_{sl}^{ab} + \sum_{s}^{n_b} V_{is}^{ab}\Pi^b\gamma_{sl}^{bb},$$

$$\gamma_{jl}^{bb} = V_{jl}^{bb} + \sum_{s}^{n_a} V_{js}^{ba}\Pi^a\gamma_{sl}^{ab} + \sum_{s}^{n_b} V_{js}^{bb}\Pi^b\gamma_{sl}^{bb}. \qquad (14)$$

Here, $i = 1, 2,...,n_a$; $j = 1, 2,...,n_b$; n_{ab} are the numbers of occupied levels in the transverse motion; l is a fixed index. We shall not consider the system (14) in general but note only that it gives, in particular, a natural result: when the number of occupied levels increases, the Debye screening becomes stronger.

If $n_a = n_b = 1$, Eq. (14) becomes identical with the system of equations derived in [6] in an investigation of the mutual influence of two different groups of electrons (for example, in alloys) on their superconducting properties. In our case, γ^{aa} and γ^{bb} remain positive (corresponding to repulsion), and

$$\gamma^{ab} = V^{ab}\,[1 - \Pi^a V^{aa} - \Pi^b V^{bb} + \Pi^a\Pi^b\,(V^{aa}V^{bb} - V^{ab}V^{ab})]^{-1}. \qquad (15)$$

The matrix elements in Eq. (15) can be found by specifying the functions $\frac{1}{\sqrt{L}}\varphi_1(z)$, which we shall assume to be $\sqrt{\frac{2}{L}}\sin\frac{\pi}{L}z$. Integrating Eq. (13), we obtain

$$V^{ab} = -\frac{e^2}{2Sk\varepsilon}\,[1 - k\,(\bar{L} + d)],\quad 2\bar{L} = L_a + L_b,$$

$$V^{aa} = \frac{e^2}{2Sk\varepsilon}\,(1 - 0.2kL_a)$$

(V^{bb} is obtained by replacing L_a with L_b).

At low values of k ($kL \ll 1$) the terms ΠV in the denominator of Eq. (15) can easily be seen to be much larger than the term containing $\Pi^a\Pi^b$ (which also occurs in the denominator) and

$$\gamma^{ab} = -\frac{e^2}{2S\varepsilon\,(k + \varkappa)}\,, \qquad (16)$$

where

$$\varkappa = \frac{(m_a + m_b)\,e^2\beta}{4\pi\varepsilon\hbar^2}\,, \qquad (17)$$

and β is a factor of the order of unity which is obtained by averaging the terms enclosed in parentheses in the expression for the polarization operator (12).

Critical Temperatures

Since in the isotropic case Δ_p is independent of the direction of $\mathbf{p}$, Eq. (6) can be averaged approximately along the directions of the vector $\mathbf{p}$. The result of such averaging of $\gamma(p-q)$ is

$$\gamma = \frac{e^2 \alpha}{2\pi S p_0 \varepsilon},$$

where α is a vector of the order of unity, which depends on the nature of the function $\gamma(k)$. If we assume that γ is of the form given by Eq. (16) for $k < \varkappa$ and that $\gamma = 0$ for $k > \varkappa$, we find that $\alpha = \ln 2$.

Assuming, as usual, that $\Delta = $ const near the Fermi boundary, we integrate Eq. (6) in a two-dimensional region $\left(\sum\limits_q \rightarrow \int \frac{2 \cdot 2\pi q \, dq \, S}{(2\pi)^2} \right)$ at $T = 0$:

$$\Delta_0 = 2\Delta E \exp\left(-\frac{2\pi^2 p_0 \varepsilon \hbar^2}{2\bar{m} e^2 \alpha}\right). \tag{18}$$

We have used here $\hbar^2 q \, dq = 2\bar{m} \, d\bar{\xi}_q$, where $\bar{m}$ is the reduced mass. The quantity ΔE is the half-width of the energy "layer" near $\bar{\xi} = 0$, in which attraction takes place, and its value is $\hbar^2 p_0 \varkappa / 2\bar{m}$. Comparing Eq. (18) with the usual expression for an energy gap, we obtain the effective interaction constant

$$g_{\text{eff}} \simeq \frac{2\bar{m} e^2}{2\pi^2 p_0 \varepsilon \hbar^2}. \tag{19}$$

The dependence $\Delta(T)$ is given by Eq. (6) in which $n^{\alpha,\beta} = \left(\exp\frac{\varepsilon^{\alpha,\beta}}{T} + 1\right)^{-1}$. The critical temperature T_c corresponds to closure of the gap $\Delta = 0$:

$$1 = \frac{\gamma}{2} \sum_p \frac{1}{|\xi|}\left[1 - \left(\exp\frac{\varepsilon^\alpha}{T_c} + 1\right)^{-1} - \left(\exp\frac{\varepsilon^\beta}{T_c} + 1\right)^{-1}\right].$$

The expression in square brackets can easily be transformed into $\frac{1}{2}\left(\text{th}\,\frac{\varepsilon^\alpha}{2T_c} + \text{th}\,\frac{\varepsilon^\beta}{2T_c}\right)$, and, using Eq. (8), we obtain

$$1 = \frac{\gamma}{2} \sum_p \frac{1}{2\xi_p}\left(\text{th}\,\frac{\xi_p^a}{2T_c} + \text{th}\,\frac{\xi_p^b}{2T_c}\right). \tag{20}$$

Replacing $\xi^{a,b}$ with $(2\bar{m}/m_{a,b})\bar{\xi}$, we integrate to find that

$$T_c = 0.57 \frac{2\sqrt{m_a m_b}}{m_a + m_b} \Delta_0. \tag{21}$$

We note that the dependence of T_c on the ratio of the effective masses is not given by Eq. (21) since ΔE may also depend on m_b/m_a. We can easily show that T_c increases when the difference between the effective masses increases for a given value of the smaller mass (this is due to the fact that $g \propto \bar{m}$).

It must be mentioned that the results obtained are valid for $\varkappa \ll p_0$, which is equivalent to the condition that the Coulomb interaction energy should be small compared with the Fermi energy. The case when the Coulomb interaction is not weak must be considered separately.

The authors are grateful to Yu. V. Kopaev, A. N. Kozlov, and V. Z. Kresin for their valuable comments.

Literature Cited

1. L. V. Keldysh and Yu. V. Kopaev, Fiz. Tverd. Tela, 6:2791 (1964).
2. A. N. Kozlov and L. A. Maksimov, Zh. Éksp. Teor. Fiz., 48:1184 (1965).
3. Yu. V. Kopaev, Fiz. Tverd. Tela, 7:2902 (1965).
4. Yu. F. Ogrin, V. N. Lutskii, and M. I. Elinson, Zh. Éksp. Teor. Fiz. Pis. Red., 3:114 (1966).
5. B. A. Tavger and V. Ya. Demikovskii, Fiz. Tverd. Tela, 5:644 (1963).
6. B. T. Geilikman, Usp. Fiz. Nauk, 88:327 (1966).

THEORY OF THE ELECTROACOUSTIC INTERACTION AT A METAL—SEMICONDUCTOR POINT CONTACT

É. G. Melikyan

Erevan' Physics Institute

A simplified theory is given for the positive electroacoustic feedback at a metal-semiconductor point contact, which has been observed experimentally by the author. A nonlinear equation of motion is derived and a method for the approximate integration of this equation is given. Expressions are obtained for the condition of self-excitation and for the amplitude of steady-state quasiharmonic oscillations.

The present author discovered [1] positive electroacoustic feedback at a metal—semiconductor point contact. An approximate theory of this feedback is given in the present communication.

As reported in [1], a metal—semiconductor point contact has a positive electroacoustic feedback — an electroacoustic interaction — sufficiently strong for the generation of electric and acoustic oscillations.

Considerable mathematical difficulties are encountered in an attempt to develop a comprehensive theory of the electroacoustic interaction.

The present paper analyzes a simplified model and gives an approximate theory of the electroacoustic interaction.

Simplified Model of the Electroacoustic Interaction

A simplified model of a metal—semiconductor point contact, suitable for the investigation of the electroacoustic interaction, is given in Fig. 1. It is assumed that the metal cylinder has no mechanical mass and that the whole mass of the metal point is concentrated at the end of the cylinder in the form of an absolutely rigid body of mass M.

We shall only consider oscillations of moderate frequencies when, on the one hand, the wavelength of these oscillations in the cylinder and the crystal exceeds the dimensions of both but, on the other hand, the thermal wavelength in the cylinder λ is considerably less than l_c. It is known that

$$\lambda = 2\pi \sqrt{\frac{2\varkappa}{\omega}},$$

46

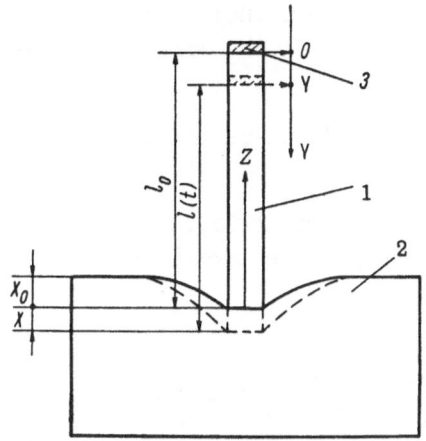

Fig. 1. Simplified model of a contact: 1) metal cylinder of cross section S and length l_c; 2) large semiconductor crystal; 3) mass M of the metal point (cylinder). The dashed curve represents the position of the cylinder at a given moment under a dynamic load.

where $\varkappa$ is the thermal conductivity of the cylinder and ω is the angular frequency of the oscillations [2].

We shall ignore the temperature dependences of the elastic constants. Moreover, we shall assume that the alternating component of the Joule heat is evolved only in the plane of the contact of the cylinder with the crystal and that this heat is transmitted only along the cylinder. In fact, this means that we shall consider solely the thermal expansion of the cylinder. An allowance for the thermal expansion of the crystal yields no additional information but simply complicates the equations. Consequently, we can ignore the thermal expansion of the crystal in an approximate theory. We shall also ignore the thermal radiation from the lateral surface of the cylinder and the temperature dependence of the resistance of the metal−semiconductor contact.

We shall not consider in detail the influence of the frictional forces and the radiation of the acoustic energy by the cylinder or crystal. We shall describe the effect of these factors phenomenologically by including an additional term in the equation of motion.

The potential difference U between the crystal and the cylinder is assumed to be independent of time, i.e., it is assumed that the load resistance in the electric circuit has no influence on the phenomena considered.

Equation of Motion

We shall use F_0 to denote the static force which presses the cylinder to the semiconductor crystal and E to denote Young's modulus of the metal cylinder. We then obtain

$$\frac{l_c - l_0}{l_c} = \frac{F_0}{ES},$$

where l_0 is the length of the cylinder in the static deformed state. The length of the cylinder at a given time t will be denoted by $l(t)$. We shall direct the Y axis as shown in Fig. 1 and place the origin at the end of the cylinder subjected to a static force F_0. The force F_0 also produces a static deformation of the semiconductor crystal, and the sag X_0 of the crystal can be represented by the formula [3]

$$X_0 = F_0 \frac{1 - \mu}{4G} \sqrt{\frac{\pi}{S}},$$

where μ is Poisson's ratio and G is the shear modulus. The position of the contact between the cylinder and the semiconductor crystal will be measured from the point X_0.

We can easily show that

$$Y(t) = X(t) + l_0 - l(t),$$

where X(t) represents the additional sag of the crystal under a dynamic load.

The instantaneous value of the alternating component of a current passing through the contact will be denoted by $\mathscr{J}(t)$. If T_0 is the average value of the temperature of the contact,

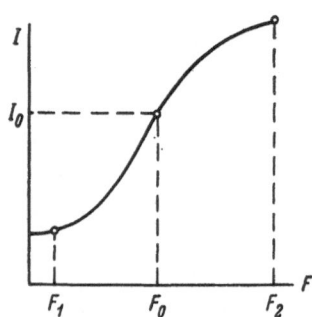

Fig. 2. Dependence of the current through the contact on the force pressing the cylinder against the crystal.

it follows that when $\mathcal{J}(t) > 0$ the current increases the temperature of the contact, but when $\mathcal{J}(t) < 0$ the current reduces the temperature below T_0.

The generality of our theory will not be affected if we assume that $T_0 = 0$.

The temperature distribution along the cylinder at a time t can be described with sufficient accuracy by the expression [4]:

$$T(z, t) = \frac{U}{KS} \sqrt{\frac{\varkappa}{\pi}} \int_0^t \frac{\mathcal{J}(t)}{\sqrt{t-\tau}} \exp\left[-\frac{z^2}{4\varkappa(t-\tau)}\right] d\tau,$$

where K is the thermal diffusivity; z is the abscissa of a point in the cylinder. We shall measure the time from the moment when the temperature of the contact is zero.

Any change in temperature alters the length of the cylinder. Denoting the increment in the length of the cylinder by Δt and its thermal expansion coefficient by α, we obtain

$$\Delta(t) = \alpha \int_0^{l_0} T(z, t)\, dz.$$

Since the amplitude of the thermal waves decreases rapidly along the cylinder axis, the upper limit in the integral can be replaced by infinity, which simplifies the expression for $\Delta(t)$:

$$\Delta(t) = \frac{\alpha \varkappa U}{KS} \int_0^t \mathcal{J}(\tau)\, d\tau.$$

The relationship between the acceleration of the mass M and the sag of the crystal X is given by the formula

$$X(t) = -MY''(t)\frac{1-\mu}{4G}\sqrt{\frac{\pi}{S}}.$$

Hence, we obtain the following equation of motion:

$$\Delta(t) + Y(t) + \left[\frac{Ml_0}{SE} + \frac{M(1-\mu)}{4G}\sqrt{\frac{\pi}{S}}\right] Y''(t) = 0.$$

We see that the equation of motion includes the current $\mathcal{J}(t)$, which can be expressed in terms of acceleration if we use the piezoresistance coefficient of the metal−semiconductor contact. Figure 2 shows the dependence of the current through the contact on the force applied to the cylinder. Near the static value of the force F_0 in the range defined by

$$F_2 - F_0 = F_0 - F_1 < \sqrt{\frac{\beta}{3\gamma}},$$

the curve in Fig. 2 can be approximated satisfactorily by the expression

$$\mathcal{J} = \beta f - \gamma f^3,$$

where $f = F - F_0$.

Substituting the value of $\mathcal{J}(t)$ into the formula for $\Delta(t)$, we obtain

$$\Delta(t) = -\frac{a\varkappa U}{KS}\beta M Y''(t) + \frac{a\varkappa U}{KS}\gamma M^3 \int_0^t (Y'')^3(\tau)\,d\tau.$$

Hence, the equation of motion assumes the following form:

$$Y'' + \omega^2 Y = \frac{a\omega^2\varkappa U}{KS}\beta M Y' - \frac{a\varkappa\omega^2 U}{KS}\gamma M^3 \int_0^t (Y'')^3(\tau)\,d\tau,$$

where the angular frequency of oscillations is

$$\omega = \sqrt{\frac{1}{M\left(\dfrac{l_0}{SE} + \dfrac{1-\mu}{4G}\sqrt{\dfrac{\pi}{S}}\right)}}.$$

We have so far ignored the frictional forces and the radiation of acoustic energy. To allow for the losses due to these causes, we shall introduce a term $-Y'\omega/Q$ on the left-hand side of the equation of motion. Here, Q is the Q factor of the mechanical oscillatory system represented by the cylinder and the crystal. We thus finally obtain the following form of the equation of motion:

$$Y'' + \omega^2 Y = \sigma\left(Y' - \varepsilon \int_0^t (Y'')^3\,d\tau\right),$$

where

$$\sigma = \frac{a\omega^2\varkappa UM}{KS}\beta - \frac{\omega}{Q},$$

$$\varepsilon = \frac{M^2\gamma}{\beta - \dfrac{KS}{a\omega^2\varkappa UMQ}}.$$

Investigation of the Equation of Motion

It is evident from the equation of motion that when $\sigma > 0$, mechanical (and electric) undamped oscillations are excited in the system. Thus, oscillations appear when the Q factor of the system is sufficiently high and the piezoresistance coefficient is sufficiently large.

At low values of the parameter σ, i.e., when

$$\frac{a\varkappa\omega UM}{KS}\beta - \frac{1}{Q} \ll 1,$$

the oscillations are nearly harmonic:

$$Y(t) = A\sin(\omega t + \varphi).$$

We shall estimate the amplitude of the oscillations using the van der Pol method [5].

It is quite easy to remove the integral from the equation of motion since in calculating the integral we can neglect the time dependence of the amplitude of the harmonic oscillations.

Without going into the details of the derivation, we shall now give the final expression for the amplitude of the steady-state quasiharmonic oscillations:

$$A = \sqrt{\frac{12}{7}\frac{1}{\varepsilon\omega^4}}.$$

Naturally, all our conclusions are valid provided that the approximate relationship $\mathscr{J}(t)$, is justified or, which is equivalent, provided that

$$MA\omega^2 < \sqrt{\frac{\beta}{3\gamma}}.$$

Hence, we obtain

$$\beta < \frac{36}{29}\frac{KS}{\omega a x U M Q}.$$

No serious difficulty is encountered in the derivation of the corresponding expressions for the higher (better) approximation in the piezoresistance coefficient formula when the current is represented by a fifth-degree polynomial

$$\mathscr{J} = \beta f - \gamma f^3 + \delta f^5.$$

However, these expressions are quite cumbersome and will not be given here.

In conclusion, we must mention that, in spite of the approximate nature of the formulas derived here, they describe quite correctly the qualitative behavior of the electroacoustic interaction.

Literature Cited

1.　　É. G. Melikyan, Radiotekhn. i Élektron., 12:1303 (1967).
2.　　H. S. Carslaw and J. C. Jaeger, Conduction of Heat in Solids [Russian translation], Izd. Nauka, Moscow (1937), p. 71. [English 2nd ed., Oxford University Press (1959).]
3.　　F. Frank and R. Mizes, Differential and Integral Equations of Mathematical Physics [in Russian], ONTI, Moscow (1937), p. 290.
4.　　H. S. Carslaw and J. C. Jaeger, Conduction of Heat in Solids [Russian translation], Izd. Nauka, Moscow (1937), p. 80. [English 2nd ed., Oxford University Press (1959).]
5.　　A. A. Andronov, A. A. Vitt, and S. É. Khaikin, Theory of Oscillations [in Russian], Fizmatgiz, Moscow (1959), p. 653.

INVESTIGATION OF THE FIELD IN A TUNNEL DIODE
WITH A HETEROJUNCTION

S. N. Dobrynin

V. I. Ul'yanov (Lenin) Leningrad Institute of Electrical Engineering

A system of equations is derived for calculating the field in a degenerate heterojunction, taking into account the presence of mobile carriers, special features of the energy bands, and the interfacial charge. Two cases are considered in these calculations: 1) heterojunctions between crystals with similar lattice constants; 2) heterojunctions in which the interfacial charge has to be taken into account because the lattice constants are different. A solution of the system for the first case gives graphs showing the deviation of the asymmetry coefficient $\beta = w_2'/w_1'$, calculated using the formulas in the present paper, from the usual definition $\beta = 1/r$. Calculations are carried out for different values of the parameters of the system and it is clear from the graphs that, in some cases, β may differ considerably from the value calculated in the usual way. Consequently, such deviations of the asymmetry coefficient must be allowed for in calculations of the characteristics of a tunnel heterodiode. When mobile charges are taken into account, the system of equations becomes more complicated. Its solution is obtained in two cases: A) when the dependence of the magnitude of the interfacial charge on the bias cannot be ignored; B) when this dependence is negligible. A method for solving the initial system of equations is given for both cases. Next, the asymmetry coefficient β is calculated taking into account the additional asymmetry solely due to the interfacial charge. This calculation is carried out for different parameters of the system. It is concluded that the asymmetry of the interfacial charge must be allowed for if correct results are to be obtained. Moreover, the position of a band of additional levels in the junction is important in computations of the interfacial charge.

Several papers on tunnel diodes with heterojunctions have been published recently [1-3]. It is reported in [2] that tunnel heterodiodes exhibit special tunneling properties observed either only in one of the crystals forming the junction or in both crystals, depending on the ratio of the impurity concentrations in the two crystals. The purpose of our investigation will be to consider the influence of mobile carriers and of the interfacial charge (appearing in the plane of the junction due to the difference between the lattice constants of the crystals [4]) on the field in a degenerate heterojunction.

We shall first consider heterojunctions of the first type, i.e., junctions between different crystals with similar lattice constants. In this case, we can ignore the charge in the plane of the junction and consider only the influence of the mobile carriers. The influence of the mobile carriers on the field in a degenerate heterojunction is manifested by the broadening of the space-charge region and, as demonstrated in [5], the appearance of an additional asymmetry in the charge layers of each crystal due to the differences between the permittivities and effective masses and to the specific features of the energy band diagrams of a heterojunction.

An investigation of the additional asymmetry is very interesting because, according to [2], the characteristics of tunnel heterodiodes depend on the relationship between the lengths of the tunneling paths in each crystal.

We shall assume that the junction is abrupt and ignore the electric dipoles which appear as a result of the different polarities of the bonds in crystals; we shall also ignore the orientation and other defects which appear due to the formation of a junction. We shall take into account the difference between the widths of the forbidden bands, electron affinities, permittivities, and effective masses in the two crystals forming a heterojunction.

Calculation of the Field in Heterojunctions

with Similar Lattice Parameters

We shall calculate the field, taking into account the mobile carriers and using a variational method developed in [6] for ordinary tunnel junctions and generalized in [5] to degenerate heterojunctions.

Following [5], we can write the system of equations for calculating the field in the form

$$w_1 = \left[\frac{2\varphi_{1c}}{\left(1 - \frac{3}{\varphi_{1c}}\mathcal{J}_1\right)}\right]^{\frac{1}{2}}, \tag{1}$$

$$w_2 = \left[\frac{2\varphi_{2c}}{r\left(1 - \frac{3}{r \cdot \varphi_{2c}}\mathcal{J}_2\right)}\right]^{\frac{1}{2}}, \tag{2}$$

$$\varphi_{1c}\frac{\sqrt{\varepsilon_1}}{w_1} = \varphi_{2c}\frac{\sqrt{\varepsilon_2}}{w_2}, \tag{3}$$

$$\varphi_c = \varphi_{1c} + \varphi_{2c}. \tag{4}$$

This system is written in terms of trial functions φ_1 and φ_2, which depend on the variational parameters w_1 and w_2, respectively, and are given by

$$\varphi_1 = \varphi_{1c}\begin{cases} 0 & \text{for } \xi_1 \leqslant -1, \\ (1 + 2\xi_1 + \xi_1^2) & \text{for } 0 > \xi_1 > -1, \\ 1 & \text{for } \xi_1 = 0, \end{cases} \tag{5}$$

$$\varphi_2 = \varphi_{2c}\begin{cases} 0 & \text{for } \xi_2 = 0, \\ (2\xi_2 - \xi_2^2) & \text{for } 0 < \xi_2 < 1, \\ 1 & \text{for } \xi_2 \geqslant 1, \end{cases} \tag{6}$$

where $w_{1,2}$ are the dimensionless variables of the thicknesses of space-charge layers in crystals I and II, given by $w_{1,2} = \dfrac{w'_{1,2}}{\left(\frac{F_n \cdot \varepsilon_{1,2}}{4\pi N_1 e^2}\right)^{1/2}}$; $w'_{1,2}$ are the dimensional thicknesses of the same space-charge layers; $\varepsilon_{1,2}$ are the permittivities of crystals I and II; $N_{1,2}$ are the impurity concentrations in crystals I and II; e is the electronic charge; F_n is the Fermi level of electrons, measured from the bottom of the conduction band of crystal I; $\xi_{1,2}$ are the dimensionless coordinates in crystals I and II, given by $\xi_{1,2} = x/w'_{1,2}$; $\Phi_{1c,2c}/e$ are potential drops in

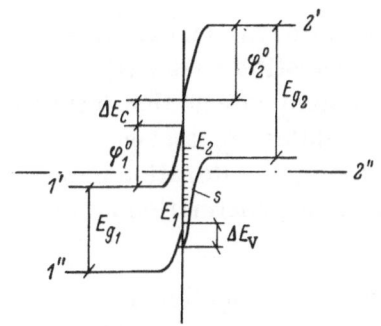

Fig. 1. Energy diagram of a
degenerate n−p heterojunction
(s is the interfacial band).

crystals I and II; $\varphi_{1c,2c} = \Phi_{1c,2c}/F_n$; Φ_c/e is the potential drop in the junction; $\varphi_c = \Phi_c/F_n$; $\varphi_c = \varphi_c^0 - \delta\varphi$; $\varphi_c^0 = \Phi_c^0/F_n$; Φ_c^0/e is the contact potential; $\delta\Phi/e$ is the applied bias; $\delta\varphi = \delta\Phi/F_n$; $r = N_2/N_1$. We also have

$$\mathscr{I}_1 = \int_{-\infty}^{0} d\xi_1 \frac{\partial\varphi_1}{\partial\xi_1} \cdot \xi_1 [v_{p_1}(\xi_1) - v_{n_1}(\xi_1)],$$

$$\mathscr{I}_2 = \int_{0}^{+\infty} d\xi_2 \frac{\partial\varphi_2}{\partial\xi_2} \cdot \xi_2 [v_{p_2}(\xi_2) - v_{n_2}(\xi_2)]. \tag{7}$$

Here, $v_{p_1} = p_1(\xi_1)/N_1$; $v_{n_1} = n_1(\xi_1)/N_1$; $v_{p_2} = p_2(\xi_2)/N_1$; $v_{n_2} = n_2(\xi_2)/N_1$; $n_{1,2}$ are the electron densities in crystals I and II and $p_{1,2}$ are the hole densities in crystals I and II.

Using the results given in [7], we can write these carrier densities in the form

$$v_{p_1} = r_2 M [F'_p - \varphi_{1c} + \varphi_1(\xi_1) - \varphi_{2c} + \Delta E'_v + \varphi']^{\frac{5}{2}}, \tag{8}$$

where $\Delta E'_v = \Delta E_v/F_n$; ΔE_v is the value band discontinuity at the junction; $\varphi' = 2kT/F_n$; $F'_p = F_p/F_n$; F_p is the Fermi level of holes, measured from the top of the valence band of crystal II; $r_2 = \left(\frac{m^*_{p_1}}{m^*_{n_1}}\right)^{\frac{3}{2}}$, where $m^*_{P_1}$ is the effective mass of a hole in crystal I and $m^*_{n_1}$ is the effective mass of an electron in crystal I; $M = \left[1 - \frac{2kT}{\left(\frac{N_1 \pi^2 \hbar^3 \cdot 15kT}{\sqrt{2}\,(m^*_{n_1})^{3/2}}\right)^{2/5}}\right]^{\frac{5}{2}}$. We also have

$$v_{p_2} = r_3 M [F'_p + \varphi' - \varphi_{2c} + \varphi_2(\xi_2)]^{\frac{5}{2}}, \tag{9}$$

where $r_3 = \left(\frac{m^*_{p_2}}{m^*_{n_1}}\right)^{\frac{3}{2}}$ and $m^*_{P_2}$ is the effective mass of a hole in crystal II. Similarly,

$$v_{n_1} = M [1 + \varphi' - \varphi_1(\xi_1)]^{\frac{5}{2}}, \tag{10}$$

$$v_{n_2} = r_1 M [1 + \varphi' - \Delta E'_v - \varphi_{1c} - \varphi_2(\xi_2)]^{\frac{5}{2}}, \tag{11}$$

where $r_1 = \left(\frac{m^*_{n_2}}{m^*_{n_1}}\right)^{\frac{3}{2}}$ and $m^*_{n_2}$ is the effective mass of an electron in crystal II; $\Delta E'_c = \Delta E_c/F_n$, where ΔE_c is the discontinuity of the conduction band at the heterojunction.

It must be mentioned that, because of the special nature of the band diagram of a heterojunction, we can neglect the electron density n_2 (Fig. 1).

The integrals in Eq. (7) can be found exactly but the expressions obtained are very cumbersome and will not be given here.

We shall now solve the system of equations (1)−(4). In general, this system cannot be solved exactly. However, in the absence of mobile carriers, the system can be solved exactly and it then yields the well-known formula obtained by Anderson [8] by solving Poisson's equation in the Schottky layer approximation.

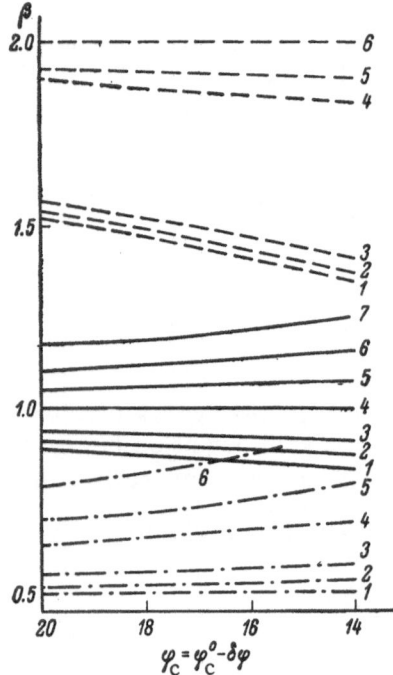

Fig. 2. Dependences of β on φ_c. Continuous curves represent the case r = 1:
1) $\varepsilon_1/\varepsilon_2 = 2$, $r_3 = 2$; 2) $\varepsilon_1/\varepsilon_2 = 2$, $r_3 = 1$;
3) $\varepsilon_1/\varepsilon_2 = 2$, $r_3 = 0.5$; 4) $\varepsilon_1/\varepsilon_2 = 1$,
$r_3 = 1$; 5) $\varepsilon_1/\varepsilon_2 = 0.5$, $r_3 = 2$; 6) $\varepsilon_1/\varepsilon_2 = 0.5$, $r_3 = 1$; 7) $\varepsilon_1/\varepsilon_2 = 0.5$, $r_3 = 0.5$.
Chain curves represent the case r = 2:
1) $\varepsilon_1/\varepsilon_2 = 1$, $r_3 = 1$ (depletion-layer calculation without mobile charges) and
$\varepsilon_1/\varepsilon_2 = 2$, $r_3 = 2$; 2) $\varepsilon_1/\varepsilon_2 = 2$, $r_3 = 1$;
3) $\varepsilon_1/\varepsilon_2 = 2$, $r_3 = 0.5$; 4) $\varepsilon_1/\varepsilon_2 = 0.5$,
$r_3 = 2$; 5) $\varepsilon_1/\varepsilon_2 = 0.5$, $r_3 = 1$; 6) $\varepsilon_1/\varepsilon_2 = 0.5$, $r_3 = 0.5$. Dashed curves represent
the case r = 0.5: 1) $\varepsilon_1/\varepsilon_2 = 2$, $r_3 = 2$;
2) $\varepsilon_1/\varepsilon_2 = 2$, $r_3 = 1$; 3) $\varepsilon_1/\varepsilon_2 = 2$, $r_3 = 0.5$; 4) $\varepsilon_1/\varepsilon_2 = 0.5$, $r_3 = 2$; 5) $\varepsilon_1/\varepsilon_2 = 0.5$,
$r_3 = 1$; 6) $\varepsilon_1/\varepsilon_2 = 0.5$, $r_3 = 0.5$ and
$\varepsilon_1/\varepsilon_2 = 1$, $r_3 = 1$ (depletion-layer calculation).

We shall use the following method in solving this system. We shall substitute Eqs. (1) and (2) into Eq. (3). We then obtain an equation (usually transcendental) which relates φ_{1c} and φ_{2c}. We shall solve this equation by the iteration method, using the depletion layer approximation as the starting point.

Since allowance for the mobile carriers yields only a correction to the results obtained in the depletion layer approximation, it follows that this approximation is fairly good. We thus obtain

$$\varphi_{2c} = \sqrt{\frac{\varepsilon_1 \varphi_{1c}}{\varepsilon_2 r}} \cdot \sqrt{\varphi_{2c}} \left[\frac{1 - \frac{3}{\varphi_{1c}} \mathscr{J}_1}{1 - \frac{3}{r\varphi_{2c}} \mathscr{J}_2} \right]^{\frac{1}{2}}. \qquad (12)$$

Denoting the zeroth (depletion-layer) approximation for φ_{2c} by $z_2 = \dfrac{\varphi_c}{1 + \frac{\varepsilon_2 r}{\varepsilon_1}}$, we shall substitute it into Eq. (12) and then use Eq. (12) in Eq. (4). Again, we obtain, in general, a transcendental equation for φ_{1c}. Applying once more the iteration method and denoting the zeroth approximation by $z_1 = \dfrac{\varphi_c}{1 + \frac{\varepsilon_1}{\varepsilon_2 r}}$, we obtain a solution for φ_{1c} in the form

$$\varphi_{1c} = \frac{\varphi_c}{1 + P_1},$$

where

$$P_1 = \sqrt{\frac{\varepsilon_1}{\varepsilon_2 \cdot r} \frac{z_2}{z_1}} \left[\frac{1 - \frac{3}{\varphi_{1c}} \mathscr{J}_1(z_1, z_2)}{1 - \frac{3}{2\varphi_{2c}} \mathscr{J}_2(z_1, z_2)} \right]^{\frac{1}{2}}. \qquad (13)$$

Then,

$$\varphi_{2c} = \frac{P_1}{1 + P_1} \varphi_c \qquad (14)$$

Substituting Eqs. (13) and (14) into Eqs. (1) and (2), we obtain formulas for the determination of the parameters w_1 and w_2.

Using this method, we shall consider the ratio $\beta = w_2'/w_1'$, representing the asymmetry of the space-charge layers for the following sets of parameters: $\varepsilon_1/\varepsilon_2 = 2$, 0.5; $r_3 = 1$, 2, 0.5; $r = 1$, 2, 0.5; T = 300°K and $N_1 = 2 \cdot 10^{19}$ cm^{-3}. We shall deal with the two most interesting cases, those when the relationships between the parameters of the system are such that holes located in the well near the heterojunction either can be ignored (first case) or must be included (second case).

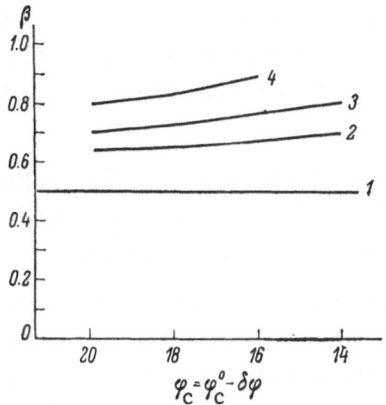

Fig. 3. Dependences of β on φ_c for r = 2 and ΔE_v = 0.2 eV: 1) $\varepsilon_1/\varepsilon_2$ = 1, r_3 = 1 (depletion-layer calculation); 2) $\varepsilon_1/\varepsilon_2$ = 0.5, r_3 = 2, r_2 = 1; 3) $\varepsilon_1/\varepsilon_2$ = 0.5, r_3 = 1, r_2 = 1 and $\varepsilon_1/\varepsilon_2$ = 0.5, r_3 = 1, r_2 = 1.5; 4) $\varepsilon_1/\varepsilon_2$ = 0.5, r_3 = 0.5, r_2 = 1.

Figure 2 shows the dependence of the asymmetry parameter β on the applied bias for the first case (holes in the well ignored). Figure 3 gives the corresponding dependence of β for the second case (taking into account the presence of holes in the well).

It is evident from Figs. 2 and 3 that, for some combinations of the parameters the coefficient β may deviate quite considerably from the values usually obtained in calculations concerned with Schottky layers.

These deviations are due to the fact that the parameters ε and m must be induced in the formulas for the calculation of β whenever allowance is made for the mobile carriers. It is known that differences between the values of ε govern differences between the potential drops across the crystals forming a heterojunction pair and, consequently, they determine carrier densities and the space charge at the interface. This gives rise to changes in the space-charge region. Depending on the relationship between $\varepsilon_1/\varepsilon_2$ and r, this change in the space-charge layer may give rise to deviations in either direction, as demonstrated in Figs. 2 and 3. The effective mass affects the positions of the Fermi levels in crystals and, therefore, the differences between the effective masses affect the mobile carrier density at the interface and they determine the asymmetry of the space-charge layers. This can be seen from the curves in Fig. 2.

Figure 3 shows the dependence of β on φ_c in the presence of interfacial holes in a well located in crystal I (Fig. 1). It is obvious that the presence of holes is equivalent to an increase of the space charge in a given crystal and that it gives rise to a corresponding reduction of the thickness of the space-charge layer.

These qualitative conclusions are fully supported by the results of the numerical calculations presented in Figs. 2 and 3.

We shall now summarize briefly the main conclusions which follow from the numerical calculations.

1. Allowance for the presence of the mobile carriers yields formulas for the calculation of the field, in which account is taken of the influence of the different effective masses and special features of the energy band diagram of a heterojunction. The coefficient β is found to depend not only on r but also on all the other parameters.

2. The values of β may, for some combinations of the parameters, deviate considerably from the values usually obtained in calculations concerned with Schottky layers. Consequently, a correct value of the tunnel current can be obtained only by taking into account the additional asymmetry of the space-charge layers in the two crystals forming a heterojunction. This must be done because such an asymmetry governs the paths of the tunneling carriers in each of the crystals, and alters considerably the characteristics of a heterojunction tunnel diode.

3. Allowance for the presence of holes in the well near the junction increases the additional asymmetry. In our case, this increase is small compared with the case in which we

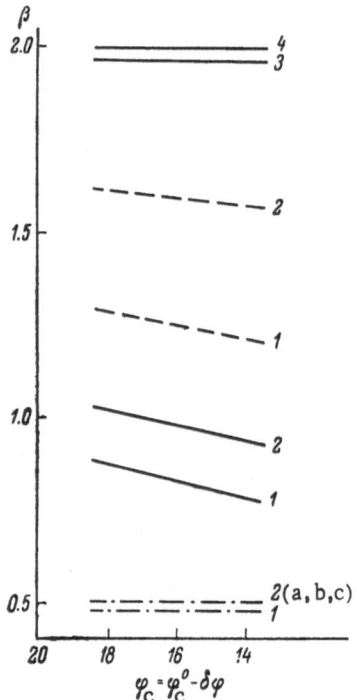

Fig. 4. Dependence of β on φ_c. Chain lines represent the results for $E_2 = -0.4$ eV, $\varepsilon_1/\varepsilon_2 = 0.5$, $r = 2$: 1) $\sigma = 10^{13}$, $E_2 - E_1 = 0.4$ eV; 2a) $\sigma = 10^{13}$, $E_2 - E_1 = 0.2$ eV; 2b) $\sigma = 10^{11}$, $E_2 - E_1 = 0.4$ eV; 2c) $\sigma = 0$. Dashed lines represent the results for $E_2 = -0.1$ eV, $\varepsilon_1/\varepsilon_2 = 2$, $r = 0.5$: 1) $\sigma = 10^{13}$, $E_2 - E_1 = 0.4$ eV; 2) $\sigma = 10^{13}$, $E_2 - E_1 = 0.2$ eV; 3) $\sigma = 10^{11}$, $E_2 - E_1 = 0.4$ eV; 4) $\sigma = 10^{11}$, $E_2 - E_1 = 0.2$ eV and $\sigma = 0$. Dashed lines 3 and 4 coincide with the corresponding continuous lines. Continuous lines represent the results for $E_2 = -0.4$ eV, $\varepsilon_1/\varepsilon_2 = 2$, $r = 0.5$: 1) $\sigma = 10^{13}$, $E_2 - E_1 = 0.4$ eV; 2) $\sigma = 10^{13}$, $E_2 - E_1 = 0.2$ eV; 3) $\sigma = 10^{11}$ for $E_2 - E_1 = 0.2$ eV and $E_2 - E_1 = 0.4$ eV; 4) $\sigma = 0$.

have ignored the presence of holes. However, at high temperatures and in the case of large values of ΔE_v, the influence of this increase may be quite considerable.

4. Allowance for the presence of the mobile carriers broadens the space-charge layer and increases correspondingly the tunneling path.

5. When dealing with specific heterojunctions it is necessary to estimate approximately the additional asymmetry using the actual parameters and, depending on the value of this asymmetry, either allow for or ignore the presence of the mobile carriers.

Calculation of the Field Taking Interfacial Charge into Account

It is known [4] that, because of the difference between the lattice constants of crystals, an additional band of energy levels appears at the heterojunction interface and this band captures electrons whose charge distorts the energy diagram of the junction.

It is interesting to consider the distortion of the energy-band diagram for different degrees of mismatch between the lattice constants and for different widths and positions of the additional-level band. We shall assume that the width and position of this band are known. Moreover, we shall postulate that the additional charge is concentrated in a very thin layer at the heterojunction interface.

Once again we shall use the variational method. However, in this case, we shall introduce an interfacial charge so that Eq. (3) becomes

$$\varphi_{1c} = \sqrt{\frac{\varepsilon_2}{\varepsilon_1}} \frac{w_1}{w_2} \varphi_{2c} + \sqrt{\frac{4\pi e^2}{F_n \varepsilon_1 N_1}} w_1 n_s, \qquad (15)$$

where n_s is the number of electrons in the additional band. The expression for n_s can be obtained easily bearing in mind that the additional band is two-dimensional:

$$n_s = \sigma \left[\frac{kT}{E_2 - E_1} \ln \left| \frac{1 + \exp\left(\frac{E_1 - F}{kT}\right)}{1 + \exp\left(\frac{E_2 - F}{kT}\right)} \right| + 1 \right], \qquad (16)$$

where E_2 and E_1 are the upper and lower edges of the additional-level band, measured from the bottom of the conduction band; σ is the number of additional levels per unit area of the interface; the value of σ depends on the crystal lattice configurations and on the mismatch between the lattice constants of the two crystals forming a heterojunction. Oldham and Milnes [4] have obtained the following formula for the calculation of σ in the case of diamond-type lattices:

$$\sigma = \frac{4}{\sqrt{3}}\left[\left(\frac{1}{\lambda_2}\right)^2 - \left(\frac{1}{\lambda_1}\right)^2\right], \tag{17}$$

where $\lambda_{1,2}$ are the lattice constants of crystals I and II, respectively.

The symbol F in Eq. (16) is the Fermi level in the additional-level band. It represents the density of the electrons localized at the additional levels and the dependence of this density on the applied bias.

In general, the value of the additional charge is governed by the following parameters: the density of the mobile carriers near the junction, recombination cross sections, and the density of the additional levels and their position in the energy diagram. Following [8], we can easily show that the expression for the number of electrons in the additional band can be written in the form

$$n_s = \frac{n_s^0 \sigma}{n_s^0 + (\sigma - n_s^0)L}, \tag{18}$$

where

$$L = \frac{\dfrac{k_{1's}n_{1C}^0}{k_{s1''}p_{1C} + k_{s2''}p_{2C}}\left(\dfrac{\sigma - n_s^0}{n_s^0}\right) + 1}{\dfrac{k_{1's}n_{1C}}{k_{s1''}p_{1C} + k_{s2''}p_{2C}}\left(\dfrac{\sigma - n_s^0}{n_s^0}\right) + \dfrac{k_{s1''}p_{1C}^0 + k_{s2''}p_{2C}^0}{k_{s1''}p_{1C} + k_{s2''}p_{2C}}}. \tag{19}$$

Here, the superscript 0 indicates the values under zero bias conditions. The subscript c indicates that the densities are taken at the contact boundary (interface). The other parameters are defined by

$$k_{1's} = \gamma_{1's}V_{n_1}, \quad k_{s1''} = \gamma_{s1''}V_{p_1}, \quad k_{2's} = \gamma_{2's}V_{n_2}, \quad k_{s2''} = \gamma_{s2''}V_{p_2}, \tag{20}$$

where γ is the cross section for the recombination of electrons and holes; V are the thermal velocities of electrons and holes, depending on the subscript. The recombination cross sections are assumed to be known. The meaning of the subscripts 1', 2', 1", 2", s is explained in Fig. 1.

Equation (18) is derived ignoring the presence of electrons in crystal II, in accordance with the special features of the band diagram of the heterojunction.

The quantity n_s^0 in Eq. (18) is found from Eq. (16), where the Fermi level F is given by

$$F = F_n - \Phi_{1C} \tag{21}$$

because the bias is zero. The carrier densities n_{1C} and p_{2C} are found from Eqs. (10) and (9) by substituting $\xi = 0$.

The hole density p_{1C} can be neglected for small values of ΔE_v and low temperatures. Then, substituting Eq. (18) into Eq. (15), we obtain a very cumbersome transcendental equation relating φ_{1C} and φ_{2C}. Solving this equation by the iteration method and using Eq. (4), we obtain φ_{1C} and then φ_{2C}. The parameters w_1 and w_2 are then found from Eqs. (1) and (2).

This method makes it possible to calculate the field in a degenerate heterojunction for various values of the bias.

However, the solution obtained is very cumbersome and, moreover, it contains coefficients k which are difficult to determine.

Obviously, the problem is greatly simplified if we can ignore the dependence of the interfacial charge on the voltage.

The problem can then be divided into two. The first problem is the determination of the additional charge under zero bias conditions, and the second is the determination, for a given charge, of the parameters w_1, w_2, φ_{1c}, and φ_{2c} as a function of the applied bias.

We shall solve the first problem using Eq. (21) and substituting Eqs. (1), (2), (4), and (16) in Eq. (15):

$$\varphi_{1c}^0 = \sqrt{\frac{\varepsilon_2 r}{\varepsilon_1}} \sqrt{\frac{\varphi_{1c}^0}{\varphi_c^0 - \varphi_{1c}^0}} \left\{ \frac{1 - \frac{3}{r\varphi_{2c}^0}\mathscr{J}_2}{1 - \frac{3}{\varphi_{1c}^0}\mathscr{J}_1} \right\}^{\frac{1}{2}} \varphi_{2c}^0 + A_1 \frac{\sqrt{\varphi_{1c}^0}}{\left[1 - \frac{3}{\varphi_{1c}^0 \mathscr{J}_1} \right]^{1/2}} [A_2 \ln|P| + 1], \tag{22}$$

where

$$A_1 = \sqrt{\frac{8\pi e^2}{F_n \varepsilon_1 N_1}\sigma}, \quad A_2 = \frac{kT}{E_2 - E_1}, \quad P = \frac{1 + \exp\left(\frac{E_1 - F_n + F_n \varphi_{1c}^0}{kT}\right)}{1 + \exp\left(\frac{E_2 - F_n + F_n \varphi_{1c}^0}{kT}\right)}.$$

Equation (22) can be solved by the iteration method, using the value of φ_{1c}^0 as the zeroth approximation and ignoring the additional charge but taking into account the mobile carriers. We note that if $E_1 > 4kT + F_n - z_1^0$, where z_1^0 is the value of φ_{1c}^0 without allowance for the additional charge at the junction interface, the band of additional levels is empty and our zeroth approximation is the exact solution.

If $E_2 < -4kT + F_n - z_1^0$, the band of additional levels is completely filled under zero bias conditions. In this case, the assumed zeroth approximation is a poor representation of the true situation and a laborious iteration procedure is required. It follows that the position of the additional-level band plays a very important role and, even when the mismatch between the lattice constants of the two crystals is large, the interfacial charge may be small. This conclusion differs from that reached by Oldham and Milnes [4], who assume that the influence of the additional charge is governed solely by the difference between the lattice constants.

After solving Eq. (22), we can easily determine the charge n_s^0 from Eq. (16) and can then proceed to solve the second problem, which is the determination of the parameters w_1, w_2, φ_{1c}, and φ_{2c} for a biased junction. As we have pointed out earlier, the quantity $A_3 = A_1[A_2 \ln|P| + 1]$ is independent of the applied bias. Bearing this in mind, we shall solve Eq. (22) for different but small values of the bias voltage. It is obvious that, once again, we have no alternative to the iteration method. In this case, we can use the solution of Eq. (22) as the zeroth approximation, ignoring the mobile carriers but taking account of the additional charge. To find this solution, we shall assume that $\mathscr{J}_1 = \mathscr{J}_2 = 0$ in Eq. (22). Then,

$$\sqrt{\varphi_{1c}} = \sqrt{\frac{\varepsilon_2 r}{\varepsilon_1}} \sqrt{\varphi_{2c}} + A_3. \tag{23}$$

Employing Eq. (4) to solve this equation, we obtain

$$\varphi_{2c} = \frac{\varphi_c}{1 + \frac{\varepsilon_2 r}{\varepsilon_1}} + A_4 \tag{24}$$

and

$$\varphi_{1c} = \frac{\varphi_c \cdot \frac{\varepsilon_2 r}{\varepsilon_1}}{1 + \frac{\varepsilon_2 r}{\varepsilon_1}} - A_4, \tag{25}$$

where

$$A_4 = \frac{A_3}{\left(1 + \frac{\varepsilon_2 r}{\varepsilon_1}\right)^2} \left\{ \left(\frac{\varepsilon_2 r}{\varepsilon_1} - 1\right) A_3 - 2 \sqrt{\frac{\varepsilon_2 r}{\varepsilon_1}} \left[\varphi_C\left(1 + \frac{\varepsilon_2 r}{\varepsilon_1}\right) - A_3^2\right]^{\frac{1}{2}} \right\}. \tag{26}$$

Using Eqs. (24) and (25) as the zeroth approximation, we can solve Eq. (22) by the iteration method for $A_3 = $ const and obtain the values of φ_{1c}, φ_{2c} and hence – using Eqs. (1) and (2) – the values of w_1 and w_2, taking into account the additional charge and mobile carriers.

We have investigated the additional asymmetry coefficient which is solely due to the presence of a charge at the junction, i.e., we have investigated solutions (24) and (25) for various combinations of the position and width of the additional band. An error appears in the dependence of $\beta = w_2'/w_1'$ because the mobile carriers are ignored and the dependence of the interfacial charge on the applied voltage is not taken into account. However, it has been our aim to study the asymmetry coefficient which is solely due to the presence of an interfacial charge. A more accurate dependence of β on the bias is obtained when the whole system (1)-(4) is solved.

Moreover, since our calculation applies to tunnel diodes, whose characteristics lie mainly at low values of the bias, it is obvious that, under these conditions, the dependence of β on the bias is not very strong. Furthermore, this asymmetry coefficient is governed by its value under zero bias conditions.

The results of the calculation of the asymmetry coefficient $\beta = \dfrac{w_2'}{w_1'} = \dfrac{1}{r}\left\{\dfrac{1 + \dfrac{A_4}{\varphi_C}}{1 - \dfrac{\varepsilon_1}{\varepsilon_2 r}\dfrac{A_4}{\varphi_C}}\right\}^{1/2}$ are given in Fig. 4.

The following conclusions can be drawn from our calculations.

1. The value of the asymmetry coefficient differs, in some cases, very considerably from the value calculated using the conventional theory which ignores the interfacial charge.

It follows that allowance for the additional (interfacial) charge is essential if the correct characteristics of a tunnel heterojunction are to be obtained.

2. We must take into account that the influence of the interfacial charge is governed not only by the difference between the lattice constants of the two crystals forming a heterojunction but also by the position of a band of additional levels, by the impurity concentration, and by the density of the mobile carriers. This is confirmed by a comparison of the dependences shown in Fig. 4.

Literature Cited

1. J. C. Marinace, IBM J. Res. Develop., 4:280 (1960).
2. M. I. Nathan and J. C. Marinace, Phys. Rev., 128:2149 (1962).
3. J. Shewchun, Phys. Rev., 141:775 (1966).
4. W. G. Oldham and A. G. Milnes, Solid State Electron., 7:153 (1964).
5. A. I. Gubanov and E. N. Dobrynin, Izv. Leningrad. Élektrotekhn. Inst., No. 64, p. 31 (1968).
6. P. E. Zil'berman, Fiz. Tverd. Tela, 5:386 (1963).
7. P. E. Zil'berman, Radiotekhn. i Élektron., 9:1270 (1964).
8. R. L. Anderson, Solid State Electron., 5:341 (1962).
9. A. I. Gubanov, Theory of the Rectifying Action of Semiconductors, GITTL, Moscow (1956).

IRREVERSIBLE CHANGES IN THE PROPERTIES OF
AN INJECTION LASER AT HIGH EXCITATION LEVELS

Yu. I. Kruzhilin

An investigation was made of gallium arsenide laser diodes prepared by the diffusion of zinc into n-type crystals. The measurements were carried out at liquid nitrogen temperature. An experimental comparison was made of the nature of the reduction in the output laser pulse power with the nature of the reduction in the continuous luminescence intensity during irreversible changes under constant forward bias conditions. This was done in order to determine the role of the shunting of a p−n junction. It was found that such shunting had little influence on the process of reduction of the injection laser efficiency. A study was also made of the watt−ampere characteristic of a laser diode before and after the application of pulses of critical amplitude causing irreversible changes. The difference between the characteristics was attributed to the formation of pits on the resonator face, which were aligned along the p−n junction. Photomicrographs were obtained of the end of a laser diode and of a diode with only one reflecting face. Pits along the p−n junction were observed also for the latter diode and, therefore, it was concluded that the pitting was not related in an unambiguous manner to the critical density of laser radiation. The appearance of pits was attributed to the thermoelastic stresses generated in a diode during a single pulse. A nonlinear watt−ampere characteristic was obtained after the application of strong current pulses. The original characteristic was linear. This was attributed to the formation of cracks which opened up only during pulses.

An irreversible reduction in the external quantum efficiency is sometimes observed for GaAs laser diodes. Studies of this phenomenon are becoming more important in view of the continuous efforts to increase the output power of laser diodes.

Deterioration of the power characteristics of a laser diode is frequently accompanied by the transformation of its current−voltage characteristic to a linear type. A study of the influence of the surface properties on the characteristics of an injection laser is reported in [1]. This study was concerned with the current−voltage and watt−ampere characteristics of a diode before and after the etching of its lateral surface. An increase in the output power of a diode was observed after it had been etched. It was concluded that this was due to a reduction in the surface recombination velocity.

An investigation of the damage to a resonator face near a p−n junction, caused by the application of strong current pulses, is reported in [2]. It was established that the critical current density (i.e., the density corresponding to the damage threshold) was independent of the dimensions and efficiency of a diode. It was found that diodes could withstand supercritical

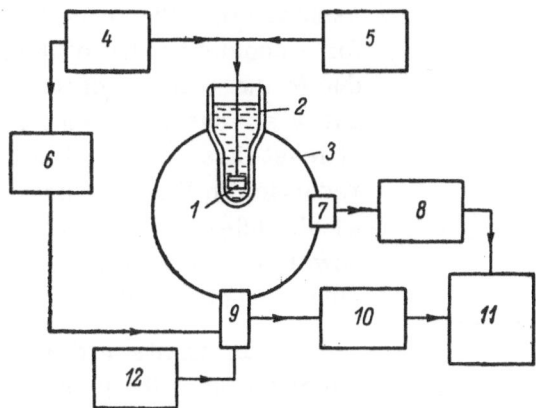

Fig. 1. Schematic diagram of the apparatus used to
determine the role of shunting in the process of deterioration of the injection laser characteristics. 1)
Investigated diode; 2) liquid-nitrogen cryostat with
transparent branch; 3) integrating sphere; 4) generator of current pulses; 5) VS-12 stabilizing rectifier; 6) G5-15 generator of square pulses; 7) silicon
photodiode; 8) V4-3 pulse millivoltmeter; 9) FÉU-28
photomultiplier; 10) U1-2 dc amplifier; 11) two-pen
automatic recorder; 12) VS-22 stabilized rectifier.

current densities when their working temperature was increased sufficiently to stop the emission of stimulated radiation. It was therefore concluded that the observed damage was the result of the emission of a critical radiation flux in the presence of randomly distributed defects
and diffusion inhomogeneities.

We investigated the role of the shunting of a p−n junction [1] by applying current pulses as
well as a constant forward bias to laser diodes. The apparatus used is shown schematically in
Fig. 1. A diode was placed in a transparent branch of a Dewar filled with liquid nitrogen. An
integrating sphere, coated internally with magnesium oxide, was used to measure the output
power of the diode under pulse conditions and the intensity of its luminescence under dc conditions. Continuous recombination radiation was recorded using an FÉU-28 photomultiplier
and a U1-2 dc amplifier. Strong laser radiation pulses saturated the photomultiplier and the
recording system. Therefore, it was necessary to employ stroboscopic illumination of the photomultiplier. A G5-15 square-pulse generator, synchronized with the current pulse generator,
was used to apply a blocking pulse ($4 \cdot 10^{-6}$ sec) to the first stage of the photomultiplier simultaneously with the application of a pumping current pulse. The resultant radiation pulses were
recorded with a silicon photodiode (operating in the photodiode mode) and a V4-3 pulse millivoltmeter. This system of instruments did not measure the constant component of the laser
radiation. The radiation observed by the integrating sphere was recorded using two selective
circuits. The readings of these circuits were applied to a two-pen automatic recorder.

A constant current of 14 mA flowed through all the diodes. The bell-shaped current pulses
of $2.5 \cdot 10^{-6}$ sec duration had a repetition frequency of 100 cps.

The diodes were prepared by the diffusion of zinc into n-type gallium arsenide. The depth
of diffusion of the zinc was 20μ. Resonator faces were obtained by cleavage. The resonator
was $6 \cdot 10^{-2}$ cm long, the diode width was $3.5 \cdot 10^{-2}$ cm and its thickness was 10^{-2} cm.

Figure 2 shows the dependence of the laser output power on the pulse current density. It
is evident from the figure that the decrease in the efficiency was discontinuous. Even if we

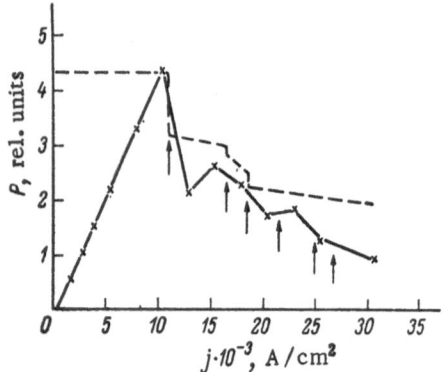

Fig. 2. Dependences, on the pulse current density, of the laser output power under pulsed pumping conditions (continuous curve) and of the luminescence power for a constant current of 14 mA (dashed line). The arrows indicate the values of the current density at which sudden changes in the laser output power took place.

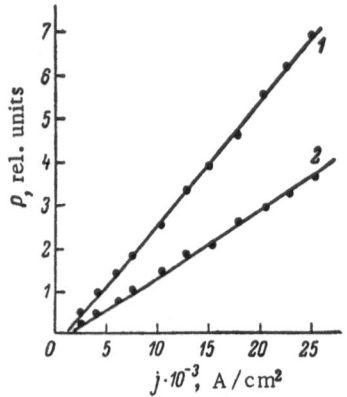

Fig. 3. Dependence of the laser output power on the current density: 1) initial characteristic; 2) characteristic after the application of strong current pulses to a diode.

assumed that the reduction in the luminescence power for a constant value of the forward current was solely due to the shunting of the p−n junction, we found that out of six cases of reduction in the laser output power at least three were not accompanied by shunting. The reduction in the luminescence power coincided to within 0.5-1 sec with the change in the laser output power (the error was due to the slow response of the measuring circuit).

The results of this experiment indicated that the shunting of the p−n junction was not of great importance in the process of rapid deterioration of the laser diode characteristics.

Apart from making a comparison of the laser and luminescence output powers, we investigated changes in the watt−ampere characteristic of a diode which occurred when the output power decreased after the application of pulses of critical amplitude. Figure 3 shows the watt−ampere characteristics of a diode before (1) and after (2) the reduction in the output power due to this cause. The threshold current density was practically constant but the differential efficiency decreased approximately by a factor of two. This could be due to the cessation of the laser emission from part of a p−n junction because of the degradation of the properties of the resonator. This was confirmed by examination of all the diodes under a microscope: when the output power decreased we found that the reflecting faces of the resonator were pitted along the p−n junction. The pit dimensions were 2-5 μ. Figure 4a shows a photomicrograph of part of the end of a laser diode whose output power was reduced by the application of pulses of critical amplitude. The damage pattern was similar to that shown in [2] but, in our case, the pit dimensions were several times smaller.

In order to determine whether the damage suffered by the p−n junction was related to the critical laser radiation density, we investigated diodes belonging to the same batch but having a low Q factor of the resonator. One of the two faces obtained by cleavage was roughened by grinding. Tests under the same conditions as those used in a study of laser diodes showed that the only reflecting face of these diodes was also pitted along the p−n junction. Figure 4b shows a photomicrograph of such a diode.

Since the output of the diode with a single reflecting surface was an order of magnitude lower but the critical current density (the density at which damage occurred) was the same as for laser diodes, we concluded that the damage could not be attributed definitely to the effects of powerful laser radiation, as was done in [2]. The damage of a diode was, in the final analysis, a mechanical phenomenon and the damage conditions depended on the properties of

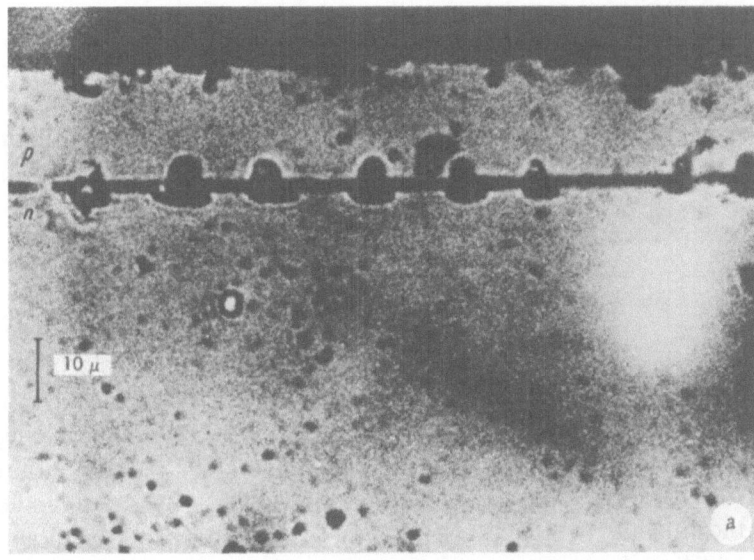

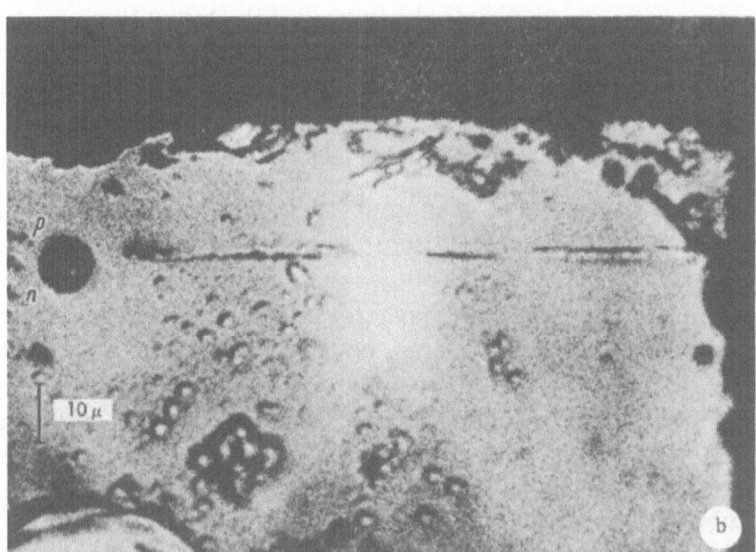

Fig. 4. Photomicrographs of damage on reflecting faces:
a) laser diode; b) diode with a single reflecting face.

the material. Therefore, the experiments carried out by Cooper et al. [2] at two different temperatures could be faulted because, in addition to altering the optical characteristics of a diode, they also changed the properties of the material (Young's modulus, expansion coefficient, thermal diffusion, mechanical strength), all of which depended strongly on the temperature. It was likely that this change in the mechanical properties of a material prevented the appearance of damage of the resonator faces at higher temperatures.

We did not determine directly the depth of the damage of the resonator faces. However, indirect experiments indicated that the depth of the pits was not very great. We studied laser diodes with four cleaved faces. The external quantum efficiency of such diodes increased strongly and irreversibly at some value of the current density. This corresponded to the moment of the appearance of pits on one of the faces of a diode, the suppression of the emission of stray laser modes, and the appearance of laser generation between one pair of mirrors. Had the pits

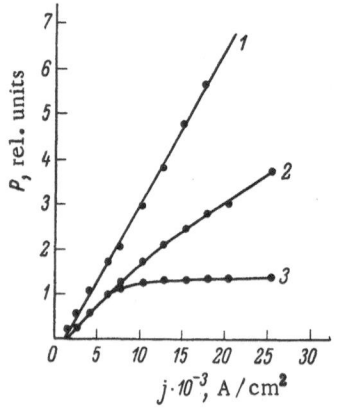

Fig. 5. Watt−ampere characteristics of a laser diode: 1)
before the pitting of the resonator face; 2) after the initial
pitting; 3) after the additional
pitting.

been deep, we would have been unable to observe the considerable increase in the efficiency.

Shallow surface pits on the reflecting faces could be
due to the bombardment of the diode surface by liquid nitrogen bubbles evaporated during each current pulse. However,
these pits were also observed on diodes investigated in a
vacuum cryostat [2].

We concluded that the observed damage was caused by
thermal stresses generated in a diode during a single pulse.
Cooper et al. [2] assumed that the damage they observed
was not due to heating since a single pulse was sufficient to
produce this damage. However, in our opinion, the temperature could rise very steeply during a single pulse. Heating
could be due to ohmic losses at the contacts and in the body
of the diode, as well as due to nonradiative recombination
and absorption of some of the intrinsic recombination radiation. The resultant temperature gradient could give rise to
mechanical stresses [3]. Since there were already stresses
in the p−n junction (due to an impurity concentration gra-
dient) and the strength in the region of the junction was lower than elsewhere in a diode, pitting
occurred preferentially along the p−n junction. We should expect diodes with less steep impurity gradients to withstand higher current densities.

The watt−ampere characteristics of many diodes became nonlinear when their output
power decreased after the application of critical pulses: the slopes of the characteristics decreased with increasing current density. This was observed most clearly in the family of characteristics shown in Fig. 5. The original characteristic 1 changed to curve 2, which exhibited
some nonlinearity. After irreversible changes, caused by powerful current pulses, had taken
place, the reproducible current−voltage characteristic assumed the form represented by curve
3. This curve is interesting for two reasons. First, curves 2 and 3 are identical up to a current
density of about $6.5 \cdot 10^3$ A/cm^2; therefore, up to this current density, we cannot say whether
the output characteristic has changed. Secondly, at current densities higher than $1.3 \cdot 10^4$
A/cm^2, curve 3 is practically horizontal. These results can be explained by the formation, on
the resonator face, of latent cracks which open up during a pulse at some value of the thermoelastic stresses and which close up again when the diode is cooled between pulses.

The author is grateful to N. V. Antonov, Yu. I. Koloskov, G. T. Pak, A. I. Petrov, G. P.
Proshko, and V. I. Shveikin for valuable discussions and for their help in this investigation.

Literature Cited

1.   O. D. Knab, V. I. Magalyas, A. S. Logginov, and A. S. Astaf'ev, Fiz. Tverd. Tela, 8:2768
     (1966).
2.   D. P. Cooper, C. H. Gooch, and R. J. Sherwell, IEEE J. Quantum Electronics, QE-2:329
     (1966).
3.   B. A. Boley and J. H. Weiner, Theory of Thermal Stresses, Wiley, New York (1960).

ELECTRICAL PROPERTIES OF
DIFFUSED p–n JUNCTIONS IN INDIUM ARSENIDE

Yu. D. Mozzhorin and V. I. Stafeev

A brief description is given of the method used to prepare diffused p–n junctions in indium arsenide. The results are reported of a study of the current–voltage characteristics (at 77 and 300°K) as a function of the electron density in the initial material and of the acceptor concentration gradient in the p–n junction. An estimate is made of the lifetime and diffusion length of the electrons.

Indium arsenide has a relatively narrow forbidden band (0.33 eV at 300°K). It is widely used in infrared detectors and sources. However, the mechanism of current rectification in indium arsenide p–n junctions has not yet been investigated sufficiently thoroughly [1, 2]. The present paper reports a study of the current–voltage characteristic of diffused p–n junctions in indium arsenide with different electron densities and different acceptor gradients in the junction region.

The investigated diodes were prepared from n-type indium arsenide single crystals grown by the Czochralski method and doped with tellurium. The initial room-temperature parameters of the material were: electron density $9.5 \cdot 10^{16}$-$1.5 \cdot 10^{19}$ cm^{-3}; electron mobility 20,000-3500 cm$^2 \cdot$ V$^{-1} \cdot$ sec^{-1}; dislocation density $5 \cdot 10^2$-10^4 cm^{-2}.

The p–n junctions were produced by the diffusion of Cd or Zn from the vapor phase in evacuated and annealed quartz ampoules. The diffusion temperature was 680-820°C, the diffusion annealing duration was $6 \cdot 10^2$ to $1 \cdot 10^4$ sec, and the p–n junction depth was 25-100 μ. The p–n junctions were parallel to the (111) planes.

The diodes were mesa structures, prepared by etching followed by electropolishing. The p–n junction area was $5 \cdot 10^{-3}$ to $5 \cdot 10^{-2}$ cm^2. Alloyed ohmic contacts were used. An In–Zn–Au alloy was used to make contact with the p-type region and In–Sn–Au alloy was used in the n-type region.

We investigated the static current–voltage characteristics in the current-density range 10^{-7}-10^2 A/cm^2 at room and liquid-nitrogen temperatures. We also determined the voltage–capacitance characteristics using a resonance method at 300 kc at liquid-nitrogen temperature. Moreover, we used the curves given in [3], which show the impurity distribution with depth in order to determine the acceptor-concentration gradients.

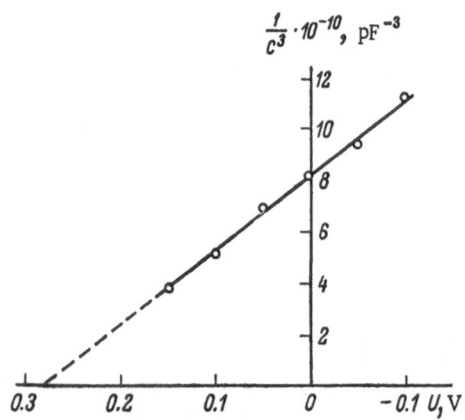

Fig. 1. Voltage−capacitance characteristic of a p−n junction in InAs
($n = 2 \cdot 10^{18}$ cm^{-3}).

Impurity Distribution in p − n Junctions and Thickness of the Space-Charge Regions

The acceptor concentration gradient in a p−n junction, determined from the curves representing the distribution of Cd or Zn with depth for samples with different initial donor concentrations, was found to lie in the range $a = 3.5 \cdot 10^{19}$ to $6 \cdot 10^{23}$ cm^{-4}. Similar values were obtained from the voltage−capacitance characteristics. The dependence of the capacitance on the voltage, obtained for InAs diodes with initial impurity concentrations up to $2 \cdot 10^{18}$ cm^{-3}, was of the $C^{-3} \propto U$ type (Fig. 1). The thickness of the space-charge region $W = \varepsilon \varepsilon_0 S/4\pi C$ (S is the p−n junction area) was $6 \cdot 10^{-5}$ to $5 \cdot 10^{-6}$ cm. In diodes with electron densities $n = (1.5-1) \cdot 10^{19}$ cm^{-3} and relatively abrupt p−n junctions, the thickness of the space-charge region did not exceed 10^{-6} cm because these diodes exhibited tunnel characteristics.

Reverse Branch of the Current − Voltage Characteristic

Figure 2 shows the reverse branches of the current−voltage characteristics obtained at room temperature, using voltages $U < U_{br}$ (U_{br} is the breakdown voltage). For diodes with a nonabrupt variation of the impurity concentration in the p−n junction region ($n = 9.5 \cdot 10^{16}$ cm^{-3}), the reverse current exhibited a saturation region in the $kT/q < U < 0.6$ V range. When the acceptor concentration and electron density gradients in the original material were increased, respectively, to $7 \cdot 10^{21}$ cm^{-4} and $2 \cdot 10^{18}$ cm^{-4}, the range of voltages in which the reverse-current saturation current was observed became narrower.

When the electron density n was increased to 10^{19} cm^{-3} and the impurity concentration gradient a was raised to 10^{22} cm^{-4}, the current at voltages $U < U_{br}$ obeyed $j \propto U^{1/\gamma}$, where $\gamma \approx 3$. Finally, for a fairly steep acceptor impurity gradient in a p−n junction ($a = 3.6 \cdot 10^{23}$ cm^{-4}) in InAs with $n = (1-1.5) \cdot 10^{19}$ cm^{-3}, we obtained tunnel diodes of the normal and backward type (Fig. 3).

The ratio of the carrier generation current in the space-charge region I_r to the generation current in the p- and n-type regions of the p−n junction, I_d, under a reverse bias, was given in [4]

$$\frac{I_r}{I_d} \approx \frac{n_n}{n_i} \frac{W}{4L}.$$

The value of this ratio increased with decreasing carrier lifetime and increasing carrier density near the boundaries of the space-charge region. A theoretical estimate showed that, for the majority of the investigated diodes, $I_r/I_d \geq 1$.

Depending on the impurity concentration gradient and on the donor concentration in the original InAs, the room-temperature reverse current was governed by the carrier generation in p- and n-type regions of the p−n junction (diodes with a saturation region in the current−voltage characteristics) or by the carrier generation in the space-charge region (diodes which obeyed $j \propto U^{1/3}$) or by the tunnel effect (ordinary and backward tunnel diodes).

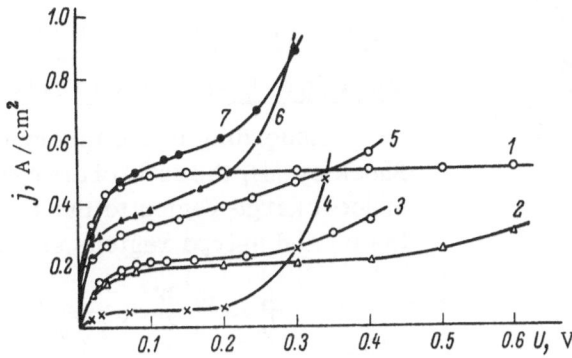

Fig. 2. Reverse branches of current−voltage characteristics at 300°K. 1) $n = 9.5 \cdot 10^{16}$ cm^{-3}, $a = 3.5 \cdot 10^{19}$ cm^{-4}; 2) $n = 5 \cdot 10^{17}$ cm^{-3}, $a = 10^{21}$ cm^{-4}; 3) $n = 6 \cdot 10^{17}$ cm^{-3}, $a = 3 \cdot 10^{21}$ cm^{-4}; 4) $n = 2 \cdot 10^{18}$ cm^{-3}, $a = 7 \cdot 10^{21}$ cm^{-4}; 5) $n = 6 \cdot 10^{17}$ cm^{-3}, $a = 1 \cdot 10^{22}$ cm^{-4}; 6) $n = 2 \cdot 10^{18}$ cm^{-3}, $a = 4 \cdot 10^{22}$ cm^{-4}; 7) $n = 1 \cdot 10^{19}$ cm^{-3}, $a = 3 \cdot 10^{21}$ cm^{-4}.

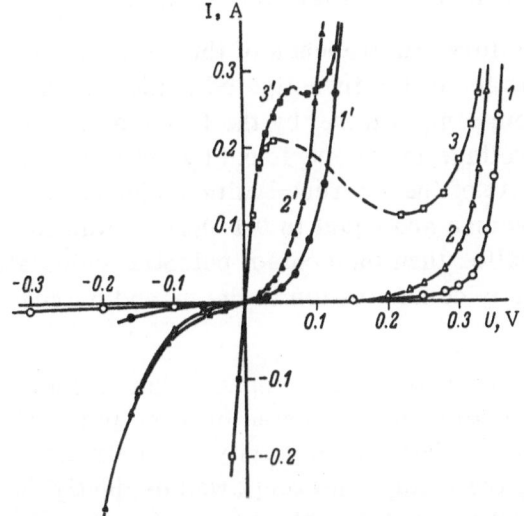

Fig. 3. Current−voltage characteristics of p−n junctions in InAs. 1, 1') $n = 1 \cdot 10^{19}$ cm^{-3}, $a = 3 \cdot 10^{21}$ cm^{-4}; 2, 2') $n = 1 \cdot 10^{19}$ cm^{-3}, $a = 3 \cdot 10^{23}$ cm^{-4}; 3, 3') $n = 1.5 \cdot 10^{19}$ cm^{-3}, $a = 6 \cdot 10^{23}$ cm^{-4}. T, °K: 1-3) 77; 1'-3') 300.

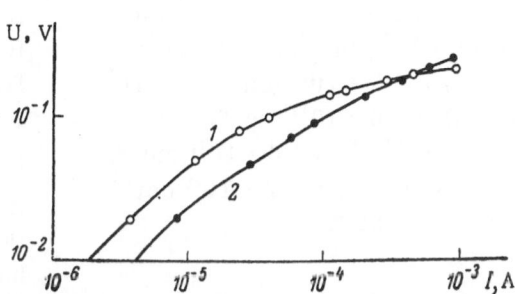

Fig. 4. Initial parts of a current−voltage characteristic ($n = 2 \cdot 10^{18}$ cm^{-3}, $a = 4 \cdot 10^{22}$ cm^{-4}): 1) forward current; 2) reverse current.

A theoretical estimate of the ratio I_r / I_d at liquid nitrogen temperature showed that the carrier generation current in the space-charge region should be a few orders of magnitude higher than the carrier generation current in the p- and n-type regions of the p−n junction. However, real currents were much higher than those calculated using the theory of Sah, Noyce and Shockley [4]. This was because of surface leakage and the tunnel effect via levels in the forbidden band. The importance of the tunnel effect was indicated by the fact that, in the initial parts of the current−voltage characteristics (U ≤ | 0.15 | V), the reverse current was usually higher than the forward current (Fig. 4).

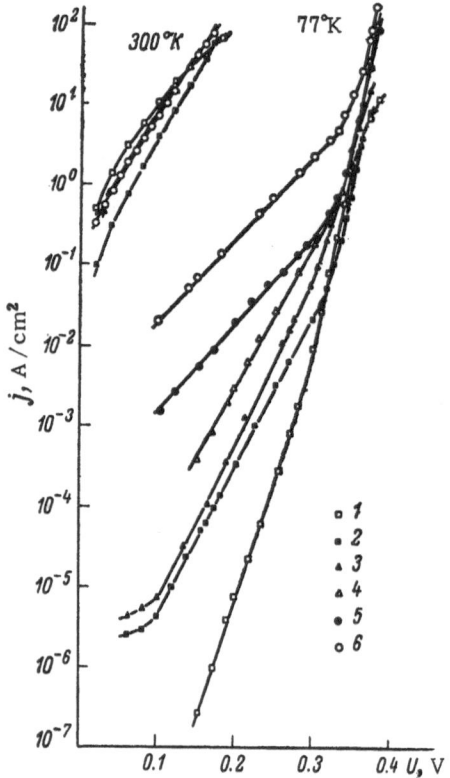

Fig. 5. Forward branches of the current−voltage characteristics obtained at 300 and 77°K: 1) n = $9.5 \cdot 10^{16}$ cm^{-3}, $a = 3.5 \cdot 10^{19}$ cm^{-4}; 2) n = $2 \cdot 10^{18}$ cm^{-3}, $a = 7 \cdot 10^{21}$ cm^{-4}; 3) n = $5 \cdot 10^{17}$ cm^{-3}, $a = 10^{21}$ cm^{-4}; 4) n = $6 \cdot 10^{17}$ cm^{-3}, $a = 3 \cdot 10^{21}$ cm^{-4}; 5) n = $6 \cdot 10^{17}$ cm^{-3}, $a = 10^{22}$ cm^{-4}; 6) n = $2 \cdot 10^{18}$ cm^{-3}, $a = 4 \cdot 10^{22}$ cm^{-4}.

Forward Branch of the Current − Voltage Characteristic

According to the theory of Sah, Noyce, and Shockley [4], the ratio of the recombination current in the space-charge region to the recombination current in the p- and n-type regions should be

$$\frac{I_r}{I_d} = \frac{n_i}{n_p} \frac{W}{2L} \times \frac{kT}{q(U_{\mathrm{br}} - U)} \exp\left(-\frac{qU}{2kT}\right).$$

Our measurements indicated that at room temperature this ratio was of the order of unity.

The forward branches of the current−voltage characteristics of diffused p−n junctions in InAs, recorded at room temperature, are shown in Fig. 5.

In the voltage range $kT/q < U < U_{\mathrm{br}}$, the forward current was an exponential function of the voltage and the argument of the exponential function was $\beta \sim 1$.

Thus, the forward branches of the current−voltage characteristics of the investigated diodes were described, at room temperature, by the Shockley theory. However, the voltage intercept, found by extrapolation of the linear part of the current−voltage characteristic to the abscissa and equal to 0.1-0.17 V, was found to be much smaller than the contact potential calculated from the impurity concentration in the p- and n-type regions.

Since the boundaries of the space-charge region in these p−n junctions were located in those parts of the semiconductor where the impurity concentration had a gradient, the voltage intercept was evidently determined by the contact potential difference between the boundaries of the space-charge region under zero bias, because the experimental value of this difference agreed with the voltage intercept.

The ratio I_r/I_d, estimated at liquid-nitrogen temperature, showed that the recombination current in the space-charge region should exceed the recombination current in the p- and n-type regions of the p−n junction up to voltages of 0.3-0.33 V. Figure 5 shows the current−voltage characteristics determined at 77°K.

At voltages 0.3 < U < 0.38 V, the current−voltage characteristics of the p−n junctions made of InAs with electron densities of $9.5 \cdot 10^{16}$ to $1.5 \cdot 10^{19}$ cm^{-3} were exponential with an argument $\beta = 1-1.3$. Bearing in mind the resistance of the contacts and the bulk of the semiconductor, which could be estimated from the linear parts of the current−voltage characteristics, we concluded that the value of β for all the diodes was close to 1. For diodes with moderately steep acceptor concentration gradients and with an initial electron density of n = $9.5 \cdot 10^{16}$ cm^{-3}, the characteristic had an exponential region with $\beta = 2$ in a wide range of voltages defined by 0.1 < U < 0.3 V. However, when the acceptor concentration gradient and the carrier density in the initial material were increased, the value of the exponent β obtained in this range of voltages increased and reached 3 for diodes made of InAs with n = 5 · 10^{17} to 2 · 10^{18} cm^{-3} and having impurity gradients 10^{21} to 7 · 10^{21} cm^{-4}. For some of these samples, a region with $\beta = 2$ was observed only in a narrow range of voltages close to U = 0.3-0.33 V. When the acceptor impurity gradient was increased still further to 10^{22} to 4 · 10^{22} cm^{-4}, the value of β for diodes made of indium arsenide with n = 5 · 10^{17} to 2 · 10^{18} cm^{-3} increased to 7. Finally, as already mentioned, when the gradient was (3-6) · 10^{23} cm^{-3} in InAs with n = 1.5 · 10^{19} cm^{-3}, we obtained ordinary and backward tunnel diodes (Fig. 3).

For the sake of comparison, Fig. 3 shows the current—voltage characteristic typical of diodes with a moderately steep impurity concentration gradient and made of InAs with n = 1 · 10^{19} cm^{-3} (curve 1 in Fig. 3).

Thus, all the diodes (with the exception of those prepared from InAs with a relatively low initial donor concentration of 9.5 · 10^{16} cm^{-3}) exhibited excess currents in the voltage range up to 0.3-0.33 V.

In order to determine the nature of the excess current, we prepared diodes with different ratios of the p—n junction area to the junction perimeter. Variations in this ratio did not have much influence on the absolute value of the excess current density or the value of β. Therefore, we concluded that the excess current was not due to surface leakage. We could assume that the carrier lifetime in the expression for the forward current given by the Sah—Noyce—Shockley theory was not constant but increased with the injection level. However, calculations showed that this assumption was justified only for those diodes for which the value of β did not exceed 3. Otherwise, we found that the initial electron lifetime would have to be unrealistically short (of the order of 10^{-17} sec for U = 0.1 V and $\beta = 7$). The observed current was evidently similar to that of the excess current in tunnel diodes, i.e., it was due to tunnel transitions through allowed states in the forbidden band. This was supported also by the observation that the excess current increased (although more slowly than the reverse current) with increasing impurity concentration gradient. When the gradient was sufficiently high the forward current in the initial part of the current—voltage characteristic could be even lower than the reverse current.

Estimate of the Diffusion Length and Lifetime
of Minority Carriers

According to the Shockley theory, the saturation current is given by the following expression:

$$I_{d_0} = q\left(\frac{n_p D_n}{L_n} + \frac{p_n D_p}{L_p}\right).$$

For a linear distribution of impurities in a p—n junction, the electron and hole densities at the boundaries of the space-charge region are approximately equal. Moreover, in the case of indium arsenide $D_p \ll D_n$ and $L_p > L_n$ so that the second term in parentheses can be neglected. According to the photomagnetic effect data [5], the diffusion length of electrons in InAs is very short and amounts to 0.3-3 μ. Therefore, we may assume that, over a path equal to the diffusion length of the electrons in the p-type region of a p—n junction, the majority carriers (holes) are distributed in accordance with the linear law $p \approx n_i + ax$. Using the law of mass action (but ignoring degeneracy), we find that the effective ("working") electron density is

$$n_{\text{eff}} = \int_0^{L_n} \frac{n_i^2 dx}{n_i + ax} = \frac{n_i^2}{L_n a}\ln\left(1 + \frac{aL_n}{n_i}\right),$$

where n_i is the intrinsic carrier density. Bearing this in mind, we find that $L_n = en_{eff} D_p / I_{d_0}$ and $\tau_n = L_n^2 / D_n$. The values of L_n and τ_n, calculated from the current−voltage characteristics using values of μ_n equal to the electron mobilities in the original material, were 1-5 μ and $2 \cdot 10^{-10}$ to $2 \cdot 10^{-9}$ sec, respectively, at T = 77°K. The saturation current, found by extrapolation of the forward branch of the current−voltage characteristic with $\beta = 1$ to the ordinate, was $7 \cdot 10^{-23}$ to $4 \cdot 10^{-22}$ A/cm². At room temperature, the diffusion length of the electrons was 1.5-2 times longer.

Conclusions

The electrical properties of diffused p−n junctions in InAs, with an initial electron density of $9.5 \cdot 10^{16}$ to $1.5 \cdot 10^{19}$ cm⁻³ and different acceptor impurity gradients in the junctions, were investigated at room and liquid-nitrogen temperatures. Tunnel diodes of the ordinary and backward types were obtained from InAs with an initial electron density of 10^{19} cm⁻³ when the acceptor impurity gradient was $3 \cdot 10^{23}$ cm⁻⁴.

The forward branches of the current−voltage characteristics of the investigated p−n junctions obeyed the Shockley theory at room temperature. At liquid-nitrogen temperature, the forward branches of the characteristics of diodes with a relatively low electron density in the initial material (up to $n \approx 10^{17}$ cm⁻³) obeyed the theory of Sah, Noyce, and Shockley.

Samples prepared from crystals with higher electron densities exhibited excess currents in the initial parts of their current−voltage characteristics (up to U < 0.3 V) and these excess currents were probably due to tunneling through impurity levels in the forbidden band.

At room temperature, the reverse branches of the current-voltage characteristics of the investigated diodes obeyed the Shockley theory for $n = 9.5 \cdot 10^{16}$ to $2 \cdot 10^{18}$ cm⁻³ and $a \leq 7 \cdot 10^{21}$ cm⁻⁴ or the Sah−Noyce−Shockley theory for $n = 5 \cdot 10^{17}$ to $1 \cdot 10^{19}$ cm⁻³, $a = 3 \cdot 10^{21}$ to $4 \cdot 10^{22}$ cm⁻⁴. At liquid-nitrogen temperatures, the reverse currents were governed by the tunnel effect.

The diffusion lengths and lifetimes of electrons, deduced from the current−voltage characteristics, were 1-5 μ and $2 \cdot 10^{-9}$ to $2 \cdot 10^{-10}$ sec at T = 77°K.

The authors take this opportunity to thank R. A. Rusakov for determining the voltage−capacitance characteristics.

Literature Cited

1. G. Lucovsky, Brit. J. Appl. Phys., 12:311 (1961).
2. N. P. Esina, N. V. Zotova, and D. N. Nasledov, Radiotekhn. i Élektron., 8:1602 (1963).
3. Yu. D. Mozzhorin and V. I. Stafeev, Fiz. Tekh. Poluprov., 1:829 (1967).
4. C. T. Sah, R. N. Noyce, and W. Shockley, Proc. IRE, 45:1228 (1957).
5. M. P. Mikhailova, Author's Abstract of Dissertation for Candidate's Degree, Leningrad (1967).

PARAMETERS OF THE ACTIVE REGION
IN AN INJECTION LASER

Yu. I. Kruzhilin

A method is suggested for the determination of the loss factor of a p−n junction and of the initial threshold current density necessary for population inversion (negative temperature state). The loss factor is given by the ratio of the slopes of the watt−ampere characteristics of a diode, recorded using two different values of the diode length or two values of the reflection coefficient of the resonator mirror. The initial threshold current density is then determined using the value of the loss factor found in this way and the threshold current densities for two values of the diode length, or for two values of the reflection coefficient. This method was employed to investigate diffused gallium arsenide diodes. The loss factor was found to be 9.6 cm^{-1} and the threshold current density was 690 A/cm^2 for a resonator with 100% negative feedback. These two parameters were then used to calculate the initial threshold current density (485 A/cm^2).

The principal parameters of an injection laser are its threshold current density and the rate of rise of the output power beyond the threshold. The threshold current density depends on the losses in the p−n junction of the laser and on the power lost in the process of emergence of radiation from the resonators (due to finite transmission of the end reflectors):

$$j_t = j_t' + \frac{\gamma}{\beta} + \frac{\ln \frac{1}{R_1 R_2}}{2\beta L} .$$

(1)

Here, j_t' is the threshold current, which ensures that a state of negative temperature (population inversion) is reached; γ is the factor representing the losses in the active region due to absorption, scattering, and diffraction; L is the resonator length; R_1 and R_2 are the reflection coefficients of the resonator mirrors; β is a coefficient representing the activity of the amplifying medium (it is known as the gain factor).

In some treatments [1, 2], the value of j_t' is assumed to be negligibly small. Therefore, it is found that the threshold current density tends to zero when $\gamma \to 0$ and $\frac{1}{2L} \ln \frac{1}{R_1 R_2} \to 0$. The results of a numerical calculation of the threshold current density for the parabolic band model are given in [3], where it is shown there that the value of j_t' may be considerable. Moreover, it is concluded that in general it is necessary to take account of j_t' for any band structure at all temperatures other than that of absolute zero.

The differential external efficiency η depends on the probability of radiative recombination in the active region η' and on the resonator-load coupling coefficient K [2-5], defined as the ratio of the radiant power emerging from the resonator to the radiant power dissipated within the resonator:

$$\eta = \eta' \frac{K}{K+1},$$
(2)

$$K = \frac{\ln \frac{1}{R_1 R_2}}{2\gamma L}.$$
(3)

The loss factor γ can be found from Eqs. (2) and (3):

$$\gamma = \left(\frac{\eta'}{\eta} - 1\right) \frac{1}{2L} \ln \frac{1}{R_1 R_2}.$$
(4)

A similar expression is derived in [3] but the value of η' used in [3] is assumed to be unity. It is suggested in [3] that the loss factor γ can be determined using absolute measurements of the output power and of the pumping power, i.e., by measuring η. However, in order to calculate γ, it is also necessary to know exactly the value of η', which is difficult to determine experimentally. Moreover, the accuracy of the absolute measurements of the pumping power and, particularly, of the output power is usually low. Therefore, direct determination of the loss factor by means of Eq. (4) is not necessarily the best method.

The difficulties just mentioned can be eliminated by making relative measurements for various reflector transmission losses $\frac{1}{2L} \ln \frac{1}{R_1 R_2}$. Making two measurements and assuming that γ and η' are constant, we obtain

$$\left(\frac{\eta'}{\eta_1} - 1\right) \frac{1}{2L} \ln \frac{1}{R_{11} R_{21}} = \left(\frac{\eta'}{\eta_2} - 1\right) \frac{1}{2L_2} \ln \frac{1}{R_{12} R_{22}}.$$
(5)

Here, the subscript (or the second subscript if there are two) indicates the number of the measurement. Equation (4) yields an expression for the probability of radiative recombination in the active region η':

$$\eta' = (1 - q) \left(\frac{1}{\eta_1} - q \frac{1}{\eta_2}\right)^1,$$
(6)

where

$$q \equiv \frac{L_1 \ln \frac{1}{R_{12} R_{22}}}{L_2 \ln \frac{1}{R_{11} R_{21}}}.$$

In order to determine η', we must know the absolute values of η_1 and η_2. The loss factor γ can be found from Eqs. (4) and (6):

$$\gamma = \left(\frac{\eta_1}{\eta_2} - 1\right) \left(1 - q \frac{\eta_1}{\eta_2}\right)^{-1} \frac{1}{2L_2} \ln \frac{1}{R_{12} R_{22}}.$$
(7)

Equation (7) contains only the ratio of the differential efficiencies η_1 / η_2 and, therefore, there is no need to carry out absolute measurements.

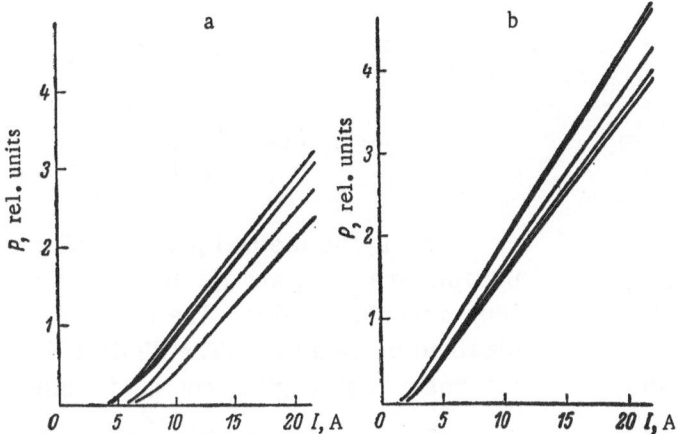

Fig. 1. Dependence of the output power P on the injection current I: a) resonator length 750 μ, diode width 730 μ; b) diode length 343 μ, diode widths 372 and 415 μ.

If, in addition to such relative measurement of the differential efficiencies, we also determine the threshold current densities in the first and second experiment (j_{t1} and j_{t2}), we can easily calculate [1, 2, 6] the gain factor β:

$$\beta = \frac{1}{2}\left(\frac{1}{L_1}\ln\frac{1}{R_{11}R_{21}} - \frac{1}{L_2}\ln\frac{1}{R_{12}R_{22}}\right)\ (j_{t1} - j_{t2})^{-1}. \tag{8}$$

Assuming that γ is known, we can determine the initial threshold current density j_t' from Eqs. (1) and (8):

$$j_t' = [j_{t2}\,(1 + p) - j_{t1}\,(q + p)]\,(1 - q)^{-1}, \tag{9}$$

where

$$p \equiv \frac{2\gamma L_1}{\ln\dfrac{1}{R_{11}R_{21}}}.$$

Thus, using Eqs. (7)-(9), we can calculate the factors γ and β and determine the value of j_t'. All that is necessary is to determine experimentally the ratio of the slopes of the watt-ampere characteristics for two values of $\frac{1}{2L}\ln\frac{1}{R_1 R_2}$ and to measure the threshold current densities under the same conditions.

The parameter $\frac{1}{2L}\ln\frac{1}{R_1 R_2}$ can be altered either by changing the reflection coefficients of the resonator mirrors or by varying the diode length. The former approach is preferable because it allows us to carry out both measurements on the same diode. This ensures that any inhomogeneities in the p−n junction have the same effect in both measurements. However, if a diode is small, it is difficult to determine the reflection and absorption coefficients of the mirrors and this may give rise to difficulties in the interpretation of the results obtained [2]. Therefore, the method of varying the diode length may be more reliable.

Let us assume that $R_{11} = R_{21} = R_{12} = R_{22} = R$. Then, instead of Eqs. (7)-(9), we obtain

$$\gamma = \left(\frac{1}{L_2}\ln\frac{1}{R}\right)\left(\frac{\eta_1}{\eta_2} - 1\right)\left(1 - \frac{L_1\eta_1}{L_2\eta_2}\right)^{-1}, \tag{10}$$

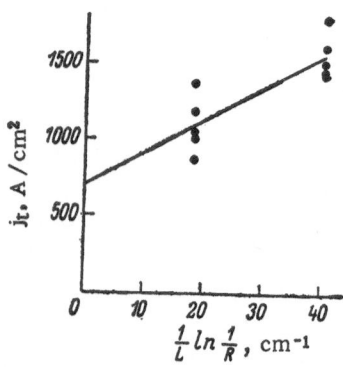

Fig. 2. Dependence of the threshold current density j_t on the reflector transmission loss.

$$\beta = (j_{t1} - j_{t2})^{-1}\left(\frac{1}{L_1} - \frac{1}{L_2}\right)\ln\frac{1}{R}, \tag{11}$$

$$j_t' = \left[j_{t2}\left(1 + \frac{\gamma L_1}{\ln\frac{1}{R}}\right) - j_{t1}\left(\frac{L_1}{L_2} + \frac{\gamma L_1}{\ln\frac{1}{R}}\right)\right]\left(1 - \frac{L_1}{L_2}\right)^{-1}. \tag{12}$$

Using the method just described, we determined the parameters γ, β, and j_t'. We used a batch of diffused gallium arsenide diodes. The resonator faces of the diodes were obtained by cleaving. The diode width was always equal to the resonator length. The diode thickness was 100 μ.

Our measurements were carried out in liquid nitrogen. A pressure holder, containing a diode, was located in a transparent branch of a Dewar flask and this branch was placed in an integrating sphere. A silicon photodiode, subjected to a reverse bias of 20 V, was used as the radiation detector.

We investigated the dependence of the output power on the injection current. The threshold current density was then determined by extrapolation of the linear part of the watt−ampere characteristic until it intersected the current axis. We also determined the threshold current from the narrowing of the emission line, using a Fabry−Perot interferometer with an air gap of $3 \cdot 10^{-2}$ cm and an image converter for the infrared radiation. The threshold current measured in this way was always lower than the current found by extrapolation of the watt−ampere characteristic, and it corresponded to the initial nonlinear part of this characteristic. This difference between the threshold currents could be explained by assuming that laser emission appeared first in a filament which had the lowest threshold. This was sufficient to cause line narrowing but insufficient for the diode to reach its maximum differential efficiency because only part of the p−n junction area participated in the emission of stimulated radiation. Since the threshold current density was used to calculate quantities averaged over the whole diode, it was more correct to determine the threshold current from the watt−ampere characteristic.

The experimental results are given in Figs. 1 and 2. Figure 1 shows a family of the watt−ampere characteristics of the investigated diodes. The threshold currents, determined by extrapolation of the linear parts of the characteristics in Fig. 1, are represented by points in Fig. 2.

We assumed that the Fresnel reflection coefficient was R = 0.25. We used the average slope of the characteristics in Fig. 1a to obtain η_1; the value of η_2 was found from Fig. 1b. The ratio of the efficiencies obtained in this way was 1.23. The value of this ratio was then used in Eq. (10) to find the loss factor $\gamma = 9.6$ cm^{-1}. The data given in Fig. 2 yielded $\beta = 4.7 \cdot 10^{-2}$ cm/A. Substituting the average values $j_{t1} = 1095$ A/cm^2 and $j_{t2} = 1561$ A/cm^2 into Eq. (12), we obtained $j_t' = 485$ A/cm^2. Using Eq. (1) or Fig. 2, we found the minimum threshold current density for the investigated p−n junction (for $\frac{1}{2L}\ln\frac{1}{R_1R_2} \to 0$), which was $j_{t0} = 690$ A/cm^2. This current density consisted of two components: 485 A/cm^2 was required to establish population inversion and 205 A/cm^2 was necessary to produce amplification which balanced out the losses represented by γ. Consequently, in this case, the threshold current density j_{t0} was governed primarily by the initial threshold current density j_t' rather than by the passive losses in the p−n junction.

The usual method [1, 2, 6] for the determination of the losses, based solely on an investigation of the dependence of the threshold current density on the parameter $\frac{1}{2L}\ln\frac{1}{R_1R_2}$ and on the approximation $j_t' = 0$, could give incorrect results. This possibly accounted for a considerable scatter of the values of the loss factor and the quantum yield reported in [2] for diodes whose external differential efficiency did not vary greatly (particularly in the case of high-efficiency diodes). In our experiments, the data in Fig. 2 give $\gamma = 32.4$ cm^{-1}, which was considerably higher than the value determined by the method described in the present paper.

This method, which we used to determine the parameters γ and j_t' of the active region, could be employed also to investigate the dependences of these parameters on the previous history of a diode or on its working conditions.

The author is grateful to N. V. Antonov and Yu. I. Koloskov for their help in this investigation.

Literature Cited

1. M. Pilkuhn and H. Rupprecht, Proc. IEEE, 51:1243 (1963).
2. N. G. Basov, P. G. Eliseev, S. D. Zakharov, Yu. P. Zakharov, I. N. Oraevskii, I. Z. Pinsker, and V. P. Strakhov, Fiz. Tverd. Tela, 8:2616 (1966).
3. V. P. Gribkovskii, Zh. Prikl. Spektrosk., 6:669 (1967).
4. W. W. Rigrod, J. Appl. Phys., 34:2602 (1963).
5. A. Yariv, Quantum Electronics (Proc. Third Intern. Congress, Paris, 1963), Dunod, Paris, and Columbia University Press, New York (1964), p. 1055.
6. E. J. Walker and A. E. Michel, J. Appl. Phys., 35:2285 (1964).

MINIMUM THRESHOLD CURRENT
FOR AN INJECTION LASER AND ITS RELATIONSHIP
WITH THE GAIN FACTOR

Yu. I. Kruzhilin

An expression is obtained for the minimum threshold current of a laser diode whose resonator length can be made as small as required. A method is described for the determination of the gain factor β, using a single measurement of the threshold current of a diode with a sufficiently short resonator. The validity of this method is demonstrated by applying it to available experimental data.

The threshold current of an injection laser is given by

$$I_t = j_t l L, \tag{1}$$

where j_t is the threshold current density; l is the diode width; L is the resonator length.

Substituting the well-known expression for j_t

$$j_t = j_{t0} + \frac{\ln \frac{1}{R_1 R_2}}{2\beta L}, \tag{2}$$

we obtain

$$I_t = I_{t0} + \frac{l \cdot \ln \frac{1}{R_1 R_2}}{2\beta}. \tag{3}$$

Here, R_1 and R_2 are the reflection coefficients of the resonator mirror; j_{t0} is the threshold current density for $R_1 = R_2 = 1$, which is equal to the sum of the initial threshold current density necessary for the establishment of a state with negative temperature (population inversion) and the current density necessary to produce a gain equal to the passive losses in the p−n junction; β is the gain factor of the p−n junction (equal to the rate of rise of the gain with increasing current in the subthreshold region), which is independent of the current density [1, 2]; $I_{t0} = j_{t0} l L$.

The value of j_{t0} is constant for a given p–n junction. Therefore, when $L \to 0$, we find that $I_{t0}/l \to 0$. Then, Eq. (3) yields an expression for the minimum threshold current density of a diode whose length can be made as short as required:

$$\frac{1}{l} I_{min} = \frac{\ln \frac{1}{R_1 R_2}}{2\beta} . \qquad (4)$$

If we assume that $R_1 = R_2 = 0.32$, $\beta = 3 \cdot 10^{-2}$ cm/A, we find from Eq. (4) that $I_{min}/l = 38$ A/cm. The condition for the smallness of L is $j_t \gg j_{t0}$.

In the same approximation, Eq. (4) yields an expression for the factor β:

$$\beta = \frac{l \cdot \ln \frac{1}{R_1 R_2}}{2 I_{min}} \qquad (5)$$

or

$$\beta = \frac{\ln \frac{1}{R_1 R_2}}{2 L j_t} . \qquad (5a)$$

If we measure I_{min} (the threshold current of a short diode), we can use such a single measurement to determine the factor β. The method used up to now requires that measurements be carried out for at least two values of the reflector transmission loss $\frac{1}{2L} \ln \frac{1}{R_1 R_2}$, A/cm, which means that either L or R_1, R_2 must be varied.

The proposed method for the determination of the gain factor β can be checked by carrying out two experiments: first, we ought to determine β in the usual way from the slope of the dependence of j_t on $\frac{1}{2L} \ln \frac{1}{R_1 R_2}$, and then we should measure I_{min} for a short diode of the same bath and calculate β using Eq. (5).

The traditional method for determination of the gain factor β was used by Basov et al. [2]. A value $\beta = 1.6$ cm/A was obtained for one of the batches of the diodes. Basov et al. quoted the parameters for a diode 28 μ long belonging to this batch. The threshold current density for this diode was $3 \cdot 10^4$ A/cm^2. They also reported that $R_1 = R_2 = 0.25$ [2]. Using Eq. (5a), we find that $\beta = 1.65$ cm/A, i.e., the calculated value is approximately equal to that obtained from a series of measurements.

Literature Cited

1.   M. P. Pilkuhn and H. Rupprecht, Proc. IEEE, 51:1243 (1963).
2.   N. G. Basov, P. G. Eliseev, S. D. Zakharov, Yu. P. Zakharov, I. N. Oraevskii, I. Z. Pinsker, and V. P. Strakhov, Fiz. Tverd. Tela, 8:2616 (1966).

SPECIAL FEATURES OF
EXPOSURE—ILLUMINANCE CHARACTERISTICS OF THE
PHOTOPOLARIZATION OF HgI$_2$ SINGLE CRYSTALS

V. F. Zolotarev, D. G. Semak, and D. V. Chepur

Uzhgorod State University

Investigations of the constant-potential curves of mercuric iodide single crystals showed that the nature of the exposure—illuminance characteristics (which showed whether the reciprocity law was obeyed) depended appreciably on the potential of the electrode through which a sample was illuminated. Thus, when a crystal was illuminated through the negative electrode, the reciprocity law was obeyed but when it was illuminated through the positive electrode, this law was not obeyed. Experimental investigations were made of the special features of the exposure—illuminance characteristics obtained by monochromatic illumination of the positive electorde. The causes of the departure from the reciprocity law were determined and a possible mechanism was suggested for the formation of the photoelectret state in mercuric iodide single crystals.

Investigations of the constant-potential curves of the photopolarization are of scientific and practical importance. They allow us to determine the relationship between the exposure—illuminance, current—illuminance, and other characteristics. Information obtained on these characteristics can be used to determine the mechanism of the formation of the photoelectret state.

The exposure—illuminance characteristic is one of the basic relationships of the photoelectret state because it shows the limits of validity of the reciprocity law, which are important in electrophotographic, television, and other applications of photoelectrets. Studies of the reciprocity law can often be used to check experimentally whether the quasi-steady-state conditions [1] are satisfied (these conditions are important in investigations of the kinetics of electron processes giving rise to the photoelectret state).

The photoelectret state in mercuric iodide was observed first by Bugrienko and Demidov [2], who used powder samples. The present paper describes an investigation of the characteristic features of the exposure—illuminance characteristics of HgI$_2$ single crystals.

Experimental Method

We investigated single crystals grown from a solution in acetone. The optically determined degree of inhomogeneity of these crystals and was less than 5%. The investigated samples were in the form of plates, $4 \times 4 \times 1$ mm. The dark resistivity of the samples was 10^{13}-10^{14} $\Omega \cdot$ cm. The photoelectric state was investigated by the capacitor method, using the measuring circuit described in [3]. A constant voltage of up to 500 V was applied to a sample placed in a cryostat. The sample holder was cooled with liquid nitrogen. During the investigations of the photopolarization and photodepolarization, a sample could be illuminated with white or mono-chromatic light (the latter supplied from a UM-2 monochromator). The photoelectret state of the HgI_2 single crystals was determined using the depolarization current method [1]. We re-corded the dependence of the polarization on the exposure time for different values of illum-inance. The initial value of the photodepolarization current, recorded using a dc amplifier, was used as a measure of the photopolarization. The sequence of operations was as follows: a sample originally stored in darkness was subjected simultaneously to an electric field and illumination. The sample was illuminated for an exposure time t; 15 sec later the field was switched off, the sample stored in darkness for 45 sec in the short-circuited state, and finally, a depolarizing light pulse applied for 5 sec. This light pulse was provided by an incandescent lamp of 30 W power. A new photopolarization cycle was started after storing the short-circuited sample for 2 min in order to eliminate the influence of the residual photoexcited state, which had been reported for these crystals [4]. The initial photopolarization current (proportional to the photoelectret charge) was found by subtracting the dark polarization from the total polariza-tion (the dark polarization was determined under the same conditions as above but in the ab-sence of illumination during the application of an external electric field).

Experimental Results

Figures 1 and 2 show the dependences of the initial depolarization current on the dura-tion of exposure to white light of different intensities during the photopolarization process. These curves were used to plot, employing a well-known method [1], the exposure—illuminance characteristics of the polarization for two polarities of the illuminated electrode. Illumina-tion through the positive electrode produced conditions which failed to satisfy the reciprocity law (curve a in Fig. 3) but illumination through the negative electrode produced a state which satisfied the reciprocity law (curve b in Fig. 3). A study of the current—illuminance character-istics showed that these were sublinear when the positive electrode was illuminated and linear when the illumination came from the negative side (Fig. 4). This agreed with the nature of the exposure—illuminance characteristics of Fig. 3 in the light of the theory given in [1].

The difference between the current —illuminance characteristics obtained by illumination through the positive and negative electrodes may be explained using one of the models for HgI_2 [5, 6].

We shall use one of these models and the experimental data deduced from an investiga-tion of the photoconductivity [5] and from a study of the photopolarization of HgI_2 single crys-tals [7-9]. We shall use the energy level scheme shown in Fig. 5. There are two types of level in this scheme and, for the sake of convenience, we shall subdivide further the levels of the first type into two groups: "trapping" levels M_3 and deep levels M_2, for which thermal excita-tion processes are negligible, i.e., which act as recombination centers at low temperatures. Levels of the second type correspond to centers denoted by M_1. The electron-capture cross section of levels of the first type (M_3 and M_2) range from 10^{-13} cm^2 for shallower levels to 10^{-15} cm^2 for deeper levels. The electron-capture cross section of the M_1 levels is about 10^3 times smaller than the hole capture cross section.

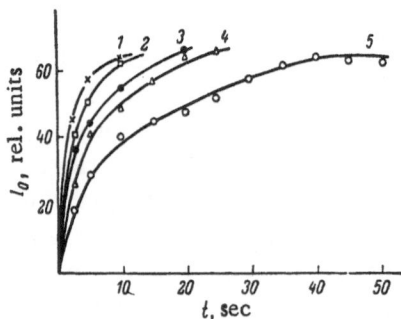

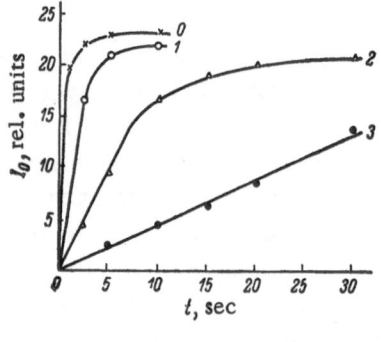

Fig. 1. Dependence of the initial depolarization current on the exposure time in the case of illumination of the positive electrode during photopolarization. The values of illuminance corresponding to curves 1-5 were in the ratio 0.25:0.06:0.015: 0.0035:0.0009.

Fig. 2. Dependence of the initial depolarization current on the exposure time in the case of illumination of the negative electrode during photopolarization. The values of illuminance corresponding to curves 0-3 were in the ratio 1:0.25:0.06:0.015.

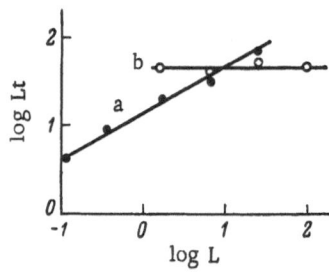

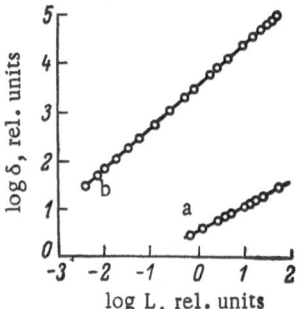

Fig. 3. Exposure−illuminance characteristics of HgI_2 single crystals: a) illumination through a transparent electrode subjected to a positive potential; b) illumination through a transparent electrode subjected to a negative potential. The exposure (product of the illuminance and exposure time) and the illuminance are in relative units.

Fig. 4. Current−illuminance characteristics of HgI_2 single crystals: a) illumination of the positive electrode; b) illumination of the negative electrode.

According to this model, the condition for neutrality is of the following form:

$$n + n_3 = p + (M_2 - n_2) + (M_1 - n_1),$$ (1)

where, as usual, n_1, n_2, and n_3 are the concentrations of electron-filled centers. Using the neutrality condition [1], we find that the dependence of p on n, obtained by solving the relevant transport equations for the steady-state case (Fig. 1) and assuming that $p \ll n \ll M_3$, is of the form

$$p = \frac{n^2 a_1}{n\,(\beta M_1 + a_1) + a_1 M_2} \cdot \frac{\beta M_3 + a_3}{\beta n + a_3},$$ (2)

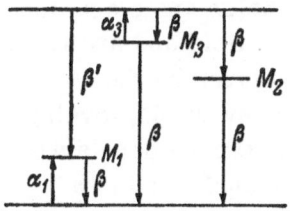

Fig. 5. Energy level scheme of HgI$_2$.

where

$$\alpha_1 = N_v \beta \exp\left(-\frac{E_1}{kT}\right), \quad \alpha_3 = N_c \beta \exp\left(-\frac{E_3}{kT}\right),$$

and β is the probability of carrier capture.

Since the M_3 centers are the trapping centers ($\beta n \ll \alpha_3$), we can distinguish two different cases. In the first case, when $(\beta M_1 + \alpha_1)n \ll \alpha_1 M_2$ and $p \propto F$, we find that

$$n \propto F^{1/2}; \tag{3}$$

in the second case, when $(\beta M_1 + \alpha_1)n \gg \alpha_1 M_2$, we obtain

$$n \propto F. \tag{4}$$

Comparing Eqs. (3)–(4) with the experimentally determined current−illuminance characteristics (Fig. 4) of HgI$_2$ single crystals, obtained by illumination through the positive and negative electrodes, we find that both cases occur in HgI$_2$, depending on whether a crystal is illuminated through the positive or negative electrode.

When a crystal is illuminated throughout the positive electrode, we obtain a sublinear current−illuminance characteristic with a power exponent of 0.5-0.6 but when it is illuminated through the negative electrode, the characteristic is linear, in agreement with Eqs. (3) and (4). This agreement is obviously due to the fact that the current is carried mainly by electrons and the free-electron densities in the illuminated part of the crystal are different in the first and second cases (it is assumed that the crystal is in an electric field).

It is shown in [7-9] that all trapping levels, which have a complex energy spectrum, take part in the formation of the photoelectret state. Moreover, the photoconductivity of HgI$_2$ is unipolar [5], i.e., when the positive electrode is illuminated, the photoconductivity is much weaker than in the case of illumination through the negative electrode. This indicates that holes are captured more readily than electrons and that the conduction is n-type. The spectral range of the photoconductivity lies in the impurity-absorption region [10, 11] since at liquid-nitrogen temperature pure single crystals are transparent in the visible region right up to the blue-green wavelengths ($\lambda \approx 4800$ Å), corresponding to a forbidden band width of $\sim$2.58 eV. The sensitivity spectrum of the photodepolarization (Fig. 6) lies in the same spectral region as the photoconductivity. The relative positions of the $\lambda_1 = 5535$ Å, $\lambda_2 = 5675$ Å, and $\lambda_3 = 5845$ Å maxima of the photodepolarization spectrum coincide with the relative energy positions of the levels deduced by the thermally stimulated current method under photoelectret conditions [7, 8].

It is worth noting that the sum of the corresponding energy levels, determined by the thermally stimulated current method under photoelectret conditions, and of the maxima in the depolarization or photoconductivity spectra is constant (2.245 eV + 0.344 eV $\approx$ 2.182 eV + 0.384 eV $\approx$ 2.120 eV + 0.450 eV $\approx$ 2.58 eV). Since, like the thermal depolarization under photoelectret conditions, the photodepolarization can be produced by infrared illumination [3], it is natural to assume that the same energy levels are active in the thermally stimulated currents, in the photoconductivity, and in the photodepolarization and that the sum of their energy positions, deduced from the thermally stimulated current spectra under photoelectret conditions and from the depolarization spectra, is equal to the forbidden band width ($\Delta E = 2.58$ eV), i.e., the mechanism of the formation of the photoelectret state is probably as follows. In the photopolarization and photoconductivity processes, the incident radiation transfers electrons from the valence band to impurity levels. Next, electrons are transferred by the thermal vibrations and the incident radiation to the conduction band. Free holes are captured mainly by the M_1 levels because of their large hole-capture cross sections and, therefore, the photoconduc-

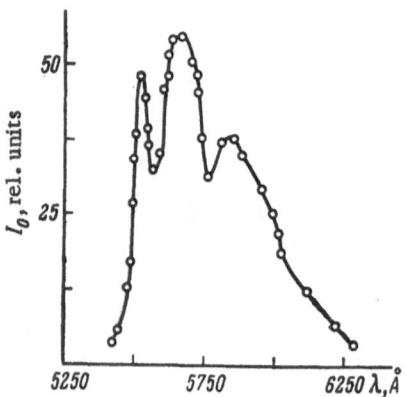

Fig. 6. Depolarization spectrum of an HgI$_2$ single crystal in the case of polarization and depolarization by illumination through the positive electrode.

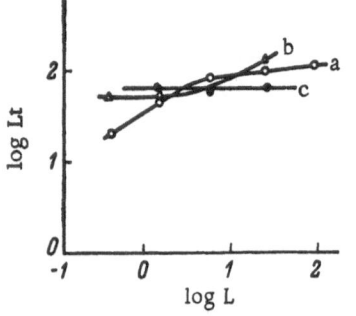

Fig. 7. Exposure−illuminance characteristics in the case of illumination with monochromatic light through the positive electrode. a) $\lambda_0 < 5535$ Å; b) $\lambda_2 = 5675$ Å; c) $\lambda_3 = 5845$ Å. The exposure (product of the illuminance and exposure time) and illuminance are given in relative units.

tivity is n-type. The remaining free holes and electrons are spread out by the external electric field and are trapped near the electrodes (electrons are captured by the M$_3$ trapping levels and holes are captured by the M$_1$ levels), giving rise to the photoelectret state. During depolarization with visible light, electrons are transferred in a similar manner from the valence to the conduction band and they neutralize the photoelectret charge. In the case of infrared illumination and thermal depolarization, the trapped electrons are transferred directly to the conduction band.

The dependence of the validity of the reciprocity law on the polarity of the electrode through which a crystal is illuminated can be explained by assuming that the concentration of electron recombination centers p$_2$ is higher than the concentration of hole recombination centers n$_2$ in darkness because the M$_2$ levels (M$_2$ = n$_2$ + p$_2$) lie above the Fermi level [12].

When a crystal is illuminated through the negative electrode, the concentration of electron recombination centers is practically unaffected by the capture of holes (electrons are assumed to move into the interior of a sample) and this is responsible for the constancy of the electron lifetime, the linearity of the current−illuminance characteristic and the validity of the reciprocity law. Conversely, when a crystal is illuminated through the positive electrode, the lifetime of the holes moving into the interior of a crystal is less because of electron capture by the M$_2$ levels. This makes the current−illuminance characteristic sublinear and leads to failure of the reciprocity law.

To check the correctness of this explanation of the failure of the reciprocity law, we determined the exposure−illuminance characteristics using monochromatic illumination through the positive electrode. Because of differences in the absorption coefficients the electron population of the recombination levels depended on the wavelength of incident light. Figure 7 shows the exposure−illuminance characteristics for monochromatic illumination with light of wavelengths $\lambda_0 < 5535$ Å, $\lambda_2 = 5675$ Å, $\lambda_3 = 5845$ Å through the positive electrode (these characteristics were plotted in the same way as for white light illumination). It is evident from curve c in Fig. 7 that the reciprocity law is satisfied for illumination with light of $\lambda_3 = 5845$ Å. This can be explained, in accordance with the suggested interpretation, by assuming that light of this wavelength penetrates to a greater depth and produces a more or less uniform absorption throughout a crystal. The reciprocity law is also satisfied at low values of the illuminance in the case of $\lambda_2 = 5675$ Å (Fig. 7b). When the illuminance is increased, the reciprocity law for this wavelength is no longer obeyed because the recombination centers in the illuminated part of the crystal have become filled. In the case of monochromatic light of $\lambda_0 < 5535$ Å, when the process of interaction between electromagnetic radiation and matter takes place in a surface layer of a crystal, the reciprocity law is not obeyed at low values of the illuminance but it is obeyed at high values of this quantity (Fig. 7a). This can be explained as follows: local surface levels do not make an appreciable contribution to the formation of the photoelectret state when light is absorbed mainly in the interior of a crystal. However, in the strong-absorption case, when the effective depth of penetration of light is small, these local levels play the dominant role and are responsible for the failure of the reciprocity law. When the intensity of monochromatic radiation is increased, the exposure−illuminance characteristic becomes parallel to the abscissa, which indicates that the reciprocity law is obeyed. This change in the characteristic can be explained by assuming that the concentration of surface levels is not sufficiently high. At relatively high values of the illuminance, these surface levels are saturated and cause no further changes in the hole lifetime. This conclusion is supported by the results of an investigation of the dependence of the cathodoluminescence intensity on the electron beam density, where again we are dealing with surface processes [13]. The dependence of the cathodoluminescence efficiency on the beam current density is sublinear at low beam densities when surface levels play an important role; however, when the beam current density is increased, the dependence becomes linear; this is followed by a sublinear region when the filling of volume recombination levels becomes important. Thus, we may assume that if the illuminance were increased, the characteristic of Fig. 7a would rise again like the characteristic obtained for monochromatic light of $\lambda_2 = 5675$ Å (Fig. 7b).

We may thus conclude that the exposure−illuminance characteristic obtained using monochromatic light is governed by short wavelengths at low illumination levels and by long wavelengths at high levels. It follows that the use of monochromatic radiation in the recording of the exposure−illuminance characteristics makes it possible to account for the nature of the corresponding characteristics obtained using monochromatic illumination.

Literature Cited

1. V. M. Fridkin, Physical Basis of the Electrophotographic Process, Énergiya, Moscow (1966).
2. V. I. Bugrienko and K. K. Demidov, Fiz. Tverd. Tela, 4:1424 (1962).
3. D. G. Semak, D. V. Chepur, and D. B. Goer, Abstracts of Papers Presented at the XIXth Scientific Conference, Physics Section, Uzhgorod, 1965, p. 47.
4. V. F. Zolotarev, D. G. Semak, and D. V. Chepur, Izv. Vuzov, Fizika, No. 6, p. 130 (1967).
5. R. H. Bube, Phys. Rev., 106:703 (1957).
6. R. H. Bube, J. Phys. Chem. Solids, 1:234 (1957).
7. D. G. Semak, D. V. Chepur, and V. F. Zolotarev, Ukr. Fiz. Zh., 12:1462 (1967).
8. D. G. Semak, D. V. Chepur, and V. F. Zolotarev, Fiz. Tverd. Tela, 9:1242 (1967).
9. V. F. Zolotarev (V. F. Zolotarjov), D. G. Semak, and D. V. Chepur, Phys. Status Solidi, 21:437 (1967).
10. D. G. Semak, Abstracts of Papers Presented at Conf. of Young Scientists on the Occasion of 20th Anniversary of Foundation of the University [in Ukrainian], Uzhgorod (1965), p. 87.
11. D. V. Chepur, Zh. Tekh. Fiz., 25:2411 (1955).
12. A. Rose, Concepts of Photoconductivity and Allied Problems, Interscience, New York (1963).
13. D. G. Semak, Abstracts of Papers Presented at Conf. of Young Scientists, Uzhgorod (1966), p. 79.

ALLOYED p–n JUNCTIONS MADE OF
Be-DOPED p-TYPE SiC

A. A. Kal'nin, V. V. Pasynkov,
Yu. M. Tairov, and D. A. Yas'kov

V. I. Ul'yanov (Lenin) Leningrad Institute of Electrical Engineering

A study was made of alloyed electroluminescent light sources made of Be-doped p-type SiC. Pure silicon was used as the electrode material. Nitrogen was employed as the dopant to produce the n-type region. A study was made of the current–voltage characteristics of the p–n junctions, of the lux–ampere characteristics, and of the temperature dependence of the electroluminescence yield. Estimates were made of the properties of these alloyed sources under pulse conditions. Typical forward voltage drops were 5-7 V for a working current of 1 mA. The lux–ampere characteristics were linear between 0.1 and 30 A/cm^2. The electroluminescence efficiency decreased to ~50% when the temperature was raised from room temperature to 60°C.

The present authors have reported earlier [1] that beryllium-doped p-type silicon carbide crystals can be used to make electroluminescent p–n junctions by alloying or by vapor-phase epitaxy. Such junctions can also be produced during the growth of crystals by sublimation. The junctions prepared by alloying Cr + Si to Be-doped p-type SiC crystals in an argon–nitrogen atmosphere are found to have satisfactory rectifying properties and they emit electroluminescence when the current is passed in the forward direction. It is assumed that the electroluminescence is emitted from the p-type region.

The present paper describes alloyed electroluminescent p–n junctions prepared using a somewhat different technique. Silicon carbide p-type crystals, used in the fabrication of these p–n junctions, were prepared by the sublimation method. They were doped with beryllium during growth. The resistivity of these crystals was 10-100 $\Omega \cdot$ cm, corresponding to a net acceptor concentration of 10^{17}-10^{18} cm^{-3}. We used crystals not smaller than 3 mm across. Each crystal had one mirror-smooth face, later used in the alloying process. Illumination with ultraviolet light produced uniform photoluminescence at room temperature and the application of a negative voltage by a point contact generated electroluminescence.

These crystals were ground down to a thickness of 0.2 mm and cut into plates of 1.5 × 1.5 mm area. Before being alloyed, they were subjected to the usual (for silicon carbide) treat-

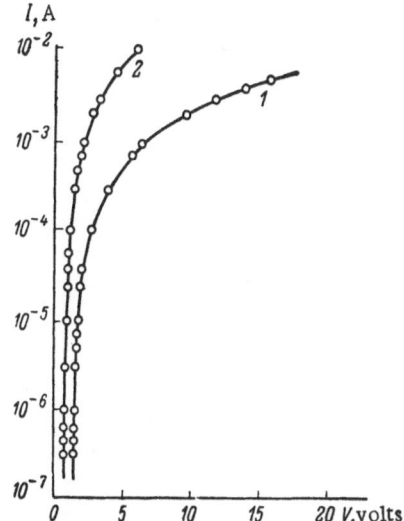

Fig. 1. Forward branch of the current−voltage characteristic of an alloyed p−n junction at two temperatures.

ment in a molten $3KNO_3$ + KOH mixture and were repeatedly washed in distilled water. In contrast to our earlier investigation [1], in which we used Cr + Si, the electrode material employed in the present study was pure silicon. The silicon wetted well the surface of the SiC, had a thermal linear expansion similar to that of the carbide, and at high temperatures dissolved the carbide without contaminating it (because it was one of the component elements of the compound). The replacement of the Cr + Si electrode alloy with pure silicon increased the efficiency of the electroluminescence of the p−n junctions and the mechanical strength of the bond between the silicon electrode and the original crystal. This increase in the efficiency of the electroluminescence could be due to the fact that the electrode alloy no longer included Cr, which was electrically active in the SiC. The electrode dimensions were selected bearing in mind two factors. An increase in the electrode size reduced the voltage drop across the p−n junction in the forward direction. However, when the p−n junction area was large, the current density for a nominal current of 1 mA through the p−n junction could be too low and the efficiency of the electroluminescence source could be reduced (this was deduced from an examination of the dependence of the yield on the current density).

Alloying was carried out in an atmosphere consisting of 98% Ar and 2% N_2. The use of nitrogen as a dopant simplified the fabrication of the p−n junction and ensured an abrupt structure because of the low diffusion coefficient of nitrogen in SiC and the alloying temperature. The alloying process was carried out at 1750-1850°C, applied for 15 sec and followed by cooling at a rate of 5-10 deg/sec. The contact with the p-type region of the junction was made by alloying with aluminum in an inert gas atmosphere. The contact with the n-type region was made via the silicon electrode and a thermally welded gold wire.

The forward branch of the current−voltage characteristic (Fig. 1) consisted of three different regions, representing the leakage current (not shown in Fig. 1), the space-charge-limited current, and the current limited by the series resistance of the crystal. The leakage current appeared at current densities below 10^{-5} A/cm^2. In the exponential part of the characteristic, the current was proportional to $\exp(eV/nkT)$. For the characteristics given in Fig. 1, we found that n = 2, but for some samples this parameter was larger than 2 even after the cleaning of the p−n junction periphery. This indicated that the structure of the p−n junctions was not perfect. The results in Fig. 1 show also that, at high applied voltages, practically all the drop was concentrated in the high-resistivity p-type region. A reverse voltage of 10 V produced an average current of 10 μA.

Figure 2 shows the dependences of the electroluminescence yield on the current density through a p−n junction, recorded using pulses of 4 μsec duration. The upper limit of the current density was 30 A/cm^2 because at higher current densities the results were distorted by heating. The minimum pulse duration of 4 μsec was sufficient to establish steady-state luminescence. At low current densities we used a circuit for modulation of the luminescence (at 74 cps), followed by amplification of the signal with a U2-6 selective amplifier and an SD-1 attachment.

In the range of current densities up to 0.1 A/cm^2, the dependence was superlinear. At low current densities, the superlinearity could be due to leakage through the p−n junction.

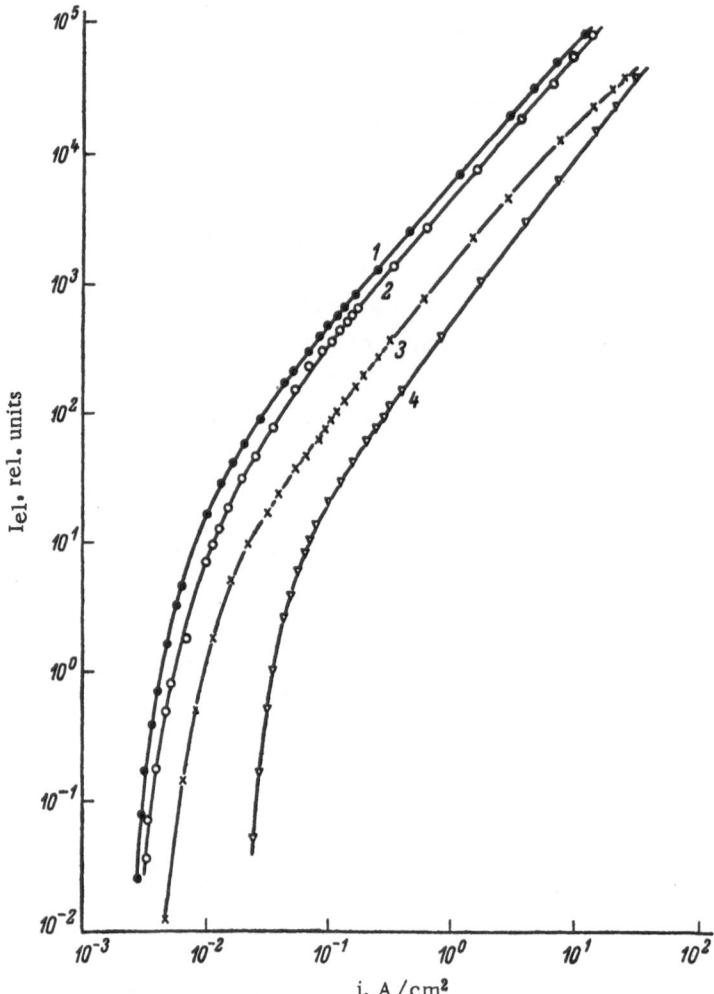

Fig. 2. Dependence of the luminescence yield on the current density through a p−n junction at 290°K. 1) Sample 11N2, p−n junction area S = 0.72 · 10^{-2} cm²; 2) sample 49N3, S = 0.55 · 10^{-2} cm²; 3) sample 11N1, S = 0.32 · 10^{-2} cm²; 4) sample 11N3, S = 0.28 · 10^{-2} cm².

The shape of the forward branch of the characteristic indicated that the leakage currents certainly appeared in the total current through the p−n junction below 10^{-5} A/cm². However, the superlinearity of the dependence of the luminescence yield was also observed at mugh higher current densities. Thus, the superlinearity could not be due to leakage through the p−n junction. The power exponent in the expression relating the electroluminescence yield to the current through the p−n junction was m = 3−4 in the superlinearity region of the curves shown in Fig. 2. The rise in the electroluminescence yield obeying a power law with an exponent larger than 3 could be due to a shift of the quasi-Fermi level of electrons in the p-type region and the consequent filling of the coactivator levels.

We assumed that the heavily doped n^+-type SiC region and the original electroluminescent p-type region were separated by a weakly luminescent compensated layer. The power exponent m could then be large if the carrier lifetime increased with increasing current density through the p−n junction. As the current density increased, deeper layers of the original electroluminescencent p-type material were excited. This assumption was based on the observed weakness of the room-temperature photoluminescence of Be-doped p-type SiC crystals with high concentrations of the compensating nitrogen impurity.

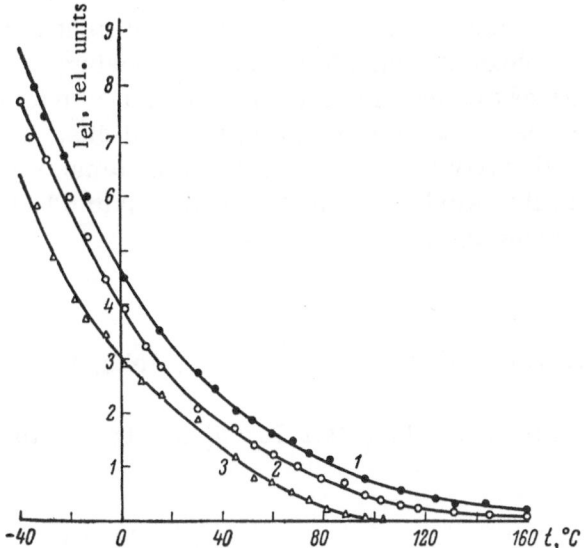

Fig. 3. Temperature dependences of the electroluminescence yield for three samples.

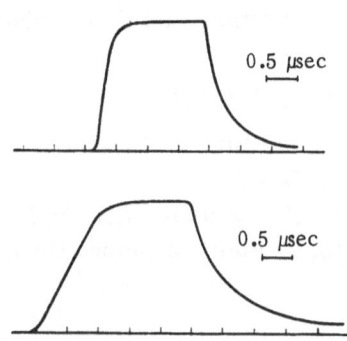

Fig. 4. Electroluminescence
relaxation curves (298°K).

Figure 3 shows the temperature dependence of the electroluminescence yield for three
samples, recorded between −40 and +160°C. Voltage pulses of up to 300 V amplitude, 10 μsec
duration, and 50 cps repetition frequency were used at low temperatures in order to avoid heating. The electroluminescence yield was estimated from the amplitude of a signal taken from
the load resistance of an FÉU-22 photomultiplier. Between room temperature and 160°C, we
used square pulses of 50 cps repetition frequency and an off-duty factor of two. The signal
from the FÉU-22 load was amplified with an SD-1 selective amplifier. The current through
the sample was maintained at a constant level of 1 mA at all temperatures.

The dependences given in Fig. 3 show that the electroluminescence yield of the majority
of the samples decreased to 40% between room temperature and 70°C. The decrease of the
electroluminescence yield throughout the whole investigated range of temperatures could be
due to the following factors: 1) thermal quenching of the luminescence due to thermal transfer
of carriers from the activation centers to the appropriate energy band; 2) increase in the
leakage current through the p−n junction at high temperatures; 3) shift of the quasi-Fermi
level of electrons with increasing temperature in the direction of the weakly luminescent compensated layer; 4) increase in the hole components of the current through the p−n junction with
rising temperature. The last factor could be due to the fact that the high value of the activation energy of the acceptor level in the p-type region (compared with the activation energy of
the donor level in the n-type region) was responsible for a more rapid fall of the resistance
of the p-type region [2].

In order to determine whether alloyed light sources can be used under pulse conditions,
we carried out preliminary estimates of their pulse characteristics. Figure 4 shows the
electroluminescence relaxation curves of two samples recorded at room temperature. These
curves were obtained using an FÉU-30 pulse photomultiplier. The time resolution of the apparatus employed was at least 15 nsec. The resolution was estimated using a source of light
pulses of nanosecond duration, with rise and decay times not exceeding 10 nsec.

The relaxation curves indicated that the rise time of the electroluminescence varied
from sample to sample and was within the range 0.5-2 μsec. The electroluminescence decay
followed a curve consisting of several exponential regions. The initial part of the decay curve
of the majority of the samples was exponential, with a time constant of about 0.24 μsec.

This investigation established that efficient electroluminescent sources of light could be produced from Be-doped p-type SiC by the technologically simple method of alloying, using pure silicon as the electrode material. The linearity of the lux−ampere characteristic up to 30 A/cm^2 should make it possible to use these sources under strong-current conditions. In spite of the fact that the electroluminescence yield fell fairly rapidly with increasing temperature, light sources of this type could be used in apparatus working at temperatures up to 60-70°C. The working voltage could be reduced by decreasing the base thickness.

Literature Cited

1.   A. A. Kal'nin, V. V. Pasynkov, Yu. M. Tairov, and D. A. Yas'kov, Fiz. Tekh. Poluprov., 1:484 (1967).
2.   G. F. Kholuyanov, Low-Voltage Electroluminescent Indicator Lamps for Transistor Circuits, Znanie, Moscow (1965).

MODULATION OF $\lambda = 3.39\mu$ LASER RADIATION
BY EXCESS CARRIERS IN A GALLIUM ARSENIDE DIODE

Yu. A. Bykovskii, I. G. Goncharov, and V. A. Maslov

Moscow Engineering Physics Institute

The $\lambda = 3.39\ \mu$ radiation emitted by an He−Ne laser was amplitude modulated by a GaAs diode placed in the resonator of the gas laser. This was done by passing current pulses through the diode in the forward direction. The modulation coefficient was 10%. The possibility of using a shutter for the Q-factor modulation of a CO_2 laser ($\lambda = 10.6\ \mu$) was examined.

1. The development of sources of coherent infrared radiation has made it possible to produce optical wide-band modulators and shutters working at these wavelengths. The modulation of infrared radiation in a germanium diode has been reported already [1-3]. The absorption of radiation by excess carriers in such a diode is observed in the transparency region of Ge ($\lambda = 2$-$10\ \mu$). Appreciable absorption in Ge is favored by the long carrier lifetime ($\tau > 10\ \mu$sec) [3] but this long lifetime limits the modulator response. Using a GaAs diode, we can vary the infrared radiation at a modulation frequency in the microwave range because such modulation is observed in recombination radiation [4]. The difficulties are to produce a free-carrier density which is sufficiently high to ensure appreciable absorption in a layer whose thickness is governed by the diffusion length of excess carriers (this layer adjoins the p−n junction) and to focus the radiation into this small region. The latter difficulty is overcome by using a laser as a radiation source.

2. The present paper describes a study of the modulation of $\lambda = 3.39\ \mu$ radiation emitted by an He−Ne laser. A gallium arsenide diode laser was used as the modulator. The dimensions of this diode were $0.4 \times 0.4 \times 0.4$ mm. The construction of the diode (pressure contacts with apertures of 0.2 mm diameter) made it possible to transmit radiation at right-angles to the p−n junction. Two faces of the gallium arsenide laser were cleaved and the faces carrying the contacts were optically polished. The thickness of the p-type region was $10\ \mu$ (the Zn concentration in this region was $6 \cdot 10^{18}\ cm^{-3}$). The absorption coefficient of the p-type region was $k > 100\ cm^{-1}$ at $\lambda = 3.39\ \mu$, due to transitions within the valence band [5]. The absorption coefficient of the n-type region (the Te concentration in this region was $n = 1.12 \cdot 10^{13}\ cm^{-3}$) was $6\ cm^{-1}$ [6]. Since the absorption by holes was stronger than that by electrons (for the same density of each carrier species), the change in the intensity of the radiation transmitted through

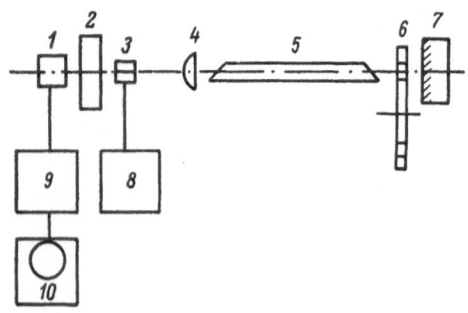

Fig. 1. Schematic diagram of the experimental layout: 1) PbS photoresistor; 2) Ge filter; 3) modulator diode; 4) quartz lens (f = 23 cm); 5) gas laser cuvette; 6) mechanical chopper; 7) Ge mirror; 8) generator of current pulses; 9) U2-4 measuring amplifier; 10) S1-16 oscillograph.

the p−n junction of a current-carrying diode was governed by the holes injected into the n-type region. By placing the diode inside the gas laser resonator (Fig. 1), we were able to produce an appreciable change in the output power of the gas laser. The modulation coefficient $M = I_a/I_0$, defined as the ratio of the alternating component of the radiation I_a to the constant component of the signal I_0, reached 10% when the diode was illuminated parallel to the p−n junction, but fell to 3% when the diode was illuminated at right angles to the junction. The modulation coefficient M was independent of the size of the illuminated region only when the light was incident at right angles to the junction; it was then governed by variations in the hole density. The dependence of the modulation coefficient on the current through the diode in the case of normal incidence of the radiation on the p−n junction is shown in Fig. 2.

When the absorption in the p−n junction increased, the reflection coefficient of the gas laser mirror decreased, which was equivalent to an increase in the losses in the resonator. A glass wedge was used to determine the dependence of the output power on the losses in the resonator due to the presence of the diode (Fig. 3). The reflection coefficient of a plane-parallel plate whose thickness is d (a multiple of $\lambda/4$) and whose volume absorption coefficient is k, is given by [7]

$$R_{max} = \frac{R[1 + \exp(-kd)]^2}{[1 + R\exp(-kd)]^2},$$ (1)

where R is the reflection coefficient at the boundary of a medium whose refractive index is n; $R = (n-1)^2/(n+1)^2$. The reflection coefficient of GaAs is R = 0.27 [8].

For a modulation coefficient of 3%, the reflection coefficient of a plate changed by 0.7% (Fig. 3), which corresponded − in accordance with Eq. (1) − to kd ≈ 0.001.

3. Modulation of the λ = 3.39 μ laser radiation is of special interest in communication because this wavelength lies in the transparency window of the atmosphere. A GaAs diode modulator does not require high voltages, it can work at room temperature, and its power

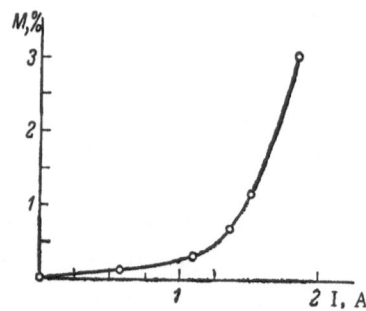

Fig. 2. Dependence of the modulation coefficient on the amplitude of current pulses passing through the modulator diode (T = 300°K).

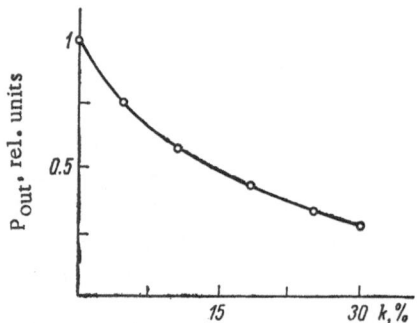

Fig. 3. Dependence of the output power of the gas laser on the absorption coefficient of a glass wedge.

consumption is not high: a change in the output power by 10% requires a pulse power of 8 W. It is known [9] that the modulation of the Q factor of CO_2 lasers increases the pulse power. We can expect the gallium arsenide modulator to be more efficient at the CO_2 wavelength of 10.6 μ because at this wavelength the absorption cross section of the holes in GaAs is an order of magnitude higher than at $\lambda = 3.39$ μ.

The modulator described in the present paper should have advantages over a rotating-mirror modulator because the diode modulator can produce radiation pulses with steep fronts and an arbitrary off-duty factor.

The authors are grateful to V. V. Nikitin and Yu. P. Zakharov for their help and comments.

Literature Cited

1. R. Newman, Phys. Rev., 91:1311 (1953).
2. A. F. Gibson, Proc. Phys. Soc. (London), B66:588 (1953).
3. Yu. N. Ukhanov, in: Semiconductor Devices and Their Applications, Sov. Radio, Moscow (1958), p. 3.
4. B. S. Goldstein and Q. M. Weigand, Proc. IEEE, 53:195 (1965).
5. R. Braunstein, J. Phys. Chem. Solids, 8:280 (1959).
6. W. G. Spitzer and J. M. Whelan, Phys. Rev., 114:59 (1959).
7. J. A. Stratton, Electromagnetic Theory, McGraw-Hill, New York (1941).
8. E. M. Voronkova, B. N. Grechushnikov, G. I. Distler, and I. P. Petrov, Optical Materials for Infrared Technology, Nauka, Moscow (1965).
9. C. K. N. Patel, Phys. Rev. Letters, 16:613 (1966).

IMPURITY PHOTOEFFECT IN GaAs p—n JUNCTIONS

A. A. Gutkin, É. M. Magerramov, D. N. Nasledov, and V. E. Sedov

A. F. Ioffe Physicotechnical Institute
Academy of Sciences of the USSR, Moscow

An investigation was made of the photo-emf spectra of GaAs p—n junctions at ~90 and 293°K in the photon energy range $\hbar\omega \geq 0.46$ eV, using modulated or constant illumination and various values of the reverse bias. Two types of sample were investigated: those prepared by the diffusion of sulfur in p-type gallium arsenide and those prepared by the diffusion of zinc in n-type crystals. The photo-emf spectra obtained using modulated illumination at $T \approx 90°K$ indicated the presence of impurities in the samples prepared by the diffusion of sulfur; these impurities had energy levels at 0.49, 0.54, 0.59, 0.65, 0.73, 0.90, and 1.05 eV. These levels were compared with the known levels of impurities usually encountered in gallium arsenide (such as copper, oxygen, and iron) and with lattice defect levels. In a certain part of the spectrum, the photosensitivity decreased with time, which was attributed to the liberation of carriers from impurity levels. It was found that the steady-state photo-emf appeared only at photon energies $\hbar\omega \gtrsim 0.5 E_g$ and it was due to the single-stage optical excitation of minority carriers in the space-charge layer. The transient photo-emf depended weakly on the reverse bias across the p—n junction but the steady-state effect was increased strongly by such a bias.

Investigations of the photo-emf of p—n junctions in the impurity absorption region can be used to detect deep-level impurities in a semiconductor, to determine the positions of these levels in the forbidden band, and to study the mechanism and kinetics of electron processes in which these levels participate.

The impurity (extrinsic) photo-emf has been investigated in cuprous oxide [1,2], germanium, and silicon [3, 4] p—n junctions. It has also been investigated in GaAs p—n junctions but only in a narrow spectral range near the fundamental absorption edge ($\hbar\omega \geq 1.26$ eV) [5, 6]. The present paper reports the results of an investigation of the photo-emf of GaAs p—n junctions in the photon energy range $\hbar\omega \geq 0.46$ eV.

Investigated Samples and Measurement Method

We investigated two types of sample. Samples of the first type were prepared from p-type GaAs ($p \approx 10^{16}$ cm^{-3}, $u_p \approx 230$ cm$^2 \cdot$ V$^{-1} \cdot$ sec^{-1}); the p—n junctions in this were prepared

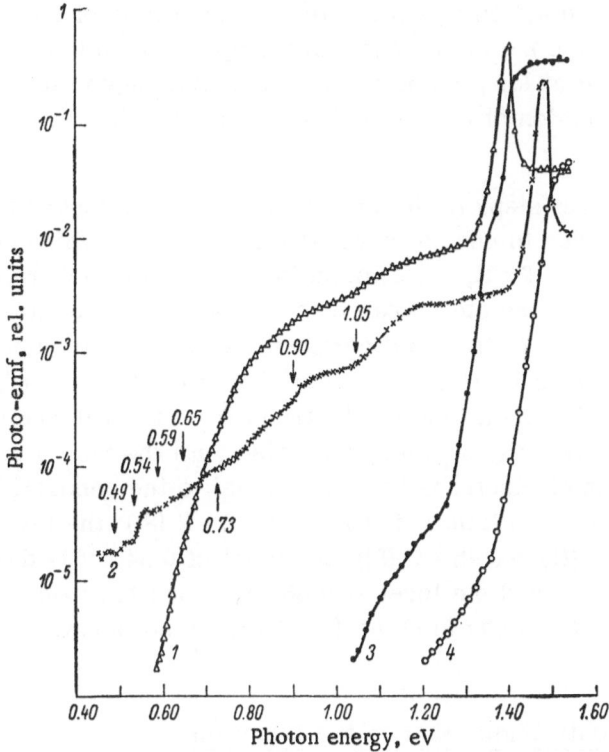

Fig. 1. Photosensitivity spectra obtained using
modulated illumination: 1, 2) sample No. 1; 3, 4)
sample No. 2. T, °K: 1, 3) 293; 2, 4) 90.

by the diffusion of sulfur at 800 or 900°C, applied for ∼30 h. Samples of the second type were
prepared from n-type GaAs ($n \approx 10^{17}$ cm^{-3}, $u_n \approx 3000$ cm$^2 \cdot$ V$^{-1} \cdot$ sec^{-1}); p–n junctions were
produced by the diffusion of zinc at 680°C, applied for 6 min. The diffusion was carried out in
sealed quartz ampoules evacuated to ∼5 $\cdot$ 10^{-6} mm Hg. No deep-level impurities were introduced deliberately into the sample. The areas of the samples were within the range 0.05-0.1
cm^2. The specific differential capacitance of samples of the first type was (4-10) $\cdot$ 10^3 pF/cm^2;
the corresponding capacitance of the samples of the second type was (1-2) $\cdot$ 10^5 pF/cm^2. The
samples were illuminated normally to the p–n junction plane on the diffused layer side (the
thickness of this layer was a few microns).

The photosensitivity spectra were determined under photovoltaic and photodiode conditions at temperatures of ∼90 and 293°K. They were recorded using an IKS-12 spectrometer
with an F1 glass prism. A germanium filter was placed at the entry slit of the spectrometer
in order to eliminate the influence of scattered light in the photon energy range $\hbar\omega < 0.7$ eV.
The measurements at 90°K were carried out in a cryostat with a fused quartz window. The
photosensitivity was determined using constant and modulated illumination (square pulses of
∼2.5 msec duration, 0.5 msec rise time, and ∼200 cps repetition frequency). An F116 photo-
compensation amplifier, with an M95 galvanometer at its output, was used to measure the photo-
signal under constant illumination conditions. Narrow-band amplifiers were employed to record
the signal produced by the modulated illumination. The load resistance was 36-100 kΩ.

Experimental Results and Discussion

1. Impurity Photo-Emf Spectra Obtained Using Modulated Illumination

Figure 1 shows the photo-emf spectra, obtained using modulated illumination, of samples
prepared by the diffusion of sulfur (sample No. 1) and zinc (sample No. 2). The impurity photo-

emf of these samples differed not only because the compositions of the samples were different. The conditions under which the p−n junctions were prepared could also have influenced the results. We could assume that at least some of the accidental impurities were introduced into sample No. 1 during the diffusion of the sulfur because of the high temperature and long duration of this process.

The existence of several steps in the low-temperature photosensitivity spectrum showed that the photoeffect was not due to one but several levels in the forbidden band. The energy positions of these levels at $T \approx 90°K$, determined to within ~ 0.02 eV, relative to the onset of the photosensitivity rise preceding the step, are given in Fig. 1. The majority of these levels can be associated with the presence of the most commonly encountered accidental impurities in GaAs: copper, oxygen, and iron [7-10]. Thus, the rise of the photo-emf, which begins at $\hbar\omega \approx 0.59$ eV, is evidently due to the excitation of electrons from the valence band to an iron level at $E_v + 0.59$ eV [7]. This level may be associated also with the rise of the photo-emf at $\hbar\omega \approx 0.90$ eV due to the excitation of electrons from this level to the conduction band. The level at 0.73 eV is evidently due to oxygen impurities ($E_c - 0.73$ eV [8]), the level at 1.05 eV is due to doubly-charged copper ions ($E_v + 0.45$ eV [9]), the level at 0.54 eV is due to iron or nickel impurities ($E_v + 0.54$ eV [7, 11]), and the level at 0.65 eV is due to lattice defects [10]. The level at 0.49 eV has been observed already in GaAs [12] but it is not known with which impurity atoms or defects it is associated.

2. Dependence of the Impurity Photo-Emf on the Duration

of Illumination

It was found that the samples prepared by the diffusion of sulfur exhibited a slow decay of the photo-emf at $T \approx 90°K$ during illumination with photons of energies $\hbar\omega = 1.05$-1.4 eV. In view of this, the spectra reported in the present paper for this range of energies were deduced from the values of the photo-emf measured during the first few seconds after beginning of illumination. Before the measurement of the signal at a given wavelength, a sample was kept in darkness for a time sufficiently long to ensure that the initial state was close to equilibrium.

These measurements showed that the dependence of the photo-emf on the duration of illumination with photons of energies $\hbar\omega \approx 1.05$-1.4 eV was exponential with a time constant of the order of 10 min. This decay of the photo-emf with time could be explained by the gradual emptying of the impurity level from which nonequilibrium carriers were excited. The emptying of this level was due to the fact that the rate of optical excitation of electrons from the deep impurity level at $E_v + 0.45$ eV to the conduction band was higher than the rate of recapture of electrons by this level.

If the density of electrons excited by light from the level being emptied is low, compared with the total electron density (this was always true in the n-type region), the dependence of the concentration of filled impurity centers m on the duration of illumination t should be given by the expression

$$m = \frac{\gamma_n M n}{\gamma_n n + q_n I} + \frac{q_n I M}{\gamma_n n + q_n I} \exp\left[-(\gamma_n n + q_n I)t\right],$$

where M is the concentration of the impurity centers; q_n is the cross section for the capture of a photon by an electron localized at the impurity level; I is the intensity of illumination; γ_n is the electron-capture coefficient of a free impurity level; n is the density of the majority carriers (electrons). Under equilibrium conditions in darkness, we should have m = M. The experimentally determined time constant for $\hbar\omega \approx 1.1$ eV was 15 min for an illumination level $I \approx 5 \cdot 10^{13}$ photons $\cdot$ cm^{-2} $\cdot$ sec^{-1}. Assuming that we could neglect γ_n^n compared with $q_n I$,

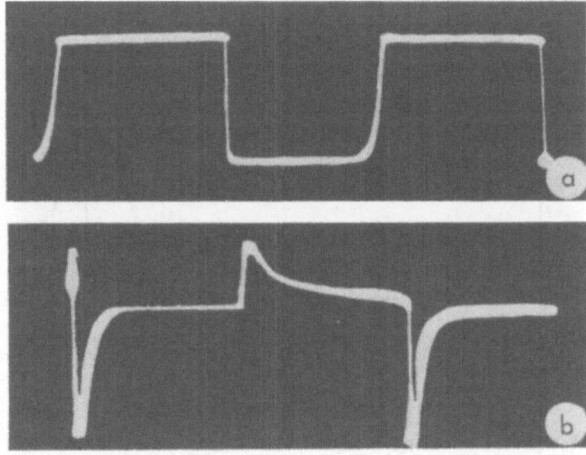

Fig. 2. Oscillograms of the photo-emf for sample 1 illuminated with square light pulses (T ≈ 90°K): a) $\hbar\omega \gtrsim E_g$; b) $\hbar\omega < 0.72$ eV. The duration of the light pulse was 12 msec and the repetition frequency was ~40 cps. The load resistance was ~0.5 MΩ.

since γ_n and n are small, we found that these results yielded $q_n \approx 2 \cdot 10^{-17}$ cm² for $\hbar\omega \approx 1.1$ eV. This order of magnitude of the cross section was expected for the absorption of photons by localized impurities in a semiconductor [13].

Such a slow establishment of the photo-emf was not observed when carriers were excited from other impurity levels. This was because the value of γ for other levels was sufficiently high and, therefore, a steady state was established rapidly so that the occupancy of the level differed little from the equilibrium value. The level at 0.45 eV differed from the others because it corresponded not to singly but to doubly charged copper ions (Cu^{2-}). Therefore, the electron-capture cross section of this level could be very small because of the Coulomb barriers surrounding these copper ions. At room temperature, the slow variation of the impurity photo-emf disappeared: this indicated that the electron-capture cross section increased strongly with increasing temperature, as expected for centers with Coulomb barriers.

A low value for the electron-capture coefficient of the copper level at low temperatures was also reported in [14] on the basis of a study of slow decay of the $\hbar\omega = 1.02$ eV electroluminescence of GaAs diodes after switching off the current.

Prolonged preliminary illumination with photons of energies $\hbar\omega = 1.05-1.4$ eV did not affect the photo-emf at $\hbar\omega < 1.05$ eV. This showed that the lifetime and diffusion length of free carriers, which governed the photo-emf, did not change greatly when the $E_v + 0.45$ eV level was emptied.

3. Steady-State and Transient Photo-emf. Mechanism of Minority Carrier

Generation by Excitation with Light of $\hbar\omega < E_g$ Energy

It is known that illumination with light of $\hbar\omega < E_g$ can produce a steady-state and a transient photo-emf in a p−n junction; the transient effect appears only at the beginning and end of illumination [4, 15]. The steady-state photo-emf is associated with the generation of minority carriers, the transient effect with the generation of majority carriers. Since the relative numbers of the majority and minority carriers generated depend on the photon energy, the steady-state and transient photo-emf spectra are different. Figure 2 shows oscillograms of the photo-

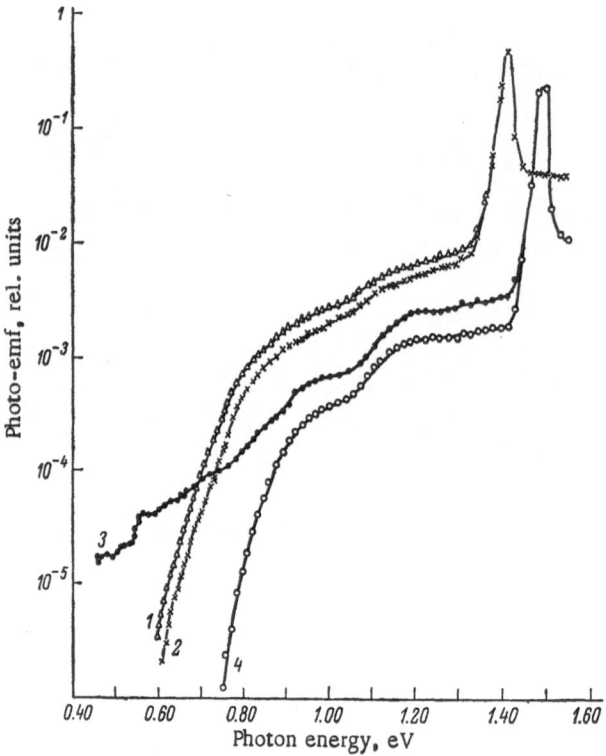

Fig. 3. Photosensitivity spectra of sample 1 obtained using modulated (1, 3) and constant (2, 4)
illumination. T, °K: 1, 2) 293; 3, 4) 90.

emf excited by square pulses of light whose energy corresponds to the fundamental absorption
region (Fig. 2a) and to the region $\hbar\omega < 0.72$ eV (Fig. 2b). In the first case, we observed the
steady-state photo-emf in the form of a pulse whose shape was the same as that of the illumination pulse. In the second case, only the transient photo-emf with a relaxation time 1-2 msec
was observed. Because the relaxation time was long, the transient photo-emf was considerable
in the measurements carried out using modulated illumination, and the contribution of this photo-
emf increased appreciably the photosensitivity, compared with its steady-state value. This is
confirmed by the curves shown in Fig. 3, which also indicate that the steady-state photo-emf
existed only at energies $\hbar\omega \gtrsim 0.5E_g$, whereas the transient effect was appreciable (at 90°K)
throughout the whole range of wavelengths investigated.

The majority carriers responsible for the transient impurity photo-emf were generated
easily at all photon energies $\hbar\omega < E_g$ both in the n-type region (by optical transitions of the
electrons from the impurity levels lying below the Fermi level to the conduction band) and in
the p-type region (by electron transitions from the valence band to unoccupied impurity levels).
We were interested in the mechanism of the minority-carrier generation which was observed
(Fig. 3) right up to photon energies $\hbar\omega \approx 0.5E_g$. It is known that if a sample is illuminated
with light of energy $\hbar\omega < E_g$, minority carriers can be generated either by single-step optical
transitions or by two-step thermo-optical or double optical transitions [3]. In the case of sufficiently deep impurity levels, the contribution of the thermo-optical transitions at low temperatures is negligible. If the minority carriers are generated by double optical transitions,
the dependence of the impurity photo-emf on the illumination intensity should be quadratic [4].
However, our measurements showed that, at all photon energies, the dependence of the steady-
state impurity photo-emf on the intensity of the illumination was slightly sublinear: $\varepsilon_{ph} \propto I^\beta$,
where $1 \geq \beta \geq 0.83$. Consequently, we concluded that double optical transitions were not the

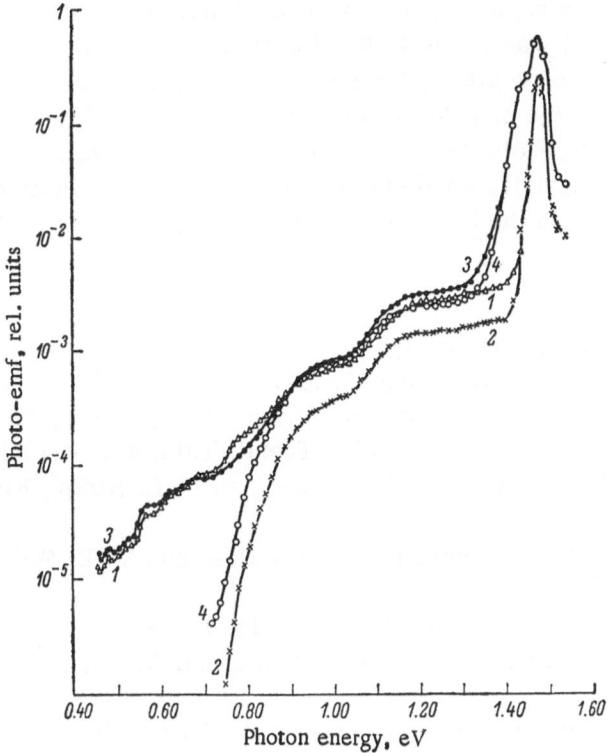

Fig. 4. Spectra of the sensitivity of modulated (1, 3) and continuous (2, 4) illumination for different values of a reverse bias U (V): 1, 2) 0; 3, 4) −20.

dominant mechanism of the minority-carrier generation. We thus found that single-step optical transitions were mainly responsible for the generation of the minority carriers by photons having energies $E_g > \hbar\omega \gtrsim 0.5E_g$. These transitions were possible only when the impurity levels near the middle of the forbidden band were empty in the n-type region and full of electrons in the p-type region. This situation existed only in the space-charge layer of the p−n junction, where the Fermi level was located deep within the forbidden band. Consequently, the minority carriers were generated by impurity-absorbed light of $\hbar\omega < E_g$ in this layer.

4. Change in the Spectral Distribution of the Photosensitivity

in the Impurity Absorption Region Due to the Application

of Reverse Bias

A reverse bias applied to a p−n junction affected in different ways the sensitivity to modulated and continuous illumination (Fig. 4). The sensitivity to the modulated illumination in the $\hbar\omega < 1.3$ eV region, where the transient photo-emf was important, was affected very little by the application of a reverse bias. The photosensitivity decreased at some wavelengths. The cause of this decrease was not clear but it was expected that a reverse bias would have little influence on the transient photo-emf because the majority carriers were generated mainly in the quasineutral p- and n-type regions, whose properties were not affected by such bias.

The sensitivity to the continuous illumination depended more strongly on a reverse bias because the minority carriers, responsible for the steady-state photo-emf in the $\hbar\omega < E_g$ region, were generated in the space-charge layer of the p−n junction. A reverse bias increased the impurity photo-emf under the continuous illumination conditions. The relative increase in the photosensitivity in the region of the long-wavelength edge of the steady-state impurity photo-emf ($\hbar\omega \approx 0.7$-0.8 eV) was greater than the increase in the region $\hbar\omega \approx 1$-1.3 eV. Therefore, the increase in the steady-state photosensitivity, caused by the application of a reverse bias, was not simply due to the broadening of the space-charge layer. The additional increase in the steady-state impurity photo-emf at its long-wavelength edge could be due to a reduction of the energy gap for the impurity level-band optical transitions in an electric field (similar to the Franz-Keldysh effect in the band−band transitions).

Literature Cited

1.   B. Lange, Physik. Z., 32:850 (1931).
2.   V. E. Lashkarev and K. M. Kosonogova, Izv. Akad. Nauk SSSR, Ser. Fiz., 5:478 (1941).
3.   F. M. Berkovskii and S. M. Ryvkin, Fiz. Tverd. Tela, 4:366 (1962).
4.   F. M. Berkovskii and S. M. Ryvkin, Fiz. Tverd. Tela, 4:376 (1962).
5.   A. A. Gutkin, M. M. Kozlov, D. N. Nasledov, and V. E. Sedov, Fiz. Tverd. Tela, 5:3617 (1963).
6.   A. A. Gutkin, É. M. Magerramov, D. N. Nasledov, and V. E. Sedov, Fiz. Tekh. Poluprov., 1:1001 (1967).
7.   V. I. Fistul' and A. M. Agaev, Fiz. Tverd. Tela, 7:3681 (1965).
8.   N. A. Vitovskii, T. V. Mashovets, S. M. Ryvkin, and R. Yu. Khansevarov, Fiz. Tverd. Tela, 5:3510 (1963).
9.   R. N. Hall and J. H. Racette, J. Appl. Phys., 35:379 (1964).
10.   D. S. Domanevskii and V. D. Tkachev, Fiz. Tekh. Poluprov., 1:377 (1967).
11.   R. W. Haisty, Appl. Phys. Letters, 7:208 (1965).
12.   M. D. Sturge, Phys. Rev., 127:768 (1962).
13.   T. S. Moss, Optical Properties of Semi-Conductors, Butterworths, London (1959).
14.   A. N. Imenkov, M. M. Kozlov, D. N. Nasledov, and B. V. Tsarenkov, Fiz. Tverd. Tela, 8:2098 (1966).
15.   S. M. Ryvkin, Photoelectric Effects in Semiconductors, Consultants Bureau, New York (1964).

1/f NOISE OF SURFACE-BARRIER DIODES

N. I. Sablina and N. B. Strokan

A. F. Ioffe Physicotechnical Institute
Academy of Sciences of the USSR, Leningrad

A study was made of the $1/f$ noise of surface-barrier diodes in various circuit configurations. A comparison was made of the noise characteristics of silicon surface-barrier diodes and of conventional germanium and silicon photodiodes. The results of noise measurements under photovoltaic conditions were analyzed on the basis of the fluctuations of the surface recombination velocity. It was found that the surface recombination velocity in "thin" photodiodes could be determined by comparing the noise under the short-circuit and open-circuit (photo-emf) conditions.

It is known [1-3] that the low-frequency noise in p−n junction semiconductor devices, characterized by a spectrum of the form $1/f^{\alpha}$ ($\alpha \approx 1$), is due either to fluctuations of the surface recombination velocity S, or to fluctuations of the leakage current. The leakage current is responsible for the noise of reverse-biased junctions and the recombination fluctuations dominate the noise in forward-biased junctions and under photovoltaic conditions of operation. According to McWhorter's theory [4], the spectral characteristic of these fluctuations is a consequence of the modulation of the recombination and generation at fast surface states by fluctuations of the population of slow states. It is assumed that these slow states are located at the outer surface of an oxide. It would appear that surface-barrier junctions should not exhibit the $1/f$ noise since the surface is in contact with a metal and electrons from the metal fill and stabilize the slow states. However, we observed the $1/f$ noise in surface-barrier junctions for all the circuit configurations.*

We shall analyze the observed noise on the basis of the fluctuations of the surface recombination velocity. We shall find the current−voltage characteristic of surface-barrier junctions, taking into account surface carrier recombination, by solving the equation of continuity in the space-charge region with the following boundary conditions:

$$\left.\begin{array}{l} \text{div } j_n = 0 \\[4pt] j_n|_{x=0} = -eI_0 + eS\,(n\,(0) - n_{0s}) \\[4pt] n|_{x=L} = n_0 \\[4pt] j_n = eEu_n n + eD_n \dfrac{dn}{dx}. \end{array}\right\} \qquad (1)$$

---

* The $1/f$ noise of a metal−semiconductor contact is reported in [5].

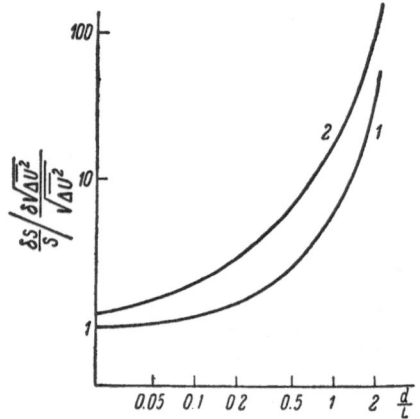

Fig. 1. Relative error in the determination of the surface recombination velocity in diodes with different relative base thicknesses D_n/L_nS: 1) 1; 2) 0.1 and 10.

We shall take the positive direction as that corresponding to the flow of the current from the surface; the generation of carriers by illumination is assumed to take place near the surface; L is the thickness of the space-charge layer; n_{0s} is the equilibrium density of the carriers at the surface; n_0 is the carrier density in the interior.

The solution is obtained in the form

$$j_n = -eI_0(1-\beta) + eSn_{0s}(1-\beta)\left(e^{\frac{eV}{kT}} - 1\right). \qquad (2)$$

The quantity β is the fraction of the current "lost" due to surface recombination and given by

$$\beta = \left(1 + \frac{D_n}{SLf(\varphi)}\right)^{-1}, \qquad (3)$$

where $f(\varphi) = e^{-\frac{\varphi(0)}{kT}} \int_0^1 e^{\frac{\varphi(\xi)}{kT}} d\xi$; $\xi = x/L$; $\varphi(0)$ is the surface

potential. We note that in order to find the total current, we must take into account its other components: the hole current and the current due to recombination-generation processes in the space-charge layer.*

The short-circuit photocurrent is i = $-eI_0(1 - \beta)A$ (A is the junction area). Assuming that the surface recombination velocity fluctuates, we find that the intensity of the fluctuations of the short-circuit current is

$$\overline{\Delta i^2} = i^2\beta^2 \frac{\overline{\Delta S^2}}{S^2}. \qquad (4)$$

The photo-emf is $\Phi = \frac{kT}{e}\ln\left(\frac{i}{i_s} + 1\right)$, where i_s is the saturation current. If the saturation current is governed mainly [Eq. (2)] by that component which is associated with surface processes, the intensity of the fluctuations of the photo-emf is given by the simple formula

$$\overline{\Delta\Phi^2} = \frac{kT}{l} \cdot \frac{\overline{\Delta S^2}}{S^2}. \qquad (5)$$

It is evident from Eqs. (4) and (5) that, by measuring the junction noise under the photo-emf and photocurrent conditions, we can estimate the surface recombination velocity and find the spectral density of its fluctuations $\overline{\Delta S_f^2}$.

A similar analysis of fluctuations under the photovoltaic conditions of operation is applicable also to conventional photodiodes with a finite base thickness, in which photocarriers are removed from the surface only by diffusion. This investigation of the noise of such photodiodes is interesting in a comparison of the noise characteristics of these two types of photo-

---

* Since the surface barrier in real metal−semiconductor contacts is difficult to penetrate, we can always ignore the flow of optically excited electrons into the metal, in exactly the same way as the contribution of such electrons to the saturation current from the metal. The difficulty of penetration of the barrier can be represented by a "permeability" which is not more than 10^{-9} for a gold−silicon contact with a barrier 4 eV high when the oxide thickness is 10 Å (assuming no additional buildup of the oxide layer) [6].

diode, and in studies of the possibility of determining the surface recombination velocity from these noise characteristics. Expressions for the intensities of fluctuations can be obtained using the formulas for the photocurrent and saturation currents given in [7, 8], which take into account surface recombination. Fluctuations of the photocurrent under short-circuit conditions are given by

$$\overline{\Delta i^2} = i^2 \left[1 + \frac{D_n}{L_n S} \operatorname{cth} \frac{d}{L_n} \right]^{-2} \frac{\overline{\Delta S^2}}{S^2},$$
(6)

where d is the base thickness; D_n is the diffusion coefficient; L_n is the diffusion length of the minority carriers. Assuming that the saturation current is governed by the minority-carrier current originating only from the illuminated part of the diode (which is true in many cases), we obtain a formula for the intensity of the fluctuations of the photo-emf

$$\overline{\Delta \Phi^2} = \left(\frac{kT}{l} \right)^2 \left[1 + \frac{D_n}{L_n S} \operatorname{th} \frac{d}{L_n} \right]^{-2} \frac{\overline{\Delta S^2}}{S^2}.$$
(7)

The surface recombination velocity S can be determined sufficiently accurately only for thin diodes: $d/L_n \lesssim 0.5$ (Fig. 1). If the saturation current has two components (currents from the base and from the unilluminated part of the diode), Eq. (7) becomes more complicated and the accuracy of the determination of S decreases. In the limit, when the saturation current is governed only by the component flowing from the unilluminated part of the junction, this method cannot be used to determine S.

Experimental Results

We investigated noise in three types of diode: surface-barrier,* specially prepared silicon diodes with exactly known parameters (d, D_n, L_n), and mass-produced FD-3 germanium photo-diodes. The investigation was carried out in the frequency range 0.08-30 kc. The sensitivity of the measuring circuit was 0.01 μV at 80 kc and 0.005 μV at frequencies above 500 cps for an amplifier bandwidth of 10 cps.[†] The diodes were illuminated with light absorbed in the fundamental region, where the absorption coefficient was high, so that the generation of carriers was concentrated near the surface. We determined experimentally the value of the square of the fluctuation voltage across a load resistance R_l.

The surface-barrier junctions were prepared from n-type silicon ($\rho = 100-300$ $\Omega \cdot$ cm). Figure 2 shows typical noise spectra under the photocurrent ($R_l \ll R_0$) and photo-emf ($R_l \gg R_0$) conditions; here, R_0 is the internal resistance of the junction, equal to $(kT/ei_s)\exp(-e\Phi/kT)$. It is quite clear from Fig. 2 that the spectral characteristic is of the $1/f$ type. The dependence of the square of the noise voltage on the current, given in Fig. 3, is in good agreement with Eqs. (4) and (5) because saturation is observed when the photodiode goes over from the short-circuit to the photo-emf state. The parameters of the diodes were such that the hole component of the saturation current could be neglected compared with the thermal generation current at the surface provided $S \gtrsim 10^3$ cm/sec. The recombination-generation processes in the space-charge layer, important in the case of a large reverse bias across the junction, were weak under the open-circuit (photo-emf) conditions. Consequently, under the open-circuit conditions, the noise in the investigated diodes should be described by Eq. (5). Measurements

---

* The surface-barrier diodes used were silicon counters for short-range particles.
† The results were obtained using the step-up transformer of a 28-IM amplifier.

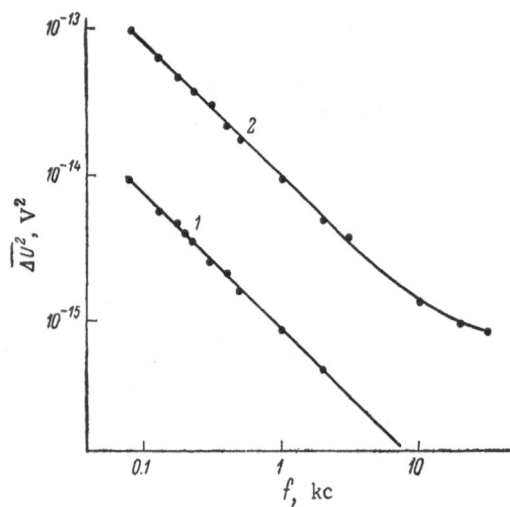
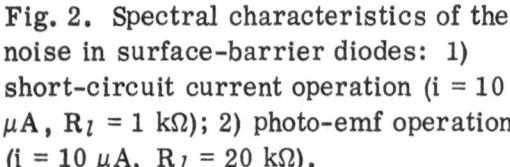

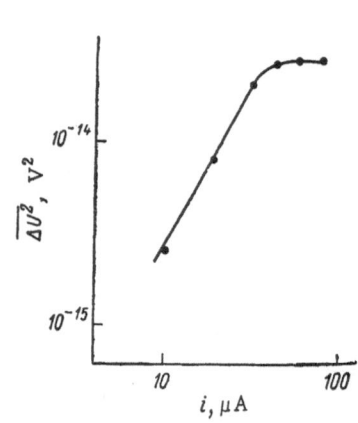

Fig. 2. Spectral characteristics of the noise in surface-barrier diodes: 1) short-circuit current operation (i = 10 μA, R_l = 1 kΩ); 2) photo-emf operation (i = 10 μA, R_l = 20 kΩ).

Fig. 3. Dependence of the square of the noise voltage on the photo-current (f = 300 cps, R_l = 1 kΩ).

carried out at 300 cps gave values of S of the order of 10^4-10^5 cm/sec and $\Delta S_f^2 = 10^{-2}$-10^{-3} cm^2/sec^2. One of the characteristic features of the surface-barrier diodes was that newly generated carriers were removed from the surface by the field and, therefore, the role played by the surface in determining the current should decrease with increasing field intensity (under reverse bias conditions). Consequently, the surface noise should decrease. We found, indeed, that the recombination component of the noise decreased by two orders of magnitude under the photodiode (photocurrent) conditions.*

Specially prepared photodiodes, made of p-type Si, were known to have small values (0.1-0.15) of the ratio $d/L_n \ll 1$ and a saturation current governed by the minority-carrier current from the illuminated p-type region. The intensity of the $1/f$ noise, proportional to the square of the current for $R_l \ll R_0$, reached saturation for $R_l \gg R_0$, in accordance with Eqs. (6) and (7). We determined the surface recombination velocity S = (1-3) $\cdot 10^3$ cm/sec and the spectral density of the fluctuations $\overline{\Delta S_f^2} = 10^{-3}$-$10^{-4}$ cm^2/sec^2 at 300 cps. The values of S were quite reasonable for etched surfaces of p-type silicon, whose resistivity was ρ = 150 $\Omega \cdot$ cm.

Similar measurements were carried out on FD-3 germanium phototodiodes. Here, we encountered a difficulty in the determination of the base thickness, which varied with the junction diameter, so that d/L_n = 0.3-0.5. The assumptions made in the derivation of Eq. (7) were again satisfied and an estimate of S gave values of 300-700 cm/sec, which were reasonable for germanium, and $\overline{\Delta S_f^2} = 10^{-6}$ cm^2/sec^2 at 300 cps, which was of the same order of magnitude as the value obtained in [2].

Since the values of the surface recombination velocity obtained by this method were reasonable, we concluded that the suggested method would be useful.

Before comparing the results obtained for conventional and surface-barrier silicon photodiodes, we must define precisely the meaning of the measured value of the surface recombina-

* Under these conditions, the dark noise (the leakage noise) had a spectral characteristic of the type $1/f^\alpha$ (α = 0.75-1.1), a linear dependence on the voltage, and a quadratic dependence on the current, just like conventional diodes [3, 9].

tion velocity. In the conventional diodes, we determine the "effective" velocity, introduced via the ratio of the recombination flux to the excess minority carrier density at the inner boundary of the space-charge layer [10]. Consequently, the value of S is governed not only by the characteristics of recombination centers but also by the majority carrier density in the interior (in this case, by the hole density p_0). In the case of the surface-barrier diodes, S is introduced as the ratio of the recombination flux to the excess density of the majority (relative to the interior) carriers on the surface. Therefore, the value of the surface recombination velocity is proportional to the density of the minority (relative to the interior) carriers on the surface, i.e., to the hole density p_{s0}. In both cases, the values of S represent recombination properties of the surface. We might expect these properties to be similar because the surfaces of the conventional and surface-barrier diodes are treated in the same etchant and have approximately equal hole densities ($p_{s0} \approx p_0$) at the surface of the p-type region. However, we found that the values of S and $\overline{\Delta S_f^2}$ for the surface-barrier diodes were much higher than for the conventional diodes. This large discrepancy between the expected and obtained values indicated that the initial assumption about the dominant role of the surface recombination processes in the current and fluctuations of the current was incorrect in the case of the surface-barrier diodes. We should seek a different source of $1/f$ fluctuations at a metal−semiconductor contact. A model of microbreakdowns at the surface [5] is in full agreement with the nature of the noise in darkness under reverse-bias conditions, but it does not explain the observed relationships governing fluctuations of the photocurrent and photo-emf.

The authors are deeply grateful to S. M. Ryvkin for his interest in this investigation and to L. V. Maslova and V. F. Afanas'ev for supplying the surface-barrier counters for short-range particles.

Literature Cited

1. A. van der Ziel, Fluctuation Phenomena in Semiconductors, Butterworths, London (1959).
2. D. N. Mirlin and Yu. S. Karpov, Fiz. Tverd. Tela, 4:700 (1962).
3. T. B. Watkins, Proc. Phys. Soc. (London), 73:59 (1959).
4. A. L. McWhorter, in: Semiconductor Surface Physics (Proc. Conf., Philadelphia, 1956, R. H. Kingston, ed.), University of Pennsylvania Press, Philadelphia (1957), p. 207.
5. R. L. Petritz, in: Semiconductor Surface Physics (Proc. Conf., Philadelphia, 1956, R. H. Kingston, ed.), University of Pennsylvania Press, Philadelphia (1957), p. 226.
6. R. J. Archer, J. Electrochem. Soc., 104:619 (1957).
7. D. E. Sawyer and R. H. Rediker, Proc. IRE, 46:1122 (1958).
8. A. A. Grinberg and N. B. Strokan, Fiz. Tverd. Tela, 2:1536 (1960).
9. Yu. S. Karpov, Fiz. Tverd. Tela, 3:1691 (1961).
10. G. L. Bir, Fiz. Tverd. Tela, 1:67 (1959).

ELECTRICAL PROPERTIES OF
α(Ge)—GaAs HETEROJUNCTIONS

Ya. A. Fedotov, V. S. Zased, and É. A. Matson

Moscow Institute of Steel and Alloys

A study was made of the electrical properties of heterojunctions formed between single crystals of a solid solution of gallium arsenide in germanium [$Ge_{0.85}(GaAs)_{0.15}$, denoted by α(Ge)] and gallium arsenide. Optical investigations showed that the forbidden band width of α(Ge) was of the order of 1 eV. The current—voltage characteristics of the p—n α(Ge)—GaAs heterojunctions were of a form typical of backward diodes. The current—voltage characteristics of the p—p α(Ge)—GaAs heterojunctions could be accounted for by the diode rectification theory. These characteristics were used to estimate the potential barrier in the p—p heterojunctions (0.2 eV), which was close to the calculated value $\psi_C = 0.25$ eV.

It is reported in [1] that a multilayer structure is formed in the contact alloying of germanium with gallium arsenide. From the point of view of heterojunctions, the most interesting is the contact between a single crystal of a solid solution of gallium arsenide in germanium [$Ge_{0.85} \cdot (GaAs)_{0.15}$, denoted by α(Ge)] and the original gallium arsenide. This contact is an abrupt heterojunction and provides a convenient object for investigating its electrical properties. Moreover, such a heterojunction has potential advantages over the Ge—GaAs heterojunction because the difference between the lattice parameters of the solid solution α(Ge) and GaAs should be less than the difference between the corresponding parameters of pure germanium and gallium arsenide. *

We used this alloying method to produce p—n and p—p α(Ge)—GaAs heterojunctions.

We determined the forbidden band width of the solid solution α(Ge) by investigating the dependence of the short-circuit photocurrent on the wavelength of the incident radiation. A typical dependence of this type, recorded for a p—n α(Ge)—GaAs heterojunction with a degenerate n-type GaAs region, is shown in Fig. 1.

The maximum of the photocurrent spectrum shown in Fig. 1 was located in the region of the absorption of light in the α(Ge) solid solution, whose forbidden band should be — according

---

* Precision measurements of the lattice parameter of α(Ge) are being carried out at present.

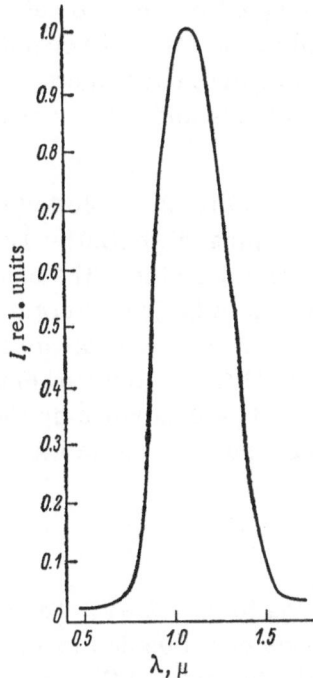

Fig. 1. Dependence of the short-circuit photocurrent of a p−n α(Ge)−GaAs heterojunction on the wavelength of incident radiation.

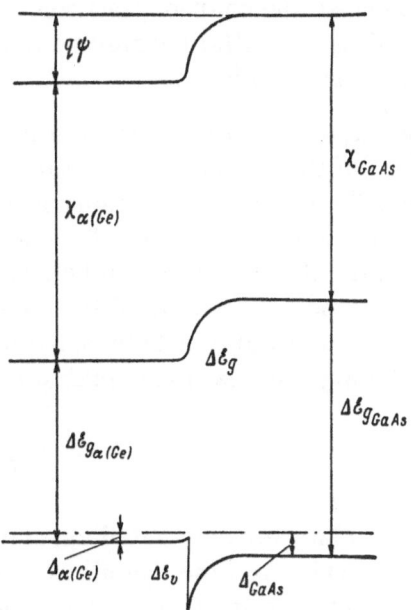

Fig. 2. Energy diagram of a p−p α(Ge)−GaAs heterojunction.

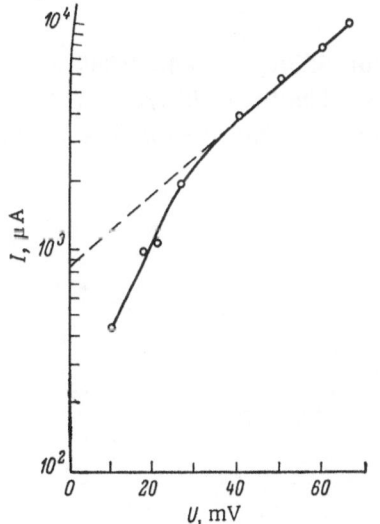

Fig. 3. Forward branch of the current−voltage characteristic of a p−p α(Ge)−GaAs heterojunction ($I_0 = 820\ \mu$A).

to this spectrum − of the order of 1 eV. The forbidden band width of the $Ge_{0.85}(GaAs)_{0.15}$ solid solution was also estimated using the results published in [2, 3]. This estimate again gave ~ 1 eV.

In order to plot the energy diagram of a heterojunction, it is necessary to know the affinities of both materials. According to the results given in [4], the difference between the electron affinities of germanium and gallium arsenide is 0.06 eV. Since the exact value of the electron affinity energy of α(Ge) was not required in our calculations, we assumed that this energy was equal to the corresponding energy of pure germanium (4.13 eV). A study of the electrical properties of the p−n heterojunctions showed that their current−voltage characteristics were typical of backward diodes. Since the concentrations of active impurities were $\sim 10^{18}$ cm^{-3} in p-type α(Ge) and $\sim 2 \cdot 10^{18}$ cm^{-3} in n-type GaAs, the main mechanisms of carrier transport were tunnel and tunnel-recombination processes, which was confirmed by the current−voltage characteristics. Since the recombination depended strongly on the number of defects in the crystal structure, the electrical processes were influenced considerably by defects.

The influence of defects on the nature of the carrier transport in the p−p heterojunctions was less marked. The presence of defects was not manifested in the form of recombination leakage because the charge was transported by the majority carriers, but defects could affect the value of the potential barrier by the formation of capture centers and dipoles at the interface between the two lattices.

The energy level diagram of a p−p heterojunction, shown in Fig. 2, is plotted using the parameters of the two semiconductors and the degree of their doping $N_A[\alpha(\text{Ge})] = 10^{18}$ cm^{-3}, $N_A(\text{GaAs}) = 5 \cdot 10^{16}$ cm^{-3} but ignoring the influence of the interfacial states. The contact potential difference, calculated from the energy diagram, was 0.25 eV. A typical current−voltage characteristic of one of the p−p heterojunctions, corrected for the voltage drop across the series resistance of the bulk of the diode, is given in Fig. 3. Analysis of the current−voltage characteristics of the p−p heterojunctions showed that they could be described by the diode rectifier theory and that the expression for the current−voltage characteristic was of the form

$$I = I_0 \left(e^{\frac{qU}{1.1kT}} - 1 \right).$$

The potential barrier calculated assuming $I_0 = 820$ μA was found to be 0.2 eV. The difference between this value and the calculated contact potential difference could be due to several effects: the distribution of the voltage between two regions of the junction, influence of the image forces, and tunneling of holes across the edge of a dip in the valence band. However, calculations showed that the corrections for these effects in the investigated p−p heterojunctions were small and their values were within the limits of the experimental error. Our study showed also that the influence of charges localized at the junction interface on the potential barrier was unimportant in the investigated p−p heterojunctions.

Literature Cited

1.   Yu. D. Chistyakov and É. A. Matson, Dokl. Akad. Nauk SSSR, 177:643 (1967).
2.   C. Kolm, S. A. Kulin, and B. L. Averbach, Phys. Rev., 108:965 (1957).
3.   D. A. Jenny, US Patent No. 2930239, applied for April 16, 1956; granted April 8, 1958.
4.   G. W. Gobeli and F. G. Allen, Phys. Rev., 137:A245 (1965).

CADMIUM TELLURIDE NUCLEAR-RADIATION COUNTER

P. S. Kireev, L. I. Kalugina, A. V. Vanyukov, and I. P. Shilo

Moscow Institute of Steel and Alloys

A brief report is given of the development of cadmium telluride nuclear-radiation detectors with p−n junctions, prepared by the diffusion of indium into a p-type material. The working area of these counters was up to 35 mm^2, their working voltage up to 20 V, and the reverse currents ranged from thousands to tenths of a microampere. The energy resolution for 5.15-MeV α particles was 6-8%. The thickness of the active region at a voltage of 10 V was about 10 μ.

Semiconductor counters are now widely used in various branches of nuclear physics and technology [1]. However, germanium counters have high reverse currents and have to be cooled in order to reduce their intrinsic noise. Silicon counters, whose spectrometric properties are excellent, are used more widely than germanium counters. Silicon counters with the p−i−n structure (prepared in our laboratory) can be used also to obtain time resolution up to 10^{-9} sec in recording γ-ray pulses. Unfortunately, the small charge of the silicon nuclei (Z = 14) means that the retardation is relatively weak, particularly for γ rays. Therefore, a search for new materials is very desirable. One of the most promising semiconductors is cadmium telluride.

Reports have appeared of the development of detectors based on n-type cadmium telluride, particularly surface-barrier detectors [2, 3]. We developed counters based on p−n junctions prepared by the diffusion of indium in p-type cadmium telluride.

The present paper gives some information on the first prototypes of these counters.

The counters had a working area up to 35 mm^2. Figure 1 shows the current−voltage characteristic of one of the first detectors. The reverse current, for voltages up to 10 V, was 10^{-7} A (the reverse currents of other detectors were within the range 10^{-9}-10^{-7} A under the same reverse voltage). The current increased rapidly when the reverse voltage was increased beyond this value. Since breakdown occurred at about 50-60 V, we concluded that the rise of the current was primarily due to leakage. This was confirmed by an additional treatment of the counter ends which increased the working voltage to 20 V. We would expect that the use of more uniform crystals and other methods of surface treatment would make it possible to increase the working voltage range still further.

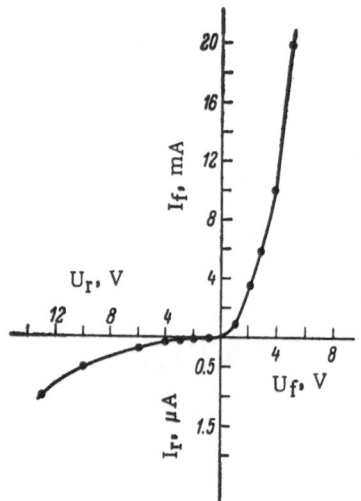

Fig. 1. Current−voltage characteristic of a counter.

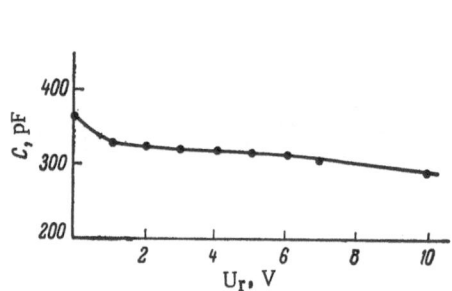

Fig. 2. Voltage−capacitance characteristic of a counter.

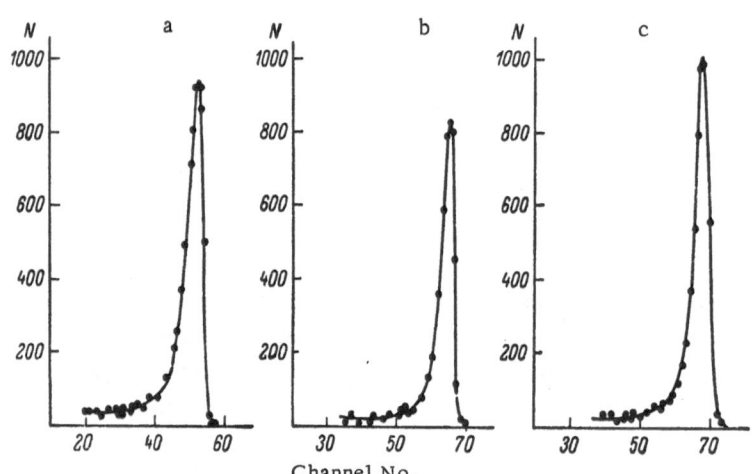

Fig. 3. Spectra of α particles of Pu[239], recorded using
different values of reverse bias (V): a) 0; b) 1.2; c) 2.5.

Figure 2 gives the voltage−capacitance characteristic of a p−n junction, showing that the sensitive region was about 10 μ thick for a voltage of 10 V and that it depended weakly on the voltage.

The counting characteristics were investigated employing 5.15-MeV α particles. Figure 3 shows the characteristics of the same counter for three different bias voltages: a) 0; b) 1.2 V; c) 2.5 V; the energy resolution for these three voltages was 8.5, 6.5, and 8% respectively. Bearing in mind that the best samples of our silicon detectors exhibited a resolution down to 0.3%, we concluded that cadmium telluride counters were still far behind silicon detectors in their characteristics. However, the results obtained for the first prototypes seem to be satisfactory and the problem is now to increase the resolving power of these counters.

Literature Cited

1.   G. Dearnaley and D. C. Northrop, Semiconductor Counters for Nuclear Radiation, 2nd ed.,
      Spon, London (1966).

2. J. W. Mayer, Nucl. Instr. Meth., 43:55 (1966).
3. E. N. Arkad'eva, L. V. Maslova, O. A. Matveev, Yu. V. Rud', and S. M. Ryvkin, Fiz. Tekh. Poluprov., 1:805 (1967).

FILAMENTATION OF THE CURRENT IN DIODES MADE OF SEMICONDUCTORS WITH DEEP IMPURITY LEVELS*

I. V. Varlamov, I. A. Sondaevskaya, and V. P. Sondaevskii

The detection and investigation of current filaments are reported for diodes made of gold-doped silicon. The investigation was made at room temperature using two methods: a scanning light probe and the sectioning of the contacts. It was found that a filament first appeared in a limited volume and then broadened with increasing voltage across the diode. An inhomogeneous distribution of an impurity in a semiconductor produced a nonuniform broadening of the filament with increasing current, and a complex current−voltage characteristic.

The possibility of the appearance of current filaments in structures with a negative differential conductivity [1] has been used by many authors [2-5] to explain the experimental data obtained in investigations of thermal and low-temperature impurity breakdown and in studies of diodes made of semiconductors with deep impurity levels. However, these authors have not reported direct evidence of the existence of current filaments in the investigated structures. The present paper describes the use of two simple methods for the detection and investigation of current filaments in diodes made of semiconductors with deep impurity levels: a scanning light probe and the sectioning of the contacts.

1. The scanning light probe method involves measurements of the dependence of the voltage photosensitivity of a diode (under constant-current conditions) on the position of a narrow illuminated strip parallel to the lines of flow of the current, when the strip is moved parallel to the contact.

In order to explain how this method can be used to detect a current filament, we must analyze the different changes in the current−voltage characteristics observed during the illumination of diodes of different geometries with strongly absorbed light. This makes it pos-

---

* A large part of this paper was presented at the Third All Union Conference on Physical Processes in p−n Junctions in Semiconductors (November, 1966, Tbilisi).

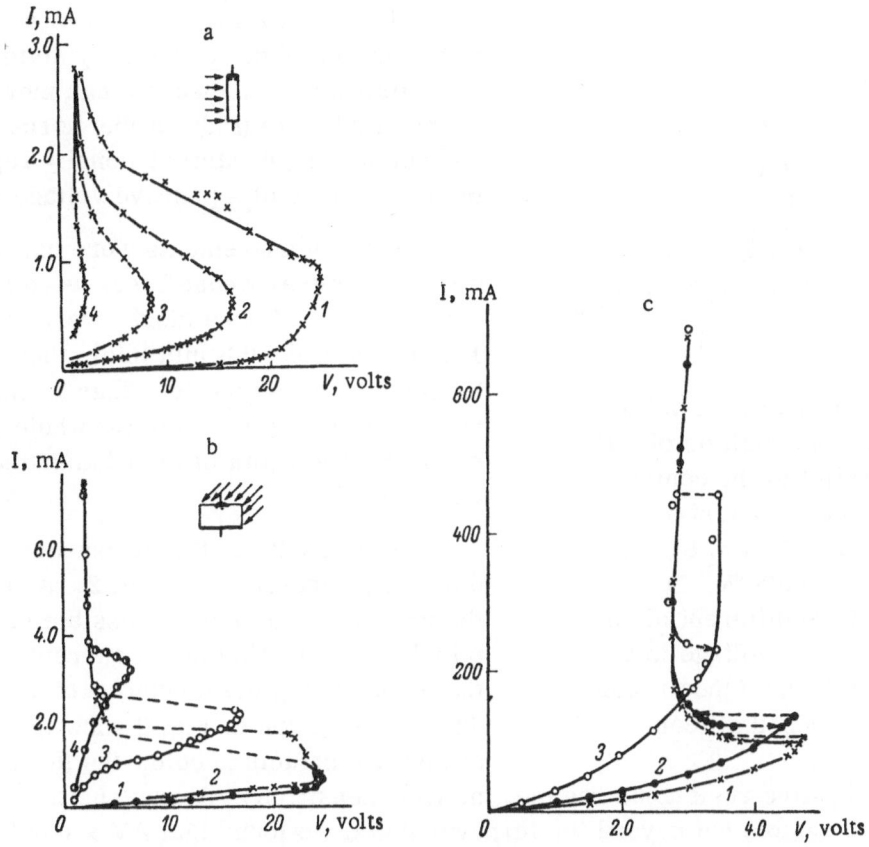

Fig. 1. Current-voltage characteristics of diodes of different
geometries illuminated with white light. a) Diode made of gold-
doped p-type germanium; cross-sectional area of diode base,
$S \approx 10^{-3}$ mm^2; p—n junction area, $S_n \approx 10^{-3}$ mm^2; distance between contacts, d = 1 mm; T = 77°K. b) Diode made of gold-
doped p-type germanium; S = 16 mm^2, $S_n \approx 1$ mm^2, d = 1 mm;
T = 77°K. c) Diode made of gold–doped n-type silicon; S = 5
mm^2, $S_n \approx 1$ mm^2; d = 0.2 mm; T = 300°K. The ascending numbers of the curves represent increasing illumination intensity.

sible to use the integrated (total) photosensitivity of the structure to deduce whether the light
probe is incident on a filament.

Figures 1a and 1b show the current—voltage characteristics of gold-doped germanium
diodes illuminated with light of various intensities. Measurements show that the effect of weak
illumination on a long diode with a sufficiently small cross-sectional area (Fig. 1a) is similar
to the effect of a change in the equilibrium conductivity in the base. Since the concentration and
population of the impurity centers are practically unaffected, the turnover current is the same —
at different illumination intensities. In this case, for any value of the current through the diode,
the difference ΔV between the voltages in darkness V_d and during illumination V_i is positive

$$\Delta V = V_d - V_i > 0.$$

Illumination of a diode whose contact area is considerably less than the cross-sectional
area of the sample (Fig. 1b) produces a high-conductivity layer near the surface. The effect
of this layer is equivalent to a shunting resistance, whose value decreases with increasing illumination intensity. In this case, the voltage photosensitivity ΔV in the negative resistance
region is negative. Similar results are obtained by the illumination of a diode structure of the
same geometry but made of gold-doped silicon (Fig. 1c).

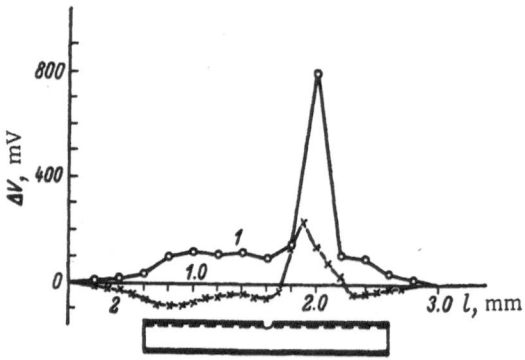

Fig. 2. The dependence of the voltage photosensitivity on the position of a light probe, moved parallel to the contacts. The width of the illuminated strip was about 0.1 mm; S = 2 × 1 mm, S_n = 2 × 1 mm, d = 0.2 mm. T = 300°K. 1) Bias voltage in the positive differential resistance region; 2) bias voltage in the negative differential resistance region. Diode made of gold-doped silicon.

Thus, if the investigated structure is switched to a high-conductivity state and a current filament is formed, we can move a narrow illuminated strip (light probe) across the base and detect a high-current-density region from the appearance of a positive voltage sensitivity ΔV.

Such measurements were carried out on diode structures whose bases were made of gold-doped silicon. These diodes were strips about 1 mm wide and ~0.3 mm thick; they differed in length, which was not less than 2 mm. The contacts were deposited over the whole 1-mm-wide surface. The width of the illuminated strip was ≲ 0.1 mm.

The results of the measurements on these diodes are presented in Fig. 2. A section across the p−n junction made it possible to detect electrically a rapid rise in the current in a part of the diode during the formation of a filament. The curves given in Fig. 2 show that, at voltages across the structure below the turnover point (the onset of the negative resistance region), the photosensitivity ΔV was positive across the whole length of the sample but beyond the turnover point, we found that $\Delta V > 0$ only in a certain narrow region where a current filament had formed.

This method has some advantages, such as simplicity, as well as important disadvantages, which include fairly rigorous restrictions imposed on the geometry of the samples (the width of the contacts must be smaller than or of the same order as the diameter of a current filament), and the fact that a current filament can be investigated only when the voltage is in the negative differential resistance region. Therefore, this method requires further development. It should be supplemented by other methods, such as that involving the sectioning of the contacts.

2. It is reported in [4, 5] that the current−voltage characteristics of large-area diode structures are complex: they can have several negative and zero differential resistance regions. It is suggested in [5] that inhomogeneities in impurity or defect distributions in a semiconductor may affect the formation of a current filament, which can broaden with increasing current not uniformly but in jumps from one inhomogeneity to another.

The formation of and changes in a conducting channel (filament) with increasing current through a structure can be investigated by the sectioning of the contacts. In this method, the distribution of the current density across a sample is measured by dividing one of the current contacts into sections of the minimum possible size with the smallest possible gaps between the sections. These sections are then rejoined via low-resistance ammeters (Fig. 4).

It must be mentioned that:

a) reduction of the dimensions of the sections increases the accuracy of the measurements and yields more information;

b) reduction of the gaps between sections reduces the inhomogeneity of the sectioned contact;

c) small (compared with the diode and load resistances) internal resistances of the instruments used to measure the current through single sections ensure that the whole sectioned contact remains at the same potential.

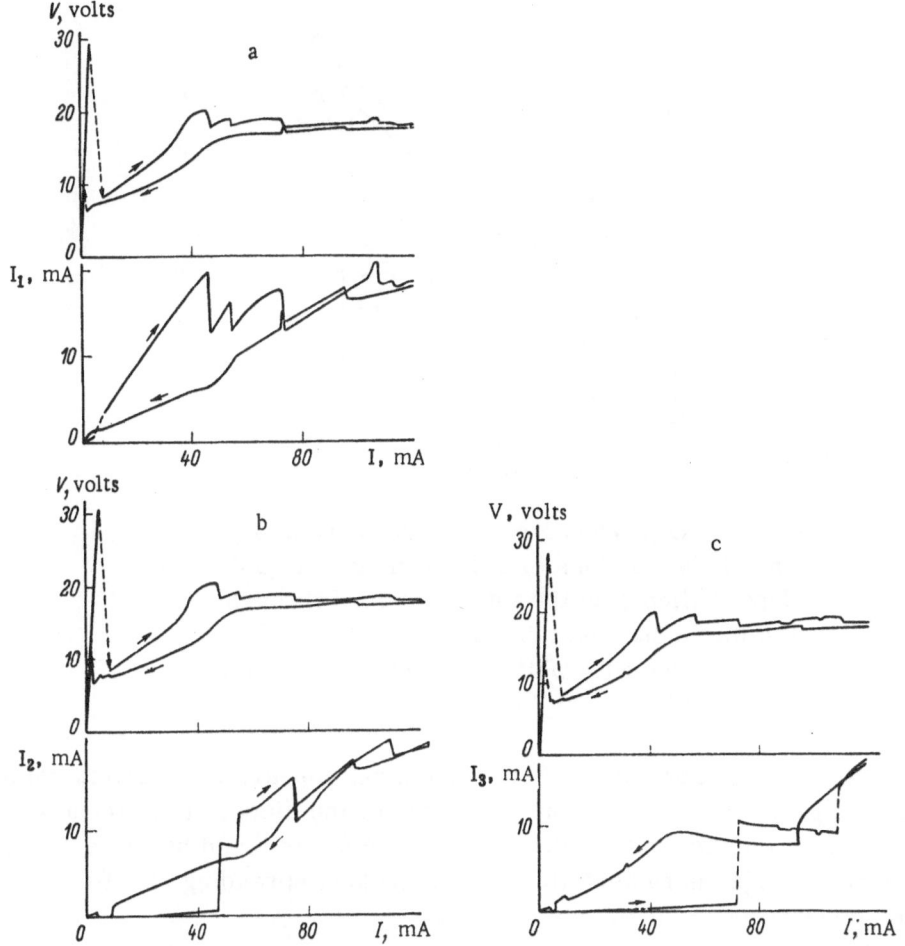

Fig. 3. Dependences of the currents I_i (i = 1, 2, 3) through single (i-th) sections and of the voltage V across the whole structure on the total current I.

Figure 3 shows the dependences of the current I_i through a given section i (i = 1, 2, 3,...) and of the voltage V across the structure on the total current I through the system. Since the current—voltage characteristics varied somewhat with time, the dependences $I_i(I)$ and $V(I)$ were determined simultaneously using two X−Y potentiometers. This eliminated the possibility of accidental agreement or disagreement between the dependences of I_i and V on I.

The curves in Fig. 3 show that a current filament appeared in the first section (Fig. 3a). This produced a strong rise in I_1 and a drop in the voltage V across the diode. The current through the other sections decreased because the voltage across the structure became lower. When the total current I through the diode was increased the filament spread to the second section (Fig. 3b) and the current through that section (I_2) increased suddenly. Further increase in the total current produced a second jump of I_2 (Fig. 3b). This indicated that the inhomogeneities in the impurity distribution had dimensions which were smaller than the size of the section (1 mm^2). Similar dependences were observed for the other sections (Fig. 3c).

Illumination of the whole structure with scattered white light altered the conditions for the formation of a filament (Fig. 4). However, once again, the filament broadened when the bias was increased and this could be followed along the six sections shown in Fig. 4.

Different dependences were obtained by increasing and decreasing the total current I (Figs. 3 and 4) because the conditions for the appearance and disappearance of a filament were not identical.

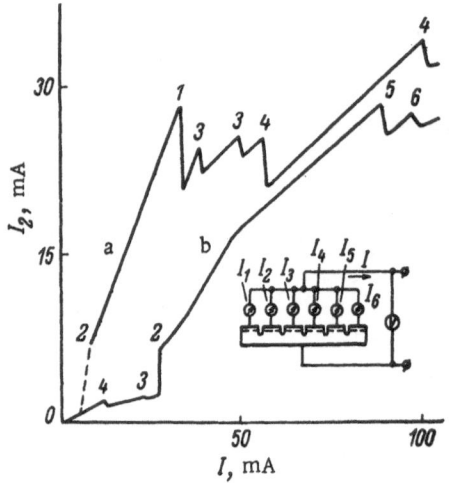

Fig. 4. Dependence of the current I_2 through one of the sections of a diode (made of gold-doped silicon) on the total current I through the whole structure in darkness (a) and during illumination with white light (b); d = 0.25 mm; $T \approx 300°K$.

Our investigations thus show that a current filament appears in a limited volume but spreads laterally and gradually occupies the total area of the diode. The formation of two or more separate filaments was not observed. A relationship was found between the complex shape of the current−voltage characteristics and the sudden spreading of a filament from one inhomogeneity to another.

Literature Cited

1. B. K. Ridley, Proc. Phys. Soc. (London), 82:954 (1963).
2. I. Melngailis and A. G. Milnes, J. Appl. Phys., 33:995 (1962).
3. P. Schnupp, Z. Angew. Phys., 19:46 (1965).
4. T. Fukami and K. Homma, Japan. J. Appl. Phys., 2:535 (1963).
5. I. V. Varlamov, I. A. Sondaevskaya, and V. P. Sondaevskii, Fiz. Tekh. Poluprov., 1:452 (1967).

HIGH-TEMPERATURE SILICON CARBIDE RECTIFIERS FOR HIGH REVERSE VOLTAGES

V. I. Pavlichenko, I. V. Ryzhikov, and T. G. Kmita

A description is given of the methods of fabrication of silicon carbide rectifiers with breakdown voltages of 100-600 V. The methods involve diffusion or alloying at relatively low temperatures, applied for long periods in order to reduce the impurity concentration gradients. Methods for the assembly and sealing of a silicon carbide diode in a metal—ceramic casing are described. The casing is evacuated before sealing and filled with an inert gas. The current—voltage and voltage—capacitance characteristics of a diode obtained in the temperature range from −70 to +500°C are given and briefly analyzed. The results indicate the presence of a wide compensated region near the p−n junctions in the silicon carbide rectifiers. The data obtained are used to determine the activation energy of aluminum (0.24-0.26 eV) and boron (0.39-0.40 eV) acceptor impurities in silicon carbide.

Various methods for the preparation of alloyed, diffused, and epitaxial p−n junctions in silicon carbide have already been described [1-6]. However, all these methods (including, diffusion from the vapor phase by the Chang method [4]) yield p−n junctions with breakdown voltages within the range 10-50 V if the usual annealing temperatures and durations are employed. Such junctions can be used as pulse sources of light and indicator lamps whose electroluminescence lies in the visible part of the spectrum.

In order to produce silicon carbide rectifiers with breakdown voltages higher than 100 V, we investigated the diffusion, alloying, and diffusion-alloy methods for the formation of p−n junctions at relatively low temperatures, applied for long periods in order to reduce impurity concentration inhomogeneities. We started with n-type silicon carbide crystals of 1-10 $\Omega \cdot$ cm resistivity.

In the diffusion processes, we used aluminum and boron dopants. The diffusion of aluminum was carried out in an atmosphere of helium of spectroscopic purity by heating the silicon carbide for 4-30 h at 1600-1800°C; boron was diffused by applying the same temperatures for 2-4 h.

Before determining their current—voltage characteristics, the crystals with diffused p-type layers and ohmic contacts were etched in a mixture of nitric and hydrofluoric acids and were then subjected to electrolysis in a solution consisting of hydrofluoric acid and glycerin. Such treatment reduced the reverse current and increased the breakdown voltage to 300-600 V.

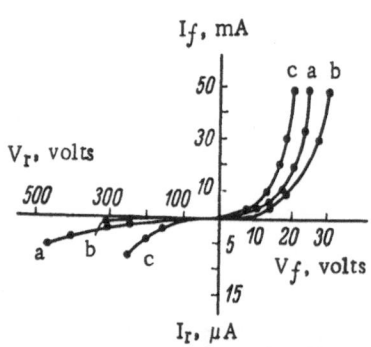

Fig. 1. Current−voltage characteristics of silicon carbide diodes at room temperature: a) diodes prepared by the diffusion of aluminum; b) diodes prepared by the diffusion of boron; c) diodes prepared by fusing together n- and p-type crystals through a layer of evaporated aluminum.

Fig. 2. Photograph of a finished silicon carbide diode.

The forward voltage for a current of 50-100 mA was 6-20 V for aluminum-doped samples and 20-50 V for boron-doped samples (Fig. 1, curves a and b).

Large-area (2 × 2 mm) alloyed junctions (Fig. 1c) were prepared from high-resistivity n-type crystals ($\rho \approx 1\text{-}10\ \Omega \cdot$ cm) and low-resistivity p-type crystals ($\rho \approx 0.1\ \Omega \cdot$ cm), which were fused together through a layer of deposited aluminum at 1500-1600°C. After this treatment, the high-resistivity n-type crystal was reduced (by grinding) to a thickness of a few tens of microns.

Prolonged annealing (lasting a few hours) of the alloyed structure of 1700-1800°C produced junctions in the n-type crystals by the diffusion of aluminum from the gap between the crystals; the aluminum was then evaporated and the n- and p-type crystals were easily separated. The diffused junctions prepared in this way had breakdown voltages of 100-150 V and the forward voltage drop was 3.5-10 V.

This method of preparing diffused p−n junctions by prolonged annealing did not cause decomposition of the surface layer of a crystal in which diffusion took place. Therefore, in contrast to the method of preparing p−n junctions by diffusion from the vapor phase, there was no need to establish an equilibrium vapor pressure of SiC around the crystal.

All the p−n junctions were provided with an ohmic contact with the n-type region: this contact, 2 × 2 mm in area, was produced by fusion-joining a crystal to a nickel-plated tungsten disk, 3 mm in diameter, at ∼1600°C. An ohmic contact with the diffused p-type layer, and with the p-type crystal in the case of alloyed junctions, was made employing a silicon−aluminum alloy (3.5% Al), which was fused at a temperature of ∼1450°C applied for a few seconds. The depth of penetration of the melt into the silicon carbide did not exceed 1 μ.

Since the rectifying and nonrectifying contacts or junctions were made using metals which oxidized in air at high temperatures (these included aluminum, tungsten, etc.), we found that prolonged operation or even measurements of the electrical parameters of silicon carbide diodes could be carried out in a wide range of temperatures only in a chemically neutral medium. Therefore, a crystal with a p−n junction was placed in a specially developed metal−ceramic casing made of oxygen-free nickel and copper. Before the final sealing, the casing was evacuated and filled with spectroscopically pure helium or argon. A photograph of a finished silicon carbide diode is shown in Fig. 2. When encased in this way, the diodes were capable of being operated for not less than 1000 h at 600°C.

The current−voltage characteristics of the diodes were determined in the temperature range from −70 to +500°C. The forward branch of the current−voltage characteristic of silicon carbide diodes consisted, in general, of four successive regions, described by:

$$I \propto V, \quad I \propto \exp \frac{qV}{ckI}, \quad I \propto (V - V_1)^4, \quad I \propto \exp \frac{qV}{nkT},$$

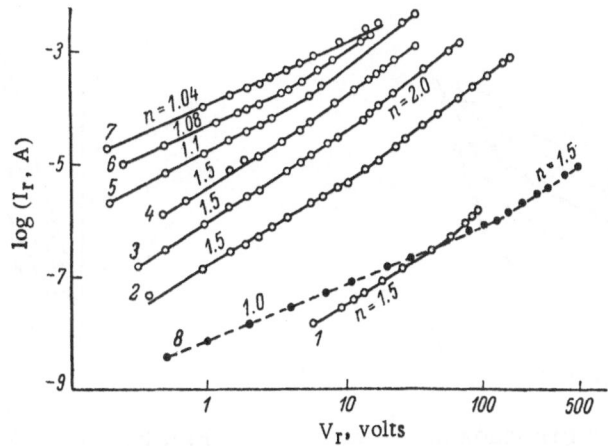

Fig. 3. Reverse branches of the current−voltage characteristics of silicon carbide diodes No. 1 (curves 1-7) and No. 2 (curve 8) prepared by the diffusion of aluminum. Temperature, °C: 1) −70; 2) 20; 3) 100; 4) 200; 5) 300; 6) 400; 7) 500. The values of the power exponent n are given alongside the curves.

(n ≫ 2), which were analyzed by us earlier [7, 9] using a theory which took into account the carrier recombination in all regions of a p^+−n−n^+ structure. The reverse current of these diodes was a power function of the applied voltage $I \propto V^n$ and consisted of three regions, shown in Figs. 3 and 4. The values of the power exponent n in these regions were, respectively, 1-1.5, 2.0, and 5-30. The capacitance of the p−n junctions, measured at a frequency of 5 Mc between +20 and +500°C, was practically independent of the applied voltage and amounted to 1-10 pF. This could probably be explained by the presence, near the p−n junction, of a high-resistivity n-type layer whose thickness ranged from a few microns to tens of microns.

Some of the diodes with breakdown voltages of 100-150 V, prepared by the diffusion of aluminum from a gap between two crystals, exhibited a dependence of the capacitance on the voltage, $C^{-3} \propto (\varphi_c - V)$, typical of p−n junctions with a linear impurity distribution in the space-charge region.

Silicon carbide diodes with high reverse breakdown voltages of up to 600 V were prepared by reducing the impurity concentration gradients near the p−n junctions either by reducing the conductivity of the initial material or increasing the duration of the diffusion annealing. The forward branches of the current−voltage characteristics of these diodes included, in addition to the leakage currents, a region described by $I \propto (V - V_1)^4$, which extended over a wide range of currents (Fig. 5). This observation was also in agreement with the theory given in [7], according to which the beginning of the power-law region in the current−voltage characteristic should shift in the direction of lower currents when the resistivity of the p-type layer was increased. Because of a decrease in the impurity concentration gradient, the thickness of the high-resistivity n-type layer decreased; according to [7], this thickness should be proportional to $\cosh^2(W/L_p)$, i.e., it should range over 3-5 orders of magnitude for the usually observed values of the ratio W/L_p.

Linear or near-linear dependences of the reverse current on the applied voltage were observed for diffused rectifiers (with high breakdown voltages) between a few tenths of a volt and 100-200 V (curve 8 in Fig. 3). The majority of the investigated diodes exhibited dependences $I \propto V^n$, where $1.4 \leq n \leq 1.6$, over a wide range of current densities. On the basis of the model

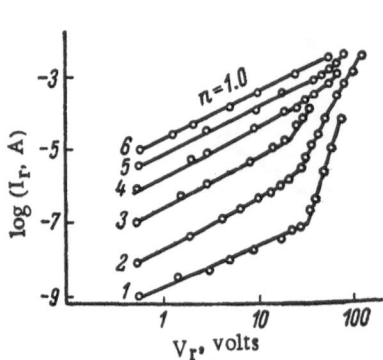

Fig. 4. Reverse branches of the current–voltage characteristics of a silicon carbide diode prepared by the diffusion of boron. Temperature, °C: 1) 20; 2) 100; 3) 200; 4) 300; 5) 400; 6) 500.

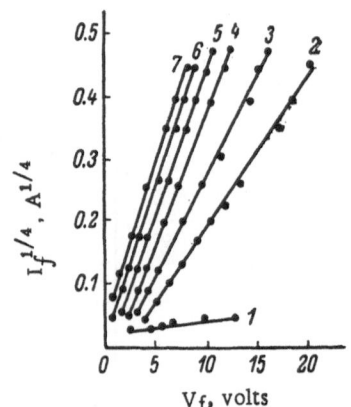

Fig. 5. Dependence of $\sqrt[4]{I_f}$ on the voltage for a diffused silicon carbide diode, recorded between −70 and +400°C. Temperature, °C: 1) −70; 2) 20; 3) 100; 4) 200; 5) 300; 6) 400.

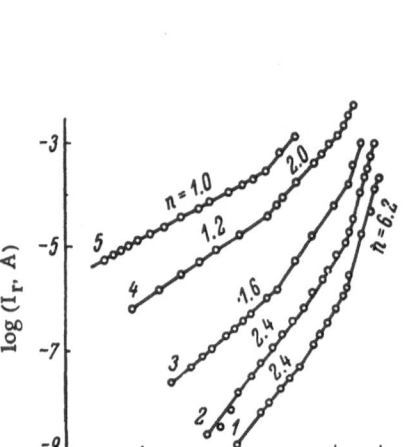

Fig. 6. Reverse branches of the current–voltage characteristics of a silicon carbide diode prepared by fusing together n- and p-type crystals through a layer of evaporated aluminum. Temperature, °C: 1) 20; 2) 90; 3) 223; 4) 300; 5) 515.

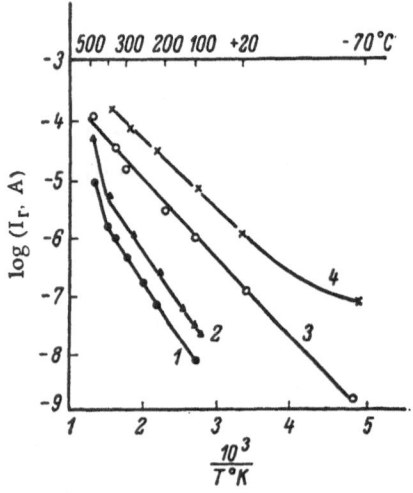

Fig. 7. Temperature dependences of the logarithm of the current in the linear part of the reverse branch of the current–voltage characteristic for diodes prepared by the diffusion of boron (1 and 2) and aluminum (3 and 4).

used in [10], the considerable widths of the regions $I \propto V^{1.0}$ and $I \propto V^{1.5}$ indicated the presence of a wide compensating n-type region, which was also confirmed by the high breakdown voltages and the relatively high values of the forward voltage drop.

The space-charge-limited region was either relatively narrow or altogether absent, but a rapid rise of the current was observed due to carrier multiplication.

The cause of the rapid rise of the electric current at breakdown voltages and of the deviations from the theoretical dependences $I \propto V^{1.0}$, $I \propto V^{1.5}$, and $I \propto V^{2.0}$, was evidently the electrostatic ionization.

The three regions of the reverse branch of the current−voltage characteristic were observed in a wide range of temperatures (Figs. 3 and 4). In the first (ohmic) region, the current at a fixed voltage was proportional to $\exp(\Delta E/kT)$, where $\Delta E = 0.24$-0.26 eV for the p−n junctions prepared by the diffusion of aluminum and $\Delta E = 0.39$-0.40 eV for the p−n junctions prepared by the diffusion of boron.

A power exponent n in the first region of the reverse branch of the current−voltage characteristic of the alloyed junctions was usually higher (Fig. 6). The exponent decreased with increasing temperature and, in the limit, reached unity. In the second region, which corresponded to the space-charge-limited current, the values of n decreased with temperature from 3-4 to 2.0.

The nature of the dependence of the reverse current on the voltage of some diodes changed at about 400-500°C (Fig. 7). Up to this temperature, the current in the first region was either linear or superlinear but above this temperature the dependence of the current on the voltage became sublinear ($I \propto V^n$, where n = 0.4-0.5). The component of the electric current associated with carrier generation in the space-charge layer of the p−n junction was probably important at these temperatures [11].

An estimate of the component of the reverse current due to carrier generation, obtained at 600°C and for a voltage of ~1 V, gave a value of 10^{-4}-10^{-3} A, which was of the same order of magnitude as the measured value. However, the thermal activation energy of the reverse current was somewhat lower than the value predicted by the Shockley−Noyce−Sah theory.

Preliminary investigations showed that silicon carbide rectifiers were more stable than germanium and silicon devices when subjected to ionizing radiation; the forward voltage drop increased after irradiation.

The authors are grateful to K. M. Kolesov for his great help in the construction of the diode casing.

Literature Cited

1.   R. N. Hall, J. Appl. Phys., 29:914 (1958).
2.   T. C. Taylor, J. Appl. Phys., 29:865 (1958).
3.   T. E. Kharlamova and G. F. Kholuyanov, Fiz. Tverd. Tela, 2:426 (1960).
4.   H. C. Chang, C. Z. LeMay, and L. F. Wallace, in: Silicon Carbide − A High-Temperature Semiconductor (Proc. Conf., Boston, 1959, J. R. O'Connor and J. Smiltens, eds.), Pergamon Press, Oxford (1960), p. 496.
5.   K. M. Hergenrother, S. E. Mayer, and A. I. Mlavsky, in: Silicon Carbide − A High-Temperature Semiconductor (Proc. Conf., Boston, 1959, J. R. O'Connor and J. Smiltens, eds.), Pergamon Press, Oxford (1960), p. 60.
6.   G. F. Lymar' and Yu. V. Polikanov, Voprosy Radioélektroniki, Ser. 2, No. 1, p. 135 (1964).
7.   I. V. Ryzhikov, V. I. Pavlichenko, and T. G. Kmita, in: Silicon Carbide, Consultants Bureau, New York (1970), p. 238.
8.   A. Yu. Leiderman and P. M. Karageorgii-Alkalaev, Radiotekhn. i Élektron., 10:720 (1965).
9.   I. V. Ryzhikov, V. I. Pavlichenko, T. G. Kmita, A. Yu. Leiderman, and P. M. Karageorgii-Alkalaev, Radiotekhn. i Élektron., 12:842 (1967).
10. L. Patrick, J. Appl. Phys., 28:765 (1957).
11. C. T. Sah, R. N. Noyce, and W. Shockley, Proc. IRE, 45:1228 (1957).

THEORY OF THE "dU/dt EFFECT" IN THYRISTORS

V. A. Kuz'min

Moscow Engineering Physics Institute

A simple theory of the dU/dt effect is developed. The effect of the capacitative current reduces to an effect of equal gate current pulses in the two base regions. A calculation is carried out on the assumption that the shape of the capacitative current pulse is rectangular. The effect of an ideal rectangular voltage pulse on a thyristor is also considered.

The reduction in the turn-on voltage of a p−n−p−n structure due to the application of a rapidly rising voltage to the anode has been discussed in [1-3]. Some experimental dependences characterizing this effect have been determined. The cause of the reduction in the turn-on voltage is the capacitative current I_C through the central junction, given by $C_2(dU/dt)$ (Fig. 1); this current increases the current gains α_p and α_n of the two transitors that can be regarded as the equivalent of the thyristor structure. However, a theory of the process responsible for the observed effect has not been given in the cited investigations. The present paper describes a simple theory of this effect.

Formulation of the Problem

The capacitative current through the central junction is the current of the majority carriers: an electron current $C_2(dU/dt)$ flows into the n-type base and an equal hole current flows into the p-type base. The application of a voltage U(t) between the anode and cathode of the thyristor (Fig. 2a) gives rise to capacitative current pulses in the two bases (Fig. 2b). The shape of these pulses is affected by the dependence of the capacitance C_2 on the voltage across the central junction (this junction will be denoted by j_2). Thus, the effect of the capacitative current through the central junction is physically equivalent to the effect of equal gate currents flowing into the two bases of the p−n−p−n structure.

If these capacitative current pulses do not exceed the static current capable of turning on the thyristor (we shall denote this current by $I_{b.on}$), the thyrsistor is not turned on. However, if $C_2(dU/dt) > I_{b.on}$, the thyristor may be turned on. The current I through the thyristor begins to increase and the current gains α_p and α_n, which depend on I, also increase. The process resembles the turn-on of a thyristor by a current pulse of finite duration applied to the circuit

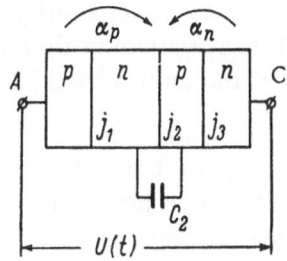

Fig. 1. Four-layer structure.

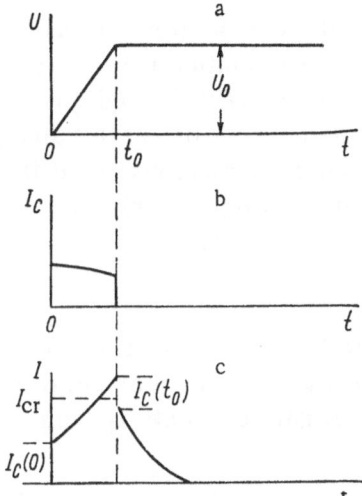

Fig. 2. Time dependences of the current and voltage: a) voltage applied between the anode and cathode; b) capacitative current through the central junction; c) current through the thyristor.

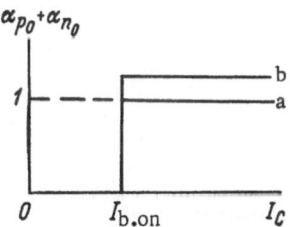

Fig. 3. Dependence of $\alpha_{p0} + \alpha_{n0}$ on the current.

of one of the bases during the delay stage [4-6], and can be calculated using similar assumptions. At $t = t_0$, the capacitative current and, therefore, the base currents I_{bp}, I_{bn} all vanish. At the same moment the current through the thyristor decreases by an amount $I_C(t_0)$.

The capacitative current pulse may be followed by one of two processes: either the current increases in an avalanche-like manner and the thyristor is turned on, or the current decreases and the thyristor remains in the off state (Fig. 2c). The current through the thyristor at the moment $t = t_0$ determines which of these two processes takes place. If I is less than a "critical" current I_{cr} the thyristor is not turned on. The "critical" current is of the same order of magnitude as the turn-off current I_{off} and there is a corresponding "critical" charge of minority carriers.

Calculation of the Current through a Thyristor

Let us assume that a voltage pulse U(t), whose amplitude is U_0, is applied to the thyristor. We shall postulate that this pulse has a flat top and that its duration is t_0. The equation for the current through the p−n−p−n structure, which takes into account the capacitative current through the central junction, is of the form [2]

$$I = \frac{I_{c0} + C_2 \dfrac{dU}{dt}}{1 - \alpha_p - \alpha_n}, \qquad (1)$$

where I_{c0} is the reverse current through the central p−n junction j_2. We must find I(t). We shall calculate this current making the following simplifying assumptions.

1. The capacitances of the outer p−n junctions j_2 and j_3 are small and can be ignored. We can also ignore the current I_{c0}.

2. The capacitative current $I_C < I_{b.on}$ cannot turn on the thyristor. For simplicity, we shall assume that $I = 0$ when $I_C < I_{b.on}$.

3. The capacitative current pulse $I_C = C_2(dU/dt)$ is rectangular and its duration is t_0. If the applied voltage pulse has a linear front, we shall assume that $I_c = \bar{C}_2 \dfrac{dU}{dt}$ where $\bar{C}_2 = \dfrac{1}{U_0} \int_0^{U_0} C_2(U)\,dU$.

4. The current gains α_p and α_n depend on the frequency in accordance with the well-known law

$$\alpha_p = \frac{\alpha_{p0}}{1 + j\omega\tau_{\alpha p}}, \qquad \alpha_n = \frac{\alpha_{n0}}{1 + j\omega\tau_{\alpha n}}.$$

This approximation is widely used for transistors and has been applied successfully in calculations dealing with transient processes in transistor-like structures, including thyristors. It describes correctly the frequency dependence of the absolute value of the current gain up to frequencies $\omega \approx 5/\tau_\alpha$.

Before the application of the current pulse $I_C > I_{b.on}$, $\alpha_p = \alpha_n = 0$. At $t \geq 0$, the values of α_{p0} and α_{n0} increase because a current begins to flow through the outer p−n junctions j_1 and j_3. This transient process corresponds to the delay stage in the process of turning on a thyristor by equal current pulses in the two bases.

The problem can be made linear by assuming the dependence $\alpha(I_C)$ shown in Fig. 3.

Such an approximation for $\alpha(I)$ was used first in [4] in a calculation of the turn-on process by a step-like current pulse applied to one of the bases. This approximation reflects the observation that the current gains α do indeed increase with increasing current through the thyristor, which is related to I_C, and that only when $I_C > I_{b.on}$ can the current increase rapidly and a four-layer structure be turned on. It is difficult to draw any other conclusions about the physical nature of this dependence. We can distinguish two cases: a) $\alpha_{p0} + \alpha_{n0} = 1$; b) $\alpha_{p0} + \alpha_{n0} > 1$.

The derivatives $\partial\alpha_{p0}/\partial I$ and $\partial\alpha_{n0}/\partial I$ vanish when $I_C > I_{b.on}$.

5. The experimentally set currents I_{bp} and I_{bn} flow parallel to the j_2 p−n junction and the current I_C flows perpendicularly to this junction. Assuming that the p−n−p−n structure is one-dimensional, we can regard the capacitive current I_C and the gate currents I_{bp} and I_{bn} as fully equivalent if $I_C = I_{bp} = I_{bn}$.

Assuming that the transforms of the current gains are

$$a_n(s) = \frac{a_{n0}}{1 + s\tau_{an}}, \quad a_p(s) = \frac{a_{p0}}{1 + s\tau_{ap}},$$

and the transform of a single pulse of duration t_0 is $1 - \exp(-st_0)$, we obtain the following operator expression for the current:

$$i(s) = \frac{W(s)}{V(s)}(1 - e^{-st_0}), \tag{2}$$

$$\left.\begin{aligned}
i(s) &= \frac{I(s)}{I_C - I_{b.on}}, \\
W(s) &= s^2 + a_1 s + a_0, \\
V(s) &= s^2 + b_1 s + b_0.
\end{aligned}\right\} \tag{3}$$

The coefficients a_1, a_0, b_1, and b_0 are related to the structure parameters:

$$a_1 = \frac{\tau_{an} + \tau_{ap}}{\tau_{ap}\tau_{an}}, \quad a_0 = \frac{1}{\tau_{ap}\tau_{an}},$$

$$b_1 = \frac{\tau_{ap}(1 - a_{n0}) + \tau_{an}(1 - a_{p0})}{\tau_{ap}\tau_{an}}, \quad b_0 = \frac{1 - a_{p0} - a_{n0}}{\tau_{ap}\tau_{an}}.$$

The current through the thyristor at $t = t_0$ is given by the expression

$$i(t_0) = i_0(t_0) - i_0(0), \tag{4}$$

where $i_0(t)$ is the original form of the function $W(s)/V(s)$. At times $t < t_0$, we have

$$i(t) \equiv i_0(t).$$

a) $\alpha_{p0} + \alpha_{n0} = 1$. This condition should be used if $C_2(dU/dt) < I_{off}$. The equality $\alpha_{p0} + \alpha_{n0} = 1$ yields $b_0 = 0$, $b_1 = (\alpha_{p0}/\tau_{\alpha n}) + (\alpha_{n0}/\tau_{\alpha p})$.

The solution for $t < t_0$ is of the form

$$i_0(t) = i(t) = \frac{a_1}{b_1} - \frac{a_0}{b_1^2} + \frac{a_0}{b_1}t + e^{-b_1 t}\left(1 - \frac{a_1}{b_1} + \frac{a_0}{b_1^2}\right), \tag{5}$$

where $i_0(0) = 1$, which corresponds to a sudden change of the current in the anode circuit by an amount $I_C - I_{b.on}$. In fact, the change in the current is equal to I_C. This discrepancy is due to the fact that we have assumed that the thyristor current is zero when $I_C < I_{b.on}$. Equation (5) can usually be approximated by a linear function when $t \geq 2/b_1$:

$$i_0(t) \simeq \frac{a_1}{b_1} - \frac{a_0}{b_1^2} + \frac{a_0}{b_1}t.$$

When $t = t_0$, we obtain

$$i(t_0) = \left(1 - \frac{a_1}{b_1} + \frac{a_0}{b_1^2}\right)(e^{-b_1 t_0} - 1) + \frac{a_0}{b_1}t_0. \tag{6}$$

When $t_0 \geq 2/b_1$, which is frequently encountered under real conditions,

$$i(t_0) \simeq \frac{a_1}{b_1} - \frac{a_0}{b_1^2} - 1 + \frac{a_0}{b_1}t_0. \tag{7}$$

Figure 2c shows the sudden change in the current at $t = 0$ and $t = t_0$. If $I(t_0)$ exceeds the "critical" current, the thyristor is turned on. The parameters of an external pulse which cannot turn on the thyristor are found from the condition

$$I(t_0) = (I_C - I_{b.on})i(t_0) < I_{cr}. \tag{8}$$

The maximum permissible rate of rise of the voltage dU/dt for a pulse whose front rises linearly and whose amplitude is U_0 can be found from the condition

$$\left(\bar{C}_2\frac{dU}{dt} - I_{b.on}\right)i\left(\frac{U_0}{\frac{dU}{dt}}\right) = I_{cr}. \tag{9}$$

The value of $I_{b.on}$ depends on the voltage U_0 and it can be found experimentally. When U_0 is reduced, $I_{b.on}$ increases somewhat. Having determined, using Eq. (9), the maximum permissible value of dU/dt for a given amplitude U_0, we can now find the minimum duration of the rise of the front $t_0 = U_0/(dU/dt)$. If the duration of the capacitative current pulse is long compared with the turn-on time, the problem reduces to the static case. The criterion for the static case is $i(t_0) \gg I_{cr}$, which gives $\bar{C}_2(dU/dt) \simeq I_{b.on}$.

b) $\alpha_{p0} + \alpha_{n0} > 1$. This approximation should be used if $I_C > I_{off}$. The solution for $t < t_0$ is of the form

$$i_0(t) = A + Be^{s_1 t} + Ce^{s_2 t},$$

where $i_0(0) = 1$. The quantities s_1 and s_2 are the roots of the denominator of Eq. (2): they have opposite signs. This means that the solution includes rising and falling components. The constants s_1, s_2, A, B, and C can be expressed in terms of the structure parameters α_{p0}, α_{n0}, $\tau_{\alpha p}$, and $\tau_{\alpha n}$.

The values of the roots s_1 and s_2 are obtained in the same way as in calculations dealing with the turn-on by a current pulse applied to one of the bases [4-6]:

$$s_1 = \frac{b_1}{2}\left[-1 + \sqrt{1 - \frac{4b_0}{b_1^2}}\right] > 0,$$

$$s_2 = \frac{b_1}{2}\left[-1 - \sqrt{1 - \frac{4b_0}{b_1^2}}\right] < 0,$$

$$A = \frac{a_0}{b_0}, \quad B = \frac{W(s_1)}{s_1(2s_1 + b_1)}, \quad C = \frac{W(s_2)}{s_2(2s_2 + b_1)}.$$

If, at $t = t_0$, we have

$$I(t_0) = (I_C - I_{b.on})(A - 1 + Be^{s_1 t_0} + Ce^{s_2 t_0}) < I_{cr}, \tag{10}$$

the thyristor is not turned on.

In our opinion, the case $\alpha_{p0} + \alpha_{n0} = 1$ is more typical than the case $\alpha_{p0} + \alpha_{n0} > 1$. When the current is rising the sum of the current gains $\alpha_{p0} + \alpha_{n0}$ increases practically from 0 to values exceeding 1. An examination of the delay stage in the turn-on process involving one base shows that the linear approximation is in best agreement with the experimental data if $\alpha_{p0} + \alpha_{n0} = 1$ [5]. We may assume that this applies also to the turn-on process involving two bases, which is considered in the present paper.

Equations (9) and (10) include the critical current I_{cr}. In general, I_{cr} can be determined from

$$\alpha_{p0}(I_{cr}, U_0) + \alpha_{n0}(I_{cr}, U_0) = 1. \tag{11}$$

For low values of U_0, $I_{cr} \approx I_{off}$.

There is no point in calculating I(t) for $t > t_0$ because the behavior of the thyristor depends on the value of the current at $t = t_0$. In fact, the condition $I(t_0) < I_{cr}$ yields the inequality $\alpha_{p0} + \alpha_{n0} < 1$ and Eq. (2) cannot have a rising current solution.

An experimental investigation of the process of turning on a thyristor by a pulse of finite duration applied to one of the bases has confirmed that when the current reaches the value I_{cr} close to I_{off}, the thyristor is turned on even after the end of the gate pulse. At working voltages considerably lower than the turn-on voltage, the value of the critical current does not depend strongly on the voltage [5].

The conditions of Eqs. (9) and (10) cannot be used when the transient process is of such short duration that the proportionality of the current and the excess charge of the minority carriers in the bases is no longer obeyed.

We shall now consider a case of this kind.

Effect of an Ideal Rectangular Pulse on a Thyristor

We shall assume that a current I_C is in the form of a δ-shaped pulse so that $dU/dt \to \infty$ and $t_0 \to 0$. The capacitance of the central p−n junction loses a charge $Q = \int_0^{U_0} C_2(U)\,dU = \bar{C}_2 U_0$, where U_0 is the amplitude of the rectangular pulse, and $\bar{C}_2 = \frac{1}{U_0}\int_0^{U_0} C_2(U)\,dU$. If the injection efficiency of the outer p−n junctions are $\gamma_1 = \gamma_3 = 1$, the base regions receive compensating minority-carrier charges $Q_p = Q_n = \bar{C}_2 U_0$. In general, $Q_p = \gamma_1 \bar{C}_2 U_0$, $Q_n = \gamma_3 \bar{C}_2 U_0$. The thyristor is not turned on if $Q_p + Q_n = \overline{C}_2 \cdot U_0 (\gamma_1 + \gamma_3) < Q_{cr}$, where Q_{cr} is the "critical" charge, i.e., when the following condition is obeyed:

$$U_0 < \frac{Q_{cr}}{\overline{C}_2 (\gamma_1 + \gamma_3)} . \tag{12}$$

Reduction of the values γ by an external shunt increases the permissible value of U_0.

The "critical" charge, corresponding to the turn-off current, is given by the relationship

$$Q_{off} = I_{off} \quad [\gamma_1 (1 - \varkappa_p) \tau_p + \gamma_3 (1 - \varkappa_n) \tau_n], \tag{13}$$

where τ_p, τ_n are, respectively, the hole lifetime in the n-type base and the electron lifetime in the p-type base; $\varkappa_p$, $\varkappa_n$ are, respectively, the transport factors representing the flow of holes through the n-type base and the flow of electrons through the p-type base. At high voltages $U_0 \sim U_{br}$, the "critical" charge should be smaller than Q_{off}. However, the dependence of the "critical" charge on the voltage should not be strong at working voltages, which are always considerably lower than the breakdown voltage U_{br}. This is confirmed, in particular, by the weak dependence of the value of $I_{b.on}$ on the voltage across the thyristor.

Equations (8)-(12) describe correctly the experimental dependence reported in [1-3]. For example, shunting one of the outer junctions increases I_{off} and, consequently, I_{cr}; it also increases $I_{b.on}$. According to Eqs. (8)-(12), shunting increases the stability of the thyristor with respect to the "dU/dt effect." When the pulse amplitude is increased but the duration of the rise of the pulse front is kept constant (i.e., when dU/dt is increased), the value of $I_C - I_{b.on}$ increases and I_{cr} decreases somewhat, which favors the turn-on of the thyristor. The most stable thyristors should be those which have the lowest value of $(I_C - I_{b.on})/I_{cr}$, i.e., those which have the lowest capacitance but high values of the currents $I_{b.on}$ and I_{off}. This is in agreement with the experimental data reported in [2].

By way of example, we shall determine the highest permissible amplitude of an ideal rectangular pulse U_{0perm} which does not turn on a low-noise asymmetrical thyristor with the following parameters: $\overline{C}_2 = 50$ pF, $\tau_p = 4 \mu$sec, $I_{off} = 2$ mA, $\gamma_1 + \gamma_3 = 2$, $Q_{cr} \approx 0.7 \cdot I_{off} \cdot \tau_p = 5.5 \cdot 10^{-9}$ C. For such a thyristor, Eq. (12) gives $U_{0 perm} < 55$ V.

Conclusions

1. The influence of the capacitative current through the central p–n junction is equivalent to two equal gate currents flowing into both bases of the structure; these currents are given by $I_{bp} = I_{bn} = C_2(dU/dt)$.

2. The time dependence of the current through a thyristor is calculated for the case when the capacitative current is in the form of a rectangular pulse. The effect of an ideal rectangular voltage pulse on the thyristor is discussed.

3. The permissible value of dU/dt is found from the condition that, at the end of the capacitative current pulse, the current through the thyristor should be lower than the "critical" current or the charge in the space regions should be lower than the "critical" charge.

The value of the "critical" current is close to the turn-off current. The most stable are those thyristors which have low values of the ratio C_2/I_{off}.

The author is grateful to V. S. Pershenkov and V. I. Stafeev for discussing this paper.

Literature Cited

1. G. E. McDuffie Jr. and W. L. Chadwell, AIEE Trans., Pt I., Commun. Electron., 79:50 (1960).
2. V. A. Kuz'min and V. A. Brazhnikov, Radiotekhn. i Élektron., 8:1193 (1963).
3. C. D. Root, Proc. IEEE, 51:1672 (1963).
4. T. Misawa, J. Electron. Control, 7:523 (1959).
5. V. A. Kuz'min and V. S. Pershenkov, Abstracts of Papers Presented at Third All-Union Conf. on p−n Junctions, Tbilisi (1966).
6. A. A. Lebedev, Dissertation for Candidate's Degree, Physicotechnical Institute, Academy of Sciences of the USSR, Leningrad (1967).

RADIATIVE TRANSITIONS IN InAs LASER DIODES

N. S. Baryshev

A hydrogen-like acceptor model is used to calculate the rate of radiative transitions of electrons from the conduction band to acceptor levels in indium arsenide. This rate is compared with the rate of band–band radiative transitions. It is shown that the transitions from the conduction band to acceptor levels should predominate in InAs laser diodes, but in the case of strong excitation of a pure sample, the band–band transitions should be dominant.

It is interesting to compare the rate of the most important (band–band and band–acceptor) radiative transitions in order to determine the nature of electron processes in InAs laser diodes.

In the hydrogen-like approximation for the ground state of acceptor centers, the absorption coefficient for the acceptor-conduction band transitions is [1]

$$\alpha_{ac} = \frac{32\pi a_v^3 N_a (2m_n)^{\frac{3}{2}} K}{\left[1 + \frac{2m_n}{\hbar^2} a_v^2 (E_a + h\nu - E_g)\right]^4} \cdot \frac{(E_a + h\nu - E_g)^{\frac{1}{2}}}{h\nu},$$

where N_a and E_a are, respectively, the concentration and ionization energy of acceptors; a_v is the effective Bohr radius; ν is the radiation frequency; E_g is the energy gap; m_n is the effective mass of electrons in the conduction band. The quantity K is governed by the matrix element of the moment of the band–band transitions. It is very difficult to obtain an accurate theoretical estimate of this quantity. However, it can be found from the experimental data on the optical absorption involving direct band–band transitions, because

$$\alpha_{vc} = 2\sqrt{2}\, K \left(\mu_1^{\frac{3}{2}} + \mu_2^{\frac{3}{2}}\right) \frac{(h\nu - E_g)^{\frac{1}{2}}}{h\nu},$$

where μ_1 is the reduced effective mass of electrons and μ_2 is the reduced effective mass of heavy and light holes.

The forbidden band width (energy gap) of pure InAs at liquid-nitrogen temperature is approximately 0.41 eV. However, the width of the forbidden band of heavily doped semiconductors is less because of shifts of the band edges caused by the presence of impurities. According to Stern and Dixon [2], an impurity concentration of $5 \cdot 10^{18}$ cm^{-3} (which corresponds approximately to the case investigated in [3]) decreases the forbidden band of InAs by 0.05 eV. Assuming that $m_n \approx m_{p2} = 0.024 m_0$, $m_{p1} = 0.41 m_0$, $E_a \approx 0.01$ eV, $a_v = 3.1 \cdot 10^{-7}$ cm, $E_g \approx 0.36$ eV, $h\nu = 0.383$ eV [3] and bearing in mind that the experimental value of the absorption coefficient for the direct band−band transitions is $\alpha_{vc} \approx 2 \cdot 10^4 (h\nu - E_g)^{\frac{1}{2}}$ cm^{-1}, we obtain $\alpha_{ac} \approx 10^{-15} N_a$ cm^{-1}. Consequently, the effective cross section for the absorption of a photon, σ_f, is approximately 10^{-15} cm^2. Using the principle of detailed balancing, the radiative capture coefficient of electrons can be written in the form [4]

$$B_{ca} \approx 408 \left(\frac{m_0}{m_n}\right)^{\frac{3}{2}} \left(\frac{300}{T}\right)^{\frac{1}{2}} \varkappa^2 \sigma_f (h\nu)^2 \quad \text{cm}^3/\text{sec},$$

where $\varkappa$ is the refractive index and the photon energy is given in electron volts. An estimate gives $B_{ca} \approx 4 \cdot 10^{-10}$ cm^3/sec at liquid-nitrogen temperature. On the other hand, the radiative recombination coefficient for the band−band transitions is $B_{cv} = 3.6 \cdot 10^{-10}$ cm^3/sec [5], i.e., in this case $B_{cv} \approx B_{ca}$. The total rates of radiative transitions from the conduction band to acceptors and between the conduction and valence bands are, respectively,

$$R_{ca} = B_{ca} n N_a^0, \quad R_{cv} = B_{cv} np,$$

where n, p, N_a^0 are, respectively, the electron density, hole density, and concentration of neutral acceptors. Under normal experimental conditions (heavy doping and relatively low temperatures), the fraction of acceptors occupied by electrons is very small. In other words, in the p-type region near the junction, we have $N_a^0 > p$ because $R_{ca} > R_{cv}$.

These estimates confirm the hypothesis put forward in [3] that radiative transitions of electrons from the conduction band to acceptor levels predominate in InAs laser diodes. It is also clear that the band−band radiative transitions should predominate in the case of strong excitation of pure InAs.

Literature Cited

1. J. Callaway, J. Phys. Chem. Solids, 24:1063 (1963).
2. F. Stern and J. R. Dixon, J. Appl. Phys., 30:268 (1959).
3. N. S. Baryshev, T. L. Maslennikova, B. P. Pyregov, P. S. Matveeva, E. E. Vdovkina, B. B. Vdovkin, I. M. Nesmelova, Yu. A. Shuba, and I. S. Aver'yanov, Opt.-Mekh. Prom., No. 4, p. 8 (1967).
4. R. N. Hall, Proc. IEE, 106B(Suppl. 17):923 (1960).
5. N. S. Baryshev, Fiz. Tverd. Tela, 6:3027 (1964).

CHARACTERISTICS OF RADIATION EMITTED
DURING AVALANCHE AND TUNNEL BREAKDOWN
OF SILICON p—n JUNCTIONS

V. G. Mel'nik

An experimental investigation was made of the spectral characteristics of recombination radiation emitted during the avalanche and tunnel breakdown. of silicon p—n junctions in the temperature range from −165 to +115°C for currents of 50-400 mA. When the temperature was lowered, the intensity of the radiation increased and the maximum of the spectral characteristic shifted in the direction of higher energies. When the current was varied, the avalanche breakdown spectrum distorted in such a way that the intensity of the long-wavelength radiation increased more rapidly with increasing current, whereas in the tunnel case it was the intensity of the short-wavelength radiation that increased more rapidly. Data are given in graphical form and a qualitative explanation of the results is suggested.

The breakdown of p—n junctions is accompanied by the emission of light resulting from the recombination of hot carriers. The radiation emitted under avalanche breakdown conditions in grown silicon junctions was first observed by Newman et al. [1]. Chynoweth and McKay [2] investigated the spectral characteristics of grown and diffused silicon junctions under avalanche breakdown conditions. A qualitative description of the emission obtained under tunnel breakdown conditions is reported by Chynoweth and McKay [3]; Migitaka [4-5] determined the recombination radiation spectra in the tunnel breakdown case. He also described the avalanche breakdown spectra, which differed from those reported by Chynoweth and McKay [2]. The difference was this: Chynoweth and McKay [2] reported that the radiation intensity increased monotonically with decreasing frequency, whereas Migitaka [4-5] found a sharp maximum at the radiation energy equal to the forbidden band width of silicon. According to Migitaka [4-5], this maximum should be observed for any breakdown mechanism.

The present paper reports an additional investigation of the radiation spectra, obtained under the avalanche and tunnel breakdown conditions, of silicon p—n junctions, as well as a study of the temperature and current dependences of these spectra.

The junctions were prepared by the diffusion and melting technique. These junctions were convenient because they did not exhibit surface breakdown and the breakdown region was not covered by the contacts. Boron was diffused into an n-type plate and a wide guard-ring

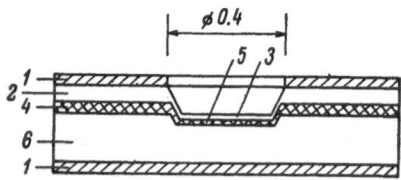

Fig. 1. Structure of a p—n junction:
1) ohmic Ni + Au contacts; 2) diffused p-type layer; 3) recrystallized
p-type layer; 4) wide diffused p—n
junction; 5) narrow alloyed p—n junction; 6) original n-type silicon.

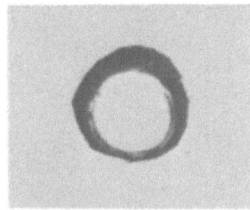

Fig. 2. Emission pattern
of a tunnel p—n junction
(the white luminous spot
in the center is surrounded
by the guard-ring area).

type diffused junction was produced. A second, working, junction with a low breakdown voltage
was produced by melting the diffused layer right across the n-type region, using an aluminum
globule. This globule was removed in hydrochloric acid and the recrystallized layer, through
which the investigated radiation was transmitted, was thinned in the following mixture: HF +
HNO_3 + CH_3COOH (2:9:4). Avalanche-type junctions were prepared from silicon with ρ = 0.09
$\Omega \cdot$ cm; tunnel junctions were prepared from silicon with ρ = 0.01 $\Omega \cdot$ cm. The breakdown
voltage of the avalanche junctions was 14-15.5 V and the tunnel junctions broke down at 3-3.2 V
(the current was 10 mA). Figure 1 shows the structure of a junction ready for measurement
and Fig. 2 is a photograph of the emission pattern of one of the tunnel junctions.

The spectra were investigated using a UM-2 monochromator whose output was detected
with a photomultiplier having a cathode of the S-1 type. The investigated sample was supplied
with an alternating current of 20 mA amplitude and 120 cps frequency, superimposed on a dc
bias of 100 mA. The junction was placed in a copper holder, which was located directly in front
of the entry slit of the monochromator (no intermediate optical system was used). The dependences of the spectra on the current were determined at three points in each spectrum. For
each point, we plotted the dependence of the output signal of the photomultiplier on the current
through the junction (in relative units). This dependence was obtained using a dc bias, which
determined the current through the sample, and superimposing an alternating current of 10 mA
amplitude (the recorded signal was due to this alternating current).

A large number of the avalanche and tunnel junctions was investigated. The observed
radiation differed from sample to sample even when the breakdown mechanism was the same.
The color of the radiation ranged from bluish-yellow to red. These variations were associated
primarily with the thickness of the recrystallized p-type layer. In the tunnel junctions, the
radiation filled uniformly the whole region of the low-voltage breakdown region but in the
avalanche junctions, the low-voltage region emitted radiation at separate spots. The samples
emitting the strongest radiation were selected for measurements and no correction was made
for the absorption in the p-type layer (it was estimated that the thickness of this layer did not
exceed 0.5-1 μ).* The reflection coefficient was also ignored because it was practically constant throughout the investigated range of wavelengths. In spite of their external differences,
all the tunnel junctions had almost identical spectra (the same was true of the avalanche junctions). Figure 3 shows typical spectra of the tunnel diodes and Fig. 4 gives the corresponding

---

* The thickness of the p-type layer was estimated (after investigation of the spectra) by dis-
  solving the junction in an etchant with a known etching rate until the p-type layer was completely removed (this was deduced from the disappearance of the radiation).

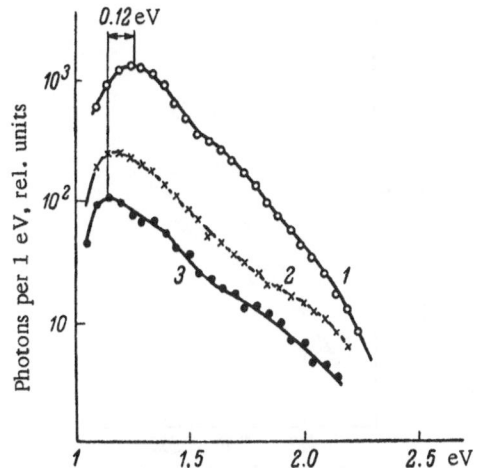

Fig. 3. Intensity of radiation as a function of the photon energy under tunnel breakdown conditions. Temperature, °C: 1) −165; 2) 40; 3) 115.

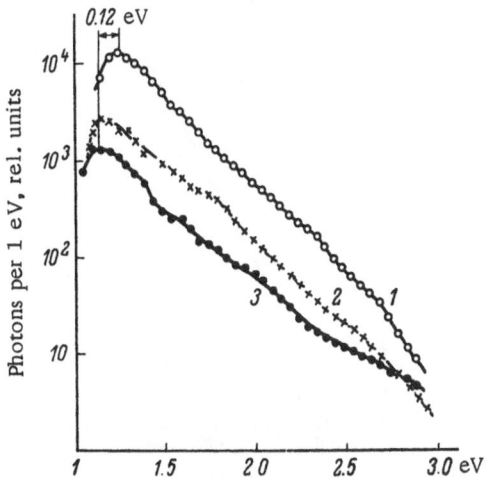

Fig. 4. Intensity of radiation as a function of the photon energy under avalanche breakdown conditions. Temperature, °C: 1) −165; 2) 40; 3) 115.

spectra for the avalanche diodes. The temperatures at which these spectra were obtained are given in the captions. The main features of all the spectra, which were independent of the breakdown mechanism and of the temperature, were very similar. When the temperature was lowered, the intensity of the radiation increased: this increase was tenfold between +115 and −165°C. The curves obtained at three different temperatures tended to approach each other at high energies. When the temperature was lowered, the maxima of the spectra shifted in the direction of higher energies.

The shift toward higher energies was 0.12 eV per 300 deg C and was independent of the breakdown mechanism.

Figures 5 and 6 show the dependences of the radiation intensity on the current passing through the junction. The behavior of the tunnel and avalanche junctions was quite different. In the tunnel breakdown case, the intensity of the radiation increased with the current more rapidly at high than at low photon energies; the reverse was true of the avalanche junctions. When the current was increased, the radiation emitted by the tunnel junctions became "whiter," whereas the radiation emitted in the avalanche breakdown became "redder." Visual observations of the investigated junctions confirmed this conclusion. In the tunnel breakdown case, the relationship between the current and intensity was rigorously linear at each of the investigated points in the spectrum but in the avalanche case the characteristics were in the form of broken lines.

These results confirmed the presence of a maximum in the recombination radiation spectra of hot carriers. There was no basic difference between the spectra observed under tunnel and avalanche breakdown conditions. The curves in Figs. 3 and 4 resembled the recombination radiation spectra under quasi-equilibrium conditions (van Roosbroeck−Shockley curves [6]) with the exception that the parts of the curves to the right of the maxima had less steep slopes and, moreover, these slopes varied with the breakdown voltage. The van Roosbroeck− Shockley curves were plotted for all three temperatures used in this investigation. The theoretical curves were in good agreement with the experimental data near the maxima in the radiation spectra.

At energies exceeding the spectral maximum by about 0.15 eV, the theoretical and experimental spectra began to differ and the difference increased rapidly with increasing photon

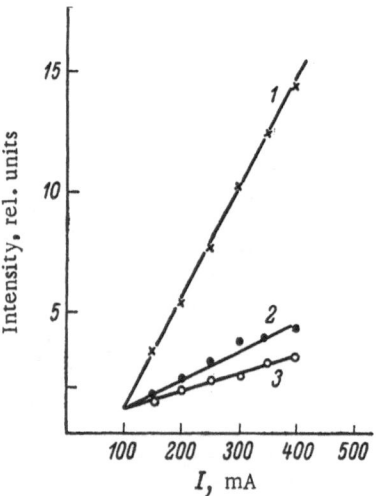

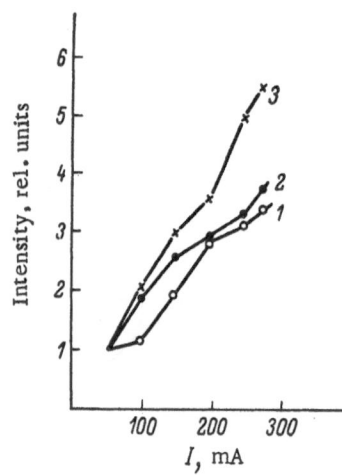

Fig. 5. Dependence of the relative radiation intensity on the current through the junction at three points in the spectrum under tunnel breakdown conditions. $h\nu$ (eV): 1) 1.58; 2) 1.26; 3) 1.12.

Fig. 6. Dependence of the relative radiation intensity on the current through the junction at three points in the spectrum under avalanche breakdown conditions. $h\nu$ (eV): 1) 1.58; 2) 1.26; 3) 1.12.

energy. This indicated that the recombination radiation maximum of hot carriers corresponded to indirect transitions between extrema in the conduction and valence bands. This was supported also by the value of the radiation energy at the spectral maximum. The discrepancy between the experimental and theoretical curves at high energies indicated that the thermodynamic assumptions made in [6] were invalid at these energies and that the exact quantum-mechanical problem should be solved. However, the solution of this problem is impossible because of the absence of data on the analytic relationship between the energy and the crystal momentum in the conduction and valence bands. Only the most energetic photons, observed under the avalanche breakdown conditions, can be deduced from Wolff's analysis of germanium [7] if it is assumed that these photons correspond to direct transitions near the point k = 0.

The shift of the energy of the radiation maximum with temperature is in good agreement with the temperature dependence of the forbidden band width of silicon. It is reported in [8] that $(\partial \Delta E / \partial t)_P = 4.1 \cdot 10^{-4}$ eV/deg, which gives 0.12 eV for a temperature difference of 300 deg C, i.e., the value obtained experimentally.

The intensity of the indirect transitions is strongly dependent on the concentration of phonons and, therefore, cooling should have reduced the radiation intensity. The experimentally observed increase in the intensity at lower temperatures may be due to exciton recombination. It is reported in [9] that, at liquid-nitrogen temperatures, the recombination radiation of silicon is mainly due to the dissociation of excitons.

As reported, the current dependences of the radiation intensity were quite different for the tunnel and avalanche breakdown mechanisms. Special experiments were carried out and calculations were made of the influence of the Planck radiation due to the Joule heating of the junction on the current-intensity characteristics. This influence was found to be negligible in the range of currents investigated. The dependence of the spectral profile on the current flowing through the tunnel junctions could be due to the "softness" of the tunnel current—voltage characteristics. An increase in the current through a tunnel junction increased the field, i.e., it produced additional heating of the electron gas. This was responsible for the spectrum becoming whiter. In the avalanche junctions the field hardly increased with increasing current and the change in the spectrum could be attributed to a change from a purely avalanche mechanism to a mixed thermal-electric breakdown. When the temperature of the junction was increased, impaction ionization was produced not only by carriers of energy greater than 2.2 eV but also by carriers of lower energies. This increased the number of ionization and recombination events in the low-energy range. The reddening of the spectrum with increasing current intensity was predicted in [10]; an analysis of this effect was given in that paper.

It was interesting to note that the spectral curves recorded at different temperatures approach each other at high photon energies. Assuming that the high-energy photons are due to vertical transitions near the point k = 0, this approach of the curves can be interpreted as a change of the dependence of the energy on the crystal momentum due to a change in the temperature.

Literature Cited

1. R. Newman, W. C. Dash, R. H. Hall, and W. E. Burch, Phys. Rev., 98:1536 (1955).
2. A. G. Chynoweth and K. G. McKay, Phys. Rev., 102:369 (1956).
3. A. G. Chynoweth and K. G. McKay, Phys. Rev., 106:418 (1957).
4-5. M. Migitaka, Solid State Electron., 8:295 (1965).
6. W. van Roosbroeck and W. Shockley, Phys. Rev., 94:1558 (1954).
7. P. A. Wolff, J. Phys. Chem. Solids, 16:184 (1960).
8. F. J. Morin and J. P. Maita, Phys. Rev., 96:28 (1954).
9. L. W. Davies, Phys. Rev. Letters, 4:11 (1960).
10. Ya. K. Balodis and T. Ya. Puritis, Izv. Akad. Nauk LatvSSR, Ser. Fiz. i Tekhn. Nauk, No. 4, p. 40 (1966).

DEPENDENCE OF THE BREAKDOWN VOLTAGE OF SILICON p—n JUNCTIONS ON THE SURFACE POTENTIAL

E. V. Ostroumova

A. F. Ioffe Physicotechnical Institute
Academy of Sciences of the USSR, Leningrad

A brief report is given of the results of an experimental investigation of the dependence of the breakdown voltage (U_{br}) of silicon p—n junctions on the surface potential (ψ_s). It was established that depletion of the majority carriers in the surface layer of a high-resistivity region of a diode increased the breakdown voltage. The dependence $U_{br} = f(\psi_s)$ was determined for silicon p—n junctions.

It is predicted theoretically in [1] that the breakdown voltage of p—n junctions should depend on the state of the surface of a high-resistivity region. Some results on silicon p—n junctions are reported in [2]: it was found experimentally that the depletion of the majority carriers in a surface layer did indeed raise the breakdown voltage. The main difficulty in the experimental investigation of the correlation between the surface potential and the breakdown voltage of p—n junctions was the selection of a method for measuring the surface potential.

In particular, the method employed in [2] was suitable only for high-resistivity materials and did not give reliable results in the case when inversion layers were formed. This disadvantage was particularly important because the greatest rise in the breakdown voltage occurred in such cases. In the present investigation, the surface potential was measured using a method based on the determination of the surface photo-emf [3]. The advantages of this method included the possibility of quite accurate measurement of the surface potential in the case of the formation of pronounced inversion layers and the possibility of measuring the potential of surfaces coated with dielectric films.

The surface potential of a semiconductor differs from the volume potential by an amount $\Psi_s = \frac{kT}{e} \ln \frac{n_0}{n_s} = \frac{kT}{e} \ln \frac{p_s}{p_0}$, where n_0, p_0, and n_s, p_s are, respectively, the equilibrium carrier densities in the interior and near the surface. When a semiconductor is illuminated, the carrier density increases. If the semiconductor is sufficiently pure, the nonequilibrium carrier densities are equal: $\Delta n = \Delta p$. An increase in the nonequilibrium carrier density reduces the dif-

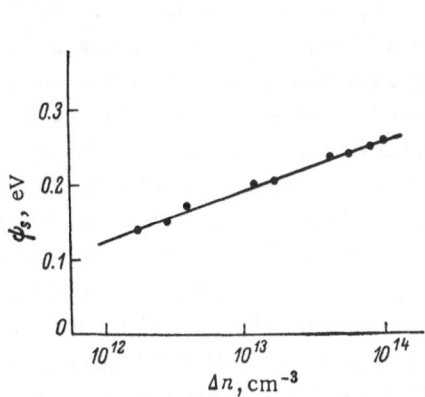

Fig. 1. Dependence of the surface photo-emf on the density of non-equilibrium carriers generated by illumination.

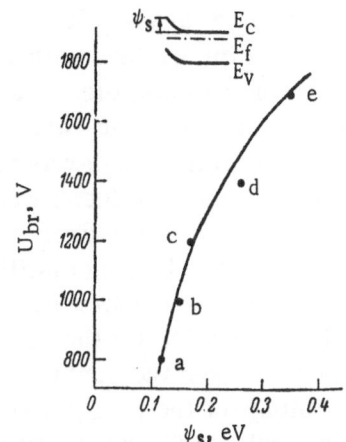

Fig. 2. Dependence of the breakdown voltage on the surface potential after various treatments: a) KO-97, alcohol; b) H_2O; c) toluene; d) KO-97 + I_2; e) etching and drying.

ference between the surface and volume potentials. A sufficiently strong illumination of the surface of a semiconductor with identical electron and hole mobilities should destroy completely the difference between the volume and surface potentials. Thus, in this case, the surface photo-emf is equal to ψ_s. In the case of a semiconductor with unequal electron and hole mobilities, it is necessary to make a correction for the Dember emf [4]. The value of Δn, necessary for the calculation of such a correction, can be found by measuring the photoconductivity due to the illumination of a sample. In practice, it is also necessary to correct ψ_s for the fact that the illumination intensity is sufficient for complete straightening of the energy bands near the surface. In the present paper, this correction was calculated using the results reported in [3]. In the majority of the experiments reported below, the nonequilibrium carrier density was never less than 10 times the equilibrium carrier density and the total values of the corrections were close to kT/e at room temperature (the experiments were carried out at this temperature).

Figure 1 shows the dependence of the surface photo-emf on the intensity of illumination, represented by the nonequilibrium carrier density. The slope of the straight line drawn through the experimental points corresponds to an energy of 0.029 eV, which is very close to kT at room temperature.

The experiments were carried out on n-type single crystals of 10-80 $\Omega \cdot$ cm resistivity. The diffusion length was $L_D \approx 0.3$-0.4 mm. The thickness of a sample was always greater than L_D. The change in the surface potential due to illumination was deduced from the change in the charge on the plates of a capacitor: one of these plates was the sample itself and the other a transparent layer of tin dioxide evaporated on a mica spacer. The surface of the sample was illuminated with single pulses of $\sim$20 msec rise time or with light modulated at 190 cps.

The breakdown voltage of samples subjected to different surface treatments was measured using silicon p−n junctions prepared by the diffusion of boron into n-type silicon of 80 $\Omega \cdot$ cm resistivity.*

* The author is grateful to O. V. Golovanova for kindly supplying these junctions.

The surface potential was altered by chemical treatments (etching in acid and alkali etchants, interaction of the surface with vapors of ethyl alcohol, water, toluene, etc.), by deposition of organosilicon enamels, etc. The breakdown voltage of the p−n junctions was measured using samples subjected to these various treatments. The experimental results are presented in Fig. 2. The breakdown voltage was found to be very sensitive to a change in the surface potential. Alcohol, water or toluene vapor reduced appreciably the surface potential and this reduction was accompanied by a drop in the breakdown voltage. A sample etched in an alkali or acid etchant, washed in boiling distilled water, and dried in filters, had a value of ψ_s amounting to 0.30-0.35 eV; treatment in water vapor reduced ψ_s to 0.10-0.15 eV and the breakdown voltage from 1700 to 800 V. On the other hand, exposure of the n-type samples to vapors of oxidizing agents increased the surface potential and the breakdown voltage. Thus, for example, the breakdown voltage of the p−n junctions coated with a thin layer (∼0.01-0.03 mm) of an organosilicon enamel KO-97 increased from 800 to 1400 V when the coating was treated in iodine vapor. Measurements of the photo-emf indicated that such treatment increased ψ_s by 0.15 eV. When the iodine evaporated completely, the quantities of U_{br} and ψ_s recovered their original values.

Thus, the experimentally observed correlation between the surface potential of the breakdown voltage of silicon diodes was found to be in agreement with theoretical predictions [1].

The author is grateful to V. E. Chelnokov for his interest in this investigation.

Literature Cited

1.   C. G. B. Garrett and W. H. Brattain, J. Appl. Phys., 27:299 (1956).
2.   T. M. Buck and F. S. McKim, J. Electrochem. Soc., 105:709 (1958).
3.   E. O. Johnson, Phys. Rev., 111:153 (1958).
4.   W. H. Brattain and J. Bardeen, Bell System Tech. J., 32:1 (1953).

SOME FEATURES OF THE TEMPERATURE DEPENDENCE
OF THE STATIC TURN-ON CURRENT OF THYRISTORS

A. N. Dumanevich, R. E. Smolyanskii, and V. E. Chelnokov

It is shown that variation of the temperature of the ambient medium can produce, in addition to a fall in the static turn-on current of a thyristor, a more complex behavior of this current: an initial descent to some minimum may be followed by a rise and another fall. This effect is atrributed to the high level of carrier multiplication in the collector p−n junction of the thyristor at a temperature corresponding to the turn-on current minimum. It is also due to the different temperature dependences of the current gains of the two transistors (which together form the equivalent circuit of a thyristor) and of the multiplication coefficients of the collector p−n junction.

It is demonstrated in [1] that, when the temperature of the ambient medium is raised, the static parameters of a thyristor (turn-on voltage and current) may change. It has been predicted that, when the temperature of the ambient medium is increased, the turn-on voltage of a thyristor may decrease or increase or it may increase at first and then decrease.

It is also shown in [1] that the turn-on current of a thyristor decreases with increasing ambient temperature because the product $L(I) M(V_c)$ increases with rising temperature, i.e., because of the condition

$$\frac{\partial}{\partial T} [L(I) M(V_c)] > 0, \tag{1}$$

where $L(I)$ is the sum of the current gains of the two transistors which together form the equivalent circuit of a thyristor (the sum of the gains is a function of the current I through the thyristor); $M(V_c)$ is the coefficient of multiplication of the minority carriers in the exhaustion layer of the collector p−n junction.

It is reported in [2] that the turn-on current obeys no regular temperature dependence. The temperature dependence of the turn-on current has also been investigated experimentally by Grekhov et al. [3] but the correct explanation of this dependence has not been given.

It should be mentioned that, in the presence of strong multiplication of the minority carriers in the collector p−n junction, the condition (1) is not satisfied throughout the whole range of temperatures in which the turn-on effect is observed in a thyristor. This is because the two factors in Eq. (1) have different temperature dependences.

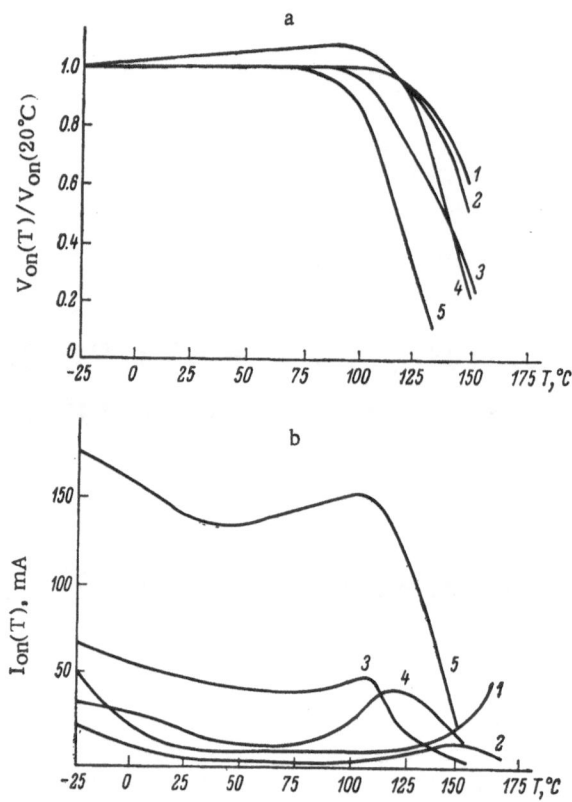

Fig. 1. Experimentally determined temperature
dependences of the turn-on voltage (a) and turn-
on current (b) of different silicon thyristors: 1, 2)
$\rho = 20\ \Omega \cdot$ cm; 3) $\rho = 32\ \Omega \cdot$ cm; 4, 5) $\rho = 80\ \Omega \cdot$ cm.

In particular, the current gain L(I) always increases with increasing ambient temperature (for a given value of I) because of the weakening of the recombination in the space-charge
layers of the emitter p—n junctions [4], and because of the saturation of the recombination centers (traps) in the bases of a thyristor. Thus, we always have $\frac{\partial}{\partial T}[L(I)] > 0$.

The multiplication coefficient $M(V_c)$ decreases with increasing ambient temperature (for
a constant value of V_c) [5]. Moreover, the turn-on voltage of a thyristor may decrease with
increasing temperature, and, consequently, $M(V_c)$ may decrease because V_c falls as the temperature T rises. Consequently, we always have $\frac{\partial}{\partial T}[M(V_c)] < 0$.

Figure 1 shows the experimentally obtained temperature dependences of the turn-on
voltage (a) and turn-on current (b) of T-300 silicon thyristors, recorded in the temperature
range between −25 and +150°C. The turn-on voltage can be seen to be practically independent
of the temperature up to 100-125°C, except for a slight rise in some cases. On the other hand,
the turn-on current first decreases, passes through a minimum, rises to a maximum (at this
temperature, the turn-on voltage begins to decrease) and, finally, the turn-on current as well
as the turn-on voltage decrease with increasing temperature.

This anomalous behavior of the turn-on current can be understood by considering the
temperature dependences of L(I) and $M(V_c)$. We shall consider these dependences in more
detail.

The expression for the turn-on current of the thyristor will be written in its general form and, for simplicity, the electron and hole multiplication coefficients of the exhaustion layer of the collector p−n junction will be assumed equal. Then,

$$I_{on} = \frac{1 - L(I) M(V_c)}{\frac{\partial}{\partial I}[L(I) M(V_c)]|_{I=I_{on}}}, \tag{2}$$

where $M(V_c) = M_p(V_c) = M_n(V_c)$.

We shall seek an extremum of I_{on} in its dependence on T. Using the condition for an extremum $\frac{\partial I_{on}}{\partial T} = 0$, we find from Eq. (2) that

$$\frac{\partial}{\partial T}[L(I) M(V_c)]|_{I=I_{on}} + I_{on}\frac{\partial}{\partial T}\left\{\frac{\partial}{\partial I_{on}}[L(I) M(V_c)]\right\} = 0,$$

and hence

$$\frac{\partial}{\partial T}[L(I) M(V_c)]|_{I=I_{on}} + \frac{\partial}{\partial T}\left\{I_{on}\frac{\partial}{\partial I_{on}}[L(I) M(V_c)]\right\} - \frac{\partial}{\partial I_{on}}[L(I) M(V_c)]\frac{\partial I_{on}}{\partial T} = 0,$$

or

$$\frac{\partial}{\partial T}\left\{I_{on}\frac{\partial}{\partial I_{on}}[L(I) M(V_c)]\right\} = 0. \tag{3}$$

Since $I_{on}\frac{\partial}{\partial I_{on}}[L(I) M(V_c)] = 1 - L(I) M(V_c)$, which follows from Eq. (2), we find that Eq. (3) is equivalent to

$$\frac{\partial}{\partial T}[1 - L(I) M(V_c)] = 0,$$

and hence

$$\frac{\partial}{\partial T}[L(I) M(V_c)] = 0. \tag{4}$$

This result shows that extremal points in the temperature dependence of the turn-on current coincide with extrema of the effective total current gain of a thyristor at the moment of turn-on. This result allows us to explain the behavior of the turn-on current.

In principle, the temperature dependence of the turn-on current may consist of four regions (Fig. 2).

Region 1. In this region, the effective current gain coefficient $L(I)M(V_c)$ increases with increasing temperature, i.e., $\frac{\partial}{\partial T}[L(I)M(V_c)] > 0$, which is equivalent to the condition*

$$\left|M(V_c)\frac{\partial}{\partial T}L(I)\right| > \left|L(I)\frac{\partial}{\partial T}M(V_c)\right|. \tag{5}$$

In this region the turn-on current decreases with increasing temperature because the rate of rise of $L(I)M(V_c)$ with the current in Eq. (2) increases and the values of $L(I)$ and $M(V_c)$ both

---

*We must mention that we always have $M(V_c)\frac{\partial}{\partial T}L(I) > 0$, and $L(I)\frac{\partial}{\partial T}M(V_c) < 0$, because $\frac{\partial}{\partial T}L(I) > 0$, $\frac{\partial}{\partial T}M(V_c) < 0$ are always true and $M(V_c)$ and $L(I)$ are positive quantities.

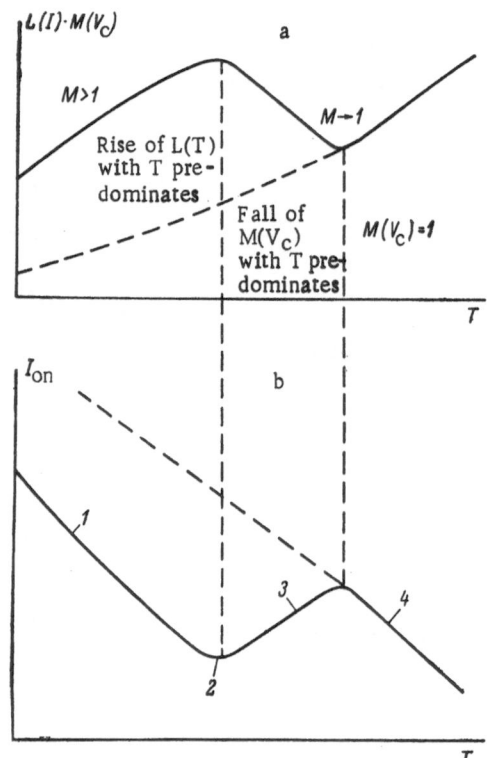

Fig. 2. Temperature dependences of the effective current gain (a) and turn-on current (b). The dashed curves represent the behavior of the two quantities when there is no carrier multiplication in the collector p−n junction of the thyristor.

increase. A characteristic feature of this region is a strong carrier multiplication in the collector p−n junction.

Region 2. The turn-on current reaches its minimum. Consequently, in accordance with Eq. (4), the following condition is satisfied:

$$\left| M\left(V_c\right)\frac{\partial}{\partial T}L\left(I\right)\right| = \left| L\left(I\right)\frac{\partial}{\partial T}M\left(V_c\right)\right|. \qquad (6)$$

The condition (6) shows that in this region the rate of change of L(I) with T (increase with increasing T) is balanced by the rate of change of $M(V_c)$ with T (decrease with increasing T because V_c decreases with increasing temperature). When the condition (6) is satisfied, the value of L(I) $M(V_c)$ and the rate of rise of this quantity with I both reach their maxima and, therefore, the turn-on current has its minimum value.

Region 3. Here, the effective current gain begins to decrease with increasing temperature because carrier multiplication in the collector p−n junction decreases rapidly and this decrease predominates over the increase of L(I) with T, i.e., $\frac{\partial}{\partial T}[L(I)\ M(V_c)] < 0$.

The following condition is satisfied in this region:

$$\left| M\left(V_c\right)\frac{\partial}{\partial T}L\left(I\right)\right| < \left| L\left(I\right)\frac{\partial}{\partial T}M\left(V_c\right)\right|. \qquad (7)$$

The rise of the turn-on current can be easily understood by considering a reduction in the "deformation" of the bell-shaped curve described in [6]: when $M(V_c)$ decreases, the current must increase in order to maintain the continuity of the current flow. At the end of this region, the multiplication coefficient M is very close to 1 and at this point the turn-on current reaches its maximum value.

Region 4. In this region, the effect current gain begins to rise with increasing temperature solely because L(I) increases with temperature. There is no carrier multiplication in this region, i.e., $M(V_c) = 1$. The drop in I_{on} is due to the same reasons as the drop in region 1, except that now $M(V_c) = 1$ and the condition for the existence of region 4 is the inequality

$$\frac{\partial}{\partial T}L(I) > 0.$$

Between regions 3 and 4 there is a temperature at which I_{on} reaches a maximum. At this point and at the minimum of the turn-on current, the condition (6) is satisfied; however, at the maximum, $M(V_c)$ slightly exceeds 1 and $\frac{\partial}{\partial T}M(V_c)$ $M(V_c)$ is very close to 0.

The temperature dependence of the static turn-on current considered here should be observed for thyristors with shunted emitter p−n junctions. Then, the turn-on at the lowest temperature (in the investigated range) takes place at voltages equal or close to the breakdown voltage of the collector p−n junction (this is accompanied by a considerable multiplication of the minority carriers in this junction).

In other cases, when M slightly exceeds 1, there are no maxima or minima in the temperature dependence of the turn on current: I_{on} simply decreases monotonically with increasing T, i.e., $\frac{\partial I_{on}}{\partial T} < 0$.

Literature Cited

1. R. E. Smolyanskii, Radiotekhn. i Élektron., 8:1615 (1963).
2. P. S. Raderecht and C. A. Hogarth, J. Electron. Control, 17:145 (1964).
3. I. V. Grekhov et al., in: Electric Drives with Semiconductor Control, Énergiya, Moscow (1964).
4. C. T. Sah, R. N. Noyce, and W. Shockley, Proc. IRE, 45:1228 (1957).
5. K. G. McKay, Phys. Rev., 94:877 (1954).
6. R. E. Smolyanskii, in: Physics of p−n Junctions, Zinatne, Riga (1966).

TUNNEL p—n JUNCTIONS IN INDIUM PHOSPHIDE

A. N. Imenkov, N. V. Siukaev, and M. K. Khadikov

A. F. Ioffe Physicotechnical Institute
Academy of Sciences of the USSR, Leningrad
K. L. Khetagurov Severo-Osetinsk State Pedagogical Institute, Ordzhonikidze

The current—voltage characteristics of indium phosphide tunnel and backward diodes were obtained and analyzed. The tunnel diodes with a negative resistance region were prepared by alloying pure tin, a mixture of tin and tellurium, or pure indium to zinc-doped InP. The backward diodes were prepared by alloying tin to cadmium-doped InP. The tunnel current density in InP diodes was found to be higher, at a given doping level, than in GaAs diodes.

Many parameters of InP and GaAs are similar: they include the band structure, forbidden band width, effective masses of electrons and holes, etc.

Investigations of InP and of the p—n junctions in this compound have been held up by the fact that the technology of the preparation of InP is not yet sufficiently developed. However, semiconducting devices made of InP may, in some respects, be superior to those made of GaAs.

For example, it is expected that InP tunnel diodes can carry higher current densities than GaAs tunnel diodes, at the same doping levels.

Observations of the tunnel effect in InP p—n junctions have been reported in [1, 2], but these papers cite very low densities of the tunnel current at low forward voltages (up to 0.3 V), at which direct interband tunnel transitions are possible. This is obviously due to the weak doping of the n- and p-type regions.

The present paper reports the preparation and a study of the electrical properties of tunnel and backward diodes made of InP.

The tunnel p—n junctions were prepared by alloying globules (0.2-0.7 mm in diameter) of tin mixed with tellurium to InP single crystals doped with zinc to give a hole density $p = 1 \cdot 10^{19}$ cm^{-3} (the hole mobility was $\mu_p = 43$ cm$^2 \cdot$ V$^{-1} \cdot$ sec^{-1} at room temperature).

The backward diodes were prepared by alloying tin globules to InP doped with cadmium to give a hole density $p = 1 \cdot 10^{18}$ cm^{-3} and a mobility $\mu_p = 60$ cm$^2 \cdot$ V$^{-1} \cdot$ sec^{-1} at room temperature. Ohmic contacts were made by alloying with pure indium.

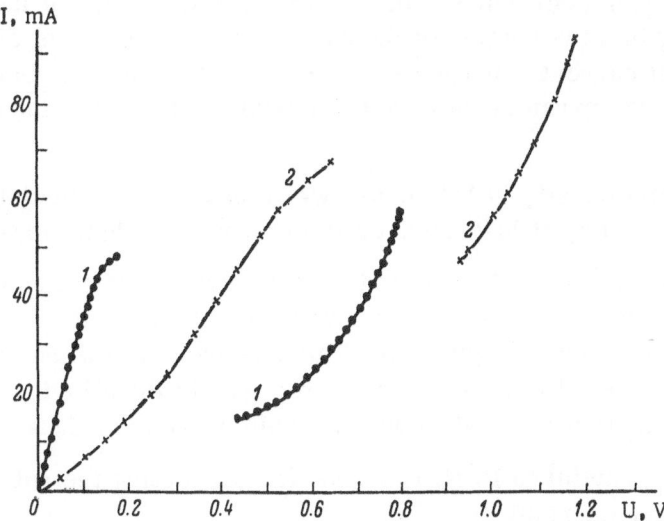

Fig. 1. Forward branches of the current—voltage characteristics of a tunnel diode prepared by alloying tin with an admixture of tellurium (1) and of a diode prepared by alloying pure indium (2).

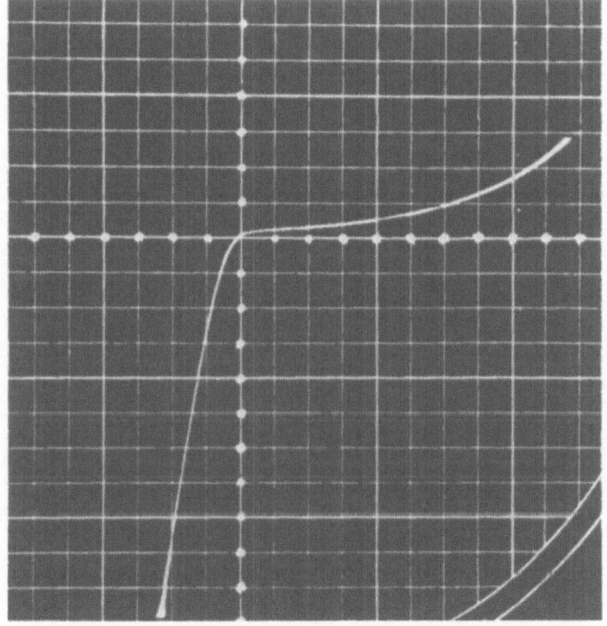

Fig. 2. Oscillogram of the current—voltage characteristic of a backward diode prepared by alloying tin to cadmium-doped indium phosphide. One division along the horizontal scale corresponds to 0.1 V and one vertical division corresponds to 10 mA.

Curve 1 in Fig. 1 is a current−voltage characteristic of one of the tunnel p−n junctions. It is worth mentioning that the scatter of the parameters of the tunnel diodes was relatively small: the peak tunnel current densities were 30-50 A/cm^2 and the peak-to-valley current ratio was $I_p/I_v \approx$ 3-4. Similar parameters have been reported for GaAs tunnel diodes with p = $4 \cdot 10^{19}$ cm^{-3}.

The tunnel current density in InP diodes was close to that calculated for direct interband transitions. The parameters of InP required in the various calculations were taken from [3].

The backward diodes (Fig. 2) exhibited a fairly high rectification coefficient (10 at U = 0.1 V and 30 at U = 0.2 V). An indium contact was found to be nonohmic. The alloying of indium to InP produced a tunnel p−n junction. This was observed when small globules of indium (considerably smaller than tin globules) were used in the alloying process. The resistance of indium contacts was found to be negligible because of their large areas.

The authors are grateful to D. N. Nasledov for suggesting the subject of this investigation and for his constant interest.

Literature Cited

1.   N. Holonyak Jr., J. Appl. Phys., 32:130 (1961).
2.   J. Shewchun and R. M. Williams, Phys. Rev. Letters, 15(4):160 (1965).
3.   O. Madelung, Physics of III−V Compounds, Wiley, New York (1964).

SOME CHARACTERISTICS OF THE CAPACITANCE OF
n'—p—n STRUCTURES WITH "DEEP" LEVELS

L. L. Makovskii, S. M. Ryvkin, N. B. Strokan,
V. P. Subasheva, and A. Kh. Khusainov

A. F. Ioffe Physicotechnical Institute
Academy of Sciences of the USSR, Leningrad

The temperature and bias dependences of the capacitance of n'—n—p structures made of germanium compensated by γ-ray generated radiation defects are explained using an equivalent circuit. It is demonstrated that the carrier energy distribution in the compensated region is of the Boltzmann type. It is shown that restrictions on the spreading of the electric field in the central region are lifted for levels with "long" times of transition to the conduction band.

There is currently much interest in n—i—p germanium and silicon structures because of their use as nuclear radiation counters. It has been reported that structures compensated by deep levels have certain advantages [3]. However, it is shown in [1, 4] that, when deep levels are used for compensation, the space-charge region of a p—n junction may be governed by the initial concentration of donors N_D^+ and not by the difference $N_D^+ - N_A^-$.*

This situation arises when the Fermi level intersects the level responsible for the compensation, which divides the space-charge region into two parts (Fig. 1a). Practically the whole voltage V applied to a p—n junction is concentrated in the first region (a), whereas in the second region (b) the voltage drop is equal to the difference $E_{F_n} - E_A$, irrespective of the value of V and of the degree of compensation [1].

The present paper proves this hypothesis, put forward in [1], and describes a study of the influence of the intersection of levels on the bias and temperature dependences of the capacitance of an n'—n—p germanium structure with a system of radiation defect levels generated by γ-ray bombardment.

---

* To be specific, we shall assume that before compensation the conductivity of the central region is n-type.

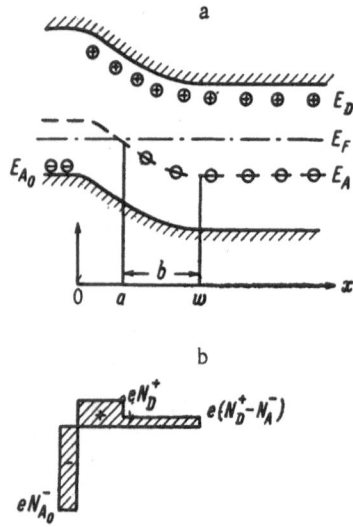

Fig. 1. a) Energy level scheme of a p−n junction in equilibrium in the case of compensation of "deep" levels. b) Assumed distribution of space charge.

Distribution of the Potential

Since, in our case, the degree of compensation is high and, therefore, the electron density is low, it would seem that under nonequilibrium conditions the Boltzmann distribution of carriers may be disturbed even at relatively low current densities. For the sake of simplicity, we shall assume that the p-type region is heavily doped (i.e., in the limiting case this region can be a metal). We shall also postulate that, at the point of intersection with the Fermi level, the population of the acceptor level drops suddenly from unity to zero and that the space-charge distribution can be represented by a step function.

The expression for the electron density in the space-charge region $n(x)$ which satisfies the boundary conditions $n(0) = \text{const}$ (at the boundary with the metal) and $n(w) = n_0$ (where $w = a + b$) is of the following form when $\varphi = 0$ in the bulk of the sample:

$$n(x) = n_0 e^{\frac{e(\varphi + V)}{kT}} - \frac{e j_n e^{\frac{e\varphi}{kT}}}{\mu_n kT} \int_0^x e^{-\frac{e\varphi}{kT}} dx. \qquad (1)$$

Under equilibrium conditions, the potential drop across the region b is $E_F - E_A$ (Fig. 1), i.e., $n(a) = n_0 \exp\left[-\dfrac{e(E_F - E_A)}{kT}\right] = n_0 \exp\dfrac{e\varphi_a}{kT}$. We shall now consider how this expression is affected by the application of a bias.

It follows from Eq. (1) that*

$$n(a) = n_0 e^{\frac{e(\varphi_a + V)}{kT}} \left\{ 1 - \frac{1 - e^{-\frac{eV}{kT}}}{1 + \dfrac{l_2 \psi\left(\dfrac{b}{l_2}\right)}{l_1 \left[\psi\left(\dfrac{\rho_2 b + \rho_1 a}{\rho_1 l_1}\right) - \psi\left(\dfrac{\rho_2 b}{\rho_1 l_1}\right)\right] \exp\left(\dfrac{b}{l_3}\right)^2}} \right\}. \qquad (2)$$

Equation (2) and Poisson's equation

$$\frac{2\pi}{\varepsilon}\left[(\rho_1 - \rho_2) a^2 + \rho_2 w^2\right] = V_c + V \qquad (3)$$

can be used together to determine, at least in principle, the dependences $a(V)$ and $b(V)$. However, in our case, it is sufficient to consider Eq. (2), which shows that the Boltzmann distribution applies in the region b when the following inequality is satisfied:

$$\frac{l_2 \psi\left(\dfrac{b}{l_2}\right)}{l_1 \left[\psi\left(\dfrac{\rho_2 b + \rho_1 a}{\rho_1 l_1}\right) - \psi\left(\dfrac{\rho_2 b}{\rho_1 l_1}\right)\right] \exp\left(\dfrac{b}{l_3}\right)^2} \leqslant 1. \qquad (4)$$

---

* The following notation is used in Eq. (2):

$$\rho_1 = e N_D^+; \quad \rho_2 = e\left(N_D^+ - N_A^-\right); \quad l_{1,2} = \sqrt{\frac{\varepsilon kT}{2\pi e \rho_{1,2}}}; \quad l_3 = l_2 \sqrt{\frac{\rho_1}{\rho_1 - \rho_2}}; \quad \varphi(x) = \int_0^x e^{z^2} dz.$$

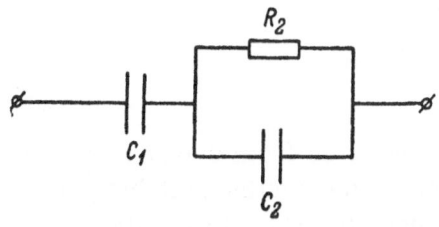

Fig. 2. Equivalent circuit of a p−n junction: C_1 is the capacitance of the space-charge layer; C_2 and R_2 are the parameters of the bulk of the sample.

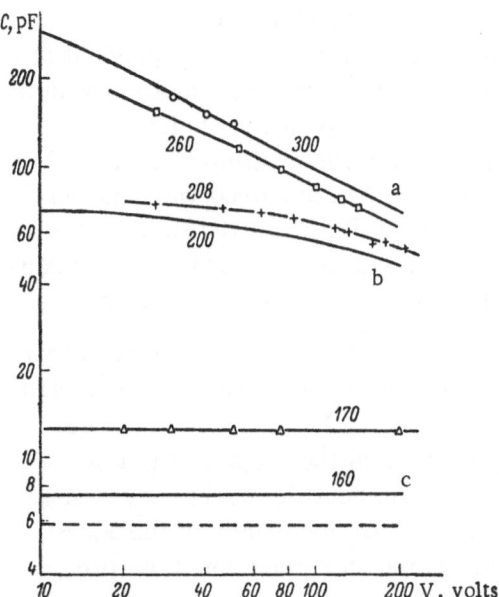

Fig. 3. Dependence of the capacitance of an n'−n−p structure on the reverse bias applied at various temperatures. a-c) Theoretical curves plotted for $N_D = 10^{13}$ cm^{-3}, $N_A = 2 \cdot 10^{13}$ cm^{-3}, $f = 2 \cdot 10^5$ cps, $V_c = 0.2$ V. The numbers alongside the curves give the temperature (in °K). The dashed line represents the geometric capacitance for a disk d = 2.5 mm thick, and the points are the experimental results.

Numerical estimates, carried out for a wide range of values of ρ_1 and ρ_2, show that this inequality is indeed well satisfied. Therefore, we can regard it as proved that the potential difference across the region b remains constant (as under equilibrium conditions) and that the applied bias is concentrated in the region a, and that we can use with confidence an expression for the capacitance of a p−n junction derived in [1].

Equivalent Circuit of a n'−n−p Structure

Apart from the possible intersection of the energy levels, the measured capacitance is affected by the relaxation processes in that part of the n-type region where there is no electric field (the bulk n-type region) [2]. Therefore, an n'−n−p structure can be represented by two RC circuits connected in series; these circuits represent the space-charge layer of the p−n junction and the bulk of the sample with a dielectric (Maxwellian) relaxation time $\tau_M = \varepsilon/4\pi\sigma$. We note that, in our case, we can neglect the active component of the p−n junction impedance since measurements are carried out mainly below room temperature. Moreover, the Fermi level may be intersected by the two extrema levels ($E_c - 0.20$ eV and $E_v + 0.27$ eV) of a system of radiation-defect levels generated by γ rays [5]. The carrier-capture cross sections of these levels are such [6] that, at T ≤ 150°K and frequencies ≥10 kc, these levels are unable to exchange electrons with the conduction band and the capacitance C_1 (Fig. 2) is described by the formula

$$C_1 = \sqrt{\frac{\varepsilon e}{8\pi}} \left[\sqrt{\frac{E_F - E_A}{e(N_D^+ - N_A^-)}} + \sqrt{\frac{V_c + V - \frac{N_A^-}{N_D^+}\left(\frac{E_F - E_A}{e}\right)}{N_D^+}} \right]^{-1}. \tag{5}$$

Our measurements were carried out using an ac bridge in which a sample was counterbalanced by an equivalent parallel RC circuit. Therefore, the measured value of the capacitance was*

---

* In principle, a rise in C can be observed with increasing bias (at a given compensation and frequency) because C_1 decreases and C_2 increases with increasing V and, therefore, the fraction on the right-hand side of Eq. (6) increases. If the increase of this fraction compensates the fall of C_1, the measured capacitance C increases. The calculated function C(V) does indeed show, in some cases, a slight rise with V.

$$C = C_1 \frac{1 + \omega^2 \tau_M^2 \left(1 + \frac{C_1}{C_2}\right)}{1 + \omega^2 \tau_M^2 \left(1 + \frac{C_1}{C_2}\right)^2}. \qquad (6)$$

Disks, 2.0-2.5 mm thick and ≈ 20 mm in diameter, were prepared from n-type germanium with $N_D = (1-2) \cdot 10^{13}$ cm^{-3}. An n'$-$n$-$p structure was produced by fabricating suitable junctions on the two faces of the disk. Acceptor levels were introduced by irradiation with Co^{60} γ rays. The maximum integral flux produced a compensation level at which the space-charge region of the p$-$n junction (at 78°K) extended to the n' contact.

Results of the Measurements

The influence of deep levels can be investigated conveniently by measuring the dependence C(V) and increasing gradually the degree of compensation of the n-type region [7]. The intersection of the levels weakens appreciably the dependence C(V) compared with the usual form $C \propto [(N_D^+ - N_A^-)/V]^{\frac{1}{2}}$ because such intersection produces the region b, which is independent of the applied bias. Figure 3 shows, by way of illustration, the capacitance of a sample in which the number of introduced acceptors is $N_A \approx 2N_D$; the capacitance is given as a function of a reverse bias applied at various temperatures. The most important observation is that the lack of a dependence of the capacitance on the bias does not mean that the measured value is the geometric capacitance C_g. Moreover, even if $C = C_g$, it does not follow that the field intensity in the n-type region is uniform. This is a well-known difficulty, which is encountered in practical applications of n'$-$n$-$p structures in, for example, spectrometric radiation counters.

The influence of the degree of compensation appears most explicitly in the temperature dependence of the capacitance (Fig. 4). When the number of acceptors introduced by bombardment is increased, the onset of the rise of C shifts in the direction of higher temperatures, a frequency dependence appears and then practically disappears, and the slope of the temperature dependence changes (this can be used to deduce information about the occupancy of the levels). The curves in Fig. 4 correspond to the following degrees of occupancy of the $E_c -$ 0.20 eV level at 78°K: a) 0; b) 0.3; c) 1.0. The experimental points for good counters (i.e., samples with a strong electric field in the n-type region) fit satisfactorily one of the theoretical curves.* At first sight, this seems surprising because the calculations show that the Fermi level intersects the $E_v + 0.27$ eV level near the p-type contact even when the contact potential difference is $V_c \geq 0.2$ eV. Measurements of C(V) and of the photo-emf generated by strong illumination of our samples, indicate that $V_c \geq 0.35$ eV. Thus, we have to assume that the intersection of the Fermi level with the $E_v + 0.27$ eV level (the compensation level) does not restrict the space-charge region.

This observation can be explained by assuming that the time constant for electron transitions from the $E_v + 0.27$ eV level to the conduction band is extremely long ($\sim 10^{23}$ sec at 78°K). Therefore, when a reverse bias is applied to the n$-$p junction, the occupancy of this level is not affected and the metastable state is retained for an indefinite time. Therefore, the potential difference across the region a in Fig. 1 decreases (this potential difference is only equal to V_c) and the field spreads into the bulk of the n-type region in accordance with the degree of compensation.

The occurrence of such "frozen" states gives rise to thermal hysteresis of the C(T) curves[†] and to a strong dependence of the capacitance on illumination. Because the differ-

---

* The discrepancy at low temperatures is due to the lesser thickness of the counter than that assumed in the calculations (2.5 mm).

[†] The results presented in Fig. 4 were obtained during the cooling of a sample when the dominant process was not the liberation of carriers but their fast capture by levels in the region b.

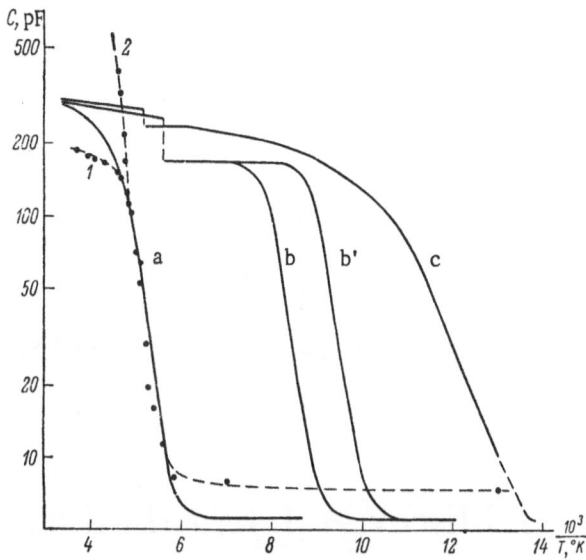

Fig. 4. Temperature dependence of the capacitance of an n'−n−p structure. a-c) Theoretical curves for $N_D = 10^{13}$ cm^{-3}, $V_c = 0.2$ V, V = 10 V, S = 1 cm^2, d = 2.5 mm, $f = 2 \cdot 10^5$ cps ($2 \cdot 10^4$ cps for curve b'). Assumed values of $N_A - N_D$ (cm^{-3}): a) 10^{13}; b, b') $5 \cdot 10^{12}$; c) -10^8. The discontinuities in curves b and c indicate the point of intersection of the Fermi and $E_v + 0.27$ eV levels due to the appearance of the region a in Fig. 1. 1) Experimental data for V = 30 V; 2) experimental data for V = 0 V.

ence between the carrier-capture cross sections of the levels at $E_c - 0.20$ eV and $E_v + 0.27$ eV is large, charge exchange takes place between these levels, in full agreement with an analysis given earlier in this paper.

Literature Cited

1. C. T. Sah and V. G. K. Reddi, IEEE Trans. Electron Dev., ED-11:345 (1964).
2. B. M. Vul and É. I. Zavaritskaya, Zh. Eksp. Teor. Fiz., 38:10 (1960).
3. S. M. Ryvkin, O. A. Matveev, N. B. Strokan, and A. Kh. Khusainov, Dokl. Akad. Nauk SSSR, 165:548 (1965).
4. C. Thompson, Arch. Elektr. Übertrag., 18:628 (1964).
5. N. A. Vitovskii, T. V. Mashovets, and S. M. Ryvkin, Fiz. Tverd. Tela, 6:1883 (1964).
6. S. M. Ryvkin and I. D. Yaroshetskii, Fiz. Tverd. Tela, 2:1966 (1960).
7. L. V. Maslova, O. A. Matveev, S. M. Ryvkin, N. B. Strokan, and A. Kh. Khusainov, in: Lithium-Drifted Germanium Detectors (Proc. Panel, Vienna, 1966), Intern. Atomic Energy Agency, Vienna (1966), p. 32.

INVESTIGATION OF DC INJECTION LASERS

P. G. Eliseev, A. I. Krasil'nikov,
M. A. Man'ko, and V. P. Strakhov

P. N. Lebedev Physics Institute
Academy of Sciences of the USSR, Moscow

A batch of injection lasers, based on epitaxial p−n junctions in gallium arsenide, was investigated under continuous working conditions using low currents at liquid-nitrogen temperature. When the laser threshold was crossed, the following effects were usually observed: the differential resistance of the laser diode became constant; the narrowing of the electroluminescence spectrum in the nonlasing modes stopped; the shift of the peak of the wide luminescence band with increasing current also stopped. These tendencies indicated uniform broadening of the emission band of an injection laser. Multimode laser emission was observed and the intensity of the radiation in the wings of the line continued to increase, and this increase was observed also at wavelengths shorter than the coherent emission wavelengths. In some cases, the voltage across a p−n junction continued to increase somewhat even under laser emission conditions. The random nature of the quantitative aspect of these deviations from uniformity confirmed an earlier suggestion of the spatial origin of the multimode emission.

Introduction

Injection lasers with a sufficiently low threshold current can be investigated under dc conditions. This makes it possible to measure the voltages and currents with a high accuracy typical of dc measurements. The present paper reports a comprehensive study of gallium arsenide injection lasers with a threshold current not exceeding 200 mA at liquid-nitrogen temperature. We were able to study the dependence of the potential drop across a p−n junction on the current in the range from low excitation levels to levels exceeding appreciably the threshold current (under virtually isothermal conditions) and to compare the results of this study with spectral investigations. The problem of the profile and uniformity of the emission band of a semiconductor laser was investigated. A hypothesis of spectral uniformity of the emission band and of the spatial origin of the deviations from complete uniformity of this band (i.e., uniformity which would guarantee the single-mode operation) was considered theoretically and experimental data were obtained to support this hypothesis. In these considerations, we used a model of the energy spectrum of the active region of a semiconductor laser consisting of an exponential distribution of the density of the upper states and a band of the lower states (as-

150

TABLE 1

Sample No.	$S \cdot 10^5$, cm^2	I_t, mA	j_t, A/cm^2	$\hbar\omega_t$, eV	R_S, Ω	$\rho_s \times 10^5$, Ω/cm^2	U_1, V
519	8.2	155	1890	1.4274	—	—	—
540	8.1	60.5	745	1.4264	0.4	3.24	1.4405
540 *	8.1	93	1150	1.4291	0.45	3.64	1.4490
543	18.8	120	640	1.4264	0.3	5.6	1.4370
544	20	155	775	1.4270	0.31	6.2	—
547	8.9	75	840	1.4266	0.37	3.3	1.4465
541	11	78	710	1.4274	0.35	3.85	1.4485
126	6.3	55.5	883	1.4268	1.84	11.8	1.4445
548	8.2	67.5	823	1.4261	0.475	3.9	1.4455
549	14.4	183	1270	1.4273	0.207	2.94	1.4500
552	16.3	105	645	1.4248	0.27	4.4	1.4554
554	10	71	710	1.4186	1.445	14.45	1.4515
555	4.9	37	755	1.4266	0.98	4.8	1.4435
556	8.1	60	740	1.4314	1.08	8.85	1.4520
557	5.2	24	462	1.4203	0.74	3.85	1.4460

Note. Data for diode 540 obtained after degradation of its characteristics.

sociated with impurity bands). The parameters used to describe the energy spectrum were derived from the experimental data.

The profile and behavior of the emission band of an injection laser were considered by Burns and Nathan [1], who reported that the intensity in the wings far from the laser line increases sublinearly with the current beyond the threshold and does not become fully saturated.

It is suggested in [2, 3] that the multimode emission may be due to spatial inhomogeneity since the emission band should be spectrally uniform.

Electrical phenomena, associated with crossing of the laser threshold, are described in [4, 5] but in both cases these phenomena were observed by connecting, in series with a p−n junction, a linear resistance R_S much larger than the nonlinear resistance of the junction so that changes in the current−voltage characteristic could be interpreted as due to a photoeffect in the layers adjoining the p−n junction [5].

In our investigation, the spectral and electrical measurements were carried out simultaneously and the electrical nonlinearity of the p−n junction was observed right up to the threshold current.

Methodological Problems

We investigated a batch of lasers prepared by liquid epitaxy (Table 1). The substrates were made of zinc-doped p-type gallium arsenide with a carrier density of $3 \cdot 10^{19}$ cm^{-3}. The substrates were oriented along the (100) planes. An epitaxial n-type layer was doped with tellurium and grown at 950-920°C in an atmosphere of purified hydrogen, using a solution of GaAs in gallium. The concentration of tellurium in the epitaxial layer was $(5-8) \cdot 10^{18}$ cm^{-3}. During the growth, which took several hours, zinc diffused from the substrate into the epitaxial layer and, therefore, the distribution of the impurity concentration near the p−n junction became linear and both the n- and p-type regions became heavily doped. The space-charge layer was approximately 0.15 μ thick.

The thickness of a plate with a p−n junction was reduced to 100 μ (the junction lay half way between the large faces of the plate). The diodes were prepared by cleaving and were placed in a spring-loaded holder. We investigated in detail a batch of lasers prepared from the same plate. The lasers with four-sided resonators had threshold current densities of 460-1900

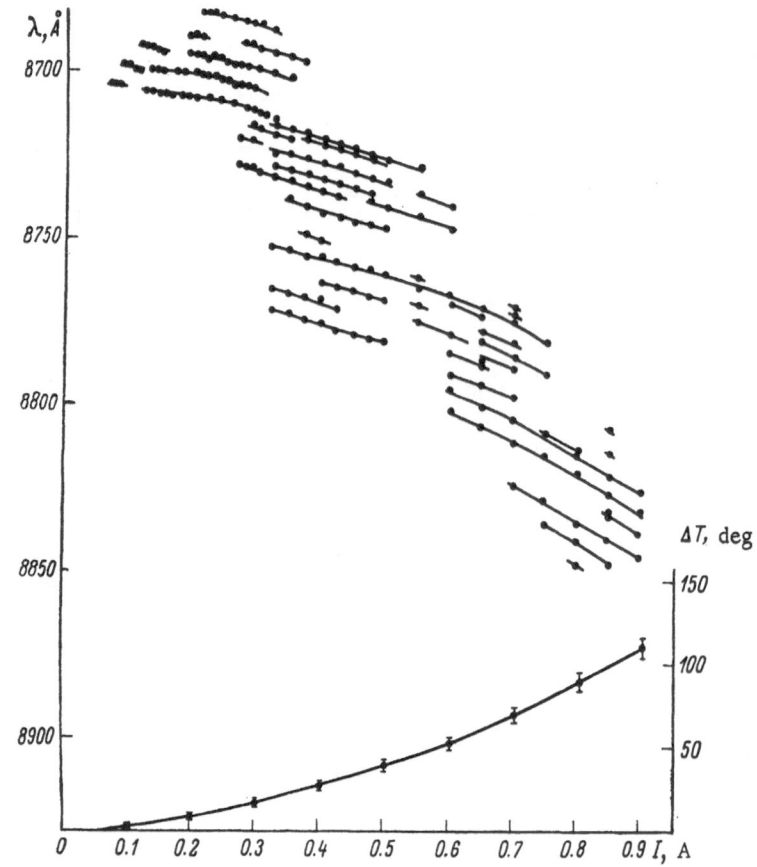

Fig. 1. Analysis of thermal conditions during the operation of an injection laser (sample 540). The upper part of the curve shows the positions of the spectral modes as a function of the direct current through the laser diode (the left-hand ordinate gives the wavelength). The lower part of the figure shows the temperature rise (the right-hand ordinate) in a p−n junction as a function of the current.

A/cm^2 at 77°K and their coherent radiation was emitted in the form of nondirectional modes with a relatively low external efficiency; however, the efficiency was sufficient to observe the onset of laser emission in the spectrum. All the measurements were carried out under dc conditions. The current and voltage were measured with a relative error not exceeding 0.2 and 0.01%, respectively. The spectral resolution was about 0.5 Å.

The series resistance of the diodes ρ_S, reduced to unit area of the p−n junction, was (in the best cases) $(3-5) \cdot 10^{-5}$ Ω/cm^2 (Table 1). Estimates of the resistance of the diode outside the p−n junction gave values of $1.5 \cdot 10^{-5}$ Ω/cm^2, i.e., about half of the total resistance or less. The contact layers made up the other half of the resistance.

The temperature of a laser during the measurements was an important parameter. Naturally, we were unable to maintain isothermal conditions in all the measurements because of the heating by the current. A calculation of the thermal resistance ρ_T, reduced to unit area of the p−n junction, i.e., a calculation of the quantity $l/\varkappa$ (l is the distance from a heat sink to a p−n junction, where most of the heat is evolved, and $\varkappa = 3$ W · cm^{-1} · deg^{-1} is the thermal conductivity of gallium arsenide at 77°K) gave a value of $8 \cdot 10^{-4}$ deg · cm^2 · W^{-1}, i.e., the heating was 10 deg/W for a junction area of $8 \cdot 10^{-5}$ cm^2. This value of the thermal resistance in-

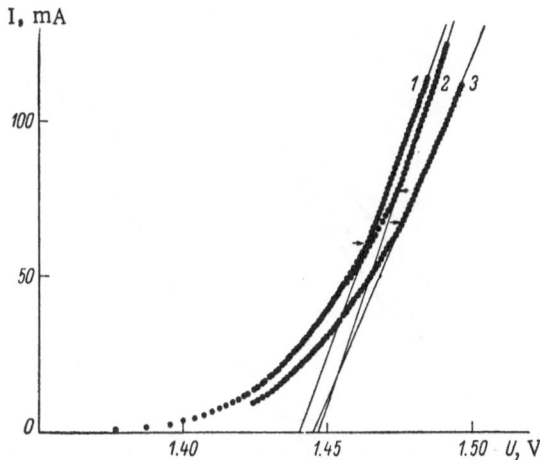

Fig. 2. Current–voltage characteristics of injection lasers. The arrows indicate the laser threshold. Samples: 1) 540; 2) 547; 3) 548.

creased with rising temperature because of the decreasing thermal conductivity. We determined experimentally the shift of the laser emission line as a function of the current flowing through a diode. Figure 1 shows the modes obtained at currents from 60 mA (the threshold current I_t) to 900 mA, as well as the temperature rise deduced from the shift of the modes.

At low values of the current, the temperature rise corresponded to a thermal resistance of $\sim$20 deg/W ($1.6 \cdot 10^{-3}$ deg $\cdot$ cm^2 $\cdot$ W^{-1} per unit area of the p–n junction), which was twice as large as the calculated value. We concluded that the discrepancy was due to the low efficiency of the heat sink at the semiconductor–metal contact.

The main measurements were carried out in the range up to 75–150 mA and the heating was deduced from the spectral shift of the modes. The characteristics obtained were isothermal to within 3°K. A correction to the forbidden band width, corresponding to this temperature difference, was 1.5 meV and the rise of the threshold current due to this rise in temperature was 5–10%.

Current – Voltage Characteristics

The investigated diodes exhibited a dependence of the current on the total applied voltage, which was nonlinear below the laser threshold and linear above the threshold (Fig. 2).

After subtracting the voltage drop across the series resistance, the subthreshold parts of the characteristics could be approximated satisfactorily by a simple exponential function

$$I = I_0 \exp \frac{e U_{pn}}{\varepsilon}, \tag{1}$$

where $\varepsilon = 16$–17 meV (for all diodes) and U_{pn} is the voltage across a p–n junction (Fig. 3). When the threshold was crossed, the nonlinear contribution of the p–n junction to the resistance disappeared. Figure 4 shows the dependence of the differential resistance dU/dI on the reciprocal of the current 1/I for diode 557. In the region where $1/I > 42$ A^{-1}, i.e., below the threshold, dU/dI was a linear function of 1/I so that, assuming Eq. (1) was obeyed, we obtained

$$U = I R_s + U_{pn} = I R_s + \frac{\varepsilon}{e} \ln I - \frac{\varepsilon}{e} \ln I_0, \tag{2}$$

and hence

$$\frac{dU}{dI} = R_s + \frac{\varepsilon}{e} \cdot \frac{1}{I}.$$

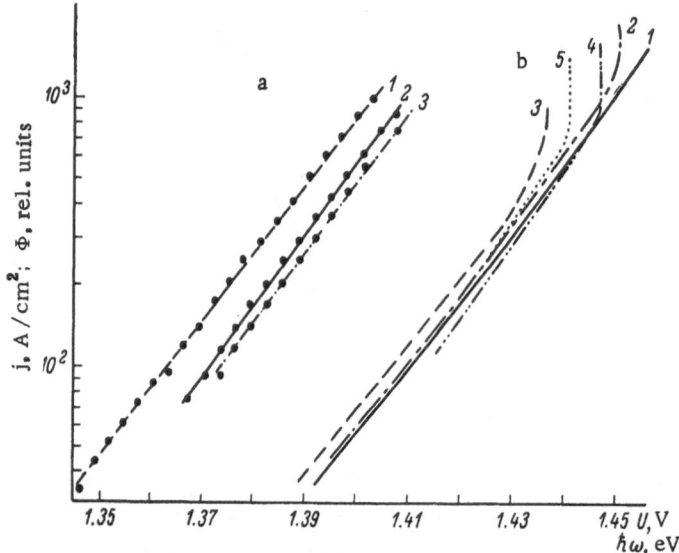

Fig. 3. Dependence of the spectral density Φ on the photon energy in the spectrum of the spontaneous radiation on the side of low values of ℏω (a) and the dependence of the current density on the external voltage applied to the p–n junction (b). Samples (a): 1) 519; 2) 557; 3) 552. Samples (b): 1) 551; 2) 549; 3) 543; 4) 547; 5) 540.

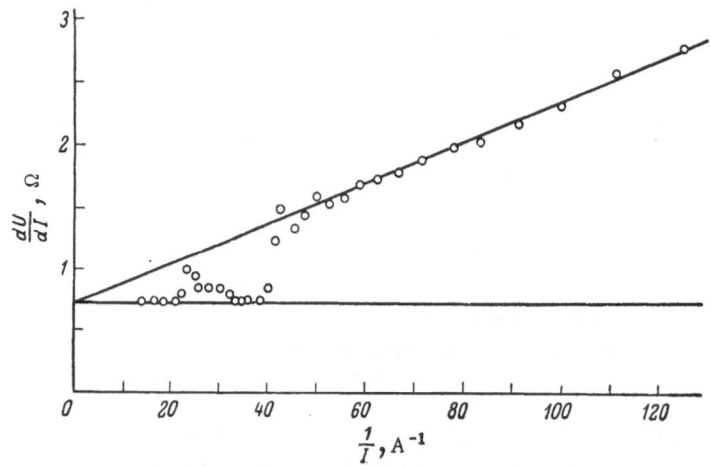

Fig. 4. Dependence of the differential resistance of an injection laser on the reciprocal of the current. Sample 557. The sloping part of the dependence (below the threshold) corresponds to $\varepsilon = 16.4$ meV. The threshold was observed at 41.7 A^{-1} and a second mode appeared at 23.5 A^{-1}. The sloping part of the dependence, extrapolated to $1/I \to 0$, intersected the ordinate at 0.74 Ω.

This indicated that the slope of the straight line in the subthreshold region was governed by the value of ε. Extrapolation of the straight line to $1/I \to 0$ gave the value of R_S. At the threshold corresponding to $1/I \approx 42$ A^{-1} the experimental points changed suddenly to the extrapolated value of R . An additional rise in the differential resistance at $1/I \approx 25$ A^{-1} will be discussed later. At still lower values of $1/I$, the differential resistance dU/dI was equal to R_S,

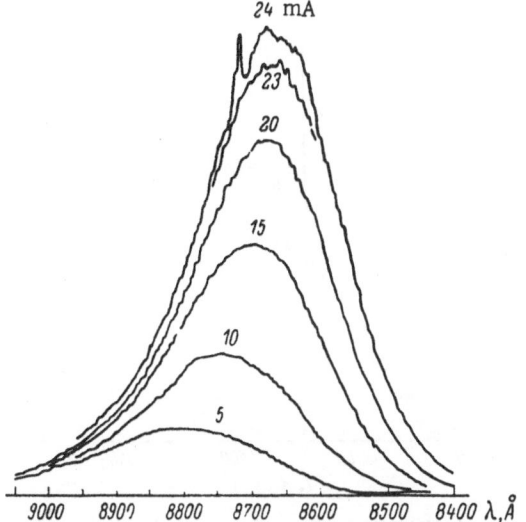

Fig. 5. Radiation spectra of sample 557. Threshold current 24 mA.

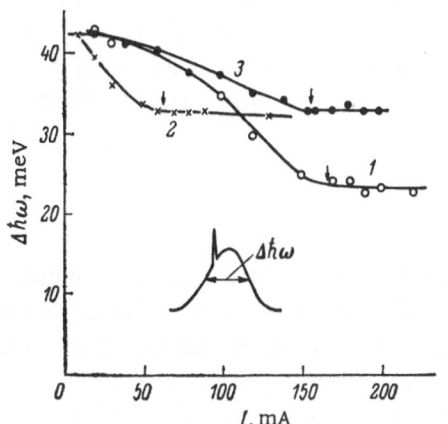

Fig. 6. Dependence of the width of the spontaneous radiation band on the current for samples 519 (1), 540 (2), and 544 (3). The arrows indicate the laser thresholds.

obtained by extrapolation of the sloping part of the dependence (Fig. 4 demonstrates most clearly the disappearance of the nonlinear resistance of the diode at the threshold). The sudden change in the resistance at the threshold point was about $3 \cdot 10^{-5} \ \Omega/cm^2$, i.e., it was comparable with ρ_S. This discontinuity was smaller for diodes with higher values of the threshold current density.

The stabilization of U_{pn} at the threshold point gave a definite meaning to the voltage intercept U_1, which was obtained by extrapolation of the linear part of the characteristic to $I \rightarrow 0$: this voltage represented the difference between the quasi-Fermi levels in the p−n junction under laser emission conditions.

Coherent and Spontaneous

Emission Spectra

Spectral investigations included a study of the profile, width $\Delta\hbar\omega$, and position of a maximum $\hbar\omega_p$ of the spontaneous radiation band and the determination of the photon energy of the laser modes $\hbar\omega_k$. In some cases, when laser emission appeared in the wing of the spontaneous radiation band, $\hbar\omega_p$ and $\Delta\hbar\omega$ could be determined at currents considerably higher than the threshold value.

A spectrum of this type was obtained for sample 557, which had the lowest threshold current of 24 mA (Fig. 5). In this spectrum, a laser line appeared in the long-wavelength wing of the spontaneous radiation band. Since laser emission was in the form of a high-Q nondirectional mode and the spontaneous radiation consisted mainly of Fabry−Perot modes, the spectral narrowing of the spontaneous radiation band was

very slight right up to the threshold current. It is evident from Fig. 6 that the narrowing of the spontaneous band slowed down or practically stopped when the threshold was reached.

Spectral oscillations of the spontaneous radiation, associated with resonance of the Fabry−Perot modes [6], were observed at right-angles to the resonator mirrors. These oscillations enabled us to estimate the magnitude and distribution of the volume (bulk) radiation losses. Oscillations in one of the spectra are shown in Fig. 7. This figure includes evidence that laser emission appeared in modes other than those of the Fabry−Perot type. At a current of 80 mA, which was slightly higher than the threshold value, a laser emission line (whose intensity extended well outside the scale of Fig. 7) appeared in a dip between two neighboring oscillation peaks, i.e., between two neighboring Fabry−Perot modes.

We shall now consider the dependence of the spectral radiation density Φ (at various wavelengths) on the current through a p−n junction. Figure 8 shows typical dependences. Laser emission was observed at a wavelength of 8688 Å when the current reached I_1. At a wavelength

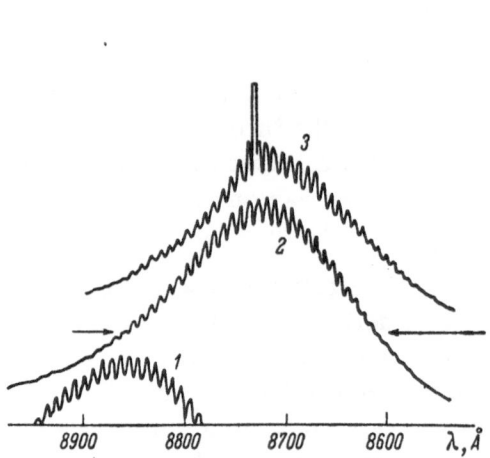

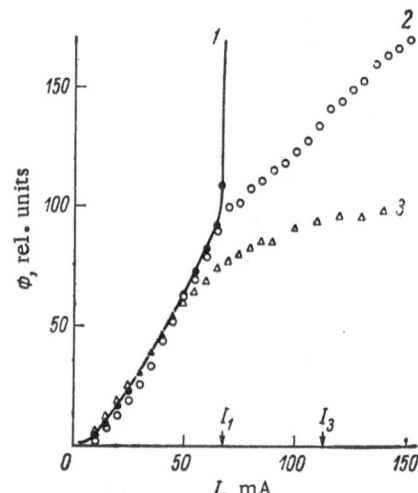

Fig. 7. Oscillations of the spontaneous radiation at 10 mA (curve 1 showing the top of the radiation band), 60 mA (2), and 80 mA (3). Sample 554. Threshold current 71 mA. The laser emission line in curve 3 is not one of the Fabry—Perot modes, whose spectral positions correspond to oscillation maxima. The curves shown here have arbitrary scales and are shifted by arbitrary amounts in the vertical direction.

Fig. 8. Dependence of the spectral density of the radiation on the current through a p—n junction in sample 548: 1) at the laser emission line; 2, 3) at wavelengths respectively shorter and longer than the laser line. The threshold current was $I_1 = 67.5$ mA. At $I_3 = 115$ mA, new modes appeared at shorter wavelengths.

of 8720 Å, i.e., at a photon energy lower than $\hbar\omega_k$, there was a tendency to saturation, which appeared before the threshold was reached. At a wavelength of 8650 Å, i.e., at a photon energy higher than $\hbar\omega_k$, there was no tendency to saturation although the rise of the intensity slowed down somewhat at the threshold current. The rise of the intensity during laser emission, particularly in the short-wavelength wing, as well as the multimode laser emission, indicated that the excess carrier density increased slightly beyond the threshold.

Only the general features of the multimode laser spectrum were similar to the spectrum described in [7] because laser emission was observed not in the form of Fabry—Perot modes but as nondirectional modes. The most interesting were the cases of multimode laser emission with a considerable spreading of the coherent spectrum in the direction of shorter wavelengths. This behavior was exhibited most clearly by sample 557 (Fig. 9). The coherent radiation spectrum of this sample was simpler than that of the other samples: between 24 and 75 mA, we observed only three laser lines, and up to 42.5 mA only a single mode was emitted (it must be pointed out that these observations were correct within the limits of resolution of the spectroscope). A second mode appeared at a photon energy 6 meV higher than the first mode and its appearance was preceded by an increase in U_{pn} of about 3 meV, so that the differential resistance increased near $1/I \sim 25$ A^{-1} (Fig. 3).

Similar behavior was observed for other diodes but an increase in U_{pn} in other cases was very slight.

Energy Spectrum of the Active Region

The experimentally observed exponential current—voltage characteristic, the exponential fall of the intensity in the long-wavelength wing of a radiation line, and the logarithmic shift of

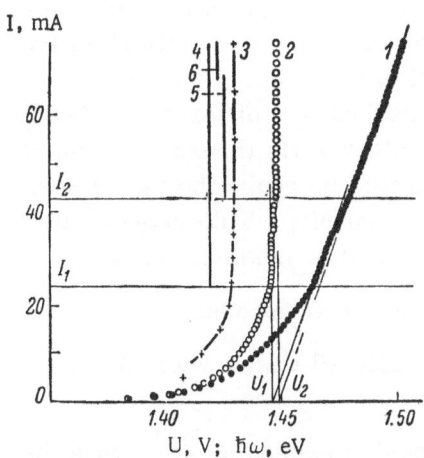

Fig. 9. Current—voltage characteristic (1), voltage across a p—n junction (2), position of the spontaneous radiation maximum (3), position of the first laser mode (4), and positions of the second and third laser modes (5 and 6). Sample 557. $I_1 = 24$ mA, $U_1 = 1.446$ V; $I_2 = 42.5$ mA, $U_2 = 1.449$ V.

the radiation line maximum with the current (a linear shift with the voltage) can all be easily explained by assuming that the density of injection-filled states increases exponentially with the energy. All these experimental dependences should have the same characteristic energy ε. We can indeed see from Fig. 3 that the fall of Φ and the dependence of the current density through a p—n junction on U_{pn} are identical. The characteristic energy for diode 557 found from the spectrum is $\varepsilon = 16.4$ meV and that found from the slope of the current—voltage characteristic (Fig. 4) is $\varepsilon = 16.5$ meV.

If the constant representing the rate of fall of the density of the upper states is 16.5 meV, the equilibrium width of a band of filled states at T = 77°K should be 36 meV for any position of the electron quasi-Fermi level in the tail. The width of the spontaneous radiation band before narrowing (this narrowing was due to stimulated emission) is 43 meV. Assuming that the difference between these two values is solely due to the finite width of the band of the lower energy levels (in the acceptor band), we can estimate the width of this energy-level band: 23.5 meV. We note that this value is practically identical with the width of the cathodoluminescence band emitted by uncompensated p-type gallium arsenide with a hole density of $3 \cdot 10^{19}$ cm^{-3}, which has been assumed to be equal to the width of the acceptor impurity band [8].

A calculation of the line profile using the formulas of Lasher and Stern [9] and the energy spectrum model just described give satisfactory agreement with the experimental curves if it is assumed that there is no conservation of the crystal momentum in radiative transitions.

Uniformity of Amplification and of the Radiation Band

Our observations will now be compared with representations of the nature of the radiation (electroluminescence) band of a p—n junction which gives rise to coherent emission. The following observations do not represent regular relationships but appear rather as definite tendencies:

A) the spontaneous radiation peak ceases to shift with the current when the threshold is reached;

B) the spectral narrowing of the radiation band with increasing current stops when the threshold of emission of nondirectional laser modes is reached;

C) the nonlinear differential resistance of a p—n junction disappears when the threshold current is reached.

We must mention also the observations of interaction in injection lasers [11], especially the quenching of laser emission by strong external illumination. In one of our earlier investigations [12], we observed strong quenching of the 8435 Å laser line by radiation of 8475 Å wavelength, i.e., the quenching of laser emission by radiation of energy 7 meV lower than the energy of the photons in the stimulated line. This shows that in the presence of a strong field an inverse population of states decreases uniformly across the spectrum, at least within an energy range of a few millielectron volts. In this case, the line may be regarded as uniformly broadened and the distribution of carriers in the energy bands as of the quasi-equilibrium type, i.e., obeying the Fermi−Dirac function. We can use the concept of the quasi-Fermi levels under laser emission conditions. When the laser threshold is reached, the difference between the quasi-Fermi levels for electrons and holes, ΔF remains constant at its threshold value because if ΔF were to continue to increase with the current, this would increase the gain above the threshold value, i.e., it would make the field amplitude rise to infinity. This explanation is correct for a spectrally uniform radiation band in the absence of spatial inhomogeneities.

We must also consider the following important experimental observations:

I) the multimode laser emission, which appears in practically all cases when the threshold current is exceeded by a sufficient amount;

II) the rise of the spectral radiation density, at photon energies exceeding $\hbar \omega_k$, with the current above the generation threshold.

These effects are very sensitive to a change in ΔF or to deformation of the spectral dependences of the spontaneous radiation and amplification.

Finally, direct measurements of the potential difference across a diode have indicated that in some cases the voltage drop across the junction U_{pn} increases when the laser threshold is reached. These observations do not agree with the completely uniform mechanism discussed earlier in this section. The multimode laser emission has been attributed in [2, 10] to a spatial inhomogeneity of the population inversion in the resonator, whose contribution can, in general, fluctuate considerably.

In order to estimate the uniform broadening in a semiconductor laser, we must take into account: a) the energy relaxation in the bands with characteristic times of the order of 10^{-12} sec; b) the selection rules for radiative transitions, which determine the states participating in transitions with a given photon energy. If the crystal momentum is conserved, a line becomes uniform only because of the energy relaxation, which gives rise to a natural level width of about 1 meV. Since the radiation band is wider, a considerable deformation of the spectral dependence of the gain is possible. A process of this kind has been considered in [13]. If the law of conservation of momentum is not obeyed (this is true of transitions in which impurities participate), the order of magnitude of the uniform broadening is governed by the width of the narrower of the two bands filled with carriers (in the upper and lower states), i.e., in our case, it is governed by the width of the acceptor band which is about 24 meV. If there are no spatial inhomogeneities, new laser modes can appear only at energies separated by intervals (which are of the order of the uniform width) from the modes being emitted. We have observed experimentally about 70 cases of the appearance of new nondirectional spectral modes, of which 45 were separated by less than 1 meV from the nearest existing mode and this separation is much smaller than the estimated uniform width of 24 meV. Obviously, we should seek an explanation of this discrepancy in spatial inhomogeneities.

The assumption of the existence of spatial inhomogeneities implies that, strictly speaking, we can no longer separate the applied voltage into U_{pn} and IR_S. In fact, if these inhomogeneities are due to an electromagnetic field, we find that, in regions where the intensity of oscillations is higher, the recombination resistance of a p−n junction is lower because the carrier lifetime is shorter [14]. This distorts the lines of flow of the current and destroys the equipotential nature of the p−n junction boundary. Consequently, ΔF can have different values in different parts of the p−n junction. This is responsible for the observed rise of ΔF beyond the threshold current.

The other important problem is the nature of the difference between eU_1 and $\hbar\omega_k$. This difference has been found to range from 10 to 30 meV, mainly because of variations in U_1. A calculation of the population of the exponential density-of-states tail with $\varepsilon = 17$ meV shows that the highest gain is obtained for transitions from levels lying about 10 meV below the quasi-Fermi level of electrons. The remainder of the energy difference $eU_1 - \hbar\omega_k$ may be due to the following factors. First, eU_1 corresponds to the limiting value of ΔF in the active region because the maximum gain corresponds to the lowest value of ΔF. Secondly, ΔF decreases because of nonradiative recombination in the p−n junction. Thirdly, the acceptor band maximum and the quasi-Fermi level of holes may be separated by a considerable energy gap, which depends on the degree of compensation and on the excitation level.

Thus, we can draw the following conclusions.

1. A laser diode loses its electrical nonlinearity at the threshold value of the current.

2. Behavior of the spontaneous radiation band is in agreement with the model of a uniformly broadened line but with definite random deviations. These deviations can be approximately related to a rise of ΔF beyond the threshold and to deformation of a maximum of the spectral dependence of the gain.

3. The cause of these deviations lies evidently in spatial inhomogeneities; in principle, single-mode laser emission is possible provided these spatial inhomogeneities are eliminated.

Literature Cited

1. G. Burns and M. I. Nathan, Proc. IEEE, 51:471 (1963).
2. H. Statz, C. L. Tang, and J. M. Lavine, J. Appl. Phys., 35:2581 (1964).
3. J. M. Lavine and A. A. Iannini, J. Appl. Phys., 36:402 (1965).
4. Y. Nannichi, Japan. J. Appl. Phys., 3:233 (1964).
5. B. I. Gladkii, D. N. Nasledov, B. V. Tsarenkov, Fiz. Tverd. Tela, 8:3282 (1966).
6. M. I. Nathan, A. B. Fowler, and G. Burns, Phys. Rev. Letters, 11:152 (1963); M. I. Nathan, G. Burns, and A. B. Fowler, Proc. Seventh Intern. Conf. on Physics of Semiconductors, Paris, 1964, Vol. 4, Radiative Recombination in Semiconductors, publ. by Dunod, Paris (1965); Academic Press, New York (1965), p. 205.
7. P. P. Sorokin, J. D. Axe, and J. R. Lankard, J. Appl. Phys., 34:2553 (1963).
8. J. I. Pankove, Proc. Eighth Intern. Conf. on Physics of Semiconductors, Kyoto, 1966, in: J. Phys. Soc. Japan, 21 (Suppl.):298 (1966).
9. G. Lasher and F. Stern, Phys. Rev., 133:A553 (1964).
10. H. Haug, Z. Physik, 194:482 (1966); 195:74 (1966).
11. A. B. Fowler, J. Appl. Phys., 35:2275 (1964); C. E. Kelly, IEEE Trans. Electron. Dev., ED-12:1 (1965).
12. P. G. Eliseev, A. A. Novikov, and V. B. Fedorov, Zh. Éksp. Teor. Fiz. Pis. Red., 2:58 (1965).
13. G. E. Pikus and A. G. Aronov, Fiz. Tverd. Tela, 7:3548 (1965).
14. H. Rieck, Solid State Electron., 8:83 (1965).

ELECTRICAL PROPERTIES OF SILICON p—n JUNCTIONS SUBJECTED TO NONUNIFORM DEFORMATION

A. L. Polyakova and V. V. Shklovskaya-Kordi

Acoustics Institute, Moscow

A study was made of the influence of nonuniform deformation, caused by the pressure of a sharp corundum needle, on the current flowing through shallow p—n junctions in silicon. Two ranges of forces applied to the needle are distinguished: weak forces, whose effects are reversible, and strong forces, whose effects are irreversible. The concept of a critical force F_{cr}, dividing these two ranges, is introduced. Comparison of the experimental data with the theory is carried out in the weak-force range.

Introduction

It was discovered in 1962 [1] that large reversible changes in the forward and reverse currents could be induced by pressing a sharp needle against the surface of a crystal with a shallow p—n junction. Since this discovery, many investigations have been made of the influence of nonuniform deformation [2-5] and of uniaxial compression [6, 7] on the p—n junction current. The results obtained can be divided into two groups. The first group comprises those changes in the currents in which the pressure-induced change in the forbidden band width plays the dominant role. The second group of effects comprises changes in the currents which may be attributed to the formation of generation—recombination centers by the application of a pressure. The first group consists mainly of reversible phenomena, whereas the second group is characterized by some residual irreversible effects.

The influence of uniaxial deformation on the p—n junction current has been investigated quite thoroughly [8, 9]. An analysis of changes in the current caused by the pressure of a needle is difficult because of the complex nature of the distribution of deformation under the needle.

The present paper describes an investigation of the influence of nonuniform deformation (produced by a needle) on the current flowing through shallow p—n junctions in silicon. We investigated both reversible effects, which appeared at low values of the force applied to a needle, and irreversible changes. The experimental data on the reversible phenomena were compared with the theory based on a calculation of a change in the forbidden band width under the needle [10].

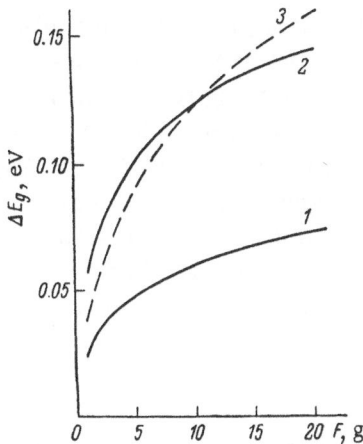

Fig. 1. Dependence of the change in the forbidden band width under a needle on the force F applied to the needle: 1) ΔE_g at the center of the contact area x = y = 0; 2) ΔE_g near the periphery of the contact area $r = r_0$; 3) results obtained using an approximate model [14]. The radius of curvature of the needle is R = 30 μ; z = 1 μ.

Material and Measurement Method

We investigated shallow p−n junctions produced by the ion bombardment method [11]. Silicon of 1 $\Omega \cdot$ cm initial resistivity was bombarded with phosphorus ions until the surface concentration of phosphorus reached 10^{20} atoms/cm^3.

Deformation was produced by a corundum needle sharpened to a point whose radius of curvature was R = 30 μ. The force applied through the needle was perpendicular to the p−n junction plane. A method was developed for smooth loading of the needle. This was done by means of an electrodynamic system using a loudspeaker magnet. The dependences of the electrical properties of a junction on the force applied to the needle were determined using an automatic X−Y recorder. The sensitivity of the apparatus was sufficient to determine a pressure-induced increase in the current $\Delta I = I_p - I_0 = 10^{-3}$ μA (in the case of a static load).

The p−n junctions were parallel to the (111) crystallographic plane. The depth of the junctions was 0.5-1.0 μ. The junction area was S $\approx 10^{-2}$ cm^2.

We investigated only the dark characteristics of p−n junctions under pressure because it was reported in [5] that illumination of the junction did not affect the pressure dependence of the current.

Influence of Deformation of the Current through a p − n Junction

1. The effective width of the forbidden band of a deformed semiconductor, defined as the difference between the energies of the lowest state in the conduction band and the highest state in the valence band, can be written in the form

$$E_g = E_{g0} - \Delta E_g, \tag{1}$$

where E_{g0} is the forbidden band width in the undeformed state. When a pressure is applied by a needle, the value of ΔE_g is found to be a complex function of the coordinates [10]: it is large in a small region immediately under the needle but it decreases rapidly away from the point of application of the force.

We shall use a system of coordinates x, y, and z whose origin is located at the point of contact of the needle with the surface of the semiconductor, directing the z axis into the semiconductor crystal. The axis of the needle (radius of curvature of the tip R) is taken to lie along the z axis. A force F is applied to the needle. We shall consider the case when the z axis coincides with the [111] crystallographic axis of the semiconductor. Calculations reported in [10] show that, in the case of germanium, ΔE_g (x, y, z) in the z = const plane has its maximum value at the center of the contact area between the needle and the semiconductor, whereas in the case of silicon the value of ΔE_g (x, y, z) has its maximum at the periphery of the contact area, i.e., where shear strains are largest. Figure 1 shows the dependence of ΔE_g on the force F, calculated for silicon using the numerical values given in [10] and assuming a fixed value of the depth z = 1 μ. Curves 1 and 2 represent, respectively, the value of ΔE_g at the center of the contact area (x = y = 0) and near the periphery of this area [at $r = \sqrt{x^2 + y^2} = r_0$, $r_0 = (FRD)^{1/3}$ is

the radius of the contact area between the needle and the semiconductor; $D = \frac{3}{4}\left[\frac{1-\sigma_2}{Y} + \frac{1-\sigma'^2}{Y'}\right]$,

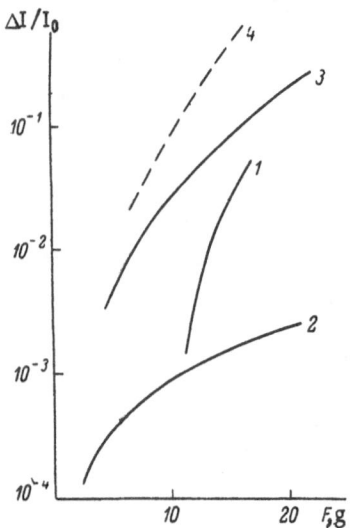

Fig. 2. Dependence of $\Delta I/I_0$, for the reverse branch of the current–voltage characteristic, on the force F applied to a needle: 1) experimental data for V = −0.2 V; 2) theoretical curve calculated on the assumption that $E_t - E_i$ = const [n_i = n_{i0}exp ($\Delta E_g/2kT$) depends on pressure but the energy level of traps is independent of pressure]; 3) theoretical curve calculated on the assumption that the energy level of the traps varies with pressure; 4) experimental curve obtained after repeated loading with a force F > F_{cr}. R = 30 μ.

where Y, Y' are Young's moduli and σ, σ' are Poisson's ratios of the semiconductor and of the needle, respectively]. These curves are calculated using the following values of the deformation potential constants: D_d = −5.4 eV [12], D_u = 9.6 eV [12], a = −2.09 eV, b = −1.36 eV [13], d = −3.09 [13]. Curve 3 in Fig. 1 shows the dependence of ΔE_g on F, calculated using a simplified model which describes the pressure under the needle [14]. In this model, the change in the forbidden band width is assumed to be

$$\Delta E_g = \alpha \frac{3F}{2\pi r_0} \frac{\sqrt{r_0^2 - r^2}}{r_0^2 + z^2},\qquad(2)$$

where α is some experimentally determined constant. Equation (2) yields a different distribution of deformation under the needle than that obtained on the basis of numerical calculations: ΔE_g now has a maximum at r = 0. Curve 3 is plotted using Eq. (2) and r = 0, but selecting the value of α in such a way that ΔE_g calculated using Eq. (2) is equal to the maximum value of ΔE_g found by numerical calculation for F = 10 g at the point r = r_0. It is evident from Fig. 1 that Eq. (2) predicts a somewhat more rapid rise of ΔE_g with increasing force than the numerical calculation carried out taking into account the spatial distribution of all the strain tensor components. The dependence of ΔE_g on z given in Eq. (2) is similar to the dependence deduced by numerical calculation and it can be used to estimate the variation of the effect of the applied force with depth.

2. We shall consider the current flowing through a p−n junction in silicon to which a voltage V is applied. Generation−recombination processes play an important role in silicon and therefore the density of the total current through the junction is the sum of two components

$$J = J_d + J_{rg},\qquad(3)$$

where J_d and J_{rg} are, respectively, the densities of the diffusion and generation−recombination currents. When a negative voltage is applied to the junction, we find that $J_{rg} \gg J_d$ [15] and when $|qV| \gg kT$, we may assume that

$$J = J_{rg} = -\frac{qWn_i}{2\sqrt{\tau_{n0}\tau_{p0}}\,\text{ch}\left[\dfrac{E_t - E_i}{kT} + \dfrac{1}{2}\ln\dfrac{\tau_{p0}}{\tau_{n0}}\right]},\qquad(4)$$

where q is the electron charge; W is the thickness of the junction; τ_{n0} and τ_{p0} are the lifetimes of minority carriers; E_t is the level of the traps at which recombination takes place; E_i is the Fermi level in an intrinsic semiconductor. Using J_P and J_0 to denote the densities of the current in the presence and absence of pressure, respectively, we can write the total current through the p−n junction subjected to pressure in the following approximate form:

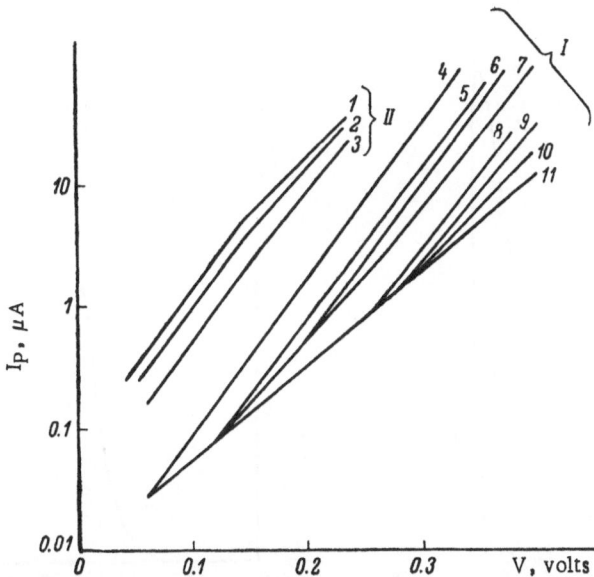

Fig. 3. Forward current—voltage character-istics of a p—n junction produced by ion bom-bardment. Applied force: I) $F < F_{cr}$; II) $F > F_{cr}$. F (g): 1) 20; 2) 18; 3) 17; 4) 16; 5) 15; 6) 14; 7) 13; 8) 12; 9) 10; 10) 8; 11) 0.

$$I_P = J_0 S + \int_\sigma J_P dx dy, \tag{5}$$

where F is the total junction area; σ is that part of the junction area which is deformed and where ΔE_g has a finite value (it is assumed that $\sigma \ll S$).

Figure 2 shows the dependence of the quantity $\Delta I / I_0 = (I_P - I_0)/I_0$, where $I_0 = J_0 S$, on the applied force F. The carrier lifetimes and masses are assumed to be independent of the applied pressure. The value of ΔE_g is found by numerical calculation and numerical integration is carried out over the region σ. Curve 3 in Fig. 2 is plotted on the assumption that the quantity $E_t - E_i$ varies with the pressure in such a way that $\frac{d(E_t - E_i)}{dF} = \frac{d\Delta E_g}{dF}$.

It is evident from Fig. 2 that the theory gives the correct order of magnitude of the ex-perimentally observed effect but the slope of the experimental curve is steeper than that pre-dicted theoretically. The theoretical slope increases somewhat when the pressure dependence of the trap level is taken into account. It must be pointed out that the scatter of the experi-mental data obtained for different points on the surface is quite large but this scatter shifts curve 1 along the ordinate, either upward or downward, without altering its slope. An analysis of the experimental data on the basis of the simplified model given in [14] does not give sat-isfactory results since the experimental curve cannot be described using a single value of the parameter α.

3. When a positive voltage is applied across the p—n junction we must take into account both components of the current density in Eq. (3). Since $J_{rg} \propto n_i \propto \exp(\Delta E_g / 2kT)$ and $J_d \propto n_i^2 \propto \exp(\Delta E_g / kT)$, the diffusion current is more sensitive to deformation. Therefore, under pressure, the diffusion current begins to predominate at those values of the voltage for which the components of the current density J_d and J_{rg} are of the same order of magnitude in the absence of pressure. The forward current—voltage characteristics, obtained experimentally using various values of the force F, are denoted by I in Fig. 3. In the absence of pressure, the current is described by the dependence $I \propto \exp(qV/2kT)$, i.e., the current is due solely to re-

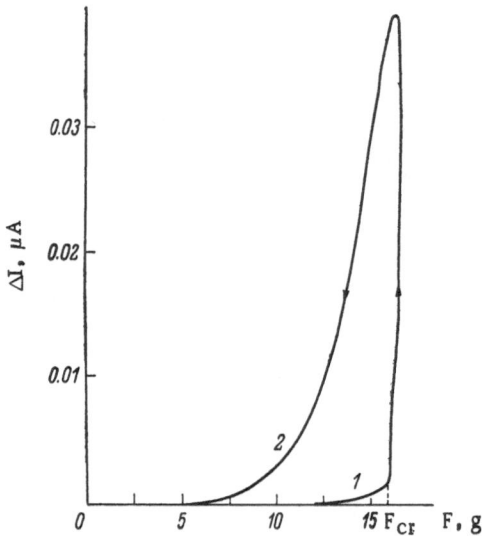

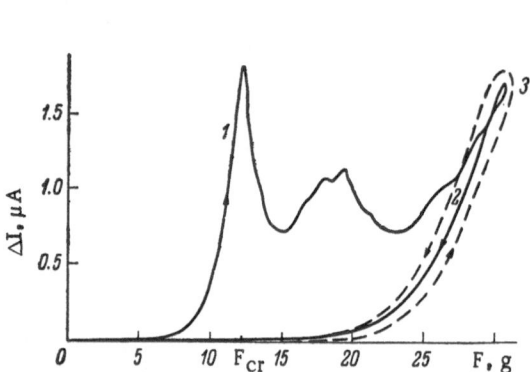

Fig. 4. Dependence of the rise of the reverse current $\Delta I = I_p - I_0$ on the force F: 1) increase of the current during the first loading using $F < F_{cr}$; 2) decrease of the current after application of a force greater than F_{cr} ($I_0 = 0.06\ \mu A$).

Fig. 5. Dependence of the rise of the forward current $\Delta I = I_p - I_0$ on the force F: 1) increase of the current during the first loading with $F < F_{cr}$; 2) decrease of the current during unloading; 3) second and subsequent loadings. V = 0.26 V, $I_0 = 0.5\ \mu A$.

combination. Under pressure, the slope of the dependences of ln I on V begins to rise at high voltages. This is due to the relatively faster rise of the diffusion current.

Reversible and Irreversible Effects

It was reported in 1962 [1] that the influence of pressure on the current was reversible as long as the force F acting on a needle did not exceed a certain critical value. It was suggested in [2, 16] that pressure created generation¬recombination centers. We investigated this process in more detail.

Figure 4 shows the dependence of the reverse current (V = −0.2 V) on the applied force. When the force is increased, the current rises approximately as $I \propto \exp(\beta F^{1/3})$ (curve 1), which is in qualitative agreement with the theoretical model based on changes in the forbidden band width in the region under the needle. At some value of the force, which we shall call the critical force F_{cr}, the current rises rapidly (nonexponentially) and when the force is reduced, the current decreases along a different curve (2). However, if the force is not allowed to exceed the critical value ($F < F_{cr}$), repeated loadings do not alter the dependence I(F) represented by curve 1. The observed hysteresis amounts to approximately 10%.

When the critical force $F = F_{cr}$ is exceeded, the dependence I(F) follows curve 2 if the loading is repeated. The hysteresis increases when the critical force is first exceeded but after repeated loadings the hysteresis of a steady-state cycle returns to 10%.

When the critical force is exceeded, the current at a given point increases. Curve 4 in Fig. 2 shows the dependence of $\Delta I/I_0$ on F obtained after repeated loadings using a force exceeding F_{cr}. It is evident from this figure that the generation−recombination current increases but the slope of the curve does not change.

The nature of the dependence of the forward current on F, observed at low voltages, does not differ from the corresponding dependence of the reverse current. However, at high values of V, when the diffusion current makes a considerable contribution, this dependence is somewhat different (Fig. 5). Curve 1 shows the results obtained for $F < F_{cr}$. When F_{cr} is exceeded, the current at first decreases with increasing force (there is a region of irregular behavior of the current) and then it begins to increase again. When the force is reduced, the current follows curve 2. During repeated loadings, the dependence I(F) is described by curve 3.

The forward current–voltage characteristics, obtained in the range of forces $F > F_{cr}$, are denoted by II in Fig. 3.

At a given depth of the p–n junction the value of F_{cr} decreases with increasing initial current I_0 and increases with increasing needle radius R.

Thus, we must distinguish two ranges of the applied force: weak forces ($F < F_{cr}$), when changes in the current are mainly reversible, and strong forces ($F > F_{cr}$), which are characterized by the appearance of irreversible effects.

The irreversible effects in a semiconductor subjected to forces $F \gtrsim F_{cr}$ are evidently due to the generation of single dislocations or groups of dislocations. Estimates show that shear strains under the needle reach values of the order of 10^{-2} and the corresponding stresses may be sufficient to generate dislocations. Each dislocation distorts the energy band structure and gives rise to additional recombination levels in the forbidden band. This increases the rate of recombination and reduces the minority carrier lifetime. Thus, the formation of dislocations explains the rise of the generation–recombination current which is observed when $F = F_{cr}$ is exceeded (see, for example, curve 4 in Fig. 2). A newly formed dislocation does not alter the value of the derivative of $\Delta I / I_0$ with respect to F, which is governed by the rate of change of the effective width of the forbidden band, but it simply increases the value of the current I_p.

To check this hypothesis, we etched our samples. Such etching was carried out after the application of different values of the force F. If $F \gtrsim F_{cr}$, we observed the characteristic triangles representing the points of emergence of dislocations on the surface, which were absent if the applied force was $F < F_{cr}$.

The authors are grateful to V. V. Zadde and A. K. Zaitseva for supplying the samples and A. B. Kuznetsov for his help in the experiments.

Literature Cited

1. W. Rindner, J. Appl. Phys., 33:2479 (1962).
2. W. Rindner and I. Braun, J. Appl. Phys., 34:1958 (1963).
3. Y. Matukura, Japan. J. Appl. Phys., 3:256, 304, 516 (1964).
4. W. Rindner, J. Appl. Phys., 36:2513 (1965).
5. A. L. Polyakova and V. V. Shklovskaya-Kordi, Fiz. Tverd. Tela, 8:208 (1966).
6. T. Imai, M. Uchida, H. Sato, and A. Kobayashi, Japan. J. Appl. Phys., 4:102 (1965).
7. Y. Matukura, Japan. J. Appl. Phys., 4:632 (1965).
8. Y. Kanda, Japan. J. Appl. Phys., 6:475 (1967).
9. J. J. Wortman, J. R. Hauser, and R. M. Burger, J. Appl. Phys., 35:2122 (1964).
10. A. L. Polyakova, Fiz. Tekh. Poluprov., 1:1164 (1967).
11. V. M. Gusev, V. V. Titov, M. I. Guseva, and V. I. Kurinnyi, Fiz. Tverd. Tela, 7:2077 (1965).
12. I. Goroff and L. Kleinman, Phys. Rev., 132:1080 (1963).
13. J. C. Hensel and G. Feher, Phys. Rev., 129:1041 (1963).
14. K. Bulthuis, J. Appl. Phys., 37:2066 (1966).
15. C. T. Sah, R. N. Noyce, and W. Shockley, Proc. IRE, 45:1228 (1957).
16. W. Bernard, W. Rindner, and H. Roth, J. Appl. Phys., 35:1860 (1964).

DEPENDENCE OF THE SENSITIVITY OF
p—n JUNCTIONS TO NONUNIFORM DEFORMATION
ON THE DEPTH OF THE JUNCTION
BELOW THE SURFACE

V. V. Zadde, A. K. Zaitseva, A. L. Polyakova,
and V. V. Shklovskaya-Kordi

All-Union Scientific Research Institute of Current Sources, Moscow
Acoustics Institute, Moscow

An investigation was made of the influence of nonuniform deformation, produced by a corundum needle pressing against a silicon p—n junction, on the current flowing through the junction. Junctions of different depths were produced by ion bombardment. When the depth of a junction below the deformed surface of a semiconductor was increased, the pressure-induced rise of the current decreased. When the junction depth was 0.1-$0.2\ \mu$, the forward and reverse branches of the current—voltage characteristics were affected by pressure (within the limits of the sensitivity of the apparatus employed). When the junction depth was greater, an appreciable sensitivity of the current to pressure was observed only in the forward branch. However, when a force F, exceeding a certain critical value F_{cr}, was applied to the needle, the reverse branch again became sensitive to pressure. The value of F_{cr} increased with increasing junction depth.

Rindner [1] reported that the rise of the current through a p—n junction, caused by the application of pressure by a needle, depended on the depth of the junction under the surface with which the needle was in contact: when the depth of the junction was increased, the pressure-induced rise of the current decreased.

The present paper describes an investigation of the influence of the depth of shallow p—n junctions (produced by ion bombardment of silicon) on the pressure-induced change in the current.

Measurement Method and Samples

The pressure was applied to a silicon p—n junction by a corundum needle whose radius of curvature at the tip was $R = 30\ \mu$. This needle was used to apply a force F normal to the plane of the junction. The method of loading the p—n junctions and of measuring their electrical properties in the deformed state was described in [2]. The sensitivity of the apparatus was such that we could measure, under static load conditions, a value of the change in the current $\Delta I = I_P - I_0$ of the order of 10^{-3} μA (I_P and I_0 are, respectively, the current under pressure and that in the absence of pressure). Those p−n junctions in which the change in the current under pressure exceeded 10^{-4} μA were arbitrarily called "pressure-sensitive."

We prepared p−n junctions at different depths below the surface of samples of silicon. Silicon plates of 1 $\Omega \cdot$ cm initial resistivity were polished mechanically and their surfaces, oriented in the (111) plane, were doped by bombardment with phosphorus ions of 30 keV energy (the total dose was $6 \cdot 10^3$ μC/cm^2). The source of these ions was an electromagnetic isotope separator. The bombardment produced p−n junctions 0.15-0.3 μ below the surface. When the surface was covered with a film of SiO$_2$, whose thickness was 700 Å, the depth of the p−n junctions was found to be 0.12-0.20 μ.

Deeper p−n junctions (up to 3 μ below the surface) were produced by annealing at 400-900°C for periods ranging from a few minutes to 25 h. The heating of doped silicon samples caused the diffusion of radiation defects and their partial annihilation. It also produced stable complexes consisting of radiation defects and various impurity atoms, which acted as generation−recombination centers. The heating also caused the diffusion of the impurity, which increased the depth of the p−n junction.

The junction depth was determined using two methods. In the first method, a polished cylindrical oblique section was decorated with copper [3]. This method was employed to determine the junction depths exceeding 1 μ. The second method, which was more accurate, was employed for shallower junctions. In this method, anodic oxidation was used to remove successive layers of silicon about 500 Å thick; at each stage of the removal procedure we determined the photo-emf [4]. We assumed that the p−n junction was reached when the photo-emf vanished.

The depths of the investigated junctions ranged from 0.1 to 3 μ.

Experimental Results and Discussion

1. A theoretical analysis reported in [2, 5] shows that a change in the forbidden band width ΔE_g due to the pressure of a needle on the surface of a semiconductor depends on the depth z in the semiconductor: if $z > r_0$, $\Delta E_g \propto 1/z^2$ [here, $r_0 = (RDF)^{1/3}$ is the radius of the contact area between the needle and the semiconductor, R is the radius of curvature of the needle tip, F is the force, D is a combination of the elastic constants of the needle and semiconductor materials]; if $z \lesssim r_0$, it is found that ΔE_g depends weakly on z. Since $\Delta I/I_0 \propto \exp(\Delta E_g/2kT)$, we expect the value of $\Delta I/I_0$ to depend on z. The experimental dependence of $\Delta I/I_0$ on the force F, observed in the forward branches of the current−voltage characteristics of p−n junctions lying at different depths, is shown in Fig. 1 for values of F not exceeding F_{cr} (the definition of F_{cr} is given in [2]). It can be seen from Fig. 1 that there is a scatter of the values of ΔI obtained for different points on the surface of a sample; the dependence of $\Delta I/I_0$ on F obtained at different points are similar but they are shifted relative to one another along the F axis. The maximum shift for a sample with a junction at a given depth is represented by a horizontal line in Fig. 1. For the sake of comparison, Fig. 1 includes a chain curve which represents the theoretical dependence of the generation−recombination current, which takes into account the shift of the trap levels under pressure [2], through a p−n junction at a depth $z = 1$ μ. All the experimental curves in Fig. 1 refer to the range of depths $z \lesssim r_0$ (F = 10 g, $r_0 = 2.5$ μ) and the weak dependence of $\Delta I/I_0$ on z for p−n junctions lying at depths of 0.1-2 μ is in qualitative agreement with the theory.

The value of $\Delta I/I_0$ in the reverse branch of the current−voltage characteristics depended more strongly on the junction depth than in the forward branch. Within the limits of the sen-

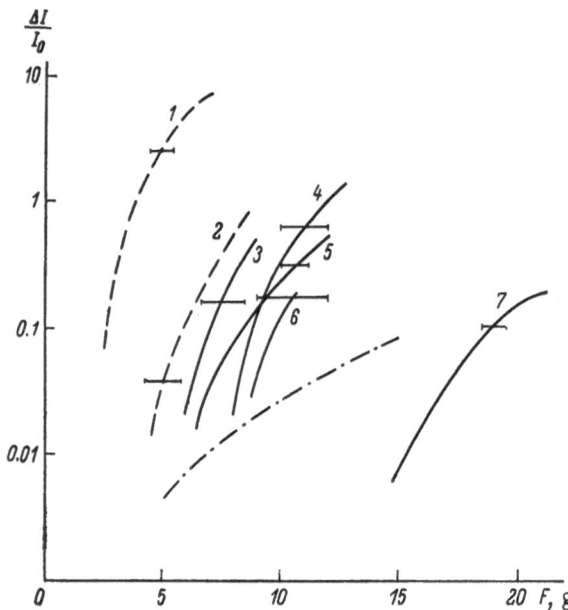

Fig. 1. Dependence of the value of $\Delta I/I_0$ in the forward branches of the current-voltage characteristics (V = 0.2 V) on the force F applied to a needle. The continuous curves represent the experimental data obtained for the samples annealed after ion bombardment; the dashed curves represent the unannealed samples; the chain curve is theoretical. Junction depth (μ): 1) 0.14-0.16; 2) 0.25; 3) 0.4; 4) 0.25; 5) 0.12; 6) 0.8; 7) 2.0.

sitivity of the apparatus, we found that the reverse current was affected by the pressure in the $F < F_{cr}$ range only in the case of the shallowest p-n junctions whose depth did not exceed $0.5\ \mu$. Deeper junctions produced measurable values of ΔI only when the applied load exceeded the critical force. In this range of forces, the values of $\Delta I/I$ for the forward and reverse branches were approximately of the same order of magnitude.

2. We investigated the effects of pressure on the reverse branch of the I(V) characteristic. The behavior of the reverse branch obeyed the theoretical dependence $I \propto \sqrt{V}$ [6] only in the case of the shallowest p-n junctions (curve 1 in Fig. 2). Deeper junctions exhibited the dependence $I \propto V$, represented by curve 3 in Fig. 2 (obviously, the leakage current was considerable). When the shallow p-n junctions, which exhibited the dependence $I \propto \sqrt{V}$, were subjected to pressure, the original dependence was retained and the inverse current increased (curves 1 and 2 in Fig. 2). In the case of deeper p-n junctions, the linear dependence of I on V was retained in the range $F < F_{cr}$ (curve 4 in Fig. 2) but in the range $F > F_{cr}$ the characteristics recorded their "correct" form $I \propto \sqrt{V}$ (curve 5 in Fig. 2).

3. The values of the critical force F_{cr}, which produced irreversible effects in a given junction [2], depended on the depth of that junction. Figure 3 shows the experimental values of F_{cr} recorded for the reverse branch (V = −0.2 V) and different depths of the p-n junction z_0. Since the sensitivity of the reverse branch to pressure was low, the measurements were carried out using the alternating component: in addition to a constant force F, we applied an alternating force $F_1 \ll F$ and measured the alternating component of the p-n junction current. The values of F_{cr} given in Fig. 3 were obtained by averaging the results obtained at a large number of

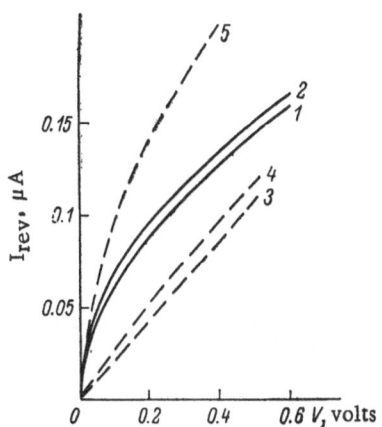

Fig. 2. Reverse branches of the current−voltage characteristics of p−n junctions lying at various depths below the surface: 1) junction at 0.14-0.16 μ, F = 0; 2) junction at the same depth, F = 5 g (F < F_{cr}); 3) junction at 0.5 μ, F = 0; 4) the same junction, F = 15 g (F < F_{cr}); 5) the same junction, F = 17 g (F > F_{cr}).

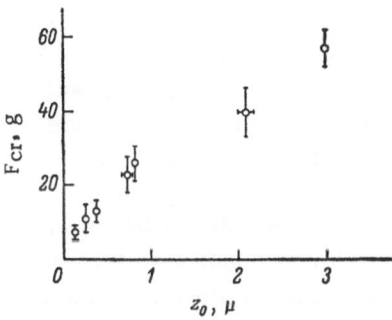

Fig. 3. Dependence of F_{cr} on the depth of p−n junctions.

points on the surface of a p−n junction at a given depth. The maximum scatter of the values of F_{cr} and of z_0 is shown, by vertical and horizontal lines, for each experimental point in Fig. 3.

4. The results obtained can be explained qualitatively using a model described in [2]. The application of forces F ≈ F_{cr} produces single dislocations or groups of dislocations, which are carrier recombination centers. The appearance of these centers increases the generation−recombination current and that is why the reverse branches of the current−voltage characteristics recover their theoretically "correct" form. Such centers play an important role only if they are in the vicinity of a p−n junction. It follows that the greater the depth of a p−n junction below the semiconductor surface the lower are the mechanical stresses produced by the needle in the junction region and the greater is the force which has to be applied before irreversible effects can take place.

This dependence of the sensitivity of the reverse branch on the junction depth makes it possible to explain the nonuniform effect of the pressure of a needle on the surface of a p−n junction, reported by many workers. This nonuniformity may be due to the variation of the depth of the p−n junction under the surface and due to local inhomogeneities of the crystal structure, such as dislocations. Estimates show that the presence of a dislocation in the region under the needle may increase considerably the value of ΔI. The strong nonuniformity of the effect in the case of p−n junctions produced by the diffusion of phosphorus in silicon (described in [7]) is evidently due to both factors. In p−n junctions prepared by the ion bombardment method, the value of ΔI is more uniform across the surface than in the case of diffused junctions. This is evidently due to the fact that the depth of such junctions is more uniform and there are no diffusion-created dislocations.

The authors are grateful to A. B. Kuznetsov for his help in the experiments.

Literature Cited

1.   W. Rindner, J. Appl. Phys., 33:2479 (1962).
2.   A. L. Polyakova and V. V. Shklovskaya-Kordi, Present Collection, p. 160.
3.   B. McDonald and A. Goetzberger, J. Electrochem. Soc., 109:141 (1962).
4.   S. P. Maminova, L. L. Odynets, A. I. Krasil'nikov, and A. N. Pechenov, Zh. Fiz. Khim., 39:531 (1965).
5.   A. L. Polyakova, Fiz. Tekh. Poluprov., 1:1164 (1967).
6.   C. T. Sah, R. N. Noyce, and W. Shockley, Proc. IRE, 45:1228 (1957).
7.   A. L. Polyakova and V. V. Shklovskaya-Kordi, Fiz. Tverd. Tela, 8:208 (1966).

"CRITICAL" TURN-ON CHARGE OF A THYRISTOR

A. I. Uvarov

A. F. Ioffe Physicotechnical Institute
Academy of Sciences of the USSR, Leningrad

A theoretical discussion is given of the conditions for turning on a thyristor by the accumulation of charge in its bases caused by short-duration external excitation or triggering pulses. The concept of a critical charge for this process is defined and justified and the concept of an effective charge accumulated in the thyristor bases is introduced. Charge "effectiveness" coefficients are estimated as a function of the thyristor parameters, the distribution of charge between bases, and the distributions of carriers within the bases. Simple approximate formulas, suitable for practical calculations, are given in the paper.

A forward-biased thyristor in the off state may be subjected to short-duration external excitation or triggering pulses capable of turning it on. These excitations may be gate current pulses of various shapes, light pulses, ionizing radiation pulses, or a rapid rise of the anode voltage. All these excitations result in the accumulation of an additional minority-carrier charge, and an equal excess majority-carrier charge in the thyristor bases. An increase in the excess carrier density enhances the injection of minority carriers by both emitters but it also increases the rate of recombination. The turn-on of the thyristor depends on which of these processes (enhanced injection or stronger recombination) predominates after the end of an external excitation pulse.

The concept of a "critical" charge is often used in discussions of the turn-on conditions in a thyristor [1]. However, a clear definition of the critical charge is still lacking and different workers interpret this concept in different ways. Some regard it as the total charge which has flowed through the anode circuit of the thyristor during the initial stage of the turn-on process [2], while others regard it as the charge which has passed through the gate circuit [3]. A more rigorous definition of the critical charge will be given in the present paper.

A rigorous theoretical discussion of the turn-on conditions in a thyristor is complicated by the need to include recombination processes which are related nonlinearly to the carrier density (ohmic emitter leakage, recombination in a space-charge layer) and which play the dominant role in the initial stage of the turn-on process. An attempt to allow for these processes has been made in [1] but the resultant nonlinear problem can be solved only for a quasi-symmetrical p−n−p−n structure which is of very limited practical importance. Moreover, although the concept of the critical charge has been introduced in [1], its value has not been determined.

We shall use a one-dimensional approximation to linearize the problem of the description of the initial stage of the turn-on process. This approximation can be used to determine exactly the magnitude of the charge in the bases which is sufficient to turn on a thyristor and to estimate the influence of the distribution of this charge between the two bases and of the distributions of carriers within the bases.

To explain the various phenomena and to justify the approximations used to solve the transient problem, we shall consider first the flow of a steady-state current through a thyristor from the point of view of the balance between charge accumulation and carrier recombination.

Steady-State Current through a Thyristor

We shall consider the model of a thyristor shown in Fig. 1. This thyristor has a thin p-type base (thickness w_1) and a thick n-type base (thickness w_2). A constant anode current I flows through the thyristor. The thicknesses of the space-charge layer can be neglected. The distribution of the excess density of the minority carriers y_1 and y_2 along the bases is assumed to have reached its steady state. The dependence of the excess density on the coordinate is taken to be governed by the diffusion lengths L_1 and L_2 in the two bases. Leakage across the emitter E_1 (or recombination in the space-charge layer of this emitter), which depends sublinearly on the minority carrier density, n_{e_1}, near the emitter, is represented by a current I_R. Leakage across the second emitter will be neglected and it will be assumed that the injection efficiency of this emitter is $\gamma_2 = 1$. We shall assume that the carrier density n_{e_1} is known. Then, the steady-state distribution of the minority-carrier density in the first base is described by a law used in transistor theory:

$$y_1 = -n_{e_1} \frac{\text{sh} \frac{x}{L_1}}{\text{sh} \frac{w_1}{L_1}}. \tag{1}$$

Using the equality of the currents flowing through the emitter E_2 and the collector C

$$I_{pe_2} = I_{pc} + I_{nc}, \tag{2}$$

which are assumed to be due to diffusion (low injection levels), and neglecting the leakage current of the collector I_{c0}, we easily obtain

$$y_2 = n_{e_1} \frac{D_1 L_2}{D_2 L_1 \, \text{sh} \frac{w_1}{L_1} \, \text{ch} \frac{w_2}{L_2} - 1} \frac{\text{sh} \frac{x}{L_2}}{}. \tag{3}$$

Here, D_1 and D_2 are the diffusion coefficients of the minority carriers in bases 1 and 2.

Integrating and adding the expressions in Eqs. (1) and (3), we obtain the total excess charge in the two bases for a given value of the minority carrier density near the first emitter n_{e_1}:

$$Q = \frac{q D_1 n_{e_1}}{L_1 \, \text{sh} \frac{w_1}{L_1}} \left(\tau_1 \frac{1 - \alpha_1}{\alpha_1} + \tau_2 \right), \tag{4}$$

where τ_1 and τ_2 are the minority carrier lifetimes in the two bases. The quantity $\alpha_1 = \text{sh}(w_1/L_1)$ is the transport factor of the minority carriers flowing across the first base; this transport factor is assumed to be constant. The transport factor differs from the current gain of the n−p−n transistor on the left-hand side because it does not include the injection efficiency of

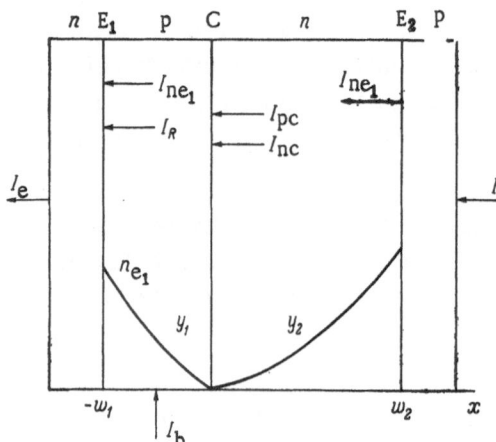

Fig. 1. Model of a thyristor.

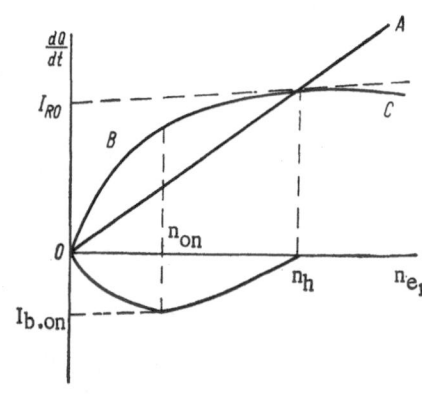

Fig. 2. Balance of the flow of
charge in the first base under
steady-state conditions.

the emitter E_1, which is allowed for separately by introducing the leakage current I_R. The first
term enclosed in parentheses in Eq. (4) refers to the charge in the first base, and the second
term to the charge in the second base.

Charge does not accumulate in the second base because of the condition (2) and because
of the steady-state variation of the carrier density with the coordinate given by Eq. (3). We
shall now consider the balance of charge flow in the first base. A current of majority carriers
("majority" refers to the first base) flows from the first to the second base:

$$I_{pc} = \frac{qD_1 n_{e_1}}{L_1 \operatorname{sh} \frac{w_1}{L_1}} \frac{\alpha_2}{1 - \alpha_2},$$

(5)

where $\alpha_2 = \operatorname{sh}(w_2/L_2)$ is the transport factor which is equal to the current gain of the $p-n-p$
transistor on the right-hand side (we recall that the injection efficiency of the emitter E_2 is
assumed to be unity). The amount of charge lost by volume recombination per unit time is

$$\frac{Q_1}{\tau_1} = \frac{qD_1 n_{e_1}}{L_1 \operatorname{sh} \frac{w_1}{L_1}} \frac{1 - \alpha_1}{\alpha_1}.$$

(6)

Consequently, if we take into account only the carrier injection and volume recombination processes in the two bases, the first base should acquire (in unit time) a charge equal to the difference between the currents defined by Eqs. (5) and (6):

$$\frac{dQ_1}{dt} = \frac{qD_1 n_{e_1}}{L_1 \operatorname{sh} \frac{w_1}{L_1}} \frac{\alpha_1 + \alpha_2 - 1}{\alpha_1 (1 - \alpha_2)}.$$

(7)

This positive flow of carriers into the first base, which depends linearly on the minority
carrier density n_{e_1} near the emitter E_1, is represented by line OA in Fig. 2. Curve OBC in the
same figure represents the leakage current I_R, which depends sublinearly on the carrier density n_{e_1}. In the ohmic leakage case [1], we have

$$I_R \propto \ln \frac{n_{e_1}}{n_p},$$

(8)

where n_p is the equilibrium minority-carrier density in the first base. In the case of recombination in the space-charge layer [4], we obtain

$$I_R \propto \sqrt{\frac{n_{e_1}}{n_p}}. \tag{9}$$

Line OA intersects curve OBC at a carrier density $n_{e_1} = n_h$, which is the carrier density corresponding to the holding current of the thyristor. We can see that when $n_{e_1} > n_h$ the balance of the flow of charge into base 1 is positive (even in the absence of the current I_b) and if the anode voltage is positive, a steady state cannot be achieved. If the carrier density n_{e_1} is somehow made larger than n_h, the thyristor is turned on. As a result of these processes the charge balance to the left of the point $n_{e_1} = n_h$ is negative because in this region a steady state can, in principle, be achieved if the charge deficit is made up by the gate (control) current I_b flowing into the first base:

$$I_b = I_R - \frac{dQ_1}{dt}. \tag{10}$$

The value of this current is represented by a curve lying below the abscissa in Fig. 2. The maximum of this curve corresponds to the minimum turn-on current $I_{b.on}$ under steady-state conditions. If the current flowing through the gate electrode is less than $I_{b.on}$, a steady-state value of the total current is reached and the carrier density n_{e_1} corresponding to this current lies between zero and n_{on}. If the constant gate current exceeds $I_{b.on}$, a steady state cannot be reached (the charge balance is positive).

However, if the gate current is maintained for a short time at a level exceeding $I_{b.on}$, the carrier density n_{e_1}, obtained under quasi-steady-state conditions after the end of the gate current pulse, lies between n_{on} and n_h. Since, in this region, the charge balance in the absence of the gate current is negative, the thyristor will eventually turn itself off. Thus, in order to turn on the thyristor by an excitation pulse of finite duration, the minority carrier density near the emitter E_1, reached under quasi-steady-state conditions, must be equal to the carrier density corresponding to the holding current of the thyristor. Substituting the value of n_h in Eq. (4), we obtain the critical turn-on charge of the thyristor in the quasi-steady-state approximation.

It is not difficult to relate the value of n_h, the turn-on value of the gate current $I_{b.on}$, and the leakage current $I_{R.h}$ corresponding to the holding current, with the value of the holding current I_h. We shall give the relevant formulas without their derivations

$$n_h = \frac{L_1 \operatorname{sh} \frac{w_1}{L_1}}{q D_1} (1 - \alpha_2) I_h, \tag{11}$$

$$I_{R.h} = \frac{\alpha_1 + \alpha_2 - 1}{\alpha_1} I_h. \tag{12}$$

For ohmic leakage through the first emitter, we have

$$I_{b.on} = I_{R.h} \left(1 - \frac{1 + \ln \ln \frac{n_h}{n_p}}{\ln \frac{n_h}{n_p}} \right), \tag{13}$$

and for recombination in the space-charge layer, we obtain

$$I_{b.on} = \frac{I_{R.h}}{4}. \tag{14}$$

Consequently, the critical turn-on charge of the thyristor in the quasi-steady-state approximation is

$$Q_{cr} = (1 - a_2)\left(\tau_1 \frac{1 - a_1}{a_1} + \tau_2\right) I_h. \tag{15}$$

We note that if the first base is thin ($w_1/L_1 \ll 1$), practically the whole charge is concentrated (under quasi-steady-state conditions) in the second (thick) base.

Linear Approximation for the Leakage Current

The excess charge introduced into the thyristor bases by a short excitation pulse may differ considerably from the steady-state charge in respect of both its distribution between the bases and its dependence on the coordinate in each base. However, in a time interval which is a fraction of the diffusion time, the charge distribution in the bases approaches closely the steady-state law [5]. Therefore, under transient conditions the turn-on charge should correspond to a carrier density n_{e_1} which is of the order of but not necessarily equal to n_h. Thus, to solve our problem we can quite justifiably approximate the leakage current I_R by a linear function of n_{e_1}, defined by the tangent to curve OBC (Fig. 2) at the point n_h, which is represented by a dashed line. This means that the leakage current of the emitter is replaced by a constant negative gate current I_{R0} and a surface recombination current which is proportional to the minority carrier density near this emitter and characterized by a surface recombination velocity s:

$$I_R = I_{R_0} + qsn_{e_1}. \tag{16}$$

Omitting the derivations, which are quite easy, we shall now give the expressions for I_{R0} and s:

$$I_{R0} = \left(1 - \frac{1}{\ln \frac{n_h}{n_p}}\right) I_{R.h}, \tag{17}$$

$$s = \frac{D_1}{L_1 \, sh \frac{w_1}{L_1}} \frac{a_1 + a_2 - 1}{a_1 (1 - a_2) \ln \frac{n_h}{n_p}} \tag{18}$$

for ohmic leakage through the emitter and

$$I_{R0} = \frac{1}{2} I_{R.h}, \tag{19}$$

$$s = \frac{D_1}{2L_1 \, sh \frac{w_1}{L_1}} \frac{a_1 + a_2 - 1}{a_1 (1 - a_2)} \tag{20}$$

for recombination in the space-charge layer of the emitter. Comparing Eqs. (17) and (19) with Eqs. (13) and (14) and bearing in mind that n_h is 6-10 orders of magnitude higher than n_p, we find that, in the ohmic leakage case, I_{R0} is within 15% of the value of $I_{b.on}$ and in the case of recombination in the space-charge layer, we have $I_{R0} = 2I_{b.on}$.

Determination of the Critical Charge under

Transient Conditions

We shall now use the approximations described in the preceding sections to consider the condition for the turn-on of a thyristor with charge accumulation in its bases under transient conditions. To determine the influence of the distribution of the charge between the bases and the distribution of carriers within the bases, we shall use the following initial conditions for the minority carrier density:

$$(y_1)_{t=0} = -\frac{Q_1}{ql_1} \frac{\operatorname{sh}\frac{x}{l_1}}{\operatorname{ch}\frac{w_1}{l_1} - 1} \tag{21}$$

and

$$(y_2)_{t=0} = \frac{Q_2}{ql_2} \frac{\operatorname{sh}\frac{x}{l_2}}{\operatorname{ch}\frac{w_2}{l_2} - 1}. \tag{22}$$

We can easily check that, for any value of the arbitrary parameters l_1 and l_2, the total excess-carrier charges in the bases are Q_1 and Q_2. By varying l_1 and l_2, we obtain different distributions of the carrier density along the bases, ranging from localization of all charge near the first emitter (l_1 and $l_2 \to 0$), which is typical of pulse injection of the charge, to a linear dependence of the carrier density (l_1 and $l_2 \to \infty$).

We shall solve the equations of continuity for the carrier densities in the two bases using the low injection level approximation, which is justified because we are considering only the initial stage of the turn-on process:

$$\frac{\partial y_{1,2}}{\partial t} = D_{1,2} \frac{\partial^2 y_{1,2}}{\partial x^2} - \frac{y_{1,2}}{\tau_{1,2}}. \tag{23}$$

We shall ignore the equilibrium value of the minority carrier density.

The boundary conditions are obtained by equating to zero the minority carrier densities near the collector:

$$(y_1)_{x=0} = (y_2)_{x=0} = 0 \tag{24}$$

and by equating the currents flowing through each p−n junction to the total current flowing through the thyristor at any given moment:

$$-qD_1 \left(\frac{\partial y_1}{\partial x}\right)_{-w_1} + I_{R0} + qs\,(y_1)_{-w_1} = -qD_1 \left(\frac{\partial y_1}{\partial x}\right)_0 + qD_2 \left(\frac{\partial y_2}{\partial x}\right)_0 = qD_2 \left(\frac{\partial y_2}{\partial x}\right)_{w_1} = I. \tag{25}$$

Solving Eq. (23) by the operational (Laplace transformation) method and using the initial and boundary conditions just given, we obtain the following operational expression for the current:

$$I\left[\left(\operatorname{ch} a_1 + \frac{\varkappa}{a_1} \operatorname{sh} a_1 - 1\right)(\operatorname{ch} a_2 - 1) - 1\right] =$$

$$= Q_1 \frac{\operatorname{ch} a_2 \left(\operatorname{ch} a_1 + \frac{\varkappa}{a_1} \operatorname{sh} a_1 - \operatorname{ch}\frac{w_1}{l_1} - \frac{\varkappa l_1}{w_1} \operatorname{sh}\frac{w_1}{l_1}\right)}{\left(\operatorname{ch}\frac{w_1}{l_1} - 1\right)\left[\left(\frac{a_1 l_1}{w_1}\right)^2 - 1\right]} + Q_2 \frac{\left(\operatorname{ch} a_1 + \frac{\varkappa}{a_1} \operatorname{sh} a_1\right)\left(\operatorname{ch} a_2 - \operatorname{ch}\frac{w_2}{l_2}\right)}{\left(\operatorname{ch}\frac{w_2}{l_2} - 1\right)\left[\left(\frac{a_2 l_2}{w_2}\right)^2 - 1\right]} - \frac{I_{R0}}{p} \operatorname{ch} a_2, \tag{26}$$

where

$$a_1 = \sqrt{1+p\tau_1}\,\frac{w_1}{L_1}, \quad a_2 = \sqrt{1+p\tau_2}\,\frac{w_2}{L_2}, \quad \varkappa = \frac{sw_1}{D_1} = \frac{w_1}{\xi L_1 \operatorname{sh}\frac{w_1}{L_1}} \cdot \frac{a_1 + a_2 - 1}{a_1(1-a_2)},$$

and $\xi = \ln(n_h/n_p)$ in the ohmic leakage case, $\xi = 2$ for recombination in the space-charge layer.

To solve the problem of the turn-on process (we are interested in the behavior of the current during a time interval considerably longer than the carrier diffusion time) we can conveniently obtain the original function by integration over a contour in a complex plane p. We can easily show that the function $\overline{I}$ has no branch points and all the singularities are simple poles of the first order.

In this case, the current, considered as a function of time, can be expanded as a series of exponential functions of the type

$$I = \sum_k A_k \exp.(p_k t), \tag{27}$$

where p_k are the roots of the transcendental equation

$$\left(\operatorname{ch} a_1 + \frac{\varkappa}{a_1}\operatorname{sh} a_1 - 1\right)(\operatorname{ch} a_2 - 1) = 1. \tag{28}$$

This equation has a single pair of real roots (a_1 and a_2), which give a positive sign of p, defined by

$$p = \frac{1}{\tau_r}. $$

The other complex roots make p a negative quantity (which represents a decrease in the current with time) and are of no interest to us. Thus, the problem of the turn-on is solved by the sign of the preexponential factor in an exponential function with a positive value of p_k, which is defined by real quantities a_1 and a_2. From now onwards, we shall regard a_1 and a_2 as the real roots of Eq. (28).

The numerical solution of a similar equation ($\varkappa = 0$) is given in [5]. This solution can be used for small values of $\varkappa$, which are often justified in practical applications. Calculating the factor in front of the exponential function with a positive argument and equating it to zero, we obtain the following condition for the turn-on of a thyristor

$$k_1 Q_1 + k_2 Q_2 = I_{R0}\tau_r, \tag{29}$$

where τ_r is the time constant of the exponential rise of the current, which can be calculated [6] or determined experimentally; in the ohmic leakage case, I_{R0}, is practically identical with $I_{b.on}$ and in the case of recombination in the space-charge layer, $I_{R0} = 2I_{b.on}$. The quantities k_1 and k_2 are given by

$$k_1 = \frac{\operatorname{ch} a_1 + \frac{\varkappa}{a_1}\operatorname{sh} a_1 - \operatorname{ch}\frac{w_1}{l_1} - \frac{\varkappa l_1}{w_1}\operatorname{sh}\frac{w_1}{l_1}}{\left(\operatorname{ch}\frac{w_1}{l_1} - 1\right)\left[\left(\frac{a_1 l_1}{w_1}\right)^2 - 1\right]}, \tag{30}$$

$$k_2 = \frac{\operatorname{ch} a_2 - \operatorname{ch}\frac{w_2}{l_2}}{(\operatorname{ch} a_2 - 1)\left(\operatorname{ch}\frac{w_2}{l_2} - 1\right)\left[\left(\frac{a_2 l_2}{w_2}\right)^2 - 1\right]}. \tag{31}$$

Interpretation of the Results of the Calculations

We must draw attention to the form of Eq. (29), which is the condition for the turn-on of a thyristor because of the accumulation of an excess charge in its bases. This condition represents the equality of the following charges. On the left, we have the sum of the charges in the bases multiplied by some coefficients k_1 and k_2, which are functions of the carrier distribution in the bases and of the symmetry of the p−n−p−n structure (ratios of the base thicknesses and carrier lifetimes). On the right, we have a charge equal to $I_{R_0}\tau_r$, which depends only on the thyristor parameters. It is natural to call this quantity the c r i t i c a l   c h a r g e of the turn-on of a thyristor:

$$Q_{cr} = I_{R_0}\tau_r. \tag{32}$$

The critical charge defined in this way can be expressed quite simply in terms of the experimentally determined parameters of a thyristor and it reflects the observation that the lower the turn-on value of the gate current and the shorter the turn-on time, the easier it is to turn on the thyristor.

The coefficients k_1 and k_2 can be called the "effectiveness" coefficients of the charges accumulated in the bases, and the quantities $k_1 Q_1$ and $k_2 Q_2$ may be called the effective charges in the bases.

The condition for turning on a thyristor by a short excitation pulse can, therefore, be formulated as follows: in order to turn on a thyristor, the sum of the effective charges accumulated in the thyristor bases at the end of an excitation pulse should be greater than the critical turn-on charge.

Analysis of the "Effectiveness" Coefficients

of Charges in Bases

Examination of the expressions (30) and (31) for the "effectiveness" coefficients shows that k_1 and k_2 increase with increasing l_1 and l_2 and that they have minimum values when $l \rightarrow 0$, i.e., when the charge is concentrated in the immediate vicinity of an emitter. The closer the initial charge is to the collector ($l \rightarrow \infty$), the stronger is its influence. Figure 3 shows, on a semilogarithmic scale, the results of a numerical calculation of k_1 as a function of the ratio of the diffusion times of the minority carriers following through the bases θ_1/θ_2 ($\varkappa = 0$ and $\tau_1 = \tau_2$). Curve I in Fig. 3 corresponds to $l_1 = 0$ (the charge is concentrated near the first emitter) and curve II represents the case $l_1 \rightarrow \infty$ (the charge is distributed linearly between the collector and the first emitter). We can see that when the first base is thinner than the second ($\theta_1/\theta_2 < 1$), right up to the quasi-symmetrical case ($\tau_1 = \tau_2$, $\theta_1 = \theta_2$), k_1 is equal to unity (to within 15%) for any distribution of carriers along the base.

Figure 4 shows, on a double logarithmic scale, the dependence of k_2 on the ratio θ_1/θ_2 for the same cases as in Fig. 3 (curve I corresponds to $l_2 \rightarrow \infty$ and curve II corresponds to $l_2 = 0$). We can see that k_2 is not greatly affected by the charge distribution in the base but it is affected more than k_1. However, the value of k_2 depends strongly on the degree of asymmetry of the p−n−p−n structure: in the quasi-symmetrical case, $k_2 = 1$ but it decreases strongly with decreasing θ_1/θ_2. In practical applications, we can use the approximation

$$k_2 \cong \sqrt{\frac{\theta_1}{\theta_2}}$$

irrespective of the carrier distribution along the base.

Thus, the turn-on condition simplifies to:

$$Q_1 + \sqrt{\frac{\theta_1}{\theta_2}} Q_2 = I_{R_0}\tau_r , \tag{33}$$

irrespective of the nature of the carrier distribution in the bases. We note that the charge accumulated in the thin base is more effective than the charge in the thick base.

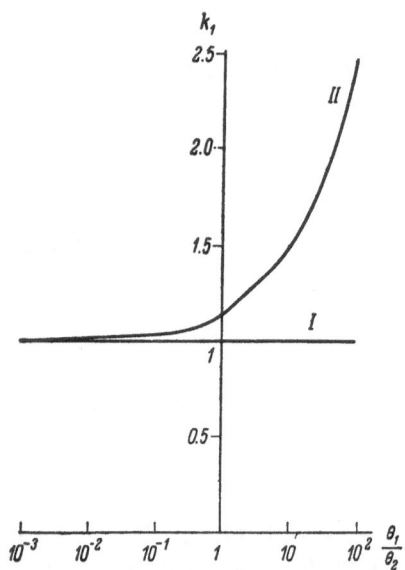

Fig. 3. Dependence of the coefficient k_1 on θ_1/θ_2.

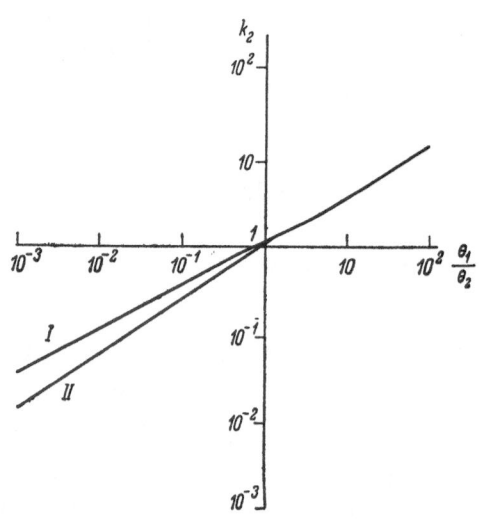

Fig. 4. Dependence of the coefficient k_2 on θ_1/θ_2.

Allowance for Leakage through the Second Emitter

In real thyristors, the leakage through the emitter adjoining the thin base is much more important because this emitter is usually separated by a low-quality p−n junction from the base and, moreover, it is normally shunted by an external resistance. Nevertheless, the leakage through the second emitter, adjoining the thick base, may be important in some low-noise thyristors.

We shall derive the turn-on conditions for a thyristor in the case of leakage through both emitters, and we shall approximate the leakage currents by the linear functions

$$I_{R_1} + qs_1 (y_1)_{-w_1}$$

for the first emitter and

$$I_{R_2} + qs_2 (y_2)_{w_2}$$

for the second emitter. We shall use only an approximate formula, which is convenient in practical calculations. In this case, the turn-on condition is formulated as follows:

$$Q_1 + \sqrt{\frac{\theta_1}{\theta_2}}\, Q_2 = \tau_r \left(I_{R_1} + \sqrt{\frac{\theta_1}{\theta_2}}\, I_{R_2} \right). \tag{34}$$

We can see that the above expression differs from Eq. (33) by the presence of an additional term which is the contribution of the leakage current of the second emitter to the critical charge; the "effectiveness" coefficient relating to this term is the same as for the charge in the second base. In the quasi-symmetrical case the charges in the bases and the leakage currents are equivalent. In asymmetrical structures the leakage current of the second emitter makes only a small contribution to the critical charge and to the effective charge in the thick base.

Unfortunately, it is difficult to relate I_{R_1} and I_{R_2} to any external characteristics of a thyristor but we can compare the right-hand side of Eq. (34) with the value of $I_{b.on}$ when the thyristor is controlled via the thin base and ohmic leakage occurs in both emitters:

$$I_{b.on} \cong I_{R_1} + \frac{a_2}{a_1} I_{R_2}. \tag{35}$$

Although this expression is not identical with the quantity enclosed in parentheses in Eq. (34), the difference between them is small and, therefore, in the case of ohmic losses in both emitters, we are quite justified in using the following turn-on condition:

$$Q_1 + \sqrt{\frac{\theta_1}{\theta_2}}\, Q_2 = \tau_r I_{b.on} \tag{36}$$

bearing in mind that the first base is thinner than or equal in thickness to the second base and that $I_{b.on}$ refers to the gate current flowing into the thin base.

Conclusions

1. A linear approximation for the leakage currents of the emitters is used to develop a mathematically rigorous but practically convenient theory of the turn-on of a thyristor by accumulation of charge in its bases under the influence of a short-duration gate pulse.

2. The concept of the original turn-on charge is defined rigorously and the value of this charge is given by the product of the time constant of the current rise during the turn-on process and of the turn-on current of the base to which the gate electrode is attached.

3. The effective charges in the bases, which determine the turn-on process, are introduced. The "effectiveness" coefficients of these charges depend weakly on the nature of the carrier distribution in the two bases. For a thin base the "effectiveness" coefficient is practically equal to unity and for a thick base this coefficient can be approximated by $(\theta_1/\theta_2)^{\frac{1}{2}}$, where θ_1 and θ_2 are the diffusion transit times of the minority carriers through the two bases.

4. The condition for the turn-on of a thyristor by the charge accumulated in the bases is formulated as an equality of the sum of the effective charges in the bases and the critical turn-on charge of the thyristor.

Literature Cited

1. A. A. Lebedev and A. I. Uvarov, Radiotekhn. i Élektron., 12:895 (1967).
2. M. A. Berg and S. A. Garyainov, Radiotekhnika, 17(1):51 (1962).
3. R. E. Smolyanskii, in: Abstracts of Papers presented at Third Conf. on Physics of p−n Junctions, Tbilisi (1966).
4. C. T. Sah, R. N. Noyce, and W. Shockley, Proc. IRE, 45:1228 (1957).
5. A. A. Lebedev, A. I. Uvarov, and V. E. Chelnokov, Radiotekhn. i Élektron., 11:1458 (1966).
6. T. Misawa, J. Electron. Control, 7:523 (1959).

SPECTRAL NARROWING OF THE RADIATION FROM InP AND InP$_{1-x}$As$_X$ INJECTION LASERS

P. G. Eliseev and I. Ismailov

P. N. Lebedev Physics Institute
Academy of Sciences of the USSR, Moscow

An investigation was made of the evolution of the radiation spectra of diffused p−n junctions in InP and InP$_{0.94}$As$_{0.06}$ in the current density range 1-10^4 A/cm^2 at 77°K. The narrowing of the spectral band was attributed to stimulated emission in the current density range 10^3-10^4 A/cm^2. Changes in the spectra at current densities 1-10 A/cm^2 were attributed to a transition from radiation originating from diagonal tunneling to injection radiation. Spectral narrowing in the stimulated emission range was used to estimate the loss and gain for the radiation propagated along the p−n junction. The loss was 3 cm^{-1} for InP and 8-16 cm^{-1} for InP$_{0.94}$As$_{0.06}$.

The present paper describes a study of the evolution of the electroluminescence spectra of diffused p−n junctions in InP and InP$_{1-x}$As$_x$ (x ≈ 0.06) in the current density range 1-10^4 A/cm^2 at 77°K. The purpose of our investigation was to determine the nature and the causes of the changes observed in the line width at various current densities.

The method used in the fabrication of the lasers has been described before [1-3]. The principal properties of the investigated samples are listed in Table 1. All the samples, except Nos. 480 and 29, emitted coherent radiation when the current density reached a sufficiently high value. Figure 1 shows the dependences of the radiation peak positions on the current density and Fig. 2 shows the dependences of the radiation band width on the current density. The results can be summarized as follows.

1. The nature of the dependence of the spectral peak position on the current density was affected by the degree of doping. The slopes of

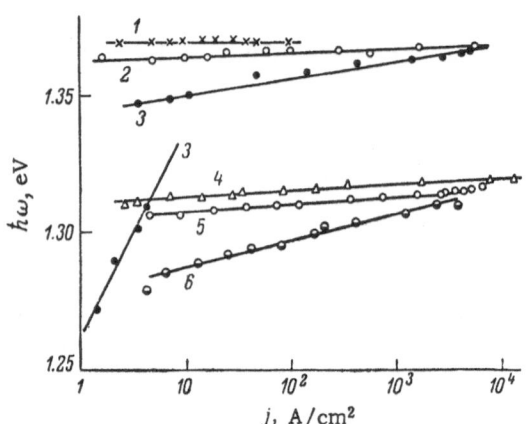

Fig. 1. Dependence of the position of the spectral peaks on the current density at 77°K. Samples: 1) 480; 2) 39; 3) 29; 4) 89; 5) 27; 6) 119B.

TABLE 1

Compound	Sample No.	Resonator length, mm	Junction area, cm²	Donor concentration, cm⁻³	Threshold current density, A/cm²	Peak shift charact. ε, meV	Radiation band width for $10 < j < 10^3$, A/cm²
InP	480	0.475	$1.08 \cdot 10^{-3}$	$5 \cdot 10^{16}$	—	0	22
	119A	1.10	$6.05 \cdot 10^{-3}$	$5 \cdot 10^{17}$	5300	0.8	30
	39	1.10	$6.05 \cdot 10^{-3}$	$5 \cdot 10^{17}$	5100	0.8	27.5
	29	0.8	$7 \cdot 10^{-3}$	$2 \cdot 10^{18}$	—	2.8	34
InP$_{0.94}$As$_{0.06}$	89	0.47	$2.8 \cdot 10^{-3}$	$6.5 \cdot 10^{17}$	10400	1.2	36
	27	0.35	$3.5 \cdot 10^{-3}$	$8 \cdot 10^{17}$	7000	1.3	32
	119B	0.47	$2.35 \cdot 10^{-3}$	$2.7 \cdot 10^{18}$	11000	3.9	40

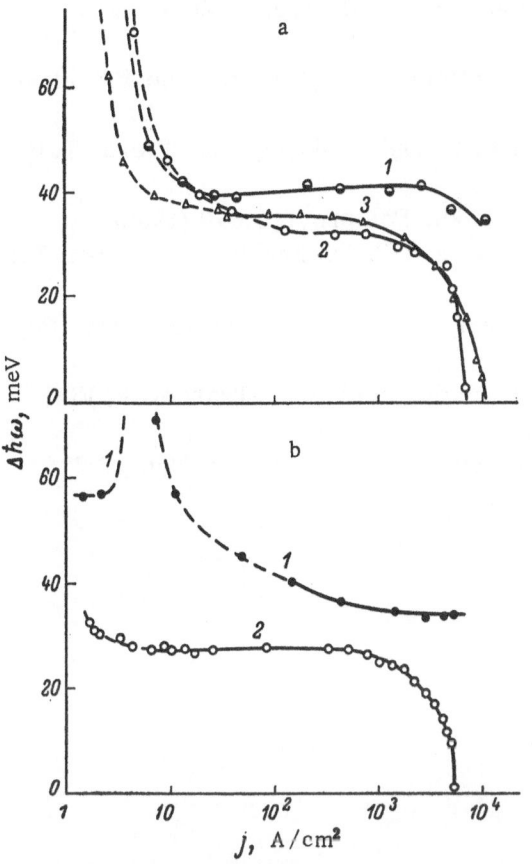

Fig. 2. Dependence of the radiation band width on the current density at 77°K. a) InP$_{0.94}$As$_{0.06}$, samples: 1) 119B; 2) 27; 3) 89; b) InP samples: 1) 29; 2) 39. Radiation was observed in the p−n junction plane.

the experimentally determined lines in Fig. 1 increased with increasing donor concentration. An energy constant ε, describing this slope, did not exceed 4 meV for $j > 10$ A/cm² (Table 1). This was important because the model of a density-of-states tail whose population decreased exponentially in the forbidden band could not be used to explain the shift of the spectral peak for $\varepsilon < kT$. Such a shift would be permissible for a Gaussian density-of-states distribution but the law governing the shift would not be exponential. Obviously, the experimentally determined exponential law was only a rough approximation, which was satisfactory because of the limited accuracy and the narrow spectral range.

2. The appearance of coherent emission from a laser with a Fabry−Perot resonator was preceded by a strong narrowing of the band, which was due to stimulated transitions. This narrowing was observed for radiation emitted in the p−n junction plane but not at right angles to this plane. The internal loss α and the gain $k = \beta j$ could be estimated from the current density at the onset of the spectral narrowing and at the laser threshold [4]. Such estimates, obtained for InP with a donor concentration of $5 \cdot 10^{17}$ cm⁻³, gave $\alpha \approx 3$ cm⁻¹, $\beta = 3 \cdot 10^{-3}$ cm/A. The corresponding parameters for different samples of InP$_{0.94}$As$_{0.06}$ were $\alpha = 8{-}16$ cm⁻¹, $\beta = (4{-}8) \cdot 10^{-3}$ cm/A. These estimates were in good agreement with our earlier measurements [2, 5].

3. Considerable changes occured in the spectra at low current densities of 1−10 A/cm². These changes were similar to those observed earlier in the spectra of GaAs diodes [6, 7] and were more likely to be associated with a change of the radiative recombination mechanism than with stimulated emission. The electroluminescence of GaAs [6] and InP [8] diodes is characterized, in this range of currents, by a transition from the diagonal tunneling mechanism to the injection radiation mechanism; in the transition region between the two mechanisms, two

bands of comparable intensities overlap and give rise to a band considerably wider than either of the component bands. This behavior is clearly exhibited by sample No. 29. Figure 1 shows the position of its radiation peak in the form of two branches with strongly differing slopes. At low values of the current density, the peak shifts rapidly and the width of the radiation band is about 56 meV (Fig. 2b). Then, the experimentally determined band width passes through a maximum and decreases to 34 meV. The higher value represents radiation due to the diagonal tunneling mechanism and the lower value represents the injection radiation. The contribution of the diagonal tunneling to radiation increases when the degeneracy is strong and when the thickness of the space-charge region of the p−n junction is small. In the investigated samples, the injection radiation predominated at $j > 10$ A/cm^2.

Literature Cited

1.   K. Weiser, R. S. Levitt, M. I. Nathan, G. Burns, and J. Woodall, Trans. Metal. Soc. AIME, 230:271 (1964).
2.   N. G. Basov, P. G. Eliseev, I. Ismailov, A. Ya. Nashel'skii, I. Z. Pinsker, and V. Yakobson, Fiz. Tverd. Tela, 8:2610 (1966).
3.   P. G. Eliseev, I. Ismailov, A. Ya. Nashel'skii, and V. Z. Ostrovskaya, Fiz. Tverd. Tela, 8:1283 (1966).
4.   P. G. Eliseev, I. Ismailov, and E. M. Kistova, Fiz. Tekh. Poluprov., 2:610 (1968).
5.   N. G. Basov, P. G. Eliseev, I. Ismailov, I. Z. Pinsker, and V. P. Strakhov, Zh. Tekh. Fiz., 37:349 (1967).
6.   R. C. C. Leite, J. C. Sarace, D. H. Olson, B. G. Cohen, J. M. Whelan, and A. Yariv, Phys. Rev., 137:A1583 (1965).
7.   T. N. Danilova, L. M. Kogan, S. S. Meskin, D. N. Nasledov, and B. V. Tsarenkov, Fiz. Tverd. Tela, 8:2462 (1966).
8.   P. G. Eliseev, I. Ismailov, A. B. Ormont, and A. É. Yunovich, Fiz. Tverd. Tela, 8:3383 (1966).

DEPARTURE FROM QUASI-EQUILIBRIUM IN SEMICONDUCTORS AND MULTIMODE LASER EMISSION

V. S. Mashkevich and V. A. Parnyuk

Physics Institute
Academy of Sciences of the Ukrainian SSR, Kiev

The method of rate equations is used to examine the influence, on the spectral composition of laser radiation, of a departure from steady-state quasi-equilibrium in the active medium, i.e., a departure from statistical equilibrium between elementary excitations in the electron subsystem (excited states of impurity centers, free carriers, and excitons). It is shown that the cause of multimode emission in the case of p−n junction lasers with typical parameters is a spatial inhomogeneity of the mode field and not a spectral inhomogeneity associated with nonuniform broadening of the luminescence line.

One of the most important problems in generation of coherent radiation is the problem of multimode emission. The cause of multimode emission under steady-state conditions (to which our analysis is restricted) is a departure from quasi-equilibrium in the active medium, i.e., a departure from statistical equilibrium between elementary excitations in the electron subsystem (excited states of impurity centers, free carriers, and excitons) [1, 2].

Two types of departure from quasi-equilibrium play the dominant role in lasers: a spatial inhomogeneity, associated with the mode field inhomogeneity, and a spectral inhomogeneity, associated with nonuniform broadening of the luminescence line.

There have been several theoretical investigations of the generation of coherent radiation in solid-state lasers with spatial inhomogeneities [3-7]. Coherent emission in the case of spectral inhomogeneities has been examined for the case of gas lasers [8], semiconductor lasers [9-11], and solid-state lasers [3, 12, 13].

We shall consider the case when only modes of a certain type, exhibiting the lowest losses, can be generated. Moreover, we shall assume that the mode spacing, $\Delta\omega$, is small compared with the width of an elementary excitation level, i.e., compared with the width in the homogeneous case. A spatial inhomogeneity gives rise to modes at neighboring frequencies ("continuous" spectrum) and a spectral inhomogeneity produces modes which are not close neighbors ("discrete" spectrum) [1, 2]. Moreover, as $\Delta\omega \to 0$, the multimode emission in the case of a spatial inhomogeneity begins at the threshold pumping level ("soft" excitation conditions) and in the case of a spectral inhomogeneity it begins at a pumping level exceeding the threshold by a finite amount ("hard" excitation conditions).

The current state of the theory of multimode emission of semiconductor lasers can be summarized as follows. An attempt to make an allowance for the spatial inhomogeneity is reported in [14]. However, the initial equations in that paper are in a form suitable for a solid-state laser (at that time a theory of semiconductor lasers was not available even for the homogeneous case). The spectral inhomogeneity is discussed in [9-11]. Unfortunately, the results reported apply to the case when the interval between the mode frequencies is large compared with the width of an elementary excitation level (this is stated clearly in [11] and implied in [9, 10]). In such a case, coherent emission is observed in modes which are not close neighbors and such emission differs basically from the case of a small frequency interval. The case of a small interval is usually realized in semiconductor lasers.

Thus, there is as yet no theory of the spatial inhomogeneity in semiconductor lasers and no analysis has been made of the most important case of the spectral inhomogeneity.

The purpose of the present paper is to fill this gap in the theory of semiconductor lasers. An allowance is made for the spatial and spectral inhomogeneities in a p−n junction laser. A modification of Pikus' theory [10, 15] is used to describe the homogeneous case.

1. Spatial Inhomogeneity

We shall give the main results of a general theory which takes into account the spatial inhomogeneity. We shall restrict our treatment to the case of a small frequency interval between modes. The main interest lies in the dependence of the number of lasing modes M on the pumping level P, i.e., on the number of elementary excitations generated per unit volume per unit time. This dependence is given by the expression

$$M = 1 + E \left\{ K \left[\frac{P - P_t}{P_t} \right]^{\frac{1}{3}} \right\}. \tag{1}$$

Here, E is used to denote the integral part of a number; P_t is the threshold pumping level;

$$K = \left(\frac{2L}{\pi \tilde{c}} \right)^{\frac{2}{3}} \left\{ \frac{4 \left[\left(\frac{d\Phi}{dy} \right)_t - G \frac{V}{v} \left(\frac{d\rho}{dy} \right)_t \right] P_t}{3 d \chi \left(y_t \right) \left(\frac{d^2\Phi}{d\omega^2} \right)_t} \right\}^{\frac{1}{3}}; \tag{2}$$

L is the resonator length; $\tilde{c}$ is the effective velocity of light in the resonator; y is a parameter describing the state of matter and the subscript "t" is used to denote that a given quantity is considered at a threshold value of y and at the emission frequency $\omega = \omega_t$; the gain of the j-th mode (per unit volume per unit time) is given by

$$B^j (\mathbf{r}, \, y) = \Phi^j (y) [1 + f^j (\mathbf{r})], \tag{3}$$

where $f^j (\mathbf{r})$ is due to an inhomogeneity of the field of the j-th mode, and

$$\int_{(v)} f^j (\mathbf{r}) \, d\mathbf{r} = 0, \tag{4}$$

$$\int_{(v)} f^j (\mathbf{r}) f^l (\mathbf{r}) \, dr = v d\delta_{jl}. \tag{5}$$

Here, v is the volume of the amplifying region; ρ is the density of elementary excitations; $G(V/v)\rho_t$ is the reciprocal of the lifetime of a mode quantum, governed by the absorption of excitation; V is the mode volume;

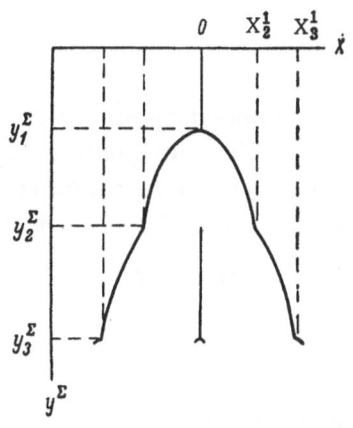

Fig. 1. Dependence of the reduced frequency on the radiation intensity.

$$\varkappa(y_t) = 4D\left(\frac{d\rho}{dy}\right)_t \left(\frac{2\pi}{\lambda}\right)^2 + \left(\frac{d\Gamma}{dy}\right)_t + \sum_i^{(H)}\left(\frac{d\Phi}{dy}\right)_t q^i + \left(\frac{d\Phi}{dy}\right)_t \frac{(P-P_t)}{\Phi_t} , \quad (6)$$

where D is the diffusion coefficient of excitations; λ is the wavelength of laser radiation; Γ is the number of excitations annihilated per unit time per unit volume in spontaneous and nonradiative transitions, summed over nonlasing modes; q^i is the number of quanta.

No difficulty is encountered in applying these results to any specific laser since the theory of the homogeneous case gives the threshold values of the quantities in Eqs. (1), (2), and (6).

2. Spectral Inhomogeneity

The most interesting is the case when the gain is a smooth function of the frequency, i.e., when the gain varies little within the limits set by the width of an elementary excitation, and the width of the luminescence line is large compared with that obtained in the homogeneous case. Then, using [16], we can deduce the dependence of the reduced frequency on the radiation intensity shown in Fig. 1. In this figure

$$X = \frac{\Omega}{\beta} , \quad (7)$$

where $\Omega = \omega - \omega_m$; $\omega_m = \omega_m(P)$ is the frequency of maximum gain, calculated ignoring regeneration; β is the half-width of the line in the homogeneous case; the line profile corresponding to an elementary excitation is assumed to be Lorentzian:

$$y^\Sigma = Cq^\Sigma = C\frac{(P-P_t)}{\Phi_t} ; \quad (8)$$

here, q^Σ is the total number of quanta in all the laser modes; C is a constant which depends on the nature of the system; X^1 is the value of X for a laser mode characterized by $X \neq 0$; the subscripts of the quantities in Fig. 1 represent the final number of modes at the end of laser emission. The dependence $X^1(y^\Sigma)$ is universal. The full expression is given in [16]. Here, we shall simply note that

$$X_2^1 = \frac{1}{2}, \quad X_3^1 \cong \frac{3}{2} ; \quad (9)$$

$$y_1^\Sigma = 8, \quad y_2^\Sigma = 18, \quad y_3^\Sigma \cong 29.4 . \quad (10)$$

3. Junction Laser. Homogeneous Case

We shall use the theory of a p−n junction laser developed for the homogeneous case by Pikus [10, 15]. The only change that will be introduced is related to the definition of the average field intensity E_0 in a p−n junction (we shall describe a p−n junction laser using the symbols employed in [10]). Since the potential in the active region is given by the linear approximation

$$\varphi(x) = -exE_0, \quad (11)$$

we can assume that the neutrality condition is satisfied in this region. The distribution of acceptors in the active region is also assumed to be linear. Therefore, we shall restrict ourselves to the linear approximation for the density of free carriers in the neutrality condition:

$$(p-n)_{\bar{x}} + \frac{\partial (p-n)}{\partial x}\Big|_{\bar{x}} (x - \bar{x}) = -N_d \gamma (x - x_0). \tag{12}$$

Here, p is the density of holes; n is the density of electrons; N_d is the concentration of donors; γ is a parameter which determines the acceptor distribution gradient; x is the coordinate along the normal to the p−n junction plane; x_0 is the point at which the acceptor and donor concentrations are equal; $\bar{x}$ is the point at which the gain has its maximum value; x = 0 is defined as in [10].

The parameter y, which describes the state of the active medium, is defined by

$$y = \Delta U = U - \left(\frac{1}{e}\right)\mathscr{E}_g, \tag{13}$$

where ΔU is the voltage drop across the p−n junction; $\mathscr{E}_g$ is the forbidden band width.

An equation for the definition of E_0 as a function of ΔU is obtained from Eq. (12):

$$\frac{\partial (p-n)}{\partial x}\Big|_{\bar{x}} = -\gamma N_d. \tag{14}$$

The results for the homogeneous case are of interest to us only as a starting point in the examination of the inhomogeneous cases. We shall need the following quantities: E_0; L_a, which is the length of the active region along the x axis, i.e., the half-width of the gain of a lasing mode considered as a function of x; ΔU_t; j_t, which is the threshold current density; $n(\bar{x})$, $p(\bar{x})$. These and other quantities are defined in the homogeneous case in the same way as in [10], except for E_0, which is given by Eq. (14).

Following Pikus [10], we shall consider only the two limiting cases of low and high temperatures:

$$\frac{e\Delta U_t}{3'kT} \gg 1, \tag{15}$$

$$\frac{e\Delta U_t}{3kT} \ll 1. \tag{16}$$

At low temperatures [Eq. (15)], we find that

$$eE_0 = \frac{\gamma N_d \pi^2 \hbar^3}{(e\Delta U)^{1/2} m_p^{*3/2} Z'}, \tag{17}$$

where

$$Z' = \left[1 + \frac{m_p^* - m_n^*}{3(m_p^* + m_n^*)}\right]^{\frac{1}{2}} \left(\frac{m_n^*}{m_p^*}\right)^{\frac{3}{2}} + \left[1 - \frac{m_p^* - m_n^*}{3(m_p^* + m_n^*)}\right]^{\frac{1}{2}}; \tag{18}$$

$$L_a = \frac{2}{3} \cdot \frac{e\Delta U}{eE_0}; \tag{19}$$

$$e\Delta U_t = \left\{\frac{\left(\frac{3}{2}\right)^{\frac{3}{2}} \pi^2 \hbar^3 \gamma N_d n^2 \mathscr{E}_g^2 a_0 a_n^*}{w_0 c^2 (\bar{m}^*)^{\frac{3}{2}} (m_p^*)^{\frac{3}{2}} Z'}\right\}^{\frac{1}{2}}; \tag{20}$$

n is the refractive index; w_0 is a constant which governs spontaneous emission [10]; α_n^* is the effective absorption coefficient [10], which takes into account the absorption by free carriers and resonator losses; a_0 is the length of the luminous region along the x axis; m_n^*, m_p^* are, respectively, the effective masses of electrons and holes; $1/m^* = (1/m_n^*) + (1/m_p^*)$;

$$j_t = \frac{4\sqrt{2}\, w_0 \bar{m}^{*1,5} (e\Delta U_t)^{\frac{5}{2}}}{15\eta E_{0t}\pi^2\hbar^3} \; ; \tag{21}$$

η is the internal quantum yield [10];

$$n(\bar{x}) = \frac{1}{3\pi^2\hbar^3}\left\{ m_n^* e\Delta U_t \left[1 + \frac{m_p^* - m_n^*}{3(m_p^* + m_n^*)} \right] \right\}^{\frac{3}{2}} ; \tag{22}$$

$$p(\bar{x}) = \frac{1}{3\pi^2\hbar^3}\left\{ m_p^* e\Delta U_t \left[1 - \frac{m_p^* - m_n^*}{3(m_p^* + m_n^*)} \right] \right\}^{\frac{3}{2}} . \tag{23}$$

At high temperatures [Eq. (16)], we obtain

$$eE_0 = 0.8\, \frac{(2\pi\hbar^2)^{\frac{3}{2}} \gamma N_d}{(kT)^{\frac{1}{2}} (m_n^*)^{\frac{3}{2}} Z''} , \tag{24}$$

$$Z'' = \left(\frac{m_p^*}{m_n^*}\right)^{\frac{3}{2}} + 1, \tag{25}$$

$$e\Delta U_t = \frac{3}{2\bar{m}^*}\left[\frac{a_0 a_n^* e E_0 n^2 \mathscr{E}_g^2}{w_0 c^2} \right]^{\frac{2}{3}} , \tag{26}$$

$$L_a = \frac{2kT}{eE_0} , \tag{27}$$

$$j_t = \frac{1.78\sqrt{2}\, w_0 (\bar{m}^*)^{\frac{3}{2}} (kT)^{\frac{5}{2}}}{\eta E_0 \pi^2\hbar^3} , \tag{28}$$

$$n(\bar{x}) = 2\left(\frac{m_n^* kT}{2\pi\hbar^2}\right)^{\frac{3}{2}} \frac{\exp\left[\frac{e\Delta U}{2kT}\left\{ 1 + \frac{m_p^* - m_n^*}{3(m_p^* + m_n^*)} \right\} \right]}{1 + 0.27\exp\left[\frac{e\Delta U}{2kT}\left\{ 1 + \frac{m_p^* - m_n^*}{3(m_p^* + m_n^*)} \right\} \right]} , \tag{29}$$

$$p(\bar{x}) = 2\left(\frac{m_p^* kT}{2\pi\hbar^2}\right)^{\frac{3}{2}} \frac{\exp\left[\frac{e\Delta U}{2kT}\left\{ 1 - \frac{m_p^* - m_n^*}{3(m_p^* + m_n^*)} \right\} \right]}{1 + 0.27\exp\left[\frac{e\Delta U}{2kT}\left\{ 1 - \frac{m_p^* - m_n^*}{3(m_p^* + m_n^*)} \right\} \right]} . \tag{30}$$

4.   Inhomogeneous Case

We shall consider first the spatial inhomogeneity. We shall start by writing down the quantities in Eqs. (2), (3), and (6) in a form appropriate to a p−n junction laser. We find that

$$\dot{P} = \frac{j}{eL_a} , \tag{31}$$

TABLE 1. Results of Calculations

j_t, A/cm²	ΔU_t, V	$n(\bar{x})$, cm⁻³	$p(\bar{x})$, cm⁻³	L_a, cm	$4D_{\text{eff}}\left(\frac{dh}{d\Delta U}\right)_t\left(\frac{2\pi}{\lambda}\right)^2 L_a$ (sec·cm²·V)⁻¹	$\frac{d\Gamma}{d\Delta U}$ (sec·cm²·V)⁻¹	$\left(\frac{d\Phi}{d\Delta U}\right)_t(j-j_t)\frac{1}{e\Phi_t}$ (sec·cm²·V)⁻¹	K	$\left(\frac{j_{\text{spl}}-j_t}{j_t}\right)_1$
Low temperatures										

Low temperatures

j_t, A/cm²	ΔU_t, V	$n(\bar{x})$, cm⁻³	$p(\bar{x})$, cm⁻³	L_a, cm	$4D_{\text{eff}}(\ldots)$	$\frac{d\Gamma}{d\Delta U}$	$(\frac{d\Phi}{d\Delta U})_t(\ldots)$	K	$(\ldots)_1$
$2.2\cdot10^2$	$4.6\cdot10^{-3}$	$1.52\cdot10^{16}$	$1.3\cdot10^{17}$	$4.1\cdot10^{-5}$	$5\cdot10^{27}$	$2\cdot10^{26}$	$1.6\cdot10^{16}$	0.33	$7.4\cdot10^5$
$1.4\cdot10^3$	$8.55\cdot10^{-3}$	$3.8\cdot10^{17}$	$3.5\cdot10^{17}$	$1.1\cdot10^{-4}$	$3.9\cdot10^{28}$	$0.3\cdot10^{27}$	$0.24\cdot10^{27}$	0.4	$3.9\cdot10^5$
$4.5\cdot10^2$	$9.25\cdot10^{-3}$	$4\cdot10^{16}$	$3.8\cdot10^{17}$	$4.9\cdot10^{-5}$	10^{28}	$1.8\cdot10^{26}$	$1.44\cdot10^{26}$	2.7	$3.46\cdot10^6$
$3\cdot10^3$	$1.76\cdot10^{-2}$	$1.1\cdot10^{17}$	10^{18}	$8\cdot10^{-5}$	$5.4\cdot10^{28}$	$7\cdot10^{26}$	$5.6\cdot10^{26}$	0.74	$3.4\cdot10^5$

High temperatures (100°K)

j_t, A/cm²	ΔU_t, V	$n(\bar{x})$, cm⁻³	$p(\bar{x})$, cm⁻³	L_a, cm	$4D_{\text{eff}}(\ldots)$	$\frac{d\Gamma}{d\Delta U}$	$(\frac{d\Phi}{d\Delta U})_t(\ldots)$	K	$(\ldots)_1$
$8.3\cdot10^3$	$5.1\cdot10^{-3}$	$8.3\cdot10^{16}$	$1.54\cdot10^{18}$	$2.5\cdot10^{-4}$	10^{30}	$\ll2\cdot10^{27}$	$\ll2\cdot10^{27}$	1.53	$1.1\cdot10^4$
$8.3\cdot10^3$	$1.2\cdot10^{-2}$	$8.3\cdot10^{16}$	$1.54\cdot10^{18}$	$2.5\cdot10^{-4}$	$0.6\cdot10^{30}$	$\ll10^{27}$	$\ll10^{27}$	0.43	$0.6\cdot10^4$
$2.1\cdot10^3$	$1.3\cdot10^{-2}$	$8.3\cdot10^{16}$	$1.54\cdot10^{18}$	$6.3\cdot10^{-5}$	$0.7\cdot10^{29}$	$\ll2\cdot10^{26}$	$\ll2\cdot10^{26}$	2.7	$2.1\cdot10^5$
$2.1\cdot10^3$	$2.8\cdot10^{-2}$	$8.3\cdot10^{16}$	$1.54\cdot10^{18}$	$6.3\cdot10^{-5}$	$3.5\cdot10^{29}$	$\ll10^{26}$	$\ll2\cdot10^{26}$	0.75	$1.37\cdot10^6$

$$\Phi = \frac{\bar{c}}{Sa_0 L_a}\int_{-\infty}^{+\infty} L_y(x)\,dx, \tag{32}$$

where S is the p−n junction area; a_0 is the length of the luminous region along the x axis; and the integral in the above equation is the gain which is used in Pikus's theory [10]. We shall ignore the sum over nonlasing modes in Eq. (6) because j_t is calculated in [10] using the spontaneous approximation, i.e., ignoring stimulated transitions corresponding to nonlasing modes. Then,

$$\Gamma = \frac{j}{eL_a}, \quad j \leqslant j_t. \tag{33}$$

We shall drop the term in Eq. (2) which contains G.

In general treatments of the spatial inhomogeneity, it is assumed that there is only one type of elementary excitation. In the case of a p−n junction, we have two types of "excitation": electrons and holes. Estimates can be made using Eq. (6) and defining D as the effective value

$$D_{\text{eff}} = D_n D_p \frac{p(\bar{x}) + n(\bar{x})}{n(\bar{x})D_n + p(\bar{x})D_p}. \tag{34}$$

The diffusion coefficient is given by

$$D = \frac{u}{e}\begin{cases} kT, & kT \gg \mu, \\ \frac{2}{3}\mu, & kT \ll \mu, \end{cases} \tag{35}$$

where u is the mobility and μ is the quasi-Fermi level measured from the bottom of a band.

Our estimates show that when the pumping level is not too high compared with the threshold value the last term in Eq. (6) is small. Moreover, K = const and the dependence of the number of lasing modes on the pumping level is given by the factor $[(j - j_t)/j_t]^{1/3}$.

We shall now consider the spectral inhomogeneity. Once again, we shall base our treatment on Pikus's theory [10], bearing in mind the special nature of the p−n junction which is manifested by the presence of two types of elementary excitation. Moreover, following the treatments given in [10, 11], we shall use an approximation in which the loss of carriers from the energy levels, caused by stimulated transitions, is compensated by the transitions between levels. The subsequent analysis follows that given in [10]. The constant C of Eq. (8) is given by

$$C = \frac{20e\bar{c}a_t^*\left(1+\dfrac{m_n^*}{m_p^*}\right)^{\frac{3}{2}}\tau_n^3\tau_p^3\,(e\Delta U_t)^{\frac{1}{2}}\,eE_0}{Sem_n^{*\frac{3}{2}}\,(\tau_n+\tau_p)^2},$$

(36)

where τ_n and τ_p are the average times between collisions of electrons and holes, respectively.

Moreover, in the approximation employed, we obtain

$$\frac{q^\Sigma}{S} = \frac{j-j_t}{ea_t^*\bar{c}}.$$

(37)

The current density j_{spl} at which the laser line is split ($y^\Sigma = y_f^\Sigma$ in Fig. 1) is given by the following expression, which is deduced from Eqs. (36), (10), and (9):

$$\frac{j_{spl}-j_t}{j_t} = \frac{8e\bar{c}a_t^*}{CSj_t}.$$

(38)

5. Numerical Estimates. Comparison with Experimental Data

We shall use the experimental data given in [10] (Table 4) and we shall assume that $w_0 = 2\cdot10^9$ sec^{-1}, $m_n^* = 0.07$, $m_p^* = 0.5$, $m^* = 0.06$, $\eta = 0.8$, $n = 3.55$, $E_g = 1.35$ eV [12], $d = 0.5$. The results of calculations of the quantities discussed in Sections 3 and 4 are listed in Table 1.

According to the data given in [10] (Table 4) the frequency interval between modes is always small. Moreover, it is evident from the last column in Table 1 that at realistic values of the pumping level (injection current density) the spectral inhomogeneity does not give rise to multimode emission.

Thus, the experimentally observed multimode emission should be attributed to the spatial inhomogeneity. We should observe a "continuous" laser emission spectrum, i.e., neighboring modes should appear in the spectrum. This conclusion is confirmed by the experimental data reported in [14]. The spectral inhomogeneity would have given rise to the pattern shown in Fig. 1, i.e., it would have produced a "discrete" laser spectrum (modes separated by large intervals compared with the mode spacing).

We shall now carry out a quantitative comparison of the theory with the experimental data.

An experimental investigation of the emission of various modes by a GaAs laser at 4.2 and 77°K was reported in [16].

The results obtained at 4.2°K are in agreement with the nature of the dependence of M on the pumping level above the threshold, given by Eq. (1). The experimental value of the threshold current at 77°K is $I_t = 0.7$ A; the second mode begins lasing at 1.2 A. Assuming that Eq. (1) is valid, we find that $K = 1.2$, which is in agreement with the results listed in Table 1 (it is not possible to determine K at 4.2°K because no data are available on the absolute values of the current at this temperature).

An experimental investigation of a GaAs laser at 1.95°K is reported in [17]: the experimental value of the threshold current is $I_t = 40$ μA and the second mode is observed when the current rises to $I = 50$ μA. Using Eq. (1), we find that $K = 1.6$, which again is in reasonable agreement with Table 1.

Thus, the theory is in quantitative agreement with the available experimental data.

Literature Cited

1. V. S. Mashkevich, Zh. Éksp. Teor. Fiz., 53:1003 (1967).
2. V. S. Mashkevich, Ukr. Fiz. Zh., 12:1814 (1967).
3. C. L. Tang, H. Statz, and G. de Mars, J. Appl. Phys., 34:2289 (1963).
4. Yu. A. Anan'ev and B. M. Sedov, Zh. Éksp. Teor. Fiz., 48:782 (1965).
5. B. L. Lifshits and V. N. Tsikunov, Zh. Éksp. Teor. Fiz., 49:1843 (1965).
6. V. S. Mashkevich, Laser Kinetics, American Elsevier, New York (1967).
7. V. S. Mashkevich and V. A. Parnyuk, Quantum Electronics, Vol. 2, Naukova Dumka, Kiev (1957), p. 45.
8. W. R. Bennett Jr., Phys. Rev., 126:580 (1962).
9. G. E. Pikus and A. G. Aronov, Fiz. Tverd. Tela, 7:3548 (1965).
10. G. E. Pikus, Fundamentals of the Theory of Semiconductor Devices, Nauka, Moscow (1965).
11. V. S. Mashkevich, Theory of Lasers and Nonlinear Properties of Materials (Rotaprint), Tartu University (1967).
12. V. S. Mashkevich, Ukr. Fiz. Zh., 8:1260 (1963).
13. A. N. Rubinov and S. A. Mikhnov, Zh. Prikl. Spektr., 5:294 (1966).
14. H. Statz, C. L. Tang, and J. M. Lavine, J. Appl. Phys., 35:2581 (1964).
15. G. E. Pikus, Fiz. Tverd. Tela, 7:3536 (1965).
16. V. S. Mashkevich, Ukr. Fiz. Zh., 12:1731 (1967).
17. R. N. Hall, G. E. Fenner, J. D. Kingsley, T. J. Soltys, and R. O. Carlson, Phys. Rev. Letters, 9:366 (1962).

AN INVESTIGATION OF PHOTODIODE RESPONSE

N. Sh. Khaikin and M. A. Trishenkov

This paper summarizes the basic laws of relaxation for photocarriers separated by a p−n junction. A method is proposed for measuring the parameters of this mechanism and determining its role in the experimentally observed transient processes. A suitable apparatus is described, together with a series of measurements on rectangular silicon photodiodes, and their results.

Introduction

The response of photodiodes is determined by the transit time of radiation-induced electron−hole pairs and by the relaxation of photocarriers separated by a space-charge layer [1, 2]. These mechanisms may occur in any proportion. It is very desirable to identify them and examine them separately in the experimental study of transient processes.

This paper describes a method of measuring the contribution of each mechanism to the form of the frequency characteristics of photodiode sensitivity.

Basic Relations

The relaxation of photocarriers separated by a space-charge layer can be quantitatively analyzed in a relatively simple manner. We shall show that the relations then obtained can serve as the basis for the proposed method.

Let us consider the operation of rectangular and circular photodiodes (Fig. 1) under negative-bias conditions with a weak signal, and assume that the base region W makes the main contribution to the resistance of the quasi-neutral parts of the device. The resistance of the collector and the conductance of the p−n junction are neglected, and we shall also assume the obvious inequalities $W \ll l$, $W \ll D$, where W, l, and D are the dimensions shown in Fig. 1a,b.

The equations governing the motion of the photocarriers separated by a space-charge layer are then simulated by a line with distributed parameters [3, 4]: the base resistances $r_b dx$ and the barrier layer capacitances $c_p dx$, where dx is an element of the coordinate x.

The relation between r_b and c_p and the dimensions and electrical parameters of the photodiode is given by the following expressions, depending on the shape of the sensitive area:

$$r_b^r = \frac{\rho_b}{bW}, \qquad c_p^r = \frac{c}{l} \tag{1a}$$

191

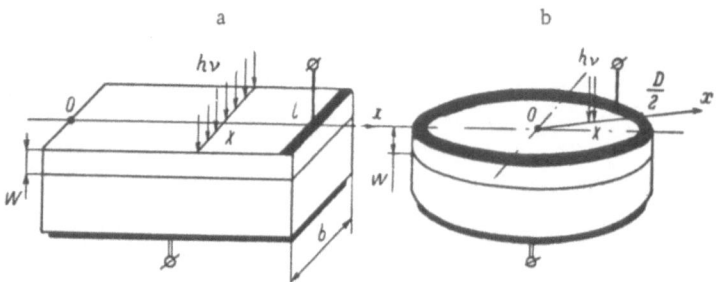

Fig. 1. Diagrams of photodiodes.

for a rectangular photodiode and

$$r_b^c(x) = \frac{\rho_b}{2\pi x W}, \qquad c_p^c(x) = \frac{8Cx}{D^2} \tag{1b}$$

for a circular one.* Here ρ_b is the resistivity of the base and C the total capacitance of the p−n junction; the remaining notation is clear from Fig. 1.

<u>Uniform Illumination of the Base.</u> In this case [5], the separation of the photocarriers by a space-charge layer is simulated by an assembly of distributed current sources $i_{ph}dx$:

$$i_{ph}^r(\omega) = \frac{I_{ph}(\omega)}{l}, \qquad i_{ph}^c(x, \omega) = \frac{8I_{ph}(\omega)\,x}{D^2}.$$

The current $I_{ph}(\omega)$ is determined by the total number of pairs produced which are separated by the p−n junction. Its dependence on the radiation modulation frequency ω is governed by the transit time of these pairs, i.e., by the first of the mechanisms mentioned above.

The relation between $I_{ph}(\omega)$ and the voltage across the base-collector contacts of a rectangular diode has been derived in [5]. We examined this problem for a circular base, using generalizations of both expressions to the case where there is a resistance R_c concentrated in the contact region:

$$U_{ph}^r(\omega) = I_{ph}(\omega) \frac{Z_l}{j\omega C\,(Z_l + R_c) + a^r l \,\text{cth}\,(a^r l)}, \tag{2a}$$

$$U_{ph}^c(\omega) = I_{ph}(\omega) \cfrac{Z_l}{j\omega C\,(Z_l + R_c) + \cfrac{a^c D}{4}\cfrac{I_0\left(a^c \frac{D}{2}\right)}{I_1\left(a^c \frac{D}{2}\right)}}, \tag{2b}$$

where Z_l is the load impedance; $\alpha = (j\omega c_p r_b)^{\frac{1}{2}}$ is a characteristic parameter of the photodiode [5]; $I_0(\alpha^c D/2)$ and $I_1(\alpha^c D/2)$ are Bessel functions of imaginary argument.

The parameter α is conveniently expressed in terms of the total capacitance C of the p−n junction and the effective resistance R_b of the base [5]:

$$R_b^r = \frac{\rho_b l}{3bW}, \qquad a^r = \frac{1}{l}\sqrt{3j\omega C R_b^r}, \tag{3a}$$

$$R_b^c = \frac{\rho_b}{8\pi W}, \qquad a^c = \frac{2}{D}\sqrt{8j\omega C R_b^c}. \tag{3b}$$

---

* All quantities with superscript r will refer to the rectangular shape, and with c to the circular shape.

It is seen from Eqs. (2) and (3) that, if C, R_b, and R_c are known, and $U_{ph}(\omega)$ is measured, the frequency dependence $I_{ph}(\omega)$ can be determined, i.e., the contributions of transit and relaxation mechanisms to the transient process in question can be experimentally distinguished. We think that the following is the most convenient way of doing this; it is valid when the load impedance (R_l) is purely resistive.

From Eqs. (2a) and (2b) we have immediately

$$\left|\frac{I_{ph}^r(\omega)}{I_{ph}^r(0)}\right| = \left|\frac{U_{ph}^r(\omega)}{U_{ph}^r(0)}\right| \cdot \left| j\eta \frac{R_l + R_c}{R_b} + \sqrt{3j\eta}\, \text{cth}\,(\sqrt{3j\eta}) \right|, \tag{4a}$$

$$\left|\frac{I_{ph}^c(\omega)}{I_{ph}^c(0)}\right| = \left|\frac{U_{ph}^c(\omega)}{U_{ph}^c(0)}\right| \cdot \left| j\eta \frac{R_l + R_c}{R_b} + \sqrt{2j\eta}\, \frac{I_0(\sqrt{8j\eta})}{I_1(\sqrt{8j\eta})} \right|, \tag{4b}$$

where $\eta = \omega R_b C$ is a dimensionless quantity which, like ω, acts as an independent variable. The values of the functions

$$\sqrt{3j\eta}\, \text{cth}\,(\sqrt{3j\eta}) = jA^r(\eta) + B^r(\eta),$$
$$\sqrt{2j\eta}\, \frac{I_0(\sqrt{8j\eta})}{I_1(\sqrt{8j\eta})} = jA^c(\eta) + B^c(\eta)$$

are shown in Table 1 for values of η between 0.01 and 10. The values of $|I_{ph}(\omega)/I_{ph}(0)|$ are calculated for frequencies corresponding to the tabulated values η_t:

$$\omega_T = \frac{\eta_T}{R_b C}. \tag{5}$$

The conversion factor in Eq. (4a) or (4b) for these frequencies can also be easily determined by means of Table 1.

The capacitance C and the total resistance $R_b + R_c$ are measured by the resonance method [6]. The value of R_b is obtained by observing the spatial distribution of sensitivity over the base at high frequencies, caused by features of the relaxation of the separated photocarriers.

Distribution of Sensitivity over the Base. It has been noted in [5] that, as the radiation modulation frequency increases, the sensitivity of the base regions farthest from the contact begins to drop. We have not found any published calculation of this effect or any description of an experimental investigation of it.

Using the method developed in [4, 5, 7], we derived expressions for the photoresponse voltage

$$U_{ph}^r(\omega, X) = I_{ph}(\omega, X) \frac{Z_l \,\text{ch}\,(a^r X)}{(Z_l + R_c)\sqrt{j\omega C/3R_b^r}\,\text{sh}\,(a^r l) + \text{ch}\,(a^r l)} \tag{6a}$$

when a rectangular diode is illuminated on a strip parallel to the contact (Fig. 1a),* X being the coordinate of the strip, and

$$U_{ph}^c(\omega, X) = I_{ph}(\omega, X) \frac{Z_l I_0(a^c X)}{(Z_l + R_c)\sqrt{j\omega C/2R_b^c}\,I_1\left(a^c \frac{D}{2}\right) + I_0\left(a^c \frac{D}{2}\right)} \tag{6b}$$

when a circular diode is illuminated at a point which moves radially, X being its distance from the center. The relative change of sensitivity over the base, when $I_{ph}(X)$ is constant, is given by

* A similar expression was derived independently by I. I. Taubkin (private communication).

TABLE 1

$\eta = \omega R_b C$	Rectangular photodiode		Circular photodiode	
	$A^r(\eta)$	$B^r(\eta)$	$A^c(\eta)$	$B^c(\eta)$
0.01	0.01	1.00	0.01	1.00
0.1	0.10	1.00	0.10	1.01
0.3	0.30	1.02	0.30	1.04
0.5	0.49	1.05	0.48	1.07
0.8	0.77	1.12	0.73	1.18
1.0	0.95	1.18	0.87	1.27
1.2	1.11	1.26	1.00	1.35
1.4	1.27	1.34	1.11	1.44
2.0	1.66	1.60	1.37	1.68
3.0	2.15	2.04	1.70	2.01
4.0	2.50	2.42	1.97	2.27
5.0	2.77	2.74	2.21	2.51
8.0	3.46	3.46	2.83	3.08
10.0	3.88	3.88	3.16	3.41

$$\frac{U_{ph}^r(\omega, X)}{U_{ph}^r(\omega, l)} = \frac{ch\,(a^r X)}{ch\,(a^r l)}, \tag{7a}$$

$$\frac{U_{ph}^c(\omega, X)}{U_{ph}^c\left(\omega, \frac{D}{2}\right)} = \frac{I_0\,(a^c X)}{I_0\left(a^c \frac{D}{2}\right)}. \tag{7b}$$

It is seen from Eqs. (7a) and (7b) that this quantity does not depend on the impedance $Z_l + R_c$. Figure 2 shows the moduli of Eqs. (7a) and (7b), with X/l and $2X/D$ as arguments and the product $\omega R_b C = \eta$ as parameter.

Figure 3 shows the quantities $|U_{ph}^r(\omega, 0)/U_{ph}^r(\omega, l)|$ and $|U_{ph}^c(\omega, 0)/U_{ph}^c(\omega, \frac{1}{2} D)|$ as functions of the parameter η.

If all the experimentally observed photodiode sensitivity distributions are in accordance with Fig. 2, the product $\omega R_b C$ can be found from Fig. 3, using the measured value of $|U_{ph}^r(\omega, 0)/U_{ph}^r(\omega, l)|$ or of $|U_{ph}^c(\omega, 0)/U_{ph}^c(\omega, \frac{1}{2} D)|$. Then, with ω, $R_b + R_c$, and C known, R_b and R_c can be determined.

Experimental Method

The measurements were made in a Kerr-cell apparatus to record the photodiode sensitivity-frequency characteristics [8, 9]. The samples were placed on a support with a micrometer gauge. With this equipment, the photodiode could be moved smoothly in the x direction and located with an accuracy of 0.05 mm. The cross section of the light probe was shaped by the Kerr-cell slit 3 × 0.2 mm. The optical system, a Jupiter-3 objective, was used to vary the slit image transmission coefficient from 0.1 to 3 without significant decrease of the radiation flux.

In order to complete the program described above, it was necessary to construct samples having the configurations shown in Fig. 1a,b. The linear dimensions of the receiving areas along the axis were made not less than 1.5 mm so as to ensure reliable probing of these areas.

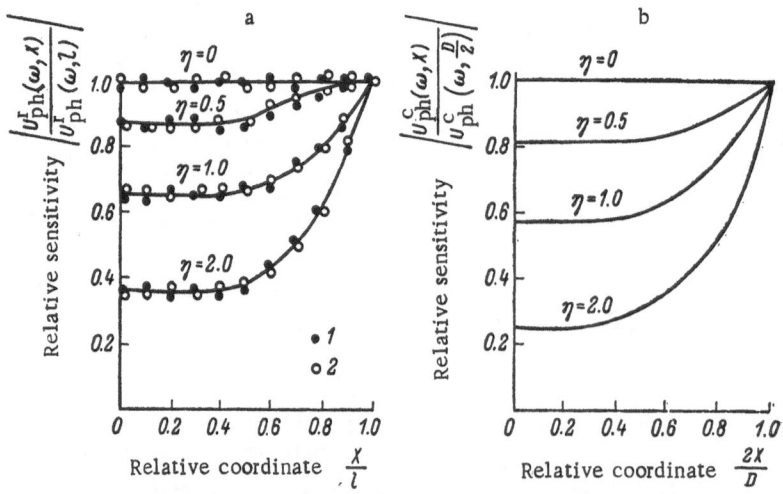

Fig. 2. Distribution of the sensitivity over the base of (a) rectangular and (b) circular diodes. $R_l(\Omega)$: 1) 150; 2) 1500.

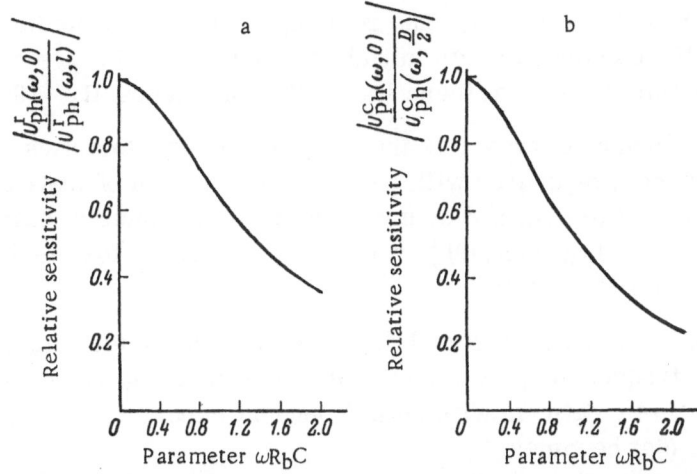

Fig. 3. Dependence of the moduli $|U_{ph}^I(\omega, 0)/U_{ph}^I(\omega, l)|$ and $|U_{ph}^C(\omega, 0)/U_{ph}^C(\omega, \frac{1}{2}D)|$ on the parameter η for (a) rectangular and (b) circular photodiodes.

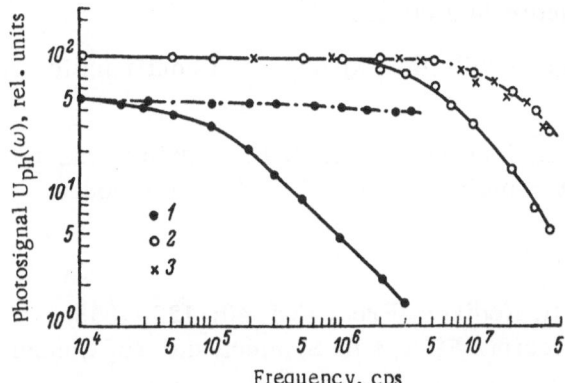

Fig. 4. Frequency dependence of photosensitivity for uniform illumination. Continuous curves: experimental; chain curves: calculated, representing the transit of radiation-induced electron−hole pairs. 1) Device No. 1; 2) device No. 2; 3) experimental points for device No. 2 with X = l and R_l = 10 Ω.

Another limitation was due to the presence of intrinsic heterogeneities which considerably alter the sensitivity distribution at high radiation modulation frequencies. In the present instance, the distribution found indicated the manner in which the samples deviated from an idealized model. The criterion of "idealness" was satisfactory agreement between all the photosensitivity distributions and the theoretical curves in Fig. 2. As an example, Fig. 2 shows the experimental points given by a rectangular silicon photodiode 7 × 7 mm for different load resistances (device No. 1). It is seen that this device satisfies the above criterion, and it was therefore possible to use it for all the measurements described above. The initial data given by the resonance method were: C = 820 pF, $R_b + R_c$ = 1900 Ω, for a working voltage U_{work} = −5 V.

The value of R_b found from Fig. 3a was 600 Ω. The contact resistance was therefore 1300 Ω. The base contact in this photodiode was a vapor-deposited gold strip, and the large value of R_c is therefore not surprising. The rectangular silicon photodiode No. 2, receiving area 2×2 mm, had alloyed contacts. Its initial parameters for $U_{work} = -4$ V where $C = 133$ pF, $R_b + R_c = 170$ Ω. For this device, we found $R_b = 170$ Ω and hence $R_c \approx 0$.

In Fig. 4, the continuous curves give the frequency characteristics of the sensitivity $| U_{ph}^r (\omega) |$ for these devices, measured with uniform illumination of their areas. The points joined by the chain curves are obtained by analyzing the experimental curves with Eq. (4a), and represent the quantity $| I_{ph}^r(\omega) / I_{ph}^r(0) |$, which is governed by the mechanism of transit of radiation-induced electron—hole pairs.

Since R_c is zero for device No. 2 and the calculated characteristic $| I_{ph}^r(\omega)/I_{ph}^r(0) |$ declines at relatively low frequencies, it is possible to measure $| I_{ph}^r(\omega) |$ directly and hence to compare the result of this treatment with an independent experiment. When $Z_l = R_l \to 0$, $R_c \to 0$, and $X = l$, Eq. (6a) becomes

$$U_{ph}^r(\omega l) = I_{ph}(\omega)R_l,$$

i.e., represents the frequency dependence of the photocurrent source. We made appropriate measurements with various load resistances. In the frequency range stated, the shape of the characteristics was independent of R_l below 20 Ω, and the experimental points (curve 3 in Fig. 4) fitted well on the theoretical curve.

This agreement, together with the experimental confirmation of Eq. (7a), shows that the method used is correct.

The authors thank S. A. Kaufman and A. I. Frimer for constant interest and assistance, and G. Z. Pis'man and O. V. Smolin for providing the photodiodes.

Literature Cited

1.   D. E. Sawyer and R. H. Rediker, Proc. IRE, 46:1122 (1958).
2.   S. M. Ryvkin, Photoelectric Effects in Semiconductors, Consultants Bureau, New York (1964).
3.   B. Ya. Moizhes, Fiz. Tverd. Tela, 2:221 (1960).
4.   M. A. Trishenkov, Radiotekh. i Élektron., 10:2046 (1965).
5.   G. Lucovsky and R. B. Emmons, IEEE Trans. Electron. Dev., 12:5 (1965).
6.   M. A. Trishenkov and O. M. Berlizova, Radiotekh. i Élektron., 13:1636 (1968).
7.   G. V. Zeveke et al., Principles of Circuit Theory, Izd. Energiya, Moscow (1965), p. 222.
8.   N. Sh. Khaikin, Prib. Tekh. Éksp., No. 3, p. 172 (1966).
9.   N. Sh. Khaikin, S. A. Kaufman, and V. A. Voronin, Fiz. Tekh. Poluprov., 1:1021 (1967).

CURRENT—VOLTAGE CHARACTERISTICS OF
p—n JUNCTIONS IN INDIUM PHOSPHIDE—GALLIUM
ARSENIDE SOLID SOLUTIONS

V. I. Osinskii, N. N. Sirota, and G. G. Shienok

Institute of Solid State and Semiconductor Physics
Academy of Sciences of the Belorussian SSR, Minsk

Current—voltage characteristics are given for diodes made from crystals of indium phosphide—gallium arsenide solid solutions, at room and liquid-nitrogen temperatures. The characteristics degenerate to straight lines at high currents. It is shown that the voltage intercepts U_i vary with the composition of the solid solution, having a maximum at 60-70% gallium arsenide.

The recombination radiation of diodes made from crystals of InP—GaAs solid solutions has been studied in [1]. It has been shown that, when forward current is passed, the diodes emit visible light over a wide range of compositions, and the color of the light varies with the composition; this may be of great practical value. It is therefore desirable to make a thorough investigation of diodes consisting of such crystals.

This paper deals with the current—voltage characteristics of p—n junctions for various compositions of InP—GaAs solid solution crystals.

The diodes were made by the customary diffusion technique, using cleaved crystals of n-type solid solutions formed by repeated zone recrystallization. Ohmic contacts were alloyed to the n-type and p-type sides of the junction under hydrogen. The quality of the junctions was tested by microscopic examination of the recombination radiation line. The diodes chosen for study were mounted in transistor holders.

The current—voltage characteristics were plotted from points obtained in steady conditions with currents up to 200-300 mA, and in pulsed operation at higher currents.

Figure 1 shows the forward and reverse static current—voltage characteristics for three diodes made from InP—GaAs solid solution crystals with various compositions, at room temperature and liquid—nitrogen temperature.

For diode S 235 (Fig. 1a), made from material close to InP, the forward and reverse branches do not differ greatly. A more noticeable asymmetry of the characteristics is seen

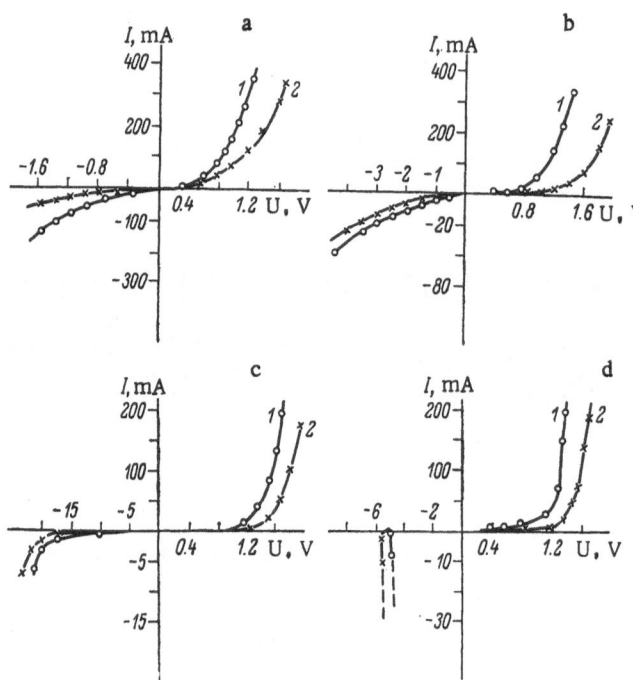

Fig. 1. Current–voltage characteristics of p–n junctions in InP–GaAs solid solution crystals. a) Sample S 235 (0.9 InP, 0.1 GaAs); b) sample P 100 (0.5 InP, 0.5 GaAs); c, d) samples H 298 and H 121 (0.1 InP, 0.9 GaAs). T, °K: 1) 300; 2) 77.

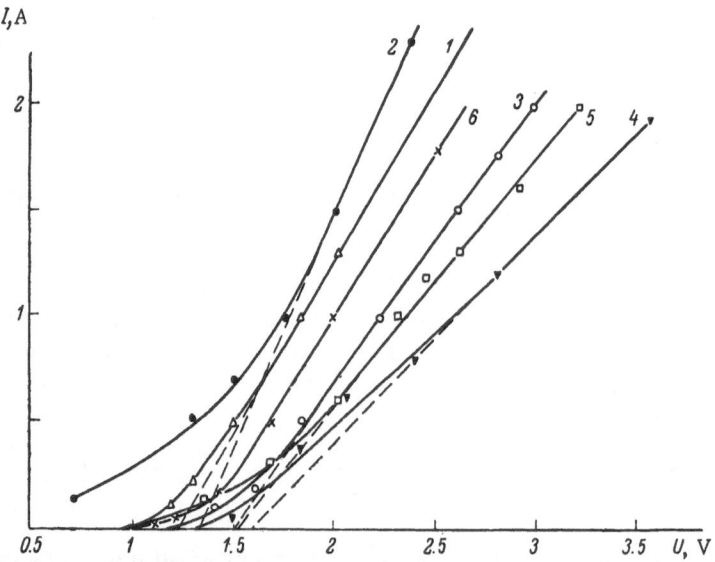

Fig. 2. Current–voltage characteristics of diodes at high currents, recorded at 300°K for samples: 1) D 61 (InP); 2) S 215 (0.9 InP, 0.1 GaAs); 3) R 127 (0.7 InP, 0.3 GaAs); 4) P 100 (0.5 InP, 0.5 GaAs); 5) H 121 (0.1 InP, 0.9 GaAs); 6) N 2 (GaAs).

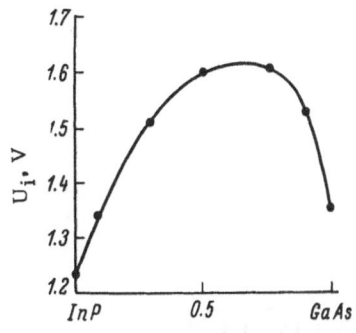

Fig. 3. Dependence of U_i on composition of InP−GaAs solid solution crystals. $T = 300°K$.

for diode P 100, made from a solid solution having an intermediate composition (Fig. 1b). The current ratio is $I_f/I_r = 40$ at 1.5 V (300°K). With reverse bias, breakdown occurs above 5 V, and the form of the curve indicates that the breakdown is thermal. For diodes made from a crystal close to GaAs, the characteristics are shown in Fig. 1c (diode H 298) and 1d (diode H 121). The ratio $I_f/I_r = 2.5 \cdot 10^3$ at 1.7 V (300°K). Breakdown occurs at 25 V for the diode having the lower impurity concentration gradient (Fig. 1c) and at 5.5 V for that having the higher gradient (Fig. 1d).

As the temperature decreases, both branches move toward higher voltages. A comparison with the corresponding characteristics for InP [2] shows that the temperature shift of the characteristics of p−n junctions in solid solutions having compositions close to InP is considerably smaller, whereas a diode having an intermediate composition shows a considerably greater temperature shift of the forward branch of the current−voltage characteristic. This appears to be due to the partial freeze-out of carriers in materials having a wide forbidden band at 77°K, leading to a considerable increase of resistivity. The diagrams shown here indicated that the forward resistance of the diodes increases rapidly with decreasing temperature.

The relatively low resistances of the diodes and the slighter temperature dependence of the current−voltage characteristics of p−n junctions in solid solutions close to InP can be correlated with the high impurity concentration and the imperfection of crystals having these compositions. The high concentration of chemical impurities and defects is also the cause of the large reverse current through diodes made from crystals having compositions close to InP.

At high currents, the characteristics degenerate to straight lines (Fig. 2). The dynamic resistance of a diode under these conditions, as determined from the slope of the line, is independent of the voltage, and is 0.5-1 Ω at room temperature and 0.6-1.3 Ω at liquid-nitrogen temperature for the diodes examined.

The voltage intercepts U_i, found by extrapolating the linear part of the curve to $I = 0$, are related to the forbidden band width of the diode material, and represent the height of the potential barrier in the p−n junction. Here, it must be remembered that the value of U_i depends on the impurity concentration, but the resulting change of U_i is small. For example, in GaAs, U_i changes by 0.01 V when the concentration increases from 10^{17} to 10^{18} cm^{-3} [3].

The curve of U_i as a function of composition (Fig. 3) has a maximum at 60-70%, in agreement with the composition dependence of the peaks of the strongest bands of recombination radiation [1].

The values of U_i are in satisfactory agreement with the corresponding values for the forbidden band width of the solid solution crystals from which the diodes were made [4].

This analysis of current−voltage characteristics of diode structures based on InP−GaAs solid solution crystals shows that the dependence of the current on the applied voltage is largely governed by the potential barrier in the region of the p−n junction.

Literature Cited

1. N. N. Sirota and V. I. Osinskii, Dokl. Akad. Nauk SSSR, 171:317 (1966).
2. V. I. Osinskii and N. N. Sirota, Izv. Akad. Nauk Beloruss. SSR, 3:93 (1965).
3. B. M. Vul, É. I. Zavaritskaya, and A. P. Shotov, Fiz. Tverd. Tela, 6:1465 (1964).
4. L. A. Makovetskaya and N. N. Sirota, Dokl. Akad. Nauk Beloruss. SSR, No. 9, 1964.

CURRENT—VOLTAGE CHARACTERISTICS OF
FORWARD-BIASED p—n_i—n SILICON DIODES

Yu. A. Bykovskii, K. N. Vinogradov,
V. F. Elesin, and V. V. Zuev

A study was made of the current—voltage characteristics of forward-biased p—n_i—n gold-doped n-type silicon diodes. S-type and N-type characteristics were found, depending on the temperature. In the N-type case, current oscillations were observed in the current pulse. The experimental data are analyzed by means of the thermal model of negative resistance.

1. In this work, we studied N-type and S-type current—voltage characteristics of diode structures under static and transient conditions.

The samples were prepared from n-type silicon whose parameters before doping were $\rho = 100 \ \Omega \cdot cm$, $\tau = 100 \ \mu sec$; forward resistance at 300°K of p—n_i—n diodes made from gold-doped n-type silicon 200 kΩ; thickness of intrinsic-conduction region 0.1-0.2 cm; junction area $\lesssim 10^{-2} \ cm^2$. A sample was placed on a heat-conducting support and put into a cryostat, which was evacuated to 10^{-5} torr. The forward-biased p—n_i—n diode was investigated when operating either as a current generator or as a voltage generator.

Figure 1b is an oscillogram of the current—voltage chracteristic of a diode at fields below 200 V/cm and T ≈ 150°K. The p—n_i—n structure clearly behaves as a diode. Figure 1a shows that an N-type negative resistance appears at temperatures below 200°K and voltages above about 300 V. At higher temperatures, only an S-type negative resistance occurs.

These negative resistances were investigated from the shape of the current in pulsed operation through a resistor in series with the sample. Figure 2 shows that the time for establishment of a steady current corresponding to the pulse amplitude is between 1 and 10 sec. As the amplitude of the applied rectangular voltage pulse increases, the "establishment" time decreases (Figs. 2b,c). The results in Fig. 2e indicate that there is a threshold voltage amplitude below which the N-type negative resistance cannot occur. When the amplitude is less than this value, the N-type negative resistance disappears. The time to restore the current corresponding to a subthreshold amplitude is of the same order as the "establishment" time (Fig. 2e).

It was also found that, when the voltage is large (450-600 V), the current pulse shows spikes when the characteristic is N-type (Figs. 2c,d), which become more noticeable and more numerous as the field amplitude increases.

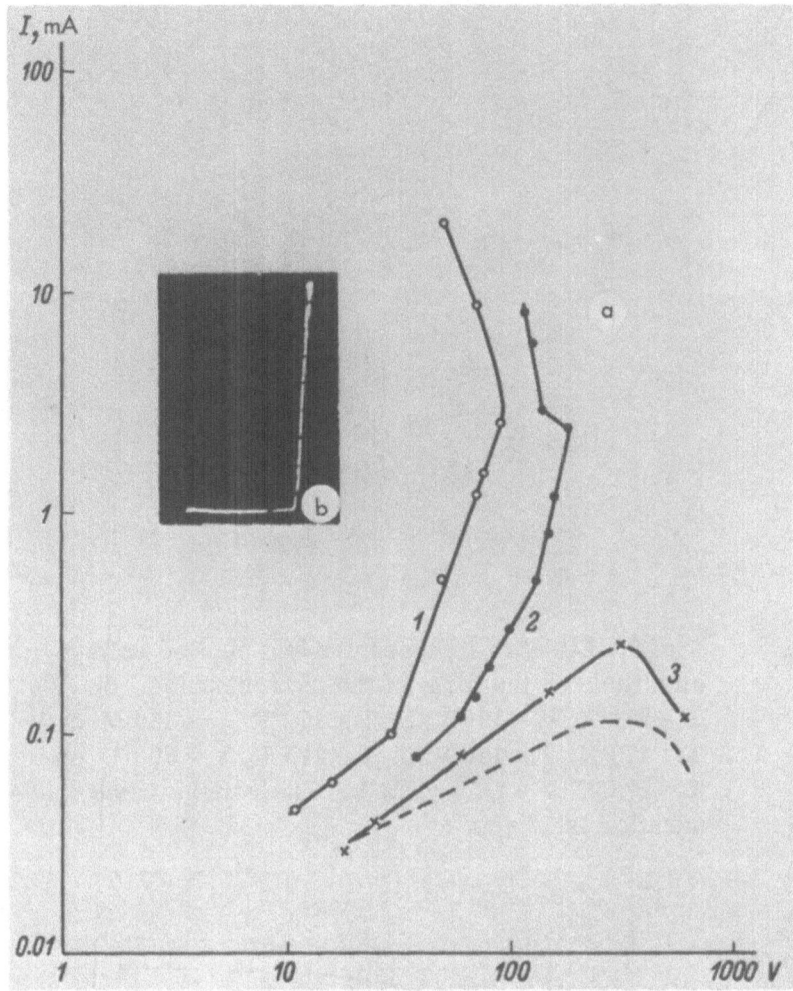

Fig. 1 Static current — voltage characteristics of a p − n_i − n diode. T, °K: 1) 293; 2) 253; 3) 173.

Monochromatic illumination with wavelengths in the range 2.5-0.3 μ had differing effects on the S-type and N-type negative resistances.

2. It is known [1] that gold can be present in silicon only in the singly charged state. Hence, the appearance of N-type characteristics in gold-doped silicon diodes cannot be ascribed to a field dependence of the capture cross section such as occurs in gold-doped or copper-doped germanium or zinc-doped silicon [2].

On the other hand, the very slight change in the conductivity of the sample between 173 and 293°K (Fig. 1) indicates that, in these samples, the occupation of the deep levels changes only slightly with temperature and that $N_{Au} < n_d$, i.e., the carrier density is independent of temperature in this range.

Thus, we can explain the experimental data using a thermal model [3, 4], which accounts both for the occurrence of N-type characteristics at low temperatures and for the long time needed to establish a steady current in pulsed measurements (Figs. 2a,b). In this model, the N-type characteristic is due to the decrease of mobility with increasing temperature at constant carrier density [$\mu = \mu_0 (T_0/T)^\gamma$]; the S-type characteristic appears at temperatures where the carrier density increases significantly with temperature.

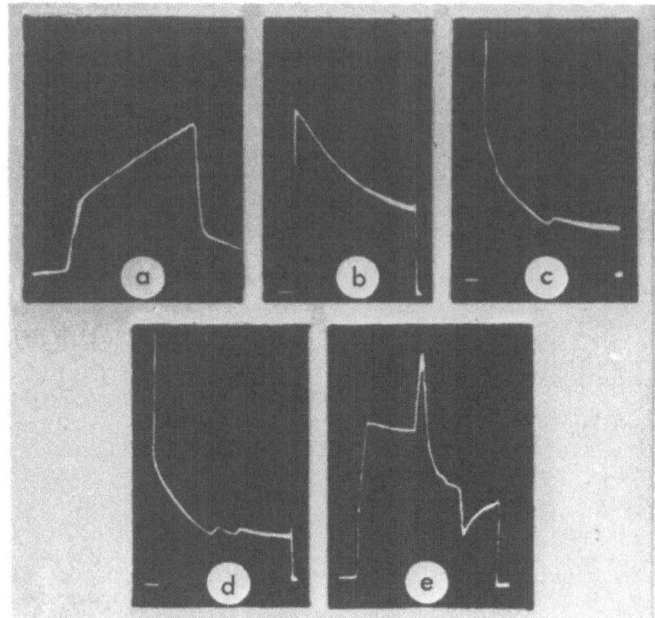

Fig. 2. Effect of temperature and applied voltage
amplitude on the form of the current pulse. a)
T = 230°K, V = 120 V; b) T = 173°K, V = 350 V; c)
T = 173°K, V = 450 V; d) T = 173°K, V = 600 V; e)
T = 173°K, V = (150 + 25) V. The voltage pulse
duration is 10 sec.

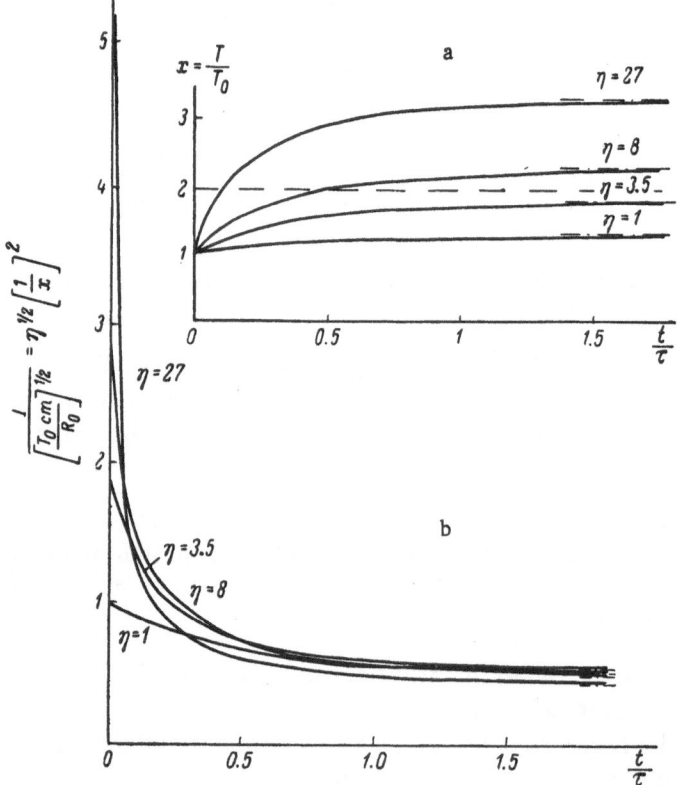

Fig. 3. Determination of the voltage necessary for the
appearance of negative resistance. See text for explanation.

3. The N-type current−voltage characteristic was calculated on the assumption that the temperature is uniform throughout the sample, starting from the equation (see, for example, [3])

$$IV = A\,(T - T_0),\tag{1}$$

where I is the current, V the voltage, A the heat-removal constant, and T_0 the temperature of the ambient medium.

In Fig. 1, the dashed line shows the theoretical curve for $\gamma = 2.6$, A = 0.3 mW/deg, which is in satisfactory agreement with the experimental curve. The value of the constant A is smaller than in [3] because the heat removal by the conducting support is less good. Nitrogen cooling, however, produces heat pockets which have an unfavorable effect on the measurements.

Compared with the work reported in [3], where n-type silicon without gold was used, our experiments involved a transition to the N-type negative resistance region at higher voltages and lower currents (higher resistance) but about the same power.

4. To describe the transient process, we consider the temperature equation

$$cm\frac{dT}{dt} = IV - A\,(T - T_0),\tag{2}$$

where c is the specific heat and m the mass of the sample. It is seen from Eq. (2) that the characteristic time of the process is, in order of magnitude,

$$\tau \sim \frac{cm}{A} \simeq 10 \text{ sec.},$$

taking c = 0.1 cal·$\deg^{-1}$·g^{-1}, m = 0.01 g, A = 0.3 mW/deg. This value of τ agrees with the time to establish a steady current in the samples examined.

The solution of Eq. (2) with $\gamma = 2.6$ is difficult. The equation has therefore been solved with $\gamma = 2$, as an illustration. Figure 3a shows $x = T/T_0$ as a function of t/τ for several values of the parameter $\eta = V^2/cmR_0T_0$. The intersection of this curve with the horizontal line $x = \gamma/(\gamma - 1)$ determines the time at which the N-type negative resistance begins. When $\gamma = 2$, the horizontal line is $x = 2$. If there is no intersection, as for the curves in Fig. 3a with $\eta = 3.5$ and 1, the applied voltage is insufficient for negative resistance to appear. Using the solution of Eq. (2), we find the current as a function of time in qualitative agreement with experiment (Fig. 3b). We also see that, in accordance with the experimental results (Figs. 2b,c,d), the current decay time decreases with increasing amplitude of the applied voltage.

5. Finally, let us consider the current oscillations which occur in strong fields (Figs. 2c,d). The form of these is similar to that described in [5], for example, where the oscillations are due to domain motion. The effect which we observed can apparently be interpreted as motion of thermal domains, since the condition for domains to occur (the presence of an N-type volume negative resistance) is satisfied.

Literature Cited

1. W. M. Bullis, Solid State Electronics, 9:143 (1966).
2. B. V. Kornilov and A. V. Anfimov, Fiz. Tekh. Poluprov., 1:340 (1967).
3. V. P. Sondaevskii and V. I. Stafeev, Physics of p−n Junctions, Zinatnu, Riga (1966).
4. E. Nebauer and E. Jahne, Phys. Status Solidi, 8:881 (1965).
5. B. K. Ridley and R. G. Pratt, J. Phys. Chem. Solids, 26:21 (1965).

SOME PROPERTIES OF A ZINC SELENIDE LASER

O. V. Bogdankevich and M. M. Zverev

P. N. Lebedev Physics Institute
Academy of Sciences of the USSR, Moscow

A description is given of some properties of a single-crystal ZnSe laser excited by a beam of fast electrons. Experimental results are reported for the threshold current densities, power, efficiency, directionality, spectrum, mode structure, and spike oscillation.

Zinc selenide is an $A^{II}B^{VI}$ compound whose edge luminescence is blue. At room temperature, the forbidden band width is $\Delta E = 2.67$ eV [1] and the band structure is of the direct type with the top of the valence band and the bottom of the conduction band being at the center of the Brillouin zone [2]. ZnSe has a cubic and a hexagonal crystal modification. The edge luminescence spectrum of ZnSe crystals at helium temperatures and at low excitation level has over 20 lines. The crystals can be divided into a number of types as regards their edge luminescence spectra, which probably correspond to different crystal lattice structures [3, 4]. Some of the spectral lines are separated by equal intervals of energy, owing to the participation of one or more optical phonons.

At liquid-nitrogen temperature the spectrum is much simpler and the intensity of the luminescence is much lower. The nature of the edge luminescence of ZnSe has not yet been fully explained. It has been suggested, on the basis of the similarity of the spectra of various $A^{II}B^{VI}$ crystals, that the edge luminescence is due to excitons, either free [3] or attached to various defects [4, 5].

$A^{II}B^{VI}$ semiconductors possess a number of features useful in the construction of lasers. They have a band structure with direct transitions, and exhibit a narrow luminescence band near the fundamental absorption edge. Almost all such compounds have now been made to lase.

The use of ZnSe single crystals as lasers has been reported previously [6]. The present paper describes some properties of a ZnSe crystal laser excited by a beam of fast electrons.

The single crystals of zinc selenide were prepared by synthesis under high pressure in a sealed vessel, followed by crystallization from the gas phase. Samples were cut in the shape of parallelepipeds, and a pair of polished opposite faces formed a Fabry–Perot resonator.

204

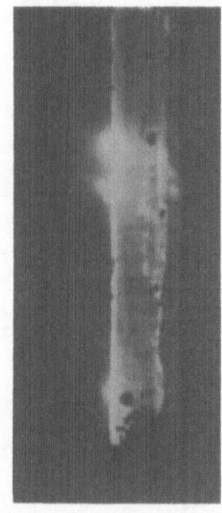

Fig. 1. Photograph of
the end face of a ZnSe
crystal under oscilla-
tion conditions.

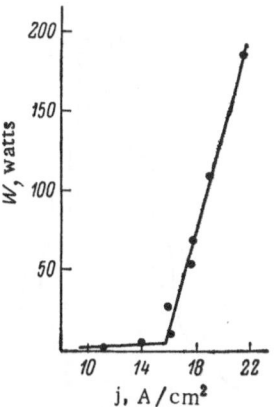

Fig. 2. Output power
as a function of pump-
ing for a ZnSe crys-
tal. Resonator length
L = 1.75 mm.

The electron irradiation was applied in a direction perpendicular to the resonator axis. The samples were placed on the end of a copper support in a liquid-nitrogen cryostat. Oscillation was brought about by a pulsed electron beam having a current density up to 30 A/cm² and energy up to 65 keV; the threshold energy for defect formation by electrons in ZnSe at 80°K is 240 keV [7]. The pulse duration was 120 nsec and the repetition frequency 50 cps. The spectra were recorded with a PGS-2 spectrograph having a first-order linear dispersion of 7.3 Å/mm in the blue region.

1. The threshold current densities for electron energy 50 keV and crystals of various sizes (minimum distance between resonator faces 72 μ, maximum 1.75 mm) were in the range from 5.5 to 20 A/cm². No detectable correlation between threshold current density and reso-nator length was found, apparently because the resonators were of insufficient quality. This was supported by the observation that the luminescence intensity was not uniform over the end of the crystal, but was emitted at isolated points which were probably due to inhomogeneity of the resonator. Figure 1 shows a typical photograph of the crystal under oscillation conditions. In a few cases the entire end face of the crystal emitted radiation.

2. The maximum output power of about 190 W (in both directions), was emitted by a crystal 1.75 mm long at 90°K. The efficiency η, defined as the ratio of the radiation power leaving the crystal to the electron-beam power supplied to it, was 1.3%. Figure 2 shows the output power as a function of the pumping current density. The following comment may be made concerning the measurement of the laser power. It will be shown later that the oscillation mode is not continuous but spiky, and the duration of a single spike may be less than 0.1 nsec. In the power measurements, a radiation-generated pulse from a coaxial photocell was mea-sured with a S1-11 oscillograph having a pass band of 100 Mc, so that the power was averaged over a large number of oscillation spikes. The output power in a single spike could be several times greater.

3. The spectrum of spontaneous radiation at 90°K has a narrow line (half-width 70-80 Å) with a peak at 4550 Å, and also two weak broad luminescence bands in the red and the green

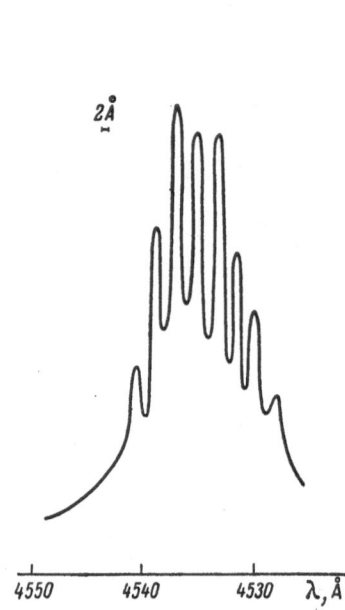

Fig. 3. Radiation spectrum of ZnSe crystal under laser conditions. Distance between resonator faces L = 72 μ; current density j = 17 A/cm².

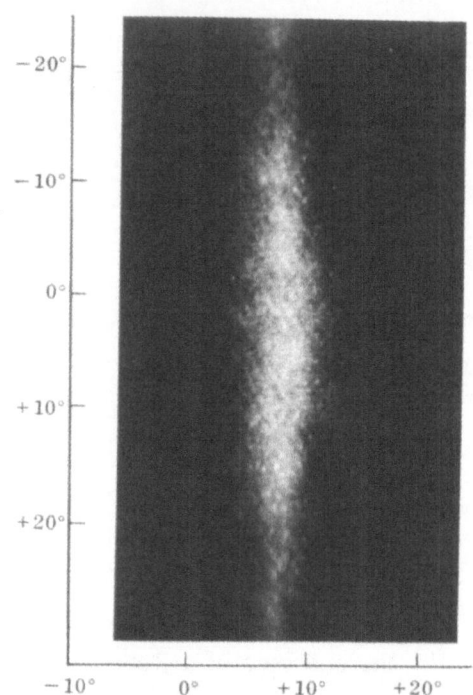

Fig. 4. Directionality pattern of ZnSe crystal laser; L = 1.75 mm; normal to crystal passes through the origin.

regions. The intensity of these bands varies only very slightly when the current density changes from 1 to 30 A/cm². These bands appear to be due to copper impurity [8]. When the threshold is reached, oscillation occurs at a wavelength of 4525-4550 Å (which varies from crystal to crystal) and the line half-width is $\Delta\lambda \approx$ 10-15 Å. As the pumping is increased further, the line broadens to occupy the range 4520-4620 Å.

Figure 3 shows the oscillation spectrum of a ZnSe crystal with L = 72 μ, where L is the distance between the resonator faces. There are obvious peaks corresponding to the axial modes of the resonator; the spacing between these modes is 2 Å.

Using the general formula

$$\Delta\lambda = \frac{\lambda^2}{2L\left(n - \lambda\,\frac{dn}{d\lambda}\right)},$$

we find

$$\left(-\lambda\,\frac{dn}{d\lambda}\right)_{\lambda=4550\,\text{Å}} \approx 4.5.$$

4. At current densities of 10-30 A/cm², the crystal is considerably heated during the pulse. This is probably the reason why the radiation pulse is considerably shorter than the current pulse. For example, when the duration of the current pulse is 100-120 nsec, that of the radiation pulse (under laser conditions) is 20-30 nsec. The duration of the spontaneous radiation pulse is equal to that of the current pulse. The pulsed heating of the crystal makes it difficult to compare the results with published data.

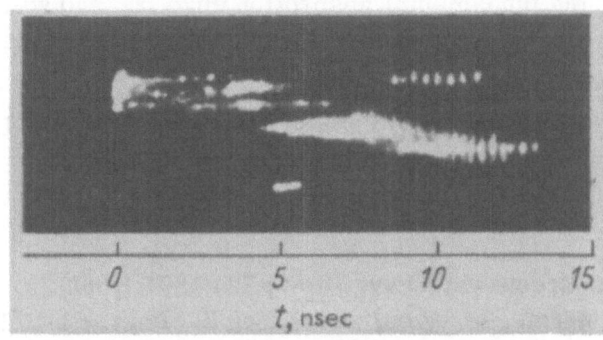

Fig. 5. Spiking in ZnSe with L = 215 μ. The ordinate represents the transverse dimension of a crystal. Nonuniformity of emission by various crystal particles due to the resonator and electron-beam nonuniformities.

Estimates show that, for a current density 30 A/cm^2, beam energy 50 keV, and pulse duration 100 nsec, the temperature of the crystal at the end of the pulse is raised by 150-170 deg (specific heat of ZnSe is C_p = 2.2 cal $\cdot$ (g-atom)$^{-1}$ $\cdot$ deg^{-1} at T = 80°K [9]). When the pulse-rise time is 20 nsec, the heating at the beginning of oscillation is 20-25 deg, i.e., the crystal temperature is 115-120°K in a cryostat filled with liquid nitrogen. At this temperature, the width of the forbidden band is 2.797 eV (exciton ground state energy E_{ex} = 2.776 eV [2] plus exciton binding energy E_b = 0.021 eV [2]). The laser transition energy (λ = 4550 Å) is E_l = 2.728 eV, which is 0.07 eV less than the forbidden band width. As the temperature increases, the radiation intensity and the efficiency decrease. The maximum temperature at which lasing was observed is 200°K.

5. Above the threshold the radiation is strongly directional, with a pattern which varies considerably with the current density and the uniformity of the exciting electron beam. Figure 4 shows a photograph of the directionality of radiation from a crystal having L = 1.75 mm, taken from a screen placed 65 mm from the crystal. In a plane passing through the normal to the electron-irradiated surface and the axis of the Fabry—Perot resonator, the angular divergence of the beam in this photograph is 3.5°, and the ratio $\lambda/d \approx$ 0.09 $\approx$ 5.1° (where d = 5 μ is the depth of the emitting layer for electron energy 50 keV); this is close to the minimum possible divergence. The angle of divergence in the other direction in the photograph is about 28°, probably due to amplification at an angle to the resonator axis. The center of the directionality pattern is 9° off the optical axis of the resonator, owing to the nonuniform depth distribution of excitation in the crystal; the same effect has previously been observed for various semiconductor lasers excited by an electron beam [10, 11]. The use of equipment having a resolving time of 5 $\cdot$ 10^{-11} sec showed that the oscillation is markedly spiky. The duration of the spikes and the intervals between them depend on the resonator length and the excitation power. It has recently been reported [12] that spiking was observed in a GaAs laser excited by an electron beam. One reason for the occurrence of spikes may be the appearance of damped or undamped oscillations of the resonator field because of the nonuniform excitation of the crystal by the electron beam [12]. Figure 5 shows a time sweep of the radiation from a ZnSe crystal having L = 215 μ. There is an obvious discontinuity, with the spike period about 0.3 nsec.

Conclusion

It can be concluded that a ZnSe crystal is a suitable substance for a semiconductor laser. The low efficiency so far achieved may be due to high losses in the material. The technique of purifying ZnSe crystals in molten zinc [13] greatly reduces the impurity concentration and the absorption coefficient near the fundamental absorption edge [2], and will probably enable more efficient ZnSe lasers to be made. A similar method has given a CdS laser having an efficiency of 26.5% [14].

We thank A. I. Krasil'nikov for providing the crystals, and N. A. Borisov, V. A. Goncharov, and B. M. Lavrushin for assisting in the work.

Literature Cited

1. M. Cardona and D. L. Greenaway, Phys. Rev., 125:1291 (1962).
2. G. E. Hite, D. T. F. Marple, M. Aven, and B. Segall, Phys. Rev., 156:850 (1967).
3. D. C. Reynolds, L. S. Pedrotti, and O. W. Larson, J. Appl. Phys., 32:2250 (1961).
4. V. V. Sobolev, Opt. Spektrosk., 18:648 (1965).
5. R. E. Halsted and B. Segall, Phys. Rev. Letters, 10:392 (1963).
6. O. V. Bogdankevich, M. M. Zverev, A. I. Krasil'nikov [Krasilnikov], and A. N. Pechonov, Phys. Status Solidi, 19:K5 (1967).
7. R. M. Detweiler and B. A. Kulp, Bull. Am. Phys. Soc., 10:321 (1965).
8. R. E. Halsted, M. Aven, and H. D. Coghill, J. Electrochem. Soc., 112:177 (1965).
9. P. V. Gul'tyaev and A. V. Petrov, Fiz. Tverd. Tela, 1:368 (1959).
10. O. V. Bogdankevich, V. A. Goncharov, B. M. Lavrushin, V. S. Letokhov, and A. F. Suchkov, Fiz. Tekh. Poluprov., 1:7 (1967).
11. V. S. Vavilov and É. L. Nolle, Fiz. Tverd. Tela, 8:532 (1966).
12. O. V. Bogdankevich, V. A. Goncharov, Yu. A. Drozhbin, B. M. Lavrushin, A. M. Mestvirishvili, and V. A. Yakovlev, Zh. Éksp. Teor. Fiz., 53:785 (1967).
13. M. Aven and H. H. Woodbury, Appl. Phys. Letters, 1:53 (1962).
14. C. E. Hurwitz, Appl. Phys. Letters, 9:420 (1966).

ELECTRON MICROPROBE STUDY OF CONCENTRATION PROFILES IN IN GaP—GaAs AND InP—InAs DIFFUSED HETEROJUNCTIONS.

T. D. Dzhafarov, T. T. Dedegkaev, and L. M. Dolginov

Institute of Semiconductors
Academy of Sciences of the USSR, Leningrad

X-ray spectroscopic microprobe analysis was used to study concentration profiles of components in linear-gradient GaP—GaAs and InP—InAs heterojunctions formed by the diffusion of phosphorus into single-crystal plates of gallium and indium arsenides.

Layer-by-layer analysis of the radioactivity, chemical or spectroscopic analysis, and methods based on layer-by-layer electrical measurements, are commonly used for the study of the impurity concentration distribution caused by diffusion in solids [1]. These methods have various disadvantages, the chief of which is that the sample has to be destroyed.

We give below the results of an x-ray spectroscopic microprobe analysis of the concentration profiles of linear-gradient heterojunctions. The measurements were made on GaP—GaAs and InP—InAs heterojunctions formed by the diffusion of phosphorus into GaAs and InAs respectively. The heterojunctions were produced in plates of GaAs ($n \approx 1 \cdot 10^{17}$ cm^{-3}) and InAs ($n \sim 10^{16}$ cm^{-3}) cut from single-crystal ingots perpendicular to the [111] direction. Phosphorus was diffused into the plates in evacuated quartz ampoules. The quantity of phosphorus used was such that its vapor pressure was 20-50 atm at the diffusion temperatures (950-1000°C for GaAs and 750°C for InAs).

The concentration profiles of the components (P, As, Ga, and In) of the GaP—GaAs and InP—InAs diffused heterojunctions were measured with a "Cameca" MS-46 electron microprobe at 25 kV, the current through the sample being 70 nA. All the heterojunction components were analyzed by means of the K series. The standards having known contents of P, As, Ga, and In were homogeneous single crystals of GaP, GaAs, and InP. Before measurements the end faces of the samples (perpendicular to the phosphorus diffusion front) were ground and carefully polished. Each sample was placed in a holder in such a way that a narrow electron beam (a microprobe having diameter $\sim 1 \mu$) was incident at right angles to the polished end surface.

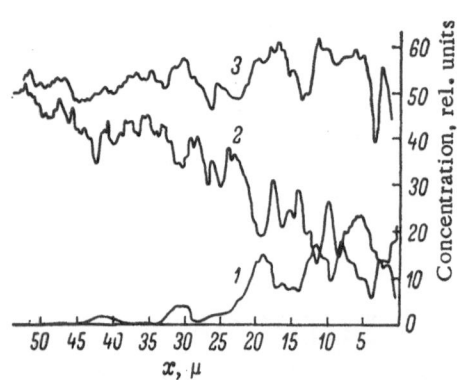

Fig. 1. Concentration distribution of phosphorus (1), arsenic (2), and gallium (3) in gallium arsenide after phosphorus diffusion at T = 1000°C, t = 20 h, p = 40 atm.

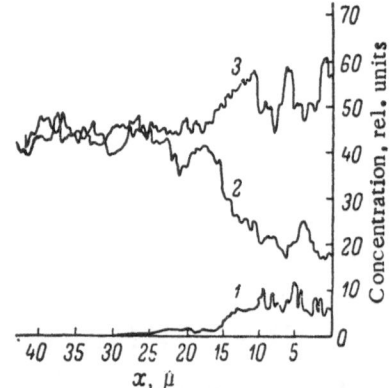

Fig. 2. Concentration distributions of phosphorus (1), arsenic (2), and indium (3) in indium arsenide after phosphorus diffusion at T = 750°C, t = 20 h, p = 30 atm.

The x rays produced were detected by a spectrometer. The electron microprobe was moved over the end face of the sample so as to determine the concentration distributions of the components of the diffused heterojunction. In some instances, where the depth of diffusion was slight, the diffusion region was widened by grinding and polishing the sample into a wedge at 5° to the phosphorus diffusion front, and the electron beam was moved along this surface during the measurements.

Figure 1 shows the measured concentration profiles of phosphorus, arsenic, and gallium in a phosphorus-diffused GaAs sample. It is seen that the general trend of the phosphorus distribution is to decrease into the sample, and that of arsenic to increase. Moreover, each curve is a sequence of alternating maxima and minima. The undulations of the distribution curves are due to microinhomogeneities of the distribution of material. This is shown by the correlation in position between the maximum phosphorus concentrations and the minimum arsenic concentrations, which occur at equal values of x. In the surface layer, Fig. 1 shows that the diffusion of phosphorus produces an almost complete series of solid solutions $GaP_\xi As_{1-\xi}$ $(0.9 > \xi \geq 0)$. The lower concentration of arsenic at the surface is due to its evaporation during diffusion. The concentration profiles of both phosphorus and arsenic differ from the familiar solutions of Fick's equation with the boundary conditions that phosphorus diffuses from a constant source and arsenic evaporates from a uniformly saturated material. This is because, in the present case, the phosphorus and arsenic are diffusing in a body whose composition varies in the x direction, and the solution of the diffusion equation demands special treatment. A rough estimate of the effective diffusion coefficients of phosphorus and arsenic from the concentration curves in Fig. 1 shows that $D_p \approx 5 \cdot 10^{-11}$ cm^2/sec and $D_{As} \approx 3 \cdot 10^{-11}$ cm^2/sec (T = 1000°C, p = 40 atm). The values found for D_p and D_{As} are several orders of magnitude greater than the diffusion coefficient of phosphorus in GaAs [2] and the self-diffusion coefficient of arsenic in GaAs [3]. This difference is probably due to differing conditions of diffusion.

Similar results were obtained in a study of concentration profiles after diffusion of phosphorus into InAs samples (Fig. 2).

Thus, the method of x-ray spectroscopic microprobe analysis can be used to determine the concentration distributions of components of GaP−GaAs and InP−InAs solid solutions, and to estimate their diffusion coefficients.

We thank B. I. Boltaks for supervising this work, and A. I. Zaslavskii and A. Ya. Nashel'-skii for their interest.

Literature Cited

1. B. I. Boltaks, Diffusion in Semiconductors, Infosearch, London (1963).
2. B. Goldstein and C. Dobin, Solid State Electronics, 5:411 (1962).
3. B. Goldstein, Phys. Rev., 121:1305 (1961).

PHOTOELECTRIC PROPERTIES OF InP p—n JUNCTIONS

V. V. Galavanov, R. M. Kundukhov,
D. N. Nasledov, and N. V. Siukaev

K. L. Khetagurov Severo-Osetinsk State Pedagogical Institute, Ordzhonikidze

A method for fabricating indium phosphide photocells is described. The spectral characteristic and the temperature dependence of the photo-emf and the short-circuit photocurrent are given.

We have not found any published results of a systematic experimental study of the photoelectric properties of InP photocells.

Rappaport has examined the photovoltaic effect in alloyed InP p—n junctions. The solar batteries used consisted of alloy junctions having an efficiency not exceeding 2% [1].

Further experiments in the technique of producing InP single crystals, the fabrication of p—n junctions and low-resistance linear contacts, and the surface treatment of these junctions have allowed a systematic experimental study of the electric and photoelectric properties of InP p—n junctions to be begun.

The present paper deals with their photoelectric properties, and continues the work reported in [2-5].

Method of Fabricating the Photocells

The photocells were fabricated from single crystals of n-type indium phosphide having electron density $\sim 10^{17}$ cm^{-3} and mobility 3500 cm$^2 \cdot$ V$^{-1} \cdot$ sec^{-1}.

The p—n junctions were made by the double diffusion method [6].

Results are given below of a study of p—n junctions formed by diffusion of cadmium under the following conditions: diffusion temperature 780°C, primary diffusion time 2 h, secondary diffusion time 2 min, temperature controlling saturated vapor pressure of cadmium 700°C, quantity of cadmium in ampoule 2 mg, quantity of phosphorus in ampoule 0.8 mg, volume of ampoule 1 cm^3. The thickness of the p-type region after the secondary diffusion was ~ 10 μ, and the photocell area 0.1 mm^2. The p-type layer was illuminated on the $(\bar{1}\bar{1}\bar{1})$ plane.

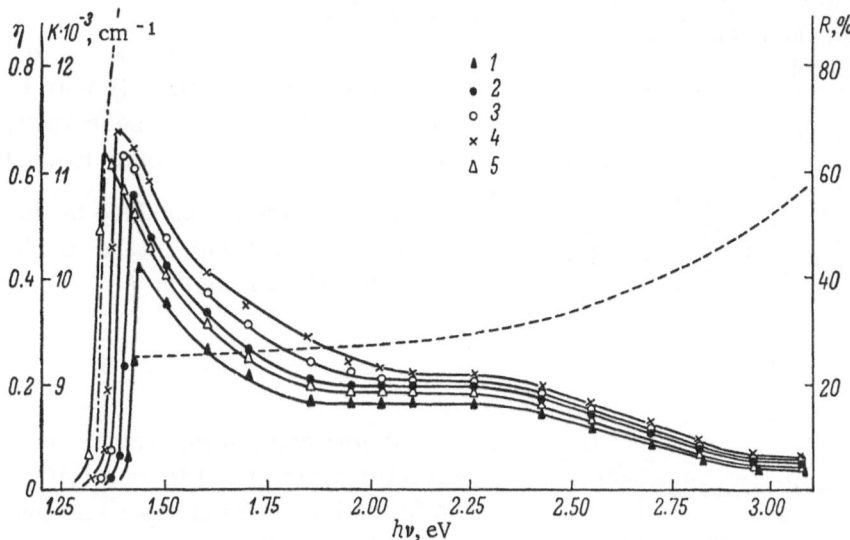

Fig. 1. Spectra of the quantum efficiency η of photocell A-17 at various temperatures, and of the absorption coefficient K (chain line) [7] and reflection coefficient R (dashed line) at room temperature. T, °K: 1) 77; 2) 194; 3) 240; 4) 295; 5) 400.

The dependence of the differential charge capacitance on the voltage is given by the empirical formula

$$C \propto (U_i^C - U)^{-\frac{1}{3}},$$

where U_i^C is the capacitative voltage intercept and U the external voltage applied to the p−n junction.

Spectral Characteristics

Figure 1 shows the spectra of the quantum efficiency of one photocell at several different temperatures, and also the absorption [7] and reflection spectra of InP at room temperature.

It is seen that the spectral characteristics have a sharp maximum at all temperatures, followed by saturation and a further drop at wavelengths of 0.45-0.55 μ.

Since, in the latter range, the decrease of the photosensitivity is more rapid than the increase of the reflection coefficient, the second drop cannot be explained by the increasing reflection coefficient in this region. It may be due either to a high surface recombination velocity or to the presence of a potential barrier near the p-type surface, which prevents the diffusion of electrons to the p−n junction and therefore reduces the photocurrent.

As the temperature increases, the low-energy edge of the photosensitivity moves to longer wavelengths. The width E(T) of the forbidden band of InP, calculated from the wavelength which corresponds to half the drop of the photocurrent in the spectral distribution curves on the long-wave side, decreases linearly with increasing temperature (Fig. 2), and E = 1.365 eV at T = 296°K. The temperature coefficient of the forbidden band width is $\alpha = 2.7 \cdot 10^{-4}$ eV/deg. These results are in good agreement with those of [7, 8].

The photocurrent corresponding to the maximum photosensitivity increases with temperature up to 300°K, and then begins to decrease.

The electron diffusion length L_n in the p-type region, determined from the spectral characteristics of a number of photocells, is 0.5-1.2 μ at room temperature.

TABLE 1. Parameters of Forward Branch of Dark Current−Voltage Characteristic of Sample A-17

$T°$, K	I_{s1}, A	β_1	I_{s2}, A	β_2
78	$3.1\cdot10^{-11}$	11.00	$8\cdot10^{-12}$	10.30
200	$1.2\cdot10^{-10}$	3.92	$7.5\cdot10^{-13}$	3.00
224	$3.8\cdot10^{-9}$	3.65	$2.1\cdot10^{-10}$	2.90
294	$1.25\cdot10^{-8}$	2.78	$2\cdot10^{-9}$	2.20
343	—	—	$2\cdot10^{-7}$	2.18
390	—	—	$2.7\cdot10^{-6}$	2.18

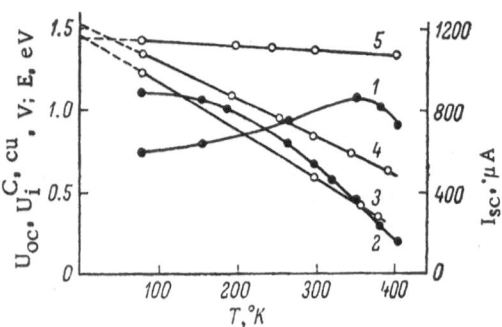

Fig. 2. Temperature dependence of short-circuit photocurrent I_{sc} (curve 1) and photo-emf U_{oc} (curve 2) on illumination with integrated light, and of capacitative voltage intercept U_i^C (curve 3) and current−voltage intercept U_i^{cu} (curve 4), and of forbidden band width E of InP (curve 5) for photocell A-17.

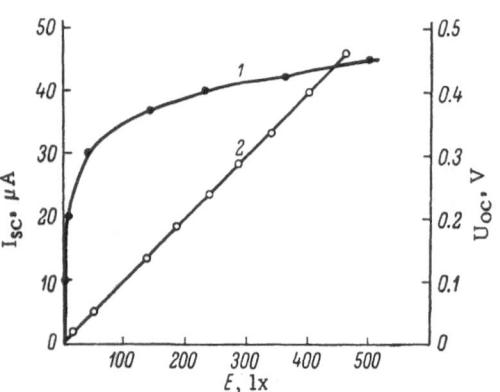

Fig. 3. Open-circuit voltage U_{oc} (curve 1) and short-circuit current I_{sc} (curve 2) of photocell A-20 as functions of illumination.

Current − Voltage Characteristics

At voltages $U \leq kT/q$, the forward and reverse currents vary linearly with the voltage (the differential resistance at zero voltage is of the order of 10^6 Ω at room temperature).

The voltage dependence of the forward current above room temperature can be represented by the empirical formula

$$I_f = I_s e^{\frac{qU}{\beta kT}}.$$

Below room temperature, current−voltage characteristics of a different type are observed. The foward branch can, in some instance, be represented as a sum:

$$I_f = I_{s1} e^{\frac{qU}{\beta_1 kT}} + I_{s2} e^{\frac{qU}{\beta_2 kT}},$$

with either $\beta_1 > \beta_2$, $I_{s1} > I_{s2}$, or $\beta_1 < \beta_2$, $I_{s1} < I_{s2}$, where I_{s1}, I_{s2}, and I_s are coefficients determined by semilogarithmic extrapolation of the linear function $I_f = f(U)$ to the current axis. At low temperatures, the dimensionless factors are β_1, $\beta_2 \gg 2$ and they are temperature-dependent.

Above room temperature, $\beta = 2.2$ and is almost independent of temperature. Table 1 shows the parameters of the forward branch of the dark current−voltage characteristic of sample A-17.

The coefficient I_s found experimentally is several orders of magnitude higher than that calculated by the theory of Sah, Noyce, and Shockley. Apparently, the forward current of the photocells, even above room temperature, is not determined solely by recombination in the region of the p−n junction. The first term in the experimental function $I_f = f(U)$ below room temperature is governed by the state of the surface at the point where the p−n junction emerges.

The reverse current for diffused p−n junctions in indium phosphide at voltages below the breakdown voltage obeys the relationship

$$I_r = kU^n + I_f.$$

The exponent n varies greatly with the treatment of the surface of the crystal containing the p−n junction; for many samples it lies between 1 and 3.5, which is several times larger than the theoretically predicted value 0.3-0.5. Apparently, the reverse current of the photocells is governed not by generation in the space-charge region, but by the state of the single-crystal surface at the point where the p–n junction emerges. The breakdown voltage U_{br} of photocells in photodiode operation does not exceed 6-8 V at room temperature.

Figure 3 shows U_{oc} and I_{sc} as functions of the illumination of photocell A-20. These results were obtained with an ambient temperature of 20°C.

As the temperature increases, the photo-emf U_{oc} decreases nonlinearly; the short-circuit photocurrent increases between 77 and 350°K, and then decreases (see Fig. 2).

Figure 2 also shows the temperature dependences of the capacitative voltage intercept U_i^C and the current–voltage intercept U_i^{cu} for photocell A-17.

It is seen that these vary linearly with temperature. The temperature coefficient of the capacitative voltage intercept is $\alpha_i^C = 4.4 \cdot 10^{-3}$ V/deg, and that of the current–voltage intercept is $\alpha_i^{cu} = 2.4 \cdot 10^{-3}$ V/deg.

Finally, the following are the parameters of photocell A-17 corresponding to solar radiation with power 62 mW/cm² and temperature 20°C: open-circuit photo-emf U_{oc} = 0.72 V; short-circuit photocurrent I_{sc} = 1.3 mA; photocurrent density 11.8 mA/cm²; maximum power delivered P_{max} = 0.49 mW, or 4.45 mW/cm² at optimum load resistance 550 Ω, voltage 0.52 V and current 940 μA; utilization factor of the load characteristic γ = 0.52; photocell efficiency 7.2%.

Literature Cited

1.   P. Rappaport, RCA Rev., 20:373 (1959).
2.   S. G. Metreveli and R. M. Kundukhov, Izv. vuzov SSSR, Fizika, No. 3, p. 151 (1966).
3.   V. V. Galavanov, R. M. Kundukhov, and D. N. Nasledov, Fiz. Tverd. Tela, 8:3402 (1966).
4.   R. M. Kundukhov, S. G. Metreveli, and N. V. Siukaev, Fiz. Tekh. Poluprov., 1:924 (1967).
5.   V. V. Galavanov, R. M. Kundukhov, and S. G. Metreveli, Fiz. Tekh. Poluprov., 1:1047 (1967).
6.   A. A. Gutkin, D. N. Nasledov, V. E. Sedov, and B. V. Tsarenkov, USSR Author's Certificate No. 156,626.
7.   W. J. Turner, W. E. Reese, and G. D. Pettit, Phys. Rev., 136:A1467 (1964).
8.   G. G. Kovalevskaya and S. V. Slobodchikov, in: Studies in Semiconductor Physics and Geophysics, Ordzhonikidze (1967).

CONDITIONS FOR TURNING ON A THYRISTOR
BY SHORT GATE CURRENT PULSES

A. I. Uvarov

A. F. Ioffe Physicotechnical Institute
Academy of Sciences of the USSR, Leningrad

Conditions for turning on a thyristor by short gate current pulses of arbitrary shape are considered in the presence and absence of an initial charge. The analysis is based on a solution of the equation of continuity of charge in the thyristor bases and on a linear approximation for the emitter leakage current. It is shown that the effective charges in the thyristor bases, produced simultaneously by various agencies, are additive when they are reduced to the initial conditions. Practical examples are given of estimates of the amplitude and duration of gate current pulses capable of turning on a thyristor. Estimates are made of the strays in the gate circuit which are too weak to turn on the thyristor.

A theoretical treatment of the conditions for turning on a thyristor by short gate current pulses is of practical value in the correct design of thyristor gate circuits and in determining the permissible values of the magnitude and duration of strays in the gate circuit which are not to turn on the thyristor. The present author [1] has formulated conditions for turning on a thyristor when there is an excess charge in its bases, accumulated during a short external pulse. However, the way in which the effective charge is related to the intensity and duration of such a pulse was not discussed.

This paper is essentially a continuation of [1]; its aim is to analyze the conditions for turning on a thyristor by a short pulse of the gate current and to estimate the effective charge in the bases when this current ceases. The calculation is based on the ideas developed in [1], and the approximations used are the same.

Formulation of the Problem

We shall consider a one-dimensional model of a thyristor having fixed base thicknesses w_1 and w_2 and zero-thickness collector space-charge layer. The origin is taken at the collector; the boundary of the left-hand emitter E_1 is at $x = -w_1$, and that of the right-hand emitter E_2 is at $x = w_2$. In the first emitter, adjoining a thin p-type base, a leakage current I_R is taken into account. This current varies sublinearly with the minority carrier density near that emitter but is approximated by a linear function:

$$I_R = I_{R_0} + qs \, (y_1)_{-w_1},$$ (1)

where y_1 is the minority carrier density in the first base; I_{R_0} and the surface recombination velocity s for ohmic leakage and recombination in the space-charge layer of the emitter are related in [1] to the thyristor parameters.

The equation of continuity of charge in the bases,

$$\frac{\partial y}{\partial t} = D \frac{\partial^2 y}{\partial x^2} - \frac{y}{\tau},$$ (2)

is solved for zero initial conditions. The equilibrium density of minority carriers in the collector and the collector leakage current are neglected, and the injection efficiency of the second emitter is taken as unity. The minority carrier density y, the diffusion coefficient D, and the lifetime τ must be different for the two bases, and will be distinguished by subscripts 1 and 2. Since we are looking at the initial stage of the turn-on process, the calculation will be made in the approximation of a low injection level.

The boundary conditions are obtained by assuming zero minority carrier density near the collector:

$$(y_1)_{x=0} = (y_2)_{x=0} = 0$$ (3)

and by applying Kirchhoff's law to the current through each p−n junction. To take a general case, we consider two gate currents I_{b_1} and I_{b_2} flowing into the two bases, which are arbitrary functions of time and cease at t_1 and t_2 respectively. Although the thyristor is actually controlled by passing a current through one electrode, the use of two gate currents is useful in discussing the conditions for turning on a thyristor by illumination or a sudden increase of the anode voltage. For the case considered, the boundary conditions are

$$-qD_1 \left(\frac{\partial y_1}{\partial x}\right)_{-w_1} + I_{R_0} + qs \, (y_1)_{-w_1} - I_{b_1} = -qD_1 \left(\frac{\partial y_1}{\partial x}\right)_0 + qD_2 \left(\frac{\partial y_2}{\partial x}\right)_0 = qD_2 \left(\frac{\partial y_2}{\partial x}\right)_{w_2} - I_{b_2} = I_c.$$ (4)

Here, I_c is the minority carrier current through the collector.

Solution and Determination of the Turn-on Conditions

The solution of Eq. (2) with zero initial conditions and the boundary conditions (3) and (4) is obtained by the Laplace-transform operational method. The expression for the transform of the collector current is

$$I_c = \frac{\left(\bar{I}_{b_1} - \dfrac{I_{R_0}}{p}\right) \mathrm{ch} \, a_2 + \bar{I}_{b_2} \left(\mathrm{ch} \, a_1 + \dfrac{\varkappa}{a_1} \mathrm{sh} \, a_1\right)}{\left(\mathrm{ch} \, a_1 + \dfrac{\varkappa}{a_1} \mathrm{sh} \, a_1\right)(\mathrm{ch} \, a_2 - 1) - \mathrm{ch} \, a_2},$$ (5)

where $\bar{I}_{b_1}$ and $\bar{I}_{b_2}$ are the transforms of the gate currents,

$$a_1 = \sqrt{1 + p\tau_1} \, \frac{w_1}{L_1}, \quad a_2 = \sqrt{1 + p\tau_2} \, \frac{w_2}{L_2}, \quad \varkappa = \frac{sw_1}{D_1}.$$

To determine I_c as a function of time [1] when inverting the transform, we integrate in the complex plane of p, taking from the series expansion of the exponential functions the only term having a positive argument, which is given by the real roots of the transcendental equation

$$\left(\operatorname{ch} a_1 + \frac{\varkappa}{a_1} \operatorname{sh} a_1\right)(\operatorname{ch} a_2 - 1) - \operatorname{ch} a_2 = 0. \tag{6}$$

In the subsequent discussion, a_1 and a_2 will denote only the real roots of Eq. (6), which give a positive value

$$p = \frac{1}{\tau_i},$$

where τ_i is the time constant of the exponential increase of current in the initial stage of the turn-on process [2].

The question of whether the thyristor will be turned on or not (i.e., whether the current will increase or decrease with time) is resolved by the sign of the factor in front of the exponential function with a positive argument. For the function φ, whose transform is unity divided by the denominator of Eq. (5), we take only the term

$$\varphi_1 = (a_{-1}) \, e^{\frac{t}{\tau_i}}, \tag{7}$$

where (a_{-1}) is the residue of the function at the pole determined by real a_1 and a_2. The original functions I_{b_1}, I_{b_2}, and I_{R_0} are unknown. Applying the convolution theorem successively to each term in Eq. (5), we obtain the positive exponential function which occurs in the series expansion of I_c

$$(I_c)_1 = (a_{-1}) \left[\operatorname{ch} a_2 \int_0^t I_{b_1}(\xi) \, e^{\frac{t-\xi}{\tau_i}} \, d\xi + \right.$$

$$\left. + \left(\operatorname{ch} a_1 + \frac{\varkappa}{a_1} \operatorname{sh} a_1\right) \int_0^t I_{b_2}(\xi) \, e^{\frac{t-\xi}{\tau_i}} \, d\xi - \operatorname{ch} a_2 \tau_i I_{R_0} e^{\frac{t}{\tau_i}} \right]. \tag{8}$$

Taking $\exp(t/\tau_i)$ outside the integrals and equating to zero the factor in the front of the exponential function, we get the turn-on condition

$$\operatorname{ch} a_2 \int_0^t I_{b_1}(\xi) \, e^{-\frac{\xi}{\tau_i}} \, d\xi + \left(\operatorname{ch} a_1 + \frac{\varkappa}{a_1} \operatorname{sh} a_1\right) \int_0^t I_{b_2}(\xi) \, e^{-\frac{\xi}{\tau_i}} \, d\xi = \tau_i I_{R_0} \operatorname{ch} a_2. \tag{9}$$

Replacing the variable ξ in the integrals by t, and using Eq. (6) and the fact that I_{b_1} ceases to flow at time t_1 and I_{b_2} at time t_2, we get the final condition for turning on the thyristor by the action of short gate current pulses:

$$\int_0^{t_1} I_{b_1}(t) \, e^{-\frac{t}{\tau_i}} \, dt + k_2 \int_0^{t_2} I_{b_2}(t) \, e^{-\frac{t}{\tau_i}} \, dt = \tau_i I_{R_0}, \tag{10}$$

where

$$k_2 = \frac{1}{\operatorname{ch} a_2 - 1} \cong \sqrt{\frac{\theta_1}{\theta_2}} \tag{11}$$

is the "effectiveness" coefficient of the charge concentrated near the emitter adjoining the wide base [1].

Analysis of the Solution and the Turn-on Conditions

when Initial Charge Is Present in the Bases

We shall compare Eq. (10) with the condition for turn-on due to accumulation of excess charge in the bases [1]

$$Q_1 + k_2 Q_2 = \tau_i I_{R_0}. \tag{12}$$

It will be assumed that the charges Q_1 and Q_2 are actually present in the bases at time t = 0. The right-hand side in each equation is a quantity defined as the critical turn-on charge of a thyristor. On the left in Eq. (12) is the effective charge present initially in the bases. The left-hand side of Eq. (10) must accordingly be regarded as the effective charge due to the gate currents.

Let us consider in more detail the physical significance of these ideas. The complete solution for the current through a thyristor as a function of time is an expansion as a series of exponentials, one of which has a positive argument and all the rest have negative arguments. After a certain delay time, all terms with the negative arguments become negligible, and the subsequent behavior of the current is governed by the positive-argument term, whose time constant is τ_i. If the left-hand side of Eq. (12) is greater than the right-hand side, the current increases exponentially, and therefore so does the excess charge present in the bases at any given time. If, on the other hand, the initial effect charge is less than the critical value, the charge will thereafter decrease according to the exponential with the positive argument. A similar effect has been noted in [3], where the transient process of turning off a thyristor by a negative gate current was discussed. The effective charge in the bases turns on the thyristor, while the emitter leakage current acts as a negative gate current, which tends to turn off the thyristor.

The left-hand side of Eq. (10) represents a certain charge, which plays the same part as the effective charge present at the beginning of the turn-on process and given by Eq. (12). Thus, it may be regarded as the effective charge in the bases of the thyristor produced by the gate currents and relating to the initial instant t = 0. This means that, under the action of short gate pulses, the behavior of the current through the thyristor for large values of t is the same as if there were no gate currents but an initial effective charge in the bases given by the left-hand side of Eq. (10). If the effective charge at the beginning and the end of each of the gate current pulses is equal to the critical value, this relation will continue to apply. The thyristor remains in a kind of unstable equilibrium state, being neither turned-on nor turned-off. This represents static conditions corresponding to a point on the negative differential resistance section of the static characteristic. In practice, under dynamic conditions, such a state does not occur, since any fluctuation of the current will cause a departure from it in one or the other direction.

Incidentally, the first term on the left of Eq. (10) corresponds to the total initial effective charge produced in both bases by the current I_{b_1}. Equation (10) does not say what fraction of the charge is in each base. Similarly, the second term on the left of Eq. (10) gives the total initial effective charge in both bases due to the gate current entering the second base.

The foregoing discussion enables us to formulate the thyristor turn-on condition for the case where, at the time when the gate pulses are applied, there is already present in the bases an excess charge due to some other process. Such an initial excess charge exists in consequence of ohmic leakage of the collector and carrier generation in its space-charge layer; it may be due to illumination of the bases or it may remain after charge dispersal during turn-off of the thyristor. In this case, clearly, the turn-on condition is governed by the critical value of the total initial effective charge:

$$Q_1 + \sqrt{\frac{\theta_1}{\theta_2}}\, Q_2 + \int_0^{t_1} I_{b_1}(t)\, e^{-\frac{t}{\tau_i}}\, dt + \sqrt{\frac{\theta_1}{\theta_2}} \int_0^{t_2} I_{b_2}(t)\, e^{-\frac{t}{\tau_i}}\, dt = \tau_i I_{R_0}. \tag{13}$$

Here, the "effectiveness" coefficients of the charge and gate current in the thick base k_2 have been written approximately in accordance with [1].

The condition (13) can be also derived by a rigorous solution of the equation of continuity with the initial condition determined by the presence of an initial charge in the base. It is found that the "effectiveness" coefficients of the initial charges are independent of the gate currents and are determined by the same quantities as in [1]. The expression for that part of the effective charge which is due to the gate currents is the same as Eq. (10). This proves the additivity of the initial effective charges no matter how they are produced.

It has been shown previously [1] that the "effectiveness" of the charges in the bases is almost independent of the carrier distribution within them. It is therefore reasonable to expect that the "effectiveness" of the gate currents will not vary greatly with the way in which these currents are applied. Consequently, Eq. (13) acquires a more general significance than was hitherto considered. It must be valid also for the turn-on conditions of a photothyristor, when the gate currents are replaced by ionization; and for the turn-on due to a rapid rise of the anode voltage from zero or when a voltage is applied after dispersal of carriers in the process of turning off. However, the treatment of these problems is beyond the scope of the present paper.

Significance of the Initial Turn-on Stage

Despite the simple form and clear physical significance of Eq. (10), its practical use to determine the effective charge may sometimes lead to a considerable error. Let us consider, for example, the effective charge produced by the gate current I_{b_1} (with $I_{b_2} = 0$) which has the various dependences on time shown in Figs. 1a,b. Figure 1a shows a rectangular pulse of current of amplitude I_0 and duration t_0. For such a pulse, Eq. (10) gives the effective charge

$$Q_{\text{eff}} = I_0 \int_0^{t_0} e^{-\frac{t}{\tau_i}}\, dt = \tau_i I_0 \left(1 - e^{-\frac{t_0}{\tau_i}} \right). \tag{14}$$

In Fig. 1b, the pulse is likewise rectangular, but shifted in time and preceded by a small current I_1 for a time t_1. Then, Eq. (10) gives

$$Q_{\text{eff}} = I_1 \int_0^{t_1} e^{-\frac{t}{\tau_i}}\, dt + I_0 \int_{t_1}^{t_1+t_0} e^{-\frac{t}{\tau_i}}\, dt = \tau_i I_1 \left(1 - e^{-\frac{t_1}{\tau_i}} \right) + \tau_i I_0 e^{-\frac{t_1}{\tau_i}} \left(1 - e^{-\frac{t_0}{\tau_i}} \right). \tag{15}$$

It is seen that, when I_1 is small and t_1 is long, the effective charge in the second case is considerably less than in the first case. At the same time, it is clear from physical arguments that a charge greater than that for the current shown in Fig. 1a must appear in the thyristor bases when the current shown in Fig. 1b has finished flowing in the gate circuit.

This discrepancy is due to our approximation for the emitter leakage current. In the calculations, we have used a linear approximation for the leakage current I_R, which varies sublinearly with the minority carrier density near the emitter E_1, replacing it by a linear function given by the tangent to the curve of I_R at the point $n_e = n_h$ (Fig. 2). In the initial turn-on stage, when the minority carrier density near E_1 has not yet increased much, our approximation greatly overestimates the recombination current, and the calculations must give too low a value for the effective charge. A corresponding correction to the critical charge would be needed. The duration of the initial stage of low minority carrier density and the error in estimating the

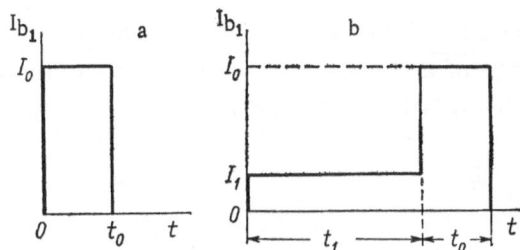
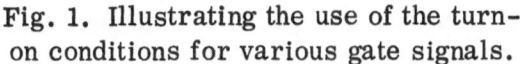
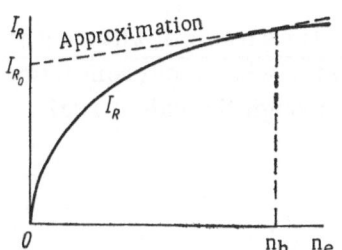

Fig. 1. Illustrating the use of the turn-
on conditions for various gate signals.

Fig. 2. Linear approximation
to emitter leakage current.

effective charge both increase with decreasing initial value of the gate current and decreasing slope of its front. The error is smaller for larger initial gate currents. If the initial value of the gate current is I_{R_0}, the maximum error for an ohmic emitter leakage does not exceed 15%, and for recombination in the space-charge layer 25%.

Thus, Eq. (10) and therefore Eq. (13) can be used if the gate current reaches I_{R_0} in a time less than τ_i. For a slowly rising gate current, the same expressions can be used in the calculation, but the initial instant $t = 0$ should be shifted to a point where the gate current becomes comparable with I_{R_0} (approximately $I_{b.on}$). The error is then within the amounts stated above. When the gate current flows into a thick base, the "effectiveness" coefficient k_2 must also be taken into account.

Some Examples of Determining the Turn-on Conditions for Gate Pulses of Various Shapes

Let us consider two examples of the determination of the amplitude and length of a gate pulse sufficient to turn on a thyristor, using Eq. (10).

Rectangular Pulse. For a rectangular gate pulse of amplitude I_0 and duration t_0, equating the effective charge given by Eq. (14) to the critical charge $\tau_i I_{R_0}$, we obtain the turn-on condition

$$I_0 \left(1 - e^{-\frac{t_0}{\tau_i}} \right) = I_{R_0}. \tag{16}$$

From this, we easily find the following relations between the amplitude and duration of the turn-on pulse:

$$I_0 = \frac{I_{R_0}}{1 - e^{-\frac{t_0}{\tau_i}}} \tag{17}$$

and

$$t_0 = \tau_i \ln \frac{I_0}{I_0 - I_{R_0}}. \tag{18}$$

An expression similar to Eq. (18) has previously been derived in [4], where a modified charge method was used to determine the critical charge buildup time in a symmetrical p−n−p−n structure with allowance for nonlinear emitter leakage. The only difference between Eq. (18) and the corresponding expression in [4] is that the latter had $I_{b.on}$ in place of I_{R_0}; $I_{b.on}$ is equal to I_{R_0} for an ohmic emitter leakage and $\frac{1}{2} I_{R_0}$ for recombination in the space-charge layer. This expression has been experimentally tested in [5] for thyristor models consisting of a p−n−p transistor and a n−p−n transistor with controllable emitter leakage. The agreement between theory and experiment is good.

RC Discharge through Gate Circuit.   Let us consider the turn-on condition for a thyristor when a current from the discharge of a capacitance C across a resistance R is passed through the gate circuit. The gate current pulse is then given by

$$I_{b_1} = I_0 e^{-\frac{t}{RC}}.$$ (19)

Although this gate current is not restricted in time, it decreases so rapidly that we can integrate to $t = \infty$ in Eq. (10). This gives the following relations between the amplitude and rate of decay of the turn-on pulse:

$$I_0 = \left(1 + \frac{\tau_i}{RC}\right) I_{R_0}$$ (20)

and

$$RC = \tau_i \frac{I_{R_0}}{I_0 - I_{R_0}}.$$ (21)

This turn-on method can be used to measure the parameters τ_i and I_{R_0} for actual thyristors. By varying RC and finding the value of I_0 which causes turn-on, we can plot a graph of $I_0 = F(1/RC)$, which should be a straight line having an intercept I_{R_0} on the axis of I_0 and a slope which easily gives the critical charge and the value of τ_i.

In the design of gate circuits in thyristorized systems, it is customary to use gate pulses whose amplitude and duration considerably exceed the minimum turn-on values, in order to ensure reliable functioning and to reduce the turn-on time, having regard to the spread of device parameter values. The estimates given above are therefore useful not so much for selecting the gate pulse values needed to turn on the device as for determining the maximum values of the external and internal strays in the gate circuit which do not cause spontaneous turn-on of the thyristor.

An analysis of Eq. (10) shows that, for very short pulses whose duration is of the order of τ_i, the effective charge is almost equal to the total quantity of electricity flowing in the gate circuit during this pulse (with the "effectiveness" coefficient k_2 for control through the thick base). For long pulses, the effective charge is a small fraction of the charge flowing. Hence, to get a very rough estimate of the acceptable magnitude of the strays in the gate circuit, we may assume that the total charge passing through this circuit as a result of the strays must not exceed the critical charge. For more precise calculations, Eq. (10) must be used, having regard to the comments in the preceding section.

Conclusions

1. The concept of the effective turn-on charge of a thyristor is used to derive the turn-on condition for short gate pulses, in the form of equality of the critical charge and the effective charge produced by the gate currents and reduced to the initial conditions.

2. It is shown that the effective charges, reduced to the initial conditions and produced by different factors acting simultaneously, are additive. A general condition is formulated for the thyristor to be turned on by a short external pulse when an initial charge is present in its bases.

3. Relations are derived between the amplitude and duration of a gate signal which causes the turn-on of the thyristor; this is done for simple gate pulses. A procedure is recommended for estimating the acceptable "noise" in the gate circuit which will not cause a spontaneous turn-on of the thyristor.

The author has pleasure in thanking Professor V. M. Tuchkevich, Professor A. I. Gubanov, A. A. Lebedev, I. V. Grekhov, V. B. Shuman, and V. E. Chelnokov for participating in discussion of this and the previous paper.

Literature Cited

1. A. I. Uvarov, present volume, p. 170.
2. A. A. Lebedev, A. I. Uvarov, and V. E. Chelnokov, Radiotekh. i Élektron., 11:1458 (1966).
3. A. A. Lebedev, A. I. Uvarov, and V. E. Chelnokov, Radiotekh. i Élektron., 13:115 (1968).
4. A. A. Lebedev and A. I. Uvarov, Radiotekh. i Élektron., 12:895 (1967).
5. A. A. Lebedev, M. V. Popova, A. I. Uvarov, and V. E. Chelnokov, Radiotekh. i Élektron., 11:1803 (1967).

THYRISTORS WITH MORE THAN ONE COLLECTOR

I. V. Grekhov and V. B. Shuman

A. F. Ioffe Physicotechnical Institute
Academy of Sciences of the USSR, Leningrad

The fabrication of p−n−p−n structures with high breakover voltages and low holding voltages in the on state is limited both by the technological feasibility of increasing the minority carrier lifetime in the thick base of the structure and by the increase in switching-off time with lifetime. These limitations are largely avoided when the structure contains 2n layers with n ≥ 3. Such a structure has n − 1 collector junctions and n emitter junctions. The breakover voltage and reverse voltage can be increased by a factor n − 1 in comparison with the four-layer structure, while forward losses are increased only slightly. The particular features of the switching on and off of multicollector structures demand that both the collector and the emitter junctions be made as p−n junctions with controllable avalanching. Experimental current−voltage characteristics are presented for a two-collector six-layer structure produced by successive diffusion.

The maximum value of the breakover voltage U_{bo} of a p−n−p−n structure with correctly chosen dimensions is governed by the breakdown voltage U_c of the collector p−n junction. The voltage U_0 which is blocked in the reverse direction is equal to the sum of the breakdown voltages of the two emitters. To increase U_{bo} and U_0, the resistivity of the thick-base material is usually increased. But, since this causes a considerable increase of the thickness of the space-charge layer, the thickness of this base also has to be increased to avoid the punch-through effect [1, 2]. In order not to increase considerably the holding voltage across the bulk of the base [3], the lifetime τ of the minority carriers in the base must also be increased. The latter process is subject to technological limitations. Moreover, increasing τ increases the switching-off time of the thyristor [4], and this is frequently undesirable.

These limitations are largely avoided in a structure (Fig. 1) which consists of six or more layers with alternating types of conductivity. When an external voltage having the polarity shown in Fig. 1 is applied to such a structure, the electrons and holes injected from the end junctions enter the layers 3 and 4 as the majority carriers and cause injection of the junction E_2. It is easily shown that, when $\gamma_{E_2} \neq 1$ and certain carrier recombination conditions exist in the bases 2, 3, 4, and 5, the whole structure is switched, i.e., the voltage on the collector junctions C_1 and C_2 changes sign. The breakover or switching voltage U_{bo} of the structure can, of course, be equal to the sum of the breakdown voltages of all the collector junctions. The voltage drop across the structure in the conducting state is

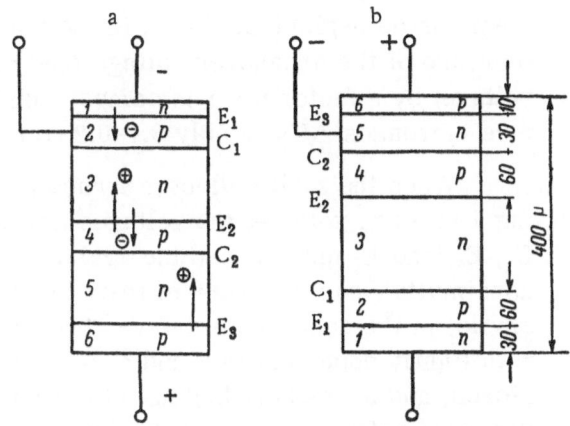

Fig. 1. Multicollector structure (a) and geometry of six-layer two-collector structure produced by successive diffusion (b).

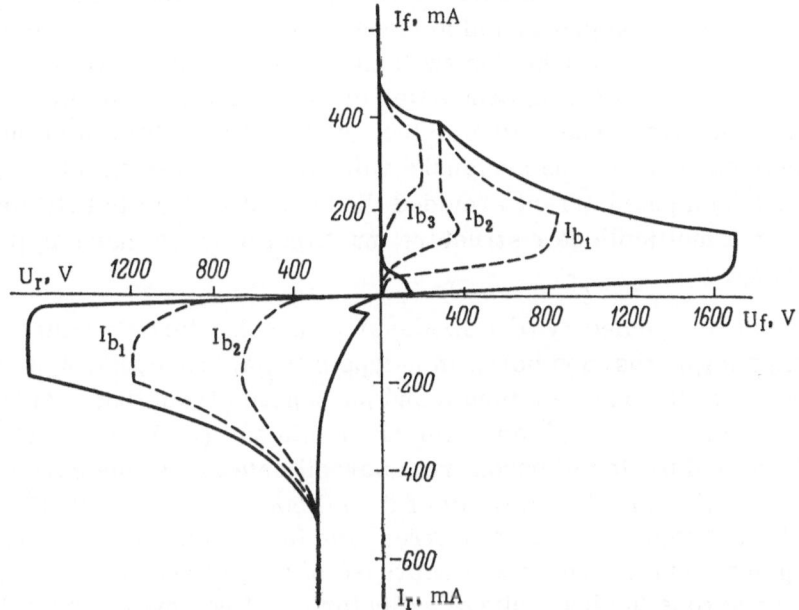

Fig. 2. Current–voltage characteristic of a six-layer structure. Continuous curve corresponds to gate current $I_b = 0$; dashed curve to $I_b > 0$; $I_{b_3} > I_{b_2} > I_{b_1}$.

$$\Delta U_h = \sum \Delta U_e - \sum \Delta U_c + \sum \Delta U_b + R_c I,$$

where ΔU_e, ΔU_c, and ΔU_b are the voltage drops across the emitter and collector junctions and across the base, and R_c is the contact resistance.

Since the number of collectors in a $2n$-layer structure is $n - 1$ and the number of emitters is n, we have $\sum \Delta U_e - \sum \Delta U_c \simeq \Delta U_e$. Hence, ΔU_h for the multicollector structure differs from that for an ordinary thyristor only in that the voltage drop across the base layers is increased because of their greater number (for correctly chosen W/L ratios, the voltage drop across the base layer is 0.1-0.2 V). Figure 1a shows that, in order to increase the breakover voltage by a factor $n - 1$ (where $2n \geq 6$) in comparison with the four-layer structure, there must be at least $n - 1$ thick and lightly doped base regions (3 and 5 in Fig. 1). The remaining layers may be

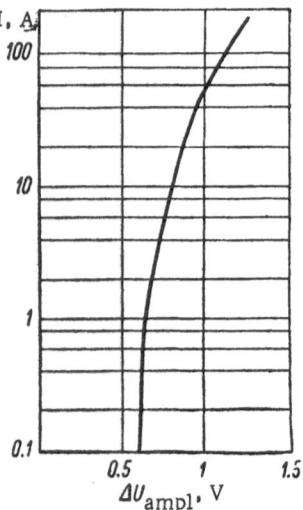

Fig. 3. Current—voltage characteristic of a two-collector thyristor in the on state.

thinner and more heavily doped, and so the voltage drop across them can be neglected. Thus, in the multicollector structure, the increase of the breakover voltage (and of the reverse breakdown voltage) by a factor $n - 1$ in comparison with the four-layer structure increases ΔU_h by only 0.1-0.2 times $(n - 2)$ volts.

When the multicollector structure is switched off by applying a reverse voltage, the voltages across the emitter junctions E_1, E_2, and E_3 may exceed the avalanche voltage because of the nonsimultaneous recovery of their blocking properties. For example, in the case of a six-layer two-collector structure (Fig. 1b) with lightly doped bases 3 and 4, these properties are first restored, and breakdown begins, at the junction E_3. Then, if E_1 recovers before E_2 and breaks down, the p−n−p−n structure consisting of layers 2, 3, 4, and 5 is switched on in the forward direction. Thus, in this case, switching-off is possible only when a reverse voltage less than the breakdown voltage for the junctions $E_1 - E_3$ is applied during the switching-off process. The switching-on process also places extra demands on the design of the p−n junctions. In switching-on by the anode voltage, because of the differing properties of the base layers adjoining the collectors, the voltage decreases sooner at one collector than at the others. This increases the voltage across the other collectors above the avalanche voltage. A similar effect occurs when the structure is switched by a gate current provided the applied voltage is sufficiently large. The collector junctions in a multicollector structure must therefore be made as p−n junctions with controlled avalanching.

Figure 1b shows the geometry of a six-layer (two-collector) structure made by the diffusion of aluminum, phosphorus, and boron in n-type silicon. Layers 1, 2, 4, and 5 were obtained by simultaneous diffusion of aluminum and phosphorus from their oxides in air [5]; the surface concentration was $(4-6) \cdot 10^{16}$ cm^{-3} for aluminum and $(1-3) \cdot 10^{18}$ cm^{-3} for phosphorus. Layer 6 was obtained by diffusion of boron from borosilicate glass, the surface concentration being about 10^{19} cm^{-3}. The initial resistivity of the silicon was $60 \, \Omega \cdot$ cm. Figure 2 shows the current—voltage characteristic of this structure. The forward branch clearly shows two negative resistance regions, corresponding to a decrease of the voltage first across the high-voltage collector C_1 and then across the low-voltage collector C_2. The reverse branch also has a negative resistance region which appears when the applied voltage exceeds the total breakdown voltage of the junctions E_1, E_2, and E_3. The presence of this region is explained, as already mentioned, by the fact that, when the junctions E_1 and E_3 break down, the four-layer structure 2-3-4-5 (Fig. 1b) is already switched on in the forward direction and then the voltage across the collectors changes sign. The holding voltage is then equal to the sum of the breakdown voltages of the junctions E_1 and E_3 (about 300 V). This voltage is approximately equal to U_{C_2}, since the junctions C_2 and E_1 are identical, and the breakdown voltage of the junction E_3 is small.

The dashed curves in Fig. 2 show the current—voltage characteristic varies under the action of a gate current of the polarity indicated in Fig. 1. In the forward branch, the voltage decreases first across C_1 and then across C_2, but not simultaneously across both collectors. In some samples, the changes take place in the opposite order. The breakover voltage in the reverse branch also decreases with increasing gate current, and the critical gate current is usually less than in the forward branch. This makes it possible to obtain, using multicollector structures having an odd number of layers, devices similar to the symmetrical thyristors [6] and having current—voltage characteristics whose forward and reverse branches are identical and can be controlled by a single electrode. Figure 3 shows the forward characteristic of such a structure having an effective area of 2.5 cm^2 in the on state.

The results given above show that the characteristics are entirely similar to the forward characteristics of ordinary single-collector thyristors, such as VKDU-150, having about the same effective area of the structure. This confirms quite well the above arguments that, in a multicollector structure in the on state, voltage inversion occurs at every collector, and the sum of the voltage drops across them is subtracted from the corresponding sum for the emitter junctions. The added voltage drop due to the layers 1 and 5 (Fig. 1b), which are additional to the single-electrode thyristor, is negligible in the samples considered here, because the layers are thin and have a relatively high majority carrier density.

Literature Cited

1. H. Lawrence and R. M. Warner, Jr., Bell System Tech. J., 39:389 (1960).
2. I. V. Grekhov, I. A. Liniichuk, V. E. Chelnokov, and V. B. Shuman, Radiotekh. i Élektron., 11:1856 (1966).
3. V. A. Kuz'min, Radiotekh. i Élektron., 8:171 (1963).
4. V. A. Kuz'min, Radiotekh. i Élektron., 9:1410 (1964).
5. I. V. Grekhov, L. N. Krylov, I. A. Liniichuk, V. M. Tuchkevich, V. B. Shuman, and V. E. Chelnokov, "Recent diffusion methods of producing silicon components for controlled and uncontrolled power rectifiers" (paper presented at a conference), Izd. Vsesoyuznogo Nauchno-Issledovatel'skogo Instituta Metrologii. Moscow (1966).
6. A. N. Dumanevich, Yu. A. Evseev, V. M. Tuchkevich, N. I. Yakivchik, and V. E. Chelnokov, Élektrichestvo, No. 5, p. 58 (1966).

p—n JUNCTION RESISTANCE OF IMPATT DIODES
AT FREQUENCIES FROM 0 TO 10 Mc

V. L. Aronov, A. I. Mel'nikov, and A. S. Tager

The results of an experimental study of p—n junction resistance in diffused germanium IMPATT diodes are compared with theoretical conclusions. It is found that, at low frequencies ($\ll 1$ Mc), the junction resistance is several times greater than the calculated value, decreasing as the frequency and current increase. It is shown that this effect is due to heating of the p—n junction by the current. A new method is proposed for measuring the thermal resistance of these diodes.

The IMPATT (impact-avalanche transit-time) diode was first constructed in 1959 in the USSR, following the discovery of coherent oscillation in avalanche breakdown of germanium-diffused microwave diodes [1-4]. In other countries, there have been many reports since 1965 [5, 6] concerning the development of IMPATT diodes and microwave devices using them.

The first theoretical analysis of the operation of these diodes is due to Read [7], who considered a simple model consisting of a narrow p—n junction with a stepped distribution of impurity and a wide i-type layer. A more general treatment, taking account of the actual distribution of impact ionization intensity in a diode having any impurity distribution, has been worked out in [3, 8] and in unpublished papers by Zakharov, as well as in a recent paper [9]. In this treatment, Read's model appears as a particular case.

The theoretical ideas developed in the publications cited above obviously need careful experimental testing and confirmation. The most direct and reliable test is to measure the impedance of the p—n junction in the operating part of the current—voltage characteristic and at various frequencies, when there are no intrinsic oscillations of the system, and to compare the results of these measurements with the relationships calculated in the theoretical model.

The two most interesting frequency ranges for study are (1) the microwave range (above 1-2 Gc), where the resistance of the p—n junction becomes negative and the impedance of the junction varies rapidly with frequency and electrical conditions, and (2) the low-frequency range (below 10 Mc), where the impedance of the p—n junction again depends on frequency and electrical conditions; these determine, in particular, the pulse characteristics of the diode. At frequencies between these ranges, the impedance of the junction under operating conditions depends only slightly on the frequency and the current.

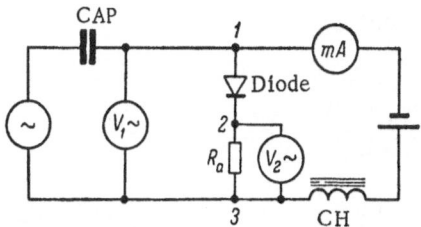

Fig. 1. Block diagram of circuit for measuring the differential resistance of an IMPATT diode.

In this paper, we report the results of an experimental study made in 1960-62 of the impedance of the p−n junctions in diffused germanium p−n IMPATT diodes having a symmetrical junction and breakdown voltage 10-40 V in the frequency range 0-10 Mc.

According to [3, 4], at frequencies much less than the characteristic frequency, the reactance of the p−n junction for a small signal is zero, while the resistance is independent of frequency and equal to the static differential resistance R_0. The isothermal resistance R_0 in the operating part of the current−voltage characteristic ($i_s \ll i^0 \ll i_m$) under avalanche breakdown conditions* must satisfy the condition [3, 4]

$$R_0 C = \eta \frac{\tau}{2}, \tag{1}$$

where C is the capacitance of the p−n junction at the breakdown voltage ($V = V_{br}$); τ is the carrier transit time across the half-width of the junction; η is a coefficient which allows for the impurity distribution in the junction (for a linear-gradient junction in germanium, $\eta = 0.43$; for an abrupt junction, $\eta = 0.62$).

The resistance R of the p−n junction in the frequency range 0-10 Mc was measured by a fairly simple method. Since the reactance of the bulk of the diode is negligible at low frequencies (at 10 Mc, the inductive reactance does not exceed $3 \cdot 10^{-2} \Omega$ and the parallel capacitive susceptance is less than $2 \cdot 10^{-5}$ mho), the diode is purely resistive and its resistance consists of the junction resistance R and the spreading resistance R_s, the latter being independent of frequency. Hence, by measuring the differential resistance R_d of the diode and subtracting R_s, whose value was found independently, it was possible to determine the resistance R of the p−n junction.

Figure 1 is a block diagram of the circuit used in the measurement. A dc voltage exceeding the breakdown value is applied to the diode, together with an ac test voltage whose amplitude does not exceed 10^{-3} V. The sources of these voltages are decoupled by an isolating capacitor CAP and a choke CH. The voltmeter readings U_{13} on V_1 and U_{23} on V_2 are noted for a certain amplitude of the ac voltage. The diode is then replaced by a standard resistor R_e of 20-100 Ω, and the ac voltage is varied until the voltmeter V_2 again reads U_{23}. The current through R_e is then the same as that through the diode, and we have

$$\frac{U_{13}}{R_e + R_a} = \frac{kU_{13}}{R_d + R_a},$$

where k is the ratio of the readings of the voltmeter V_1 in the two measurements. Hence, we find the differential resistance R_d:

$$R_d = R + R_s = kR_e + (k-1) R_a.$$

Figures 2 and 3 show typical results of these measurements. It is seen from Fig. 2 that the p−n junction resistance R does not remain constant with increasing frequency, but slowly decreases by a factor between 2 and 10 as far as a frequency of about 1-10 Mc. Figure 3 indicates the change of shape of the current−voltage characteristic of the p−n junction at various

---

*i_s is the saturation current, i^0 the diode current, i_m the limiting current at which the space charge of the mobile carriers neutralizes the impurity charge.

TABLE 1. Comparison of Experimental and Calculated Values of R_0C

Diode	V_{br}, V	$C \cdot 10^{12}$, F	R_0, Ω	$(R_0 \cdot C)_{exp} \cdot 10^{12}$, sec	$(R_0 \cdot C)_{calc} \cdot 10^{12}$, sec
I-85	14	0.27	11	3	1.6
A-11	14	0.29	14	4	1.6
I-63	14	0.30	10	3	1.6
I-1	15	0.24	14	3.2	1.7
S-7	20	0.25	16.5	4.2	2.3
S-9	20.5	0.19	19	3.6	2.3
S-46	20	0.18	24	4.4	2.3
S-55	19	0.24	14.5	3.5	2.2
S-21	21	0:21	23	5	2.4
13	23	0.25	18	4.5	2.8
19	25	0.08	55	4.1	3
23	23	0.05	70	3.8	2.8
20	26	0.10	57	5.4	3.2
99	26	0.16	29	4.5	3.2
92	27	0.10	53	5.3	3.3
97	27.5	0.13	38	4.7	3.4

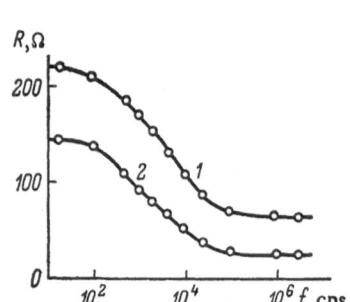

Fig. 2. Frequency dependence of the differential resistance of IMPATT diodes. 1) Diode T-9 (V_{br} = 27.5 V, C = 0.13 · 10^{-12} F, i^0 = 5 mA); 2) diode T-2 (V_{br} = 27.5 V, C = 0.21 · 10^{-12} F, i^0 = 5 mA).

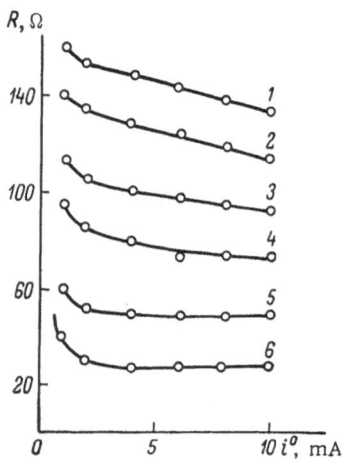

Fig. 3. Current dependence of the differential resistance of IMPATT diode T-2 at various frequencies. V_{br} = 27.5 V, C = 0.21 · 10^{-12} F. f, kc: 1) 0.02; 2) 0.1; 3) 1; 4) 2; 5) 10; 6) 100.

frequencies. At low frequencies, the resistance R decreases more or less steadily as the current increases, whereas at frequencies above 0.1 Mc it decreases only in the initial section of the characteristic, thereafter remaining constant.

The value of R exceeds that given by Eq. (1) at all frequencies, but the limiting value at frequencies above 1 Mc is close to the calculated value for some diodes (see Table 1), whereas the value at low frequencies is always greater than the calculated value.

This effect is due to heating of the p−n junction by the current and the accompanying increase of the breakdown voltage. The heating or cooling time of the junction (i.e., the thermal relaxation time) depends on the conditions of heat removal from the parts of the semiconductor, solder, pin, etc., adjoining the junction.

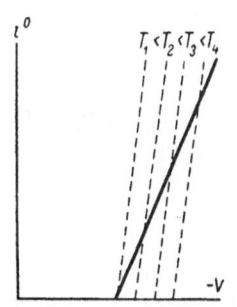

Fig. 4. To illustrate the frequency dependence of the differential resistance of IMPATT diodes.

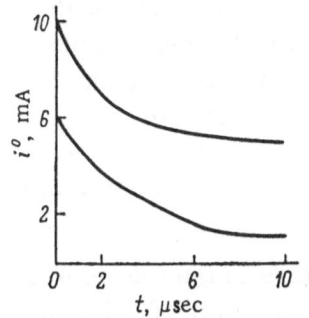

Fig. 5. Shape of current pulse on application of a rectangular voltage pulse to IMPATT diode M-8. V_{br} = 31.5 V, C = $0.22 \cdot 10^{-12}$ F.

As the temperature of the p−n junction rises, its current−voltage characteristic moves towards higher voltages, as shown in Fig. 4. If the current changes rapidly (i.e., in a time considerably shorter than the thermal relaxation time), the p−n junction temperature increases but does not change in one period, and the differential resistance R is the same as R_0; the change of voltage corresponds to one of the "isothermal" dashed curves T = constant in Fig. 4. In the opposite limiting case, where the period of variation of the current is long compared with the thermal relaxation time, the temperature of the p−n junction at any instant depends on the instantaneous value of the current. The current−voltage curve obtained under these "adiabatic" conditions is shown by the continuous curve in Fig. 4. The corresponding differential resistance R_{st}, measured under static conditions or at very low frequencies (≪ 1 Mc), is greater than the isothermal resistance R_0:

$$R_{st} = R_0 + \frac{dV_{br}}{di^0}. \qquad (2)$$

Since $V_{br}(T) = V_{br}(T_0)[1 + \alpha(T - T_0)]$, and, for germanium diodes, $\alpha \approx 10^{-3}$ deg^{-1}, and putting $T - T_0 = R_t i^0 V_{br}$, we find

$$R_{st} = R_0 + \alpha V_{br}^2 R_t, \qquad (3)$$

where R_t is the thermal resistance of the diode. Thus, at low frequencies, we measure the resistance R_{st}, and at high frequencies (>1 Mc), the resistance R_0. At intermediate frequencies, some intermediate quantity $R(\omega)$ is evidently measured. Its form depends on the way in which thermal equilibrium is established between the p−n junction and the adjoining regions of the crystal and the bulk of the diode; a precise calculation of the function $R(\omega)$ is complicated and not really necessary. In practice, all that is important is to know the thermal equilibrium establishment time, which may be estimated from the experimental form of $R(\omega)$. Figure 2 shows that this time is about 1-5 μsec for typical p−n junctions.

The experimental functions $R(\omega)$ are in good agreement with the diode current establishment curves (Fig. 5) when the diode receives pulses from a dc voltage source. In Fig. 5, the abscissa is the time from the beginning of the voltage pulse, and the ordinate is the diode current. The decrease of the current is due to the increase of V_{br} as the p−n junction is heated by the current. The steady current is reached almost exponentially, with a relaxation time of the order of a few microseconds.

From the measured values of R_{st} and R_0, we can easily find an important parameter of the diode, its thermal resistance R_t, which governs the maximum permissible value of the dc. According to Eq. (3),

$$R_t = \frac{R_{st} - R_0}{\alpha V_{br}^2}. \qquad (4)$$

In the derivation of Eq. (3), it has been assumed that the function $V_{br}(T)$ is linear (α = const). However, the function $V_{br}(T)$ is actually nonlinear, and α decreases with increasing temperature (for diffused germanium diodes with V_{br} between 15 and 60 V, α decreases by

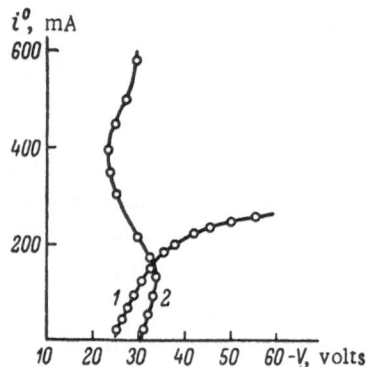

Fig. 6. Current−voltage characteristics of IMPATT diodes in the avalanche breakdown range, recorded in pulsed operation. 1) Diode T-41 (V_{br} = 24.5 V, C = 0.09 · 10^{-12} F); 2) diode M-8 (V_{br} = 31.5 V, C = 0.22 · 10^{-12} F).

30-50% when the temperature rises from 20 to 70°C). Consequently, the measured value of R_t is somewhat less than the true value. The decrease of α explains, in particular, the experimentally observed steady decrease of the differential resistance of the p−n junction with increasing diode current at low frequencies (Fig. 3).

At high frequencies (above 1 Mc), the resistance R is constant except in a small initial part of the current−voltage characteristic, where it increases rapidly with decreasing current. The latter effect is due to the gradual development of breakdown (when $i^0 \rightarrow i_s$, the intensity of impact ionization is mainly determined by the directly applied voltage, and depends only very slightly on the electron and hole space-charge density).

We can now compare the experimental results with Eq. (1). It follows from the foregoing discussion that the differential resistance in this expression is the experimental value of R_0 at frequencies above 1 Mc and small currents [for $R_0(i^0)$ = const]. The relation between the experimental and theoretical results is shown by Table 1, which contains data for sixteen arbitrarily selected IMPATT diodes that have been found to oscillate.

It is seen from Table 1 that the experimental values of R_0C are close to the theoretical values, though always somewhat higher, the difference being as much as a factor of 1.5-2 for certain diodes. This difference appears to be due to the fact that the effective area S_1 of the p−n junction, through which the current flows, is not the same as its total area S_0. Since the value of R_0 is independent of the distribution of current over the effective part of the p−n junction, we can immediately find the effective area of the p−n junction from the measured values of R_0C:

$$\frac{S_1}{S_0} = \frac{\frac{\tau}{2}}{(R_0C)_{\exp}}. \tag{5}$$

So far, we have considered the shape of the current−voltage characteristic of the diode in the range of fairly small currents, where the diode can be tested continuously ($i_s \ll i^0 \ll i_m$). The current densities in this range are relatively small (about 5 · 10^2 to 10^3 A/cm²), and some tens of times less than the maximum value i_m = eNv, which is 2.5 · 10^4 A/cm² for V_{br} = 30 V and 1 · 10^5 A/cm² for V_{br} = 12 V. It is interesting to examine the shape of the diode current−voltage characteristic at considerably higher current densities, up to values close to i_m, where theory would indicate that the characteristic is no longer a straight line and that the differential resistance of the p−n junction increases rapidly [3, 4]:

$$R_0 = \eta \frac{\tau}{2C} \frac{1}{1 - \dfrac{i^0}{i_m}}. \tag{6}$$

Such measurements have been made with a pulsed supply to the diode, the pulse duration being much less than the thermal relaxation time of the p−n junction. Figure 6 shows typical results of measurements using rectangular current pulses of 0.5 μsec duration.

It is seen that the current–voltage characteristics of the p–n junction have several interesting features.

1. The current increases linearly with voltage only up to about 100 mA. The differential resistance in this range is equal, as we should expect, to R_0 measured at frequencies above 1 Mc by the method described above.

2. At currents above 100 mA, some diodes show the theoretically predicted "droop" of the current–voltage characteristic (curve 1), but others have regions with negative slope at high currents (curve 2). The reason for this is not yet fully understood; these may be instances where the contacts are not ohmic and minority carriers are injected into the crystal, possibly leading to the occurrence of regions with negative slope in the current–voltage characteristic.

Where this phenomenon does not occur, the shape of the current–voltage characteristic curve is close to the theoretical one.

We thank A. V. Skidan and O. V. Semina for help with the experiments.

Literature Cited

1. A. I. Mel'nikov, A. S. Tager, G. P. Kobel'kov, and A. M. Tsebiev, USSR Author's Certificate No. 185,965 (1959).
2. A. I. Mel'nikov, A. S. Tager, G. P. Kobel'kov, and A. M. Tsebiev, Invention License No. 24 (1959).
3. A. S. Tager, Generation and Amplification of Microwave Oscillations by Diodes Having Negative Dynamic Resistance (Abstract of Thesis), Moscow (1962).
4. V. M. Val'd-Perlov, A. V. Krasilov, and A. S. Tager, Radiotekh. i Élektron., 11:2008 (1966).
5. R. L. Johnston, B. C. De Loach, Jr., and B. G. Cohen, Bell System Tech. J., 44:369 (1965).
6. C. A. Lee, R. L. Batdorf, W. Wiegmann, and G. Kaminsky, Appl. Phys. Letters, 6:89 (1965).
7. W. T. Read, Jr., Bell System Tech. J., 37:401 (1958).
8. A. L. Zakharov, Proceedings of Conference on Impact Ionization and Tunnel Effect in Semiconductors, Institute of Physics, Academy of Sciences of the Azerbaidzhan SSR, Baku (1962).
9. T. Misawa, IEEE Trans. Electron. Devices, 13:137, 143 (1966).

INFLUENCE OF CADMIUM VAPOR PRESSURE
ON THE DIFFUSION OF INDIUM IN CdTe

L. V. Maslova, O. A. Matveev,
Yu. V. Rud', and A. K. V. Sanin

A. F. Ioffe Physicotechnical Institute
Academy of Sciences of the USSR, Leningrad

The influence of cadmium vapor pressure on the diffusion of indium in p-type CdTe was studied in the range 700-1000°C (the cadmium source temperature was T_{Cd} = 1000-600°C). It was found that there was an anomalous increase in the rate of diffusion at temperatures $T_{Cd} \gtrsim 650°C$. At $T_{Cd} < 650°C$, the diffusion rate could be explained quantitatively using a published account of diffusion through the cadmium sublattice. At cadmium source-temperatures $T_{Cd} \gtrsim 650°C$, a different mechanism of diffusion became predominant.

An experimental study of the diffusion of indium in CdTe [1] led the authors to the conclusion that diffusion takes place through the cadmium sublattice in this compound.

It is known that the diffusion of substitutional impurities through one sublattice depends on the vacancy concentration in it, which in turn is governed by the pressure of the relevant component in the gas phase. In the present paper, we have in fact studied the dependence of the diffusion of indium in CdTe on the cadmium vapor pressure p_{Cd}.

The diffusion was investigated by determining the depth at which the p-n junction was situated when indium (which was a donor impurity for CdTe) diffused into homogeneous p-type crystals. The crystals were grown by the horizontal Bridgman method and zone recrystallization under a cadmium vapor pressure which specifically ensured a shift away from the stoichiometric composition in the direction of excess cadmium. Thus, the p-type conduction in the crystals was due to acceptor impurities, not to the presence of vacancies in the cadmium sublattice. The original crystals had a free hole density $p \approx 2 \cdot 10^{13}$ cm^{-3} and a Hall mobility of 50 cm$^2 \cdot$ V$^{-1} \cdot$ sec^{-1} at room temperature.

Experiments of two kinds were performed on the diffusion of indium into CdTe from the vapor phase. In one, the procedure used in [1] was repeated; in the other, the diffusion took place at a fixed vapor pressure of cadmium. The diffusion process was conducted in a two-

TABLE 1

Samples	T_{cr}, °C	T_{Cd}, °C	T, h	300°K		77°K	
				n, cm^{-3}	U_n, cm^2/V·sec	n, cm^{-3}	U_n, cm^2/V·sec
2N-1	1000	850	2	$4.2 \cdot 10^{17}$	105	$4.2 \cdot 10^{17}$	80
2N-2	1000	850	0.5	$6.6 \cdot 10^{17}$	350	$6.9 \cdot 10^{17}$	220
30K-1	1000	800	4	$7.3 \cdot 10^{16}$			
30K-2	1000	800	1	$3.7 \cdot 10^{16}$	360	$2.9 \cdot 10^{16}$	370
30K-3	1000	750	4	$1.0 \cdot 10^{17}$	450	$1.1 \cdot 10^{17}$	330
30K-4	1000	725	4	$5.7 \cdot 10^{16}$	210	$1.2 \cdot 10^{16}$	790
30K-5	1000	700	4	$2.8 \cdot 10^{17}$	60	$1.6 \cdot 10^{17}$	40
2N-3	1000	700	4	$4.5 \cdot 10^{17}$	180	$5.0 \cdot 10^{17}$	30
32N-5	1000	700	4	$5.3 \cdot 10^{17}$	70	$4.5 \cdot 10^{17}$	30

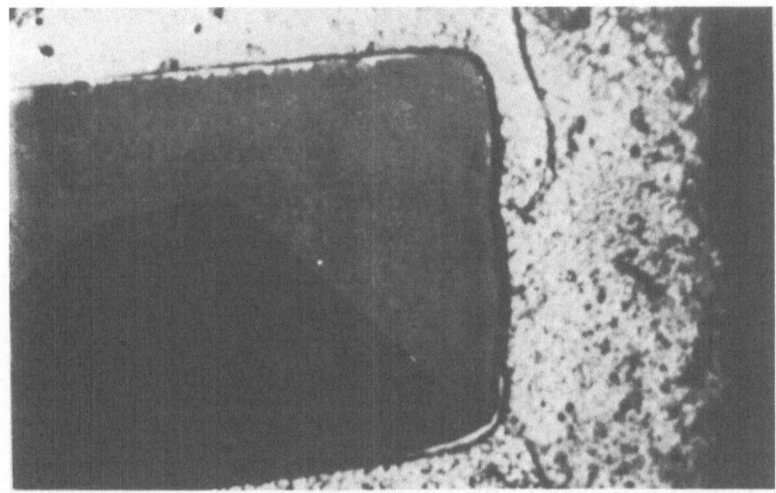

Fig. 1. Pattern of indium diffusion in CdTe at T_{Cd} = 650°C and T_{cr} = 1000°C (τ = 4 h, magnification 70).

temperature furnace. The crystals and the impurity elements were placed in quartz ampoules evacuated to 10^{-3} torr. After heating, they were quenched in water. The p—n junction was detected by using a feature of the etching of n-type and p-type CdTe which we had discovered earlier [2]. The depth at which the p—n junction appeared was measured with a MIM-7 microscope (accuracy ±5 μ). The samples used in electrical measurements were cut from the plates into which diffusion had taken place.

The experiments on indium diffusion at a high temperature (T_{cr} = 1000°C) and a high vapor pressure of cadmium (source temperature T_{Cd} = 1000°C) gave an unexpected result. The crystal was very rapidly converted completely into n-type material. We estimated that the rate of diffusion was at least an order of magnitude above what would be expected from the results of [1]. A very long (50 h) anneal of CdTe crystals, under similar conditions but with only cadmium present, caused no change in the type of conduction, even in the surface layer. Thus, we concluded that this very rapid conversion of the type of conduction in the crystals was entirely due to the diffusion of indium under a high vapor pressure of cadmium.

Diffusion annealing of crystals at a fixed temperature in the presence of indium vapor alone, between 600 and 1000°C, confirmed the results of [1]. In these experiments, it was not possible to control the cadmium pressure in the gas phase; the partial pressures of cadmium and tellurium above the crystal depended on the ampoule volume and the temperature. Equi-

TABLE 2

Samples	T_{cr} °C	T_{Cd}, °C	T, h	Type of conduction	300° K	
					n, cm^{-3}	u, cm^2/V·sec
30K-1	1000	800	4	n	$7.3 \cdot 10^{16}$	—
2N-4	900	850	0.5	n	$2.0 \cdot 10^{17}$	360
30K-6	800	800	2	n	$1.1 \cdot 10^{15}$	350
2N-6	700	700	2	p	$2.0 \cdot 10^{13}$	20

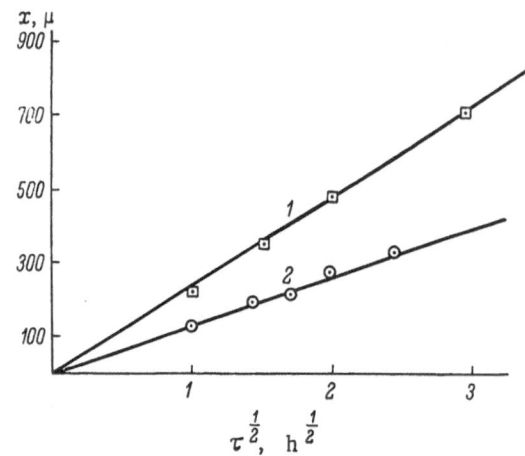

Fig. 2. Time dependence of position of p−n junction. 1) Indium diffusion in CdTe (T = 1000°C); 2) indium diffusion in CdTe under cadmium pressure (T_{cr} = 1000°C, T_{Cd} = 650°C).

librium was established by a change in the composition of the crystal, accompanied, of course, by the appearance of a certain concentration of corresponding vacancies. Therefore, the conclusion in [1] on the sublattice mechanism of indium diffusion under these conditions was entirely logical.

Let us now see what happens if the cadmium vapor pressure is gradually increased. The vacancy concentration in the cadmium sublattice will decrease and finally reach some equilibrium value for a given temperature of the crystal. If the sublattice mechanism of diffusion is maintained during this process, experiments should show a decrease of the diffusion rate. Actually, a high cadmium vapor pressure, which prevents the cadmium vacancy concentration from exceeding the equilibrium value, acts as a kind of stimulus, producing an anomalously high rate of diffusion of indium. It seems that, under these conditions, some other mechanism (such as interstitial diffusion) predominates.

Table 1 shows the results of some experiments on diffusion annealing in cadmium and indium vapors. At T_{cr} = 1000°C and T_{Cd} = 700-850°C, the crystals undergo a change of the type of conduction. The free electron density in the samples range from $3.7 \cdot 10^{16}$ to $7 \cdot 10^{17}$ cm^{-3} at room temperature. The electrical properties of these crystals are the same as of those prepared by doping during growth [3]. A similar electron density in CdTe crystals was found in [4], where the samples were doped with indium during growth, and then annealed at various cadmium vapor pressures. The results in Table 1 serve to confirm the conclusions of [4]. Above a certain value of p_{Cd}, the electron density is independent of p_{Cd}.

When the temperature T_{Cd} falls below 700°C (T_{cr} = 1000°C), vacancies form in the cadmium sublattice and the diffusion is qualitatively similar to that which occurs in the absence of a controlled pressure of cadmium. Figure 1 shows a p−n junction produced under these conditions.

Figure 2 shows the position of the p−n junction as a function of the indium diffusion annealing time, for the two kinds of experiments. In both cases, the time dependence of the diffusion is normal. The quantitative difference in the depth of penetration may be ascribed to the different indium vapor pressures, since with controlled cadmium vapor pressure, the p−n junction is formed at $T_{Cd} \lessgtr 650°C$, which also limits the indium vapor pressure.

The similarity of the results of the two experiments at $T_{Cd} = 650°C$ and $T_{cr} = 1000°C$ indicates that the number of vacancies in the cadmium sublattice is sufficient for the sublattice diffusion mechanism to begin to predominate.

By determining, at a given temperature of the crystal, the cadmium vapor pressure at which the indium diffusion rate becomes anomalously high, we can find information about equilibrium in the solid CdTe−vapor phase system.

We also studied the dependence of the conduction conversion effect on the temperature in the crystal (Table 2). As T_{cr} decreases, so does the electron density in the samples, apparently because of the temperature dependence of the solubility of indium in CdTe. Finally, when $T_{cr} = 700°C$, the properties of the original sample remain almost unchanged. It is possible that very long annealing times are necessary for the conversion of a crystal under these conditions.

The strong rise of the diffusion rate found when $T_{Cd} > 650°C$ (for $T_{cr} = 1000°C$) demonstrates very clearly the important role of thermodynamic equilibrium in the crystal−vapor phase system for CdTe, not only in crystal growth but also in diffusion.

The authors thank Prof. S. M. Ryvkin for his continued interest in this work.

Literature Cited

1. H. Kato and S. Takayanagi, Japan. J. Appl. Phys., 2:250 (1963).
2. O. A. Matveev, Yu. V. Rud', and K. V. Sanin, Izv. Akad. Nauk SSSR, Neorg. Mater., 5:372 (1969).
3. E. N. Arkad'eva, L. V. Maslova, O. A. Matveev, S. M. Ryvkin, and Yu. V. Rud', Fiz. Tekh. Poluprov., 2:279 (1968).
4. D. de Nobel, Philips Res. Rep., 14:361 (1959).

RADIATIVE RECOMBINATION
IN Ge-DOPED GaAs DIODES

C. Constantinescu, G. Popovici,
P. Mihailovici, and I. Petrescu

Institute of Physics
Academy of Sciences of the Socialist Republic of Rumania, Bucharest

The spectral and current dependences of the luminescence of Ge-doped GaAs diodes were studied. A relation between the luminescence intensity J and the current I was found: $I \propto J^{1/n}$, where n = 5-7 for small forward currents and n = 1 for large forward currents. The exact relationship between the luminescence and the current is discussed bearing in mind the competition between possible recombination processes.

There have been many publications on radiative recombination in GaAs p−n junctions at low current densities, but several aspects of the spectral and current dependences of the luminescence are not yet clarified.

In most papers, the spectrum has been found to show several peaks, some of which move to higher energies as the current increases.

The relation between the integrated luminescence intensity J and the forward current I through the p−n junction can be written in the form $I \propto J^{1/n}$, where the value of n differs according to various authors: n = 1 [1], n = 2 [2], or n > 2 [3-5].

Luminescence of p−n junctions under a reverse bias has been studied in [6], where it was shown that the spectrum is similar to that obtained for a forward bias.

In the present work, we studied the luminescence of p−n junctions in Ge-doped GaAs at forward and reverse bias. The current and voltage dependences of the luminescence intensity and the current−voltage characteristics, are explained from the point of view of competition between possible recombination processes.

Experimental

Ge-doped GaAs single crystals were grown from the molten solution of Ga by slow cooling from 900°C. It is known that Ge is an amphoteric impurity in GaAs. In crystals grown

TABLE 1

Crystal	$p(300°K)$, cm^{-3}
D	$1.1 \cdot 10^{17}$
C	$5.1 \cdot 10^{17}$
F	$2.2 \cdot 10^{18}$
G	$4.5 \cdot 10^{18}$
H	$1.2 \cdot 10^{19}$

from a stoichiometric mixture, Ge replaces Ga and behaves as a donor with low ionization energy.

In growth from a solution containing excess Ga, as in the present work, Ge is trapped mainly at As vacancies and produces an acceptor level in the forbidden band of GaAs, 0.04 eV above the valence band.

The hole density measured by the Hall effect in p-type GaAs crystals had values as shown in Table 1.

Our p−n junctions were made by alloying Sn at 500°C. The n-type regions of the junctions, recrystallized after alloying, were degenerate.

The current−voltage and spectral characteristics were recorded at 77 and 300°K. The luminescence spectra at various currents were measured with a Zeiss mirror monochromator, a ÉPP-09 recorder, and a M-13 photomultiplier or germanium photocell. The spectral resolution of this system was 0.01 eV.

Results and Discussion

When the diodes were biased in the forward direction, a single luminescence peak was observed for all the currents studied, varying with the equilibrium hole density but situated in the range 1.38-1.46 eV, i.e., at energies lower than the forbidden band of GaAs. Such spectra have been reported in [7]. This difference from the other authors' observations of two or more peaks in the luminescence spectrum is probably due to our particular method of growing crystals from the Ga solution. In [8], a single luminescence peak was likewise found for material grown in the same manner.

Figure 1 shows a typical current−voltage characteristic and the dependence of the luminescence intensity J on the voltage V for type D diodes.

Similar current−voltage characteristics have been reported in [5] for Te-doped GaAs diodes with carrier density $2 \cdot 10^{16}$-$3 \cdot 10^{18}$ cm^{-3}.

At low voltages the excess current predominated in all our diodes; it may be written as $I = I_0 \exp(\alpha V)$. Since α is almost independent of temperature, we may assume that this current is due to a two-stage process: tunneling followed by trapping at local levels in the forbidden band.

At 77°K, the injection current becomes comparable with the excess current at about 1.4 V, as has also been observed in [5, 9], etc. (In Fig. 1, the injection region is distorted because of the series resistance of the diode.)

It is seen from Fig. 1 that, in our case, the excess current is nonradiative. Radiative recombination becomes appreciable only at voltages such that the injection current outweighs the excess current.

For diodes made from a material having a high hole density, a shift of the maximum luminescence was observed in a narrow range of voltages close to the onset of injection. This shift shows rapid saturation at the energies indicated above. Such a shift is usually related to phonon-assisted diagonal tunneling [10, 11]. However, for diodes made from material having a low hole density, no shift of the peak was observed, because the electric field at low densities is insufficient for the diagonal tunneling.

Figure 2 shows the current dependence of the luminescence intensity for diodes made from materials having various hole densities, and Fig. 3 shows the same relationship for type

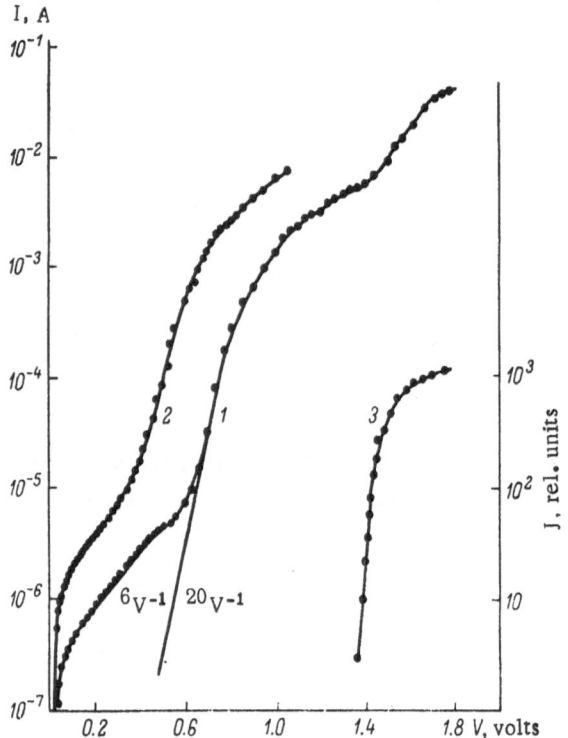

Fig. 1. Current−voltage characteristics of forward-biased diodes (curves 1 and 2) and voltage dependence of luminescence (curve 3) for diodes D_7 ($p_{300°K} = 5.1 \cdot 10^{17}$ cm^{-3}). T, °K: 1, 3) 77; 2) 300.

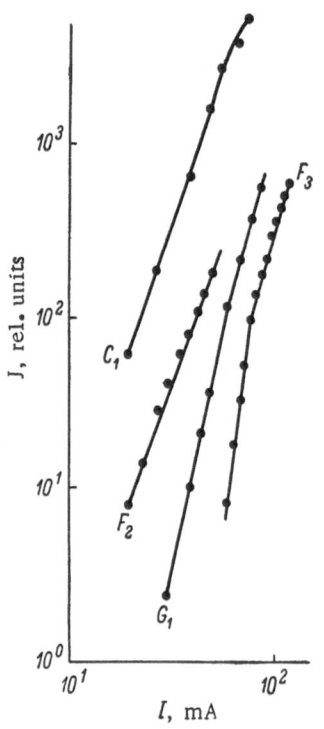

Fig. 2. Current dependence of the luminescence intensity for diodes having various hole densities under a forward voltage bias. T = 77°K.

D diodes. For all the diodes at low currents, the relation is of the form $I \propto J^{1/n}$ with n = 4-6; this means that only a part of the total current I_{p-n} through the p−n junction is radiative. For type D diodes at high currents, the relation becomes $I \propto J$ (n = 1).

It follows from these results that the total current through the junction may be written as $I_{p-n} = I_{nonrad} + I_{rad}$, where the second term, at low densities, is due only to injection radiative recombination, and at high densities to a small component of radiative tunneling.

I_{nonrad} represents all possible radiationless processes, including tunneling (which, in our diodes, is mainly radiationless). Then, the strong dependence of the luminescence on the current which we observe in the region of radiative and nonradiative transitions appears to be due to a decrease in the number of nonradiative transitions and a corresponding increase in radiative ones, until the whole current becomes radiative, and $I \propto J$, n = 1, in the injection region.

The decrease in the number of nonradiative transitions may be due to the distribution of local levels in the forbidden band, which governs the excess currents, and also to the fact that the probability of phonon-assisted nonradiative transitions, unlike that of radiative transitions, decreases rapidly with increasing dissipated energy [9].

We also investigated the luminescence of diodes under a reverse bias.

Unlike Ge and Si, they have a fairly narrow spectrum, similar to that found under a forward bias and having the same luminescence peak energy.

For some diodes, however, the peak is at lower energies than under a forward bias.

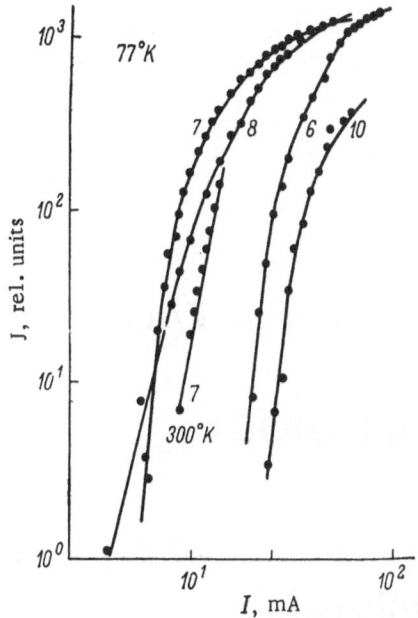

Fig. 3. Current dependence of the luminescence intensity for type D diodes under a forward bias.

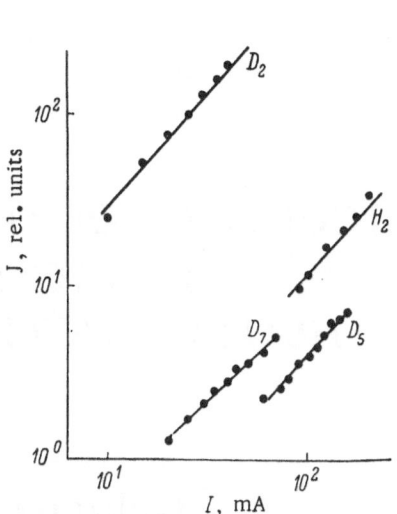

Fig. 4. Current dependence of luminescence intensity with reverse bias (77 °K).

Figure 4 shows the luminescence intensity as a function of the reverse current. The relation is linear: $I \propto J$. The results indicate that, in this case, we observe the usual interband recombination of carriers thermalized at the band edges by interaction with optical phonons in the junction itself [6].

Literature Cited

1. B. M. Vul, V. S. Vavilov, G. N. Galkin, and E. A. Bobrova, Fiz. Tverd. Tela, 8:908 (1966).
2. A. N. Imenkov, M. M. Kozlov, S. S. Meskin, D. N. Nasledov, V. N. Ravich, and B. V. Tsarenkov, Fiz. Tverd. Tela, 7:634 (1965).
3. M. I. Nathan, T. N. Morgan, G. Burns, and A. E. Michel, Phys. Rev., 146:570 (1966).
4. M. Pilkuhn and H. Rupprecht, J. Appl. Phys., 37:3621 (1966).
5. Yu. P. Demidov, M. N. Zargar'yants, and A. A. Kiselev, Physics of p−n Junctions [in Russian], Riga (1966), p. 296.
6. A. E. Michel, M. I. Nathan, and J. C. Marinace, J. Appl. Phys., 35:3543 (1964).
7. C. Constantinescu, G. Popovici, and P. Mihailovici, Proceedings of the Internal Conference on Luminescence, Budapest (1966).
8. M. B. Panish, H. J. Queisser, L. Derick, and S. Sumski, Solid State Electronics, 9:311 (1966).
9. C. A. Morato de Andrade, Solid State Electronics, 9:901 (1966).
10. A. É. Yunovich and A. B. Ormont, Zh. Éksp. Teor. Fiz., 51:1292 (1966).
11. T. N. Morgan, M. Pilkuhn, and H. Rupprecht, Phys. Rev., 138:A1551 (1965).

CALCULATION OF THE CURRENT—VOLTAGE
CHARACTERISTIC FOR A
HETEROJUNCTION TUNNEL DIODE

A. I. Gubanov and S. N. Dobrynin

V. I. Ul'yanov (Lenin) Leningrad Institute of Electrical Engineering

The WKB method is used to analyze the expression for the interband tunneling by comparing the arguments of an exponential function using attenuated electron functions and the effective-mass method. A method is thus developed for calculating a tunnel transition between bands of different crystals. An expression is also derived for calculating the current—voltage characteristic of a heterojunction tunnel diode and used to make this calculation for Ge—GaAs. It is shown that the value of the current may depend considerably on the asymmetry of the space-charge layers. Moreover, a calculation for the most favorable band diagram shows a very large increase of the current.

The development of heterojunction tunnel diodes [1, 2] has made necessary a theoretical study of the properties of such devices. In [3-5], the field in a degenerate heterojunction was calculated and the band diagrams were analyzed. The object of the present work is to consider interband tunneling in heterojunctions and to derive a theoretical current—voltage characteristic.

In general, the solution of the first problem is extremely complex, because the interface between two materials reflects the flow of tunneling carriers and because the tunneling path passes through forbidden bands of both crystals. Assuming that the field is uniform, and that Anderson's model of an abrupt heterojunction [6] is correct, we show in Fig. 1 the two principal possible types of band diagrams for various cases of doping, which exhibit all the features of the problem. There is a paper by Price [7] giving the relation between the wave functions of the different crystals at the interface and a quasiclassical formula for the probability of interband tunneling through the heterojunction:

$$T \approx \frac{\operatorname{sh} \eta_1 \operatorname{sh} \eta_2 \exp\left(-\lambda_1 - \lambda_2\right)}{\operatorname{sh}^2\left[\frac{1}{2}\left(\eta_1 + \eta_2 + \ln|P|\right)\right]}, \tag{1}$$

242

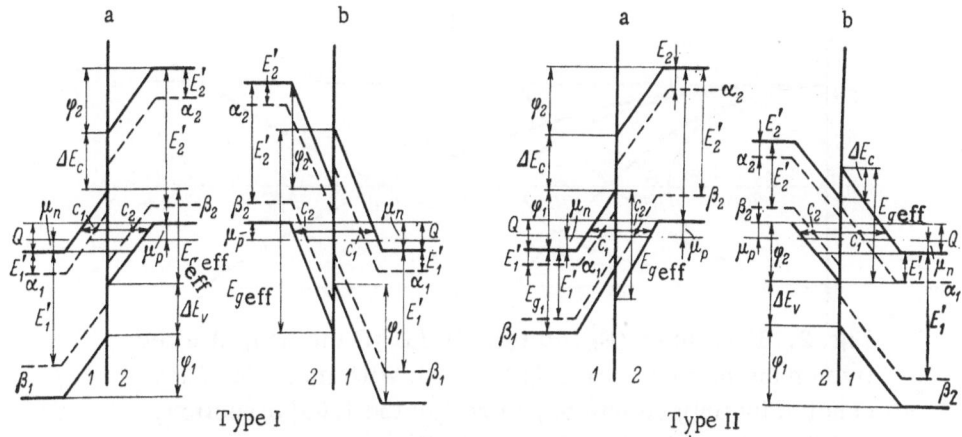

Fig. 1. Principal types of band diagrams of degenerate heterojunctions. The dashed line $\alpha_{1,2}$ shows the energy level at which the curves $\varkappa(E)$ intersect (these curves are calculated by the effective-mass method for $m_p^* < m_n^*$ in crystals 1 and 2); the dashed lines $\beta_{1,2}$ are similar to $\alpha_{1,2}$, but they apply to the case $m_p^* > m_n^*$.

where

$$\eta_{1,2} = \varkappa_{1,2} a_{1,2}, \quad \lambda_1 = 2 \int_{x_1}^{0} \varkappa_1(x)\, dx, \quad \lambda_2 = 2 \int_{0}^{x_2} \varkappa_2(x)\, dx, \quad p = \frac{H_1(1)\, H_2(1)}{H_{12}^2},$$

$$H_{1,2}(1) = \int w_{1,2}(x)\, H_{1,2} w_{1,2}(x - a_{1,2})\, dx,$$

$$H_{1,2} = \int w_1\left(x + a_1 \frac{1}{2}\right) H_{\text{junc}}\, w_2\left(x - \frac{1}{2} a_2\right) dx, \tag{2}$$

$a_{1,2}$ are the lattice constants of the two crystals; $\varkappa_{1,2}$ functions describing the behavior of carriers in passing through the forbidden bands of the two crystals; $w_{1,2}$ the Wannier functions.

Price did not specify the form of the function $\varkappa$. In the effective-mass approximation, it may be written

$$\varkappa_{1,2} = \frac{1}{\hbar} \sqrt{2 m_{1,2}^* \left[V_{1,2}(x) - E \right]}. \tag{3}$$

This expression cannot be used directly for calculations with Eq. (1), however, since the concept of the effective mass is valid only at the boundary of an energy band. Moreover, in the known expressions for calculating interband tunneling by the WKB method which makes use of the effective-mass approximation, it is not clear which mass should be taken. Some authors, as in [8], propose the effective mass over the whole width of the forbidden band, taking it equal to one of the two masses (valence band or conduction band) when these are similar in value, or to the reduced mass given by $2/m_r^* = (1/m_n^*) + (1/m_p^*)$ when they are not similar. Ivanchik [9] has proposed an expression which takes account of the change in the effective mass at the middle of the forbidden band, but again it is not clear which effective mass should be taken. All these proposals are dubious, since it is known that the use of each effective mass is valid only at the boundary of the corresponding band, and that the effective mass changes sign on passing through the forbidden band.

In the present paper, we shall attempt to clarify these problems by a specific calculation using attenuated electron functions obtained from an analysis of the band structure of the crystal, which relate the branches $E(k)$ of the valence and free bands and determine the nature of an

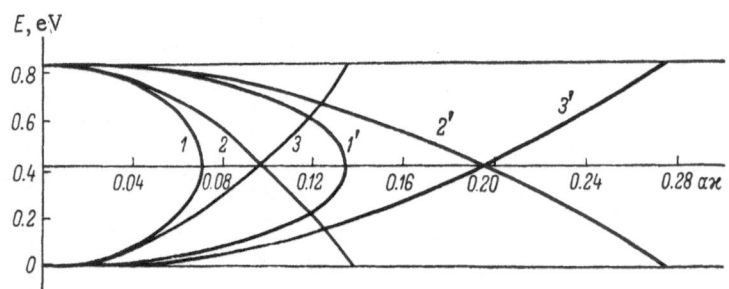

Fig. 2. Function $\varkappa(E)$ for Ge. 1) LCAO method; 2) effective-mass method with m_n^*; 3) same with m_p^*. The curves denoted by primed numbers are for the [100] direction, and the others for the [111] direction.

interband transition between specific points of k space. Next, we shall compare this calculation with the effective-mass method.

In [10], similar calculations were made by the LCAO method for germanium along the [100] direction. A comparison with the effective-mass method showed that there is an appreciable difference over about a third of the forbidden band, the maximum difference being about 40%. In order to reach more general conclusions, let us consider similar calculations for germanium and silicon in the [111] and [100] directions.

Using Eqs. (11) and (12) from [10], we obtain the function $E(\varkappa)$ along the [100] direction for k = 0

$$E_{[100]_{k=0}} = 2\gamma_0' \pm \left[4\gamma_0'^2 + \gamma^2 \pm \sqrt{2} \; \gamma\gamma_0' \left(1 + 4\,\mathrm{ch}^2\,a\varkappa + 3\,\mathrm{ch}\,a\varkappa\right)^{\frac{1}{2}} \right]^{\frac{1}{2}} \tag{4}$$

and $E(\varkappa)$ along the [111] direction for k = 0

$$E_{[111]_{k=0}} = 2\gamma_0' \pm \left[4\gamma_0'^2 + \gamma^2 \pm \sqrt{2} \; \gamma\gamma_0' \left(16\,\mathrm{ch}^4\,a\varkappa + 4\,\mathrm{ch}^3\,a\varkappa - 16\,\mathrm{ch}^2\,a\varkappa - \mathrm{ch}\,a\varkappa + 5\right)^{\frac{1}{2}} \right]^{\frac{1}{2}}, \tag{5}$$

where γ_0' and γ are the resonance interaction integrals. Knowing the experimental values for the valence and forbidden band widths at k = 0 in germanium and silicon, we can find the basic quantities $\gamma\gamma_0'$ and $4\gamma_0'^2 + \gamma^2$ needed to plot $\varkappa(E)$. These are -6.86 eV^2 and 27.615 eV^2 respectively for germanium, -29.015 eV^2 and 117.724 eV^2 for silicon. Then, using Eqs. (4) and (5), we plot graphs for the two elements along the [111] and [100] directions, as shown in Figs. 2 and 3.

We can now find the effective masses of Ge and Si along the [111] and [100] directions at the point k = 0, in accordance with the expressions for $E(k, \varkappa)$ given in [10]:

$$m^* = \frac{\hbar^2}{a^2}\,(\pm A), \tag{6}$$

where A is 0.0445 eV^{-1} for Ge along the [100] direction and 0.0114 eV^{-1} along the [111] direction; the corresponding values for Si are 0.0323 eV^{-1} and 0.0082 eV^{-1}. The plus sign refers to the conduction band and the minus sign to the valence band.

Substituting Eq. (6) in the formula for finding $\varkappa$ by the effective-mass method, we obtain

$$a\varkappa = \frac{a}{\hbar}\,\sqrt{2m^*E} = \sqrt{2AE}, \tag{7}$$

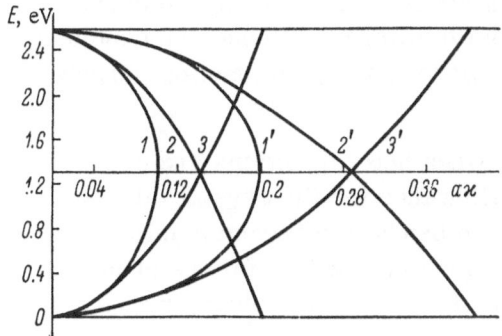

Fig. 3. Function $\varkappa(E)$ for Si. 1) LCAO method; 2) effective-mass method with m_n^*; 3) same with m_p^*. The curves denoted by primed numbers are for the [100] direction, and the others for the [111] direction.

or

$$a\varkappa = \sqrt{2A\,(E_g - E)}, \qquad (8)$$

where the first of these expressions refers to the conduction band and the second to the valence band.

Figures 2 and 3 show the functions $\varkappa(E)$ for Ge and Si along the [100] and [111] directions, obtained from Eqs. (7) and (8).

The results may be analyzed as follows.

Curves 1 and 1' in Figs. 2 and 3 were obtained using attenuated functions and the LCAO method; within the accuracy of this method, the area between the curves and the E axis represents the probability of tunneling for a triangular barrier. There are also the curves 2, 2' and 3, 3', obtained by the effective-mass method, which also represent the transition probability. A comparison of the areas bounded by the curves 1 and 2 or 3, or the corresponding curves designated by primed numbers, shows that the effective-mass method as generally employed involves a large error. It is seen from Figs. 2 and 3 that this arises mainly from neglecting the change in sign of the effective mass when the carrier passes through the forbidden band. The error is eliminated by using the effective mass m_n^* and Eq. (7) in the upper half of the forbidden band, and m_p^* and Eq. (8) in the lower half. Ivanchik's result gives the best approximation, since it allows both for the invalidity of the effective-mass method in the middle of the forbidden band and for the change of sign. If $\varkappa(E)$ is plotted from Ivanchik's expression, the deviations from curve 1 are so small that it is difficult to show them in a diagram on this scale.

The whole of this analysis applies to the case when the effective masses are equal. In practice, however, they are usually quite different. Ivanchik's expression then becomes inapplicable, since it does not make clear which mass is to be used. The analysis is complicated by the fact that there is no comparison curve similar to the curve 1 or 1' for the case of equal masses; such a curve could be obtained using more precise methods of computing the band diagram, which lead to different effective masses. But, applying the basic results to the case of equal effective masses, we can say that the best approximation will be one which takes into account the change in sign of the effective mass, and also the fact that, near the band edges, the true attenuation curve must be close to those given by the effective masses of the relevant bands.

On this basis, we can propose the following method for determining $\varkappa$. We first find the point of intersection of the curves for different effective masses (taking the edge of the conduction band as the zero of energy):

$$E' = \frac{E_g m_p^*}{m_n^* + m_p^*}. \qquad (9)$$

Then, $\varkappa$ is given in the range from 0 to E' by

$$\varkappa = \frac{1}{\hbar}\,\sqrt{2m_n^*\,[V\,(x) - E]}, \qquad (10)$$

and in the range from E' to E_g by

$$\varkappa = \frac{1}{\hbar}\,\sqrt{2m_p^*\,[E_g - (V\,(x) - E)]}. \qquad (11)$$

To complete the analysis, we must note that, as is shown by Figs. 2 and 3, the curves $\varkappa(E)$ plotted for different crystallographic directions are quite different. Hence, we conclude that, to improve tunnel diodes, it is necessary to make the plane of the p−n junction perpendicular to the direction of least attenuation in the crystal.

The above discussion becomes still more significant when heterojunctions are considered. The method given here for determining $\varkappa$, with allowance for the crystal interface, enables us to calculate the probability of a transition between bands of different semiconductors, with no uncertainty in the choice of effective mass, and with allowance for the change in sign of the latter. The following are some further features of tunneling in heterojunctions.

Figure 1 shows that the tunneling paths for carriers having different energies are not the same, and this results in an energy dependence of the transition probability even in a uniform electric field.

Evidently, tunneling in the first crystal is determined by the curve of $\varkappa_1(E)$ up to the interface, and then in the second crystal by the curve of $\varkappa_2(E)$. When a carrier passes through the boundary, $\varkappa_1 \neq \varkappa_2$ at the interface, unless there is equality by chance. Hence, the imaginary crystal momentum must change in order to bring about the transition. This change occurs when the particle interacts with the interface, which is taken into account by the transmission coefficient in Eq. (1). The position is similar to that for ordinary indirect transitions, and one can therefore say that all transitions are indirect in tunneling through a heterojunction.

The allowance for orientation and the doping conditions of the crystals is still more important when heterojunctions are used in the production of tunnel diodes. For example, in a Ge−Si junction, if the Ge is n-type and the Si is p-type, a junction in the (111) plane is best, but if the doping is reversed, the (100) plane is better, since these planes correspond to the directions of the principal minima of the conduction bands.

Let us now consider the quantity $\lambda_1 + \lambda_2$ in Eq. (1), in the uniform-field approximation. We shall allow for its dependence on $p_\perp$, the crystal momentum perpendicular to the tunneling direction. For tunneling in ordinary semiconductors, this has been taken into account by Keldysh [11]. From the above analysis, it is clear that the allowance can also be made in the WKB method. To do so, we need only increase the forbidden band width by $p_\perp^2 / 2(m^*)_{n,p}$ at each edge when $p_\perp \neq 0$. Then, using Eqs. (1), (9), (10), and (11), we obtain an expression for $\lambda_1 + \lambda_2$. Here, there are four possible cases, depending on the relationship of effective masses and forbidden bands. In the analysis, we consider the points c_1 and c_2, whose significance is clear from Fig. 1.

1. If c_1 and c_2 are outside the band Q [energies from 0 to $(\mu_n + \mu_p - qV)$], i.e., if the inequalities $E_1' > \varphi_1$ and $E_2' < E_{g_2} - \varphi_2$ hold, then

$$(\lambda_1 + \lambda_2) = \frac{4}{3} \frac{\sqrt{2m^*_{n_1}}}{\hbar q \mathcal{E}_1} \left\{ \left(\varphi_1 - E + \frac{p_\perp^2}{2(m^*_\perp)_{n_1}} \right)^{\frac{3}{2}} + \sqrt{\frac{m^*_{p_2}}{m^*_{n_1}} \frac{\varepsilon_2}{\varepsilon_1}} \left(E_{g_2} - \varphi_1 \mp \Delta E_c + E + \frac{p_\perp^2}{2(m^*_\perp)_{p_2}} \right)^{\frac{3}{2}} \right\}. \tag{12}$$

2. If only c_1 lies in Q, i.e., if $E_1' < \varphi_1$ and $E_2' < E_{g_2} - \varphi_2$, then, for energies $E \leq \varphi_1 - E_1'$

$$\lambda_1 + \lambda_2 = \frac{4}{3} \frac{\sqrt{2m^*_{n_1}}}{\hbar q \mathcal{E}_1} \left\{ \left(E_{g_1} \frac{m^*_{p_1}}{m^*_{n_1} + m^*_{p_1}} + \frac{p_\perp^2}{2(m^*_\perp)_{n_1}} \right)^{\frac{3}{2}} + \sqrt{\frac{m^*_{n_1}}{m^*_{n_1}}} \left[\left(E_{g_1} \frac{m^*_{n_1}}{m^*_{p_1} + m^*_{n_1}} + \frac{p_\perp^2}{2(m^*_\perp)_{p_1}} \right)^{\frac{3}{2}} - \right. \right.$$
$$\left. \left. - \left(E_{g_1} + E + \frac{p_\perp^2}{2(m^*_\perp)_{p_1}} - \varphi_1 \right)^{\frac{3}{2}} \right] + \sqrt{\frac{m^*_{p_2}}{m^*_{n_1}} \frac{\varepsilon_2}{\varepsilon_1}} \left(E_{g_2} - \varphi_1 \mp \Delta E_c + E + \frac{p_\perp^2}{2(m^*_\perp)_{p_2}} \right)^{\frac{3}{2}} \right\}; \tag{13}$$

for energies $E \geq \varphi_1 - E_1'$, we have Eq. (12).

3. If, on the other hand, only c_2 is in Q, i.e., if $E_1' > \varphi_1$ and $E_2' > E_{g_2} - \varphi_2$, then, for energies $E \geq \varphi_1 \pm \Delta E_c - E_2'$,

$$\lambda_1 + \lambda_2 = \frac{4}{3}\frac{V\sqrt{2m_{n_1}^*}}{\hbar q \mathscr{E}_1}\left\{\left(\varphi_1 - E + \frac{p_\perp^2}{2\left(m_\perp^*\right)_{n_1}}\right)^{\frac{3}{2}} + \sqrt{\frac{m_{n_2}^*}{m_{n_1}^*}\frac{\varepsilon_2}{\varepsilon_1}}\left[\left(E_{g_2}\frac{m_{p_2}^*}{m_{n_2}^* + m_{p_2}^*} + \frac{p_\perp^2}{2\left(m_\perp^*\right)_{n_2}}\right)^{\frac{3}{2}} - \right.\right.$$

$$\left.\left. - \left(\varphi_1 \pm \Delta E_c - E + \frac{p_\perp^2}{2\left(m_\perp^*\right)_{n_2}}\right)^{\frac{3}{2}} + \sqrt{\frac{m_{p_2}^*}{m_{n_2}^*}}\left(E_{g_2}\frac{m_{n_2}^*}{m_{n_2}^* + m_{p_2}^*} + \frac{p_\perp^2}{2\left(m_\perp^*\right)_{p_2}}\right)^{\frac{3}{2}}\right]\right\}; \tag{14}$$

for energies $E \leq \varphi_1 \pm \Delta E_c - E_2'$, we have Eq. (12).

4. If both points c_1 and c_2 lie in Q, i.e., if $E_1' < \varphi_1$ and $E_2' > E_{g_2} - \varphi_2$, then, for energies $\varphi_1 \pm \Delta E_c - E_2' < E < \varphi_1 - E_1'$,

$$\lambda_1 + \lambda_2 = \frac{4}{3}\frac{V\sqrt{2m_{n_1}^*}}{\hbar q \mathscr{E}_1}\left\{\left(E_{g_1}\frac{m_{p_1}^*}{m_{n_1}^* + m_{p_1}^*} + \frac{p_\perp^2}{2\left(m_\perp^*\right)_{n_1}}\right)^{\frac{3}{2}} + \sqrt{\frac{m_{p_1}^*}{m_{n_1}^*}}\left[\left(E_{g_1}\frac{m_{n_1}^*}{m_{p_1}^* + m_{n_1}^*} + \frac{p_\perp^2}{2\left(m_\perp^*\right)_{p_1}}\right)^{\frac{3}{2}} - \right.\right.$$

$$\left. - \left(E_{g_1} + E + \frac{p_\perp^2}{2\left(m_\perp^*\right)_{p_1}} - \varphi_1\right)^{\frac{3}{2}}\right] + \sqrt{\frac{m_{n_2}^*}{m_{n_1}^*}\frac{\varepsilon_2}{\varepsilon_1}}\left[\left(E_{g_2}\frac{m_{p_2}^*}{m_{n_2}^* + m_{p_2}^*} + \frac{p_\perp^2}{2\left(m_\perp^*\right)_{n_2}}\right)^{\frac{3}{2}} - \right.$$

$$\left.\left. - \left(\varphi_1 \pm \Delta E_c - E + \frac{p_\perp^2}{2\left(m_\perp^*\right)_{n_2}}\right)^{\frac{3}{2}} + \sqrt{\frac{m_{p_2}^*}{m_{n_2}^*}}\left(E_{g_2}\frac{m_{n_2}^*}{m_{n_2}^* + m_{p_2}^*} + \frac{p_\perp^2}{2\left(m_\perp^*\right)_{p_2}}\right)^{\frac{3}{2}}\right]\right\}; \tag{15}$$

for energies $E > \varphi_1 \pm \Delta E_c - E_2'$ and $E > \varphi_1 - E_1'$, we have Eq. (14), and for $E < \varphi_1 \pm \Delta E_c - E_2'$ and $E < \varphi_1 - E_1'$, Eq. (13).

The subscripts 1 and 2 denote here quantities pertaining to the two crystals. The quantities $E_{g\,1,2}$, $\varphi_{1,2}$, ΔE_c are as shown in Fig. 1. The upper sign before ΔE_c refers to band diagram a in Fig. 1, and the lower sign to diagram b; $m_{n_1}^*$ and $m_{p_2}^*$ are the effective masses in the conduction and valence bands; the subscript $\perp$ denotes the effective mass in a direction perpendicular to that of tunneling; q is the electron charge; $\mathscr{E}_1$ is the field strength in crystal 1; $\varepsilon_{1,2}$ the permittivities of the two crystals; V the applied bias.

It is seen from Eqs. (12)–(15) that, as already mentioned, the tunneling probability depends on E and $p_\perp$. It is clear that the asymmetry of the junction is important, as has been emphasized in [4]. A very interesting fact is that, when we change to an ordinary tunnel diode in Eqs. (13) and (14) by altering the appropriate parameters, the argument of the exponential function becomes

$$\lambda \approx \frac{4}{3}\frac{V\sqrt{2\bar{m}_r^*}}{\hbar q \mathscr{E}}E_g^{\frac{3}{2}}\left(1 + \frac{3}{4}\frac{p_\perp^2}{\bar{m}_r^* E_g}\right), \tag{16}$$

where $\bar{m}_r^*$ is the reduced mass given by $(1/m_n^*) + (1/m_p^*) = 1/\bar{m}_r^*$, which differs only slightly in the numerical factors $(4/3 \rightarrow \pi/2\sqrt{2}$ and $3/4 \rightarrow 1)$ from the expression derived by Keldysh [12] for direct transitions.

Let D denote the exponential part of Eq. (1), and D' the remainder. Expressions for D' are easily obtained, but are lengthy and will not be given here. The tunneling probability is then

$$T' = D'D. \tag{17}$$

The difference between heterojunction and ordinary tunnel diodes lies in the specific nature of interband tunneling. Hence, to calculate the current, we use the expression given in [12], simply substituting Eq. (17) in it in place of the expression for the tunneling probability in an ordinary p−n junction. Then,

$$I = \frac{qS}{2\pi^2\hbar^3} \int\limits_{0}^{\Delta\mu - qV} [f_1(E) - f_2(E)] \int\limits_{0}^{p_{n_1} \text{ or } p_{p_2}} T' p_\perp \, dp_\perp \, dE, \tag{18}$$

where S is the junction area, and $f_{1,2}(E)$ are the Fermi functions for electrons in the two crystals; $p_{n_1}^2 = 2m_{n_1}^* E$; $p_{p_2}^2 = 2m_{p_2}^* (\Delta\mu - qV - E)$; $\Delta\mu = \mu_n + \mu_p$.

The dependence of T' on $p_\perp$ is seen from Eqs. (12)–(15) to be very complicated, and the integral cannot be evaluated. But, since the quantity $p_\perp^2 / 2(m_\perp^*)_{n,p}$, which appears in $\lambda_1 + \lambda_2$ is small in comparison with the other terms in the parentheses that are raised to the power 3/2, we can expand the expressions in these parentheses in series, after which the integration with respect to $p_\perp$ is simple. The result is

$$I = \frac{qS}{2\pi^2\hbar^3} \int\limits_{0}^{\Delta\mu - qV} [f_1(E) - f_2(E)] D' D_1 dE, \tag{19}$$

where D_1 is the expression found by integrating over $p_\perp$. There are four such expressions, corresponding to Eqs. (12)–(15). All are easily derived, but are extremely lengthy and therefore will not be given here.

The integral in Eq. (19) has to be calculated numerically; the analysis of the expression itself is extremely laborious, and it can evidently be applied to exact calculations of the current−voltage characteristic.

For a further investigation, we shall write down a simplified expression based on the work of Esaki [13]. To do so, we change [as in the derivation of (18)] only the expression for the tunneling probability, using Eq. (17) and setting $p_\perp = 0$, since Esaki did not take into account the conservation of crystal momentum.

Then,

$$I = A \int\limits_{0}^{\Delta\mu - qV} T' \rho_1(E) \rho_2(E) [f_1(E) - f_2(E)] \, dE, \tag{20}$$

where ρ_1 and ρ_2 are the densities of states in the two crystals. Since the transmission coefficient D' depends only very slightly on the energy in the range from 0 to $\Delta\mu - qV$, in comparison with the other quantities, and mainly affects the numerical value of the current, it can be taken outside the integral. Next, using the approximation given in [14] for the Fermi function in a heavily doped semiconductor, we can write $[f_1(E) - f_2(E)] \approx qV/4kT$. Then, combining in a parameter A' all the quantities which depend only slightly or not at all on E and V, we obtain

$$I = A'qV \int\limits_{0}^{\Delta\mu - qV} e^{-(\lambda_1 + \lambda_2)} \sqrt{E(\Delta\mu - qV - E)} \, dE. \tag{21}$$

The chief difference between this and Esaki's formula is that the tunneling probability is energy-dependent and is within the integral. We studied Eq. (21) for the case of a degenerate heterojunction between n-type Ge and p-type GaAs, calculating the integrals numerically, for

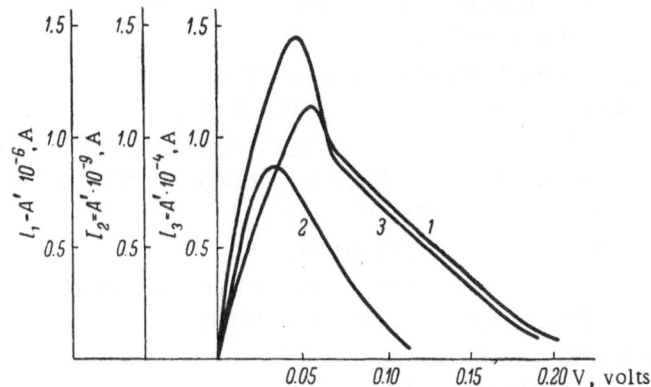

Fig. 4. Dependence of tunnel component of current on bias. 1) Basic tunneling path in GaAs: $n_{1\,Ge} = 3 \cdot 10^{19}$ cm^{-3}, $p_{2\,GaAs} = 1 \cdot 10^{19}$ cm^{-3}; 2) basic tunneling path in Ge: $n_{1\,Ge} = 1 \cdot 10^{19}$ cm^{-3}, $p_{2\,GaAs} = 3 \cdot 10^{19}$ cm^{-3}; 3) for a type Ia band diagram: $n_{1\,Ge} = 1 \cdot 10^{19}$ cm^{-3}, $p_{2\,GaAs} = 3 \cdot 10^{19}$ cm^{-3}.

various doping levels (Fig. 4). The data relating to the band diagram, and the expressions for calculation, were taken from the work of Anderson.

Figure 4 shows that curves 1 and 2 are very different. This difference occurs because the change of doping levels causes the basic tunneling path to lie in the forbidden band of either Ge or GaAs, which have different effective masses. Since that of GaAs is the smaller, the current will evidently be larger, as is in fact seen on comparing curves 1 and 2 in Fig. 4. From the above discussion, we can understand the importance of allowing for the field asymmetry as in [4], and it is obvious that even quite small deviations will noticeably affect the current — voltage characteristic.

In [5], the hypothesis has been foward that a band diagram of type Ia (Fig. 1) might give improved tunnel diode characteristics. Using Eq. (21), we can obtain a theoretical current — voltage characteristic for this case. Since it has not been possible to find satisfactorily studied pairs of semiconductors having a band diagram of this type, the calculation has been made for a pair of crystals having the same effective masses and forbidden bands as Ge and GaAs, but a band diagram of type Ia with $\Delta E_v = 0.1$ eV and $\Delta E_c = 0.96$ eV. In this way, we can compare with the ordinary band diagram for a Ge — GaAs heterojunction having the same doping, and see the advantages of the artificial system considered. In Fig. 4, the curves to be compared are 1 and 3. It can be seen that the current for curve 3, which refers to the band diagram Ia, is considerably greater than the peak current in the current — voltage characteristic for an ordinary band diagram with this pair of crystals.

Finally, it may be noted that we have found only one experimental current — voltage characteristic [1] for a tunnel heterodiode using n-type Ge and p-type GaAs. Unfortunately the parameters needed for a meaningful theoretical analysis are not stated (these parameters include μ_n, μ_p, the area of the junction, and definite values of the impurity concentrations); moreover, the concentrations used are not entirely suitable for calculations relating to the cases analyzed in the present paper. No detailed comparison with experiment has been possible, therefore.

Literature Cited

1.   J. C. Marinace, IBM J. Res. Develop., 4:280 (1960).

2. M. I. Nathan and J. C. Marinace, Phys. Rev., 128:2149 (1962).

3. A. I. Gubanov and S. N. Dobrynin, Izv. Leningr. Élektrotekh. Inst. (1967).

4. S. N. Dobrynin, present collection, p. 51.

5. S. N. Dobrynin, Fiz. Tekh. Poluprov., 3:16 (1969).

6. R. L. Anderson, Solid State Electronics, 5:341 (1962).

7. P. J. Price, Proc. Sixth Intern. Conf. on Physics of Semiconductors, Exeter, 1962, Publ. by Institute of Physics and Physical Society, London (1962), p. 99.

8. A. Shibata, J. Phys. Soc. Japan, 17:770 (1962).

9. I. I. Ivanchik, Fiz. Tverd. Tela, 3:103 (1961).

10. A. I. Gubanov and S. N. Dobrynin, Fiz. Tekh. Poluprov., 1:323 (1967).

11. L. V. Keldysh, Zh. Éksp. Teor. Fiz., 33:994 (1957).

12. G. E. Pikus, Fundamentals of the Theory of Semiconductor Devices, Moscow (1965).

13. L. Esaki, Phys. Rev., 109:603 (1958).

14. J. Karlovský, Phys. Rev., 127:419 (1962).

EXPERIMENTAL STUDY OF PHOTODIODE RESPONSE
USING THE PHOTOCURRENT SHOT NOISE SPECTRUM

D. V. Tarkhin and N. N. Armencha

A. F. Ioffe Physicotechnical Institute
Academy of Sciences of the USSR, Leningrad

The photocurrent shot noise was measured by a modulation method in the frequency range from 0.15 to 400 Mc. The spectral characteristics of noise were investigated for FD-3 and FDK-1 commercial photodiodes and also for germanium photodiodes of two types (thin-base diffused and coaxial surface-barrier diodes). Measurements of the frequency characteristics were also made on the same diodes, using the radiation flux from a helium−neon laser ($\lambda = 0.63\ \mu$) modulated by a Kerr cell. It was found that, for the FD-3 and FDK-1 photodiodes, the relation between the limiting frequencies, given by the frequency characteristic and spectral characteristic of noise, is governed by the relation between the carrier diffusion time in the base and the time constant of the RC circuit. The limiting frequencies determined from the two characteristics are in satisfactory agreement for the thin-base diffused photodiodes. For the surface-barrier photodiodes the spectral and frequency characteristics have no descending part in the frequency range covered by the measurements.

The noise spectra of photodiodes have been studied in [1-4], but the frequency range used did not go beyond 35 Mc, and the photodiodes examined had slow response. It is shown in [4] that the spectral and frequency characteristics of photodiodes need not in general coincide. The limiting frequency determined from the noise spectrum is greater than that determined from the frequency characteristic recorded using a modulated radiation signal. This difference depends on the diffusion time of the nonequilibrium minority photocarriers in the photodiode base ($t_{diff} = W^2/2D$).

The corresponding difference for thin-base diodes is very small [4]. The shot noise spectra of photodiodes then give information about their frequency characteristics, whereas the direct methods of measuring response are more complicated, since they require microwave modulation of the radiation flux. We made a study of thin-based diffused photodiodes and of coaxial surface-barrier photodiodes, in which there is no carrier diffusion caused by radiation.

At high frequencies, the noise in most photodiodes is shot noise. In photodiodes whose excess noise extends to high frequencies, the hf component can be eliminated by illuminating

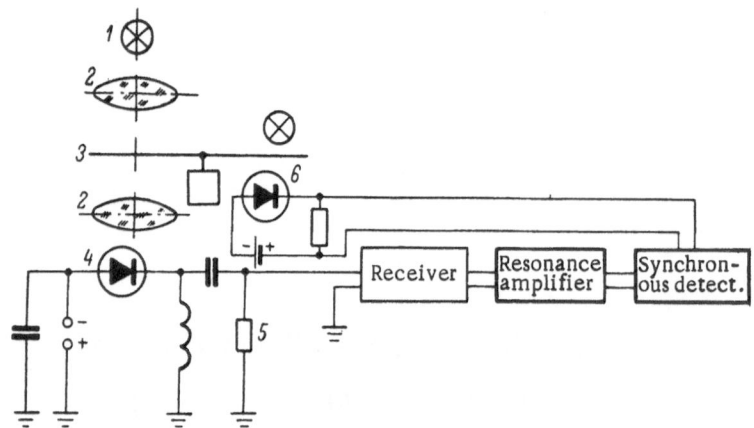

Fig. 1. Block diagram of the apparatus. 1) Incandescent lamp; 2) focusing lenses; 3) modulator; 4) photodiode under examination; 5) load; 6) photodiode whose load supplied the reference voltage.

the photodiode and "drowning" this component in the photocurrent shot noise. The spectral distribution of the shot noise voltage $F(f)$, when the fall of the spectral curve is governed by relaxation in the RC circuit (as was observed in our photodiodes), is given by the expression

$$F(f) = \left[2eI \, \frac{1}{1 + (2\pi f)^2 \, \tau^2} \right]^{\frac{1}{2}} R_l,$$

where I is the current through the photodiode; f is the limiting frequency; $\tau = RC$; $R = R_l + R_{pd}$; R_l is the load resistance; R_{pd} is the photodiode base resistance; C is the barrier capacitance.

The thermal noise was eliminated by a modulation method which yielded only the noise due to the illumination of the photodiode. Figure 1 shows a block diagram of the apparatus.

In order to exclude strays and minimize unwanted capacitances, the photodiode and load were placed in a specially designed coaxial holder.

The range from 0.15 to 400 Mc was covered by three superheterodyne receivers: from 0.15 to 35 Mc by a V6-1 selective microvoltmeter with 10 kc pass band, from 18 to 150 Mc by a P5-1 receiver with 20 kc pass band, and from 150 to 400 Mc by a P5-2 receiver with 1 Mc pass band. The load resistance was 75 Ω in order to match the input of the P5-1 and P5-2 receivers to the load of the photodiode under examination.

The radiation flux from the incandescent lamp was focused on the photodiode in order to obtain a photocurrent of about 1 mA. This value was determined by the need to produce a sufficiently large signal with a small load.

In order to raise the sensitivity of the apparatus, a K3-2 synchronous detector was used. The photodiode dc bias circuit was almost shorted, so as to exclude modulation of the photodiode capacitance at the radiation interruption frequency (800 cps) owing to the change of the voltage drop across it.

In our case, the load photocurrent noise, which constituted the signal, was considerably less than the intrinsic noise of the receiver. It was found that the noise signal leaving the apparatus was then proportional to the photocurrent: $U_{ns} \propto I_p$. Let the radiation flux be modulated

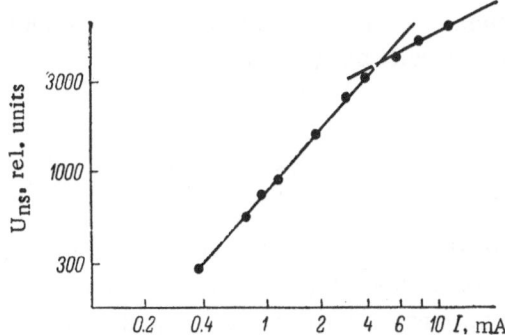

Fig. 2. Dependence of output signal on noise diode current.

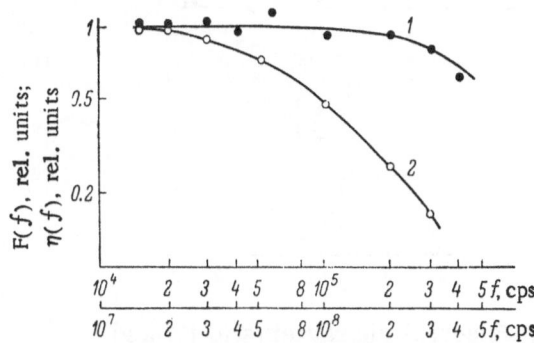

Fig. 3. Spectral characteristic (1) and frequency characteristic (2) of an FD-3 photodiode. Abscissa: upper scale for frequency characteristic, lower scale for spectral characteristic.

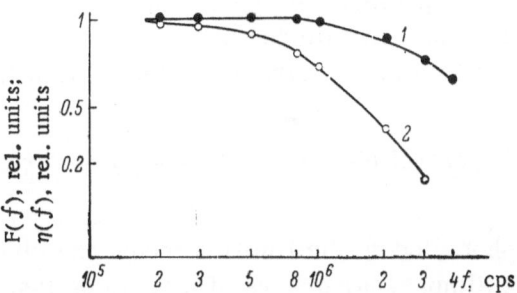

Fig. 4. Spectral characteristic (1) and frequency characteristic (2) of an FDK-1 photodiode.

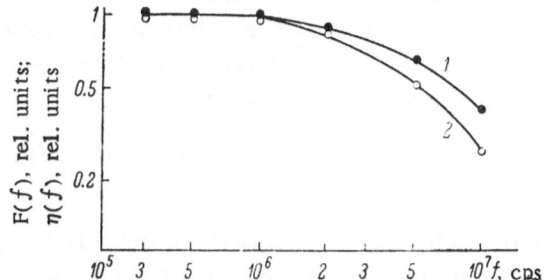

Fig. 5. Spectral characteristic (1) and frequency characteristic (2) of a thin-base diffused photodiode.

by rectangular pulses of frequency ω. Using the first two terms of an expansion in harmonic series, we can write the photocurrent noise voltage across the photodiode load as

$$U_{\text{pn}} = \sqrt{2eI_p \Delta f R_l^2} \left(\frac{1}{2} + \frac{2}{\pi} \sin \omega t \right),$$

where I_p is the photocurrent and Δf the amplifier transmission bandwidth.

The total noise voltage at the receiver input is

$$U_\Sigma = \sqrt{U_{\text{rn}}^2 + U_{\text{pn}}^2} = \left[P_{\text{rn}} R_l + 2eI_p \Delta f R_l^2 \left(\frac{1}{2} + \frac{2}{\pi} \sin \omega t \right)^2 \right]^{\frac{1}{2}} = \sqrt{P_{\text{rn}} R_l} \left[1 + \frac{eI_p \Delta f R_l}{2P_{\text{rn}}} \left(1 + \frac{4}{\pi} \sin \omega t \right)^2 \right]^{\frac{1}{2}},$$

where P_{rn} is the intrinsic noise power of the receiver. Hence, since in our case

$$\frac{eI_p \Delta f R_l}{2P_{\text{rn}}} \leqslant 1,$$

we have

$$U_\Sigma \approx \sqrt{P_{\text{rn}} R_l} \left[1 + \frac{eI_p \Delta f R_l}{4P_{\text{rn}}} \left(1 + \frac{8}{\pi} \sin \omega t + \frac{16}{\pi^2} \sin^2 \omega t \right) \right].$$

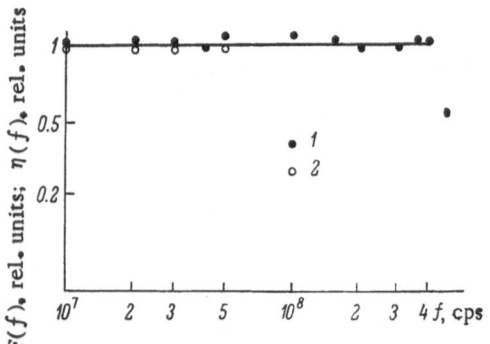

Fig. 6. Spectral characteristic (1) and frequency characteristic (2) of a surface-barrier photodiode.

The output signal from the apparatus, when the resonance amplifier is tuned to the modulation frequency ω, will be given by

$$U_{ns} \sim \frac{2eI_p\Delta f R_l^{\frac{3}{2}}}{\sqrt{P_{rn}}\,\pi}\sin\omega t.$$

The linear relation between U_{ns} and I_p was observed in the photodiodes examined, and was tested with the apparatus calibrated by means of a 2D2S noise diode whose anode received a voltage at the modulation frequency, as proposed in [3].

Figure 2 shows that the linear relation between the output signals and the 2D2S diode current is maintained until the load noise voltage becomes comparable with the intrinsic noise of the receiver (for a noise diode current of 8 mA). At high diode currents, $U_{ns} \propto I_p^{\frac{1}{2}}$. We measured the shot noise spectra of the FD-3 and FDK-1 photodiodes, germanium diffused photodiodes, and germanium surface-barrier photodiodes. The same diodes were also used for measurement of the frequency characteristics, using the radiation flux from a helium−neon laser ($\lambda = 0.63\ \mu$) modulated by a Kerr cell.

Figures 3 through 6 show frequency $\eta(f)$ and spectral characteristics $F(f)$ of various types of photodiodes.

It is seen from Fig. 3 that, in the FD-3 type photodiodes, the limiting frequency given by the frequency characteristic is $5 \cdot 10^4$ cps, which is almost four orders of magnitude less than the limiting frequency given by the spectral characteristic. This is in accordance with the ratio of the photodiode response time ($\sim 10^{-5}$ sec) and the time constant of the RC circuit ($\sim 10^{-9}$ sec).

In the FDK-1 type photodiodes (Fig. 4), the limiting frequencies given by the frequency characteristic and the spectral characteristic differ by a factor of 3.3, which is considerably less. The explanation probably lies in the different ratio of the carrier diffusion time in the base to the time constant of the RC circuit.

The limiting frequencies given by the frequency and spectral characteristics for a thin-base germanium diffused photodiode are in satisfactory agreement (Fig. 5). Figure 6 shows that the frequency and spectral characteristics of a germanium surface-barrier photodiode have no descending part in the frequency range where the measurements were made (up to 50 Mc for the frequency characteristic and up to 400 Mc for the spectral characteristic).

Literature Cited

1.   L. Ya. Pervova, Radiotekh. i Élektron., 4:330 (1959).
2.   T. M. Lifshits and L. Ya. Pervova, Radiotekh. i Élektron., 4:1541 (1959).
3.   S. A. Kaufman and K. M. Kulikov, Fiz. Tverd. Tela, 7:3132 (1965).
4.   S. M. Kozel, N. N. Kolachevskii, and A. M. Noginov, Radiotekh. i Élektron., 11:1616 (1966).

SELF-MODULATION OF MICROWAVE OSCILLATIONS
GENERATED IN GUNN DIODES

I. I. Abkevich

Scientific-Research Institute of Semiconductor Device Manufacture, Moscow

A theoretical and experimental study was made of the self-modulation in Gunn diodes (i.e., the generation by these diodes of a 100% modulated microwave signal under certain conditions). The phenomenon is due to a sudden drop of the mean diode current, when the microwave generation voltage is reached, and to the excitation of oscillations, because of the sudden current drop, in the low-frequency oscillatory circuit. An ac voltage appears in the Gunn diode circuit and this voltage modulates the microwave signal. Relations are derived between the amplitude, shape, and frequency of self-modulation and the parameters of the diode in the circuit. The condition for the occurrence of self-modulation is $2|R_e|(L/C)^{1/2} > 1$, where $R_e = RR_l/(R + R_l)$; R is the differential resistance of the diode above the critical voltage; R_l, L, and C are respectively the resistance, inductance, and capacitance of the parallel low-frequency circuit to which the diode is connected.

Introduction

The generation of an amplitude-modulated microwave signal is sometimes observed in measurements of the parameters of Gunn diodes. This has been reported in [1, 2], and is also mentioned in the review [3]. The observed modulation frequencies are of the order of tens of megacycles. It is suggested in [1] that the phenomenon is due to negative conductance of the diode. Fleming [2] considers that the modulation is due to the excitation of oscillations in the low-frequency oscillatory circuit because of negative differential conductance of the diode. A low-frequency ac voltage appears in the Gunn diode circuit and this voltage modulates the microwave signal. A criterion for the occurrence of modulation is given in [3]: the negative resistance of a domain must exceed the resistance of the crystal.

In the present paper, the physical nature of the modulation is further elucidated, and the relations are derived between the modulation voltage and the diode and circuit parameters. It is shown that the depth of modulation is always 100%. In accordance with the accepted terminology, we shall use the term self-modulation for this phenomenon [2, 3].

The diodes examined were encapsulated crystals of gallium arsenide 40-150 μ thick, with contact areas 10^{-3}-10^{-4} cm^2. The low-voltage resistance of the diodes was about 5-50 Ω. The diodes were tested in various circuits: at microwave frequencies we used either a broad-

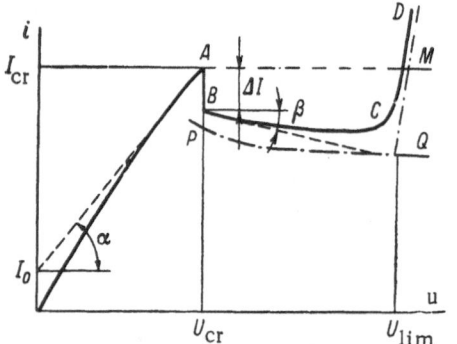

Fig. 1. Current–voltage characteristics of a Gunn diode. Continuous
curve: low-frequency characteristic; chain curve: high-frequency
characteristic; dashed curve: approximations used in the calculations.

band ~3-Ω resistance or a resonance load; at low frequencies we used either a circuit having lumped R_l,
L, and C components or a coaxial microwave section,
which acted also as a lf oscillatory circuit having distributed inductance and capacitance in the form of seg-
ments of a coaxial line. In all these circuits, when Q
was sufficiently high, the lf circuit showed self-modulation, usually in the form of nonsinusoidal diode voltage
and current oscillations in the range from megacycles
to hundreds of megacycles, and a correspondingly
modulated microwave signal.

The following are two possible causes of self-
modulation: (a) excitation of the lf circuit because of
nonlinearity of the diode in this circuit; (b) oscillations caused by some internal processes in the crystal,
e.g., motion of domains at the speed of sound [4, 5].
In our measurements, the self-modulation frequency
did not vary appreciably with the dimensions, orienta-
tions, or lengths of the GaAs crystals but could be altered considerably (by several orders of
magnitude) by changing the construction of the coaxial section or the resonance frequency of
the lf circuit having lumped components. Hence, we concluded that automanipulation oscillations appeared to be due to cause (a), the interaction of the diode with lf oscillatory circuit.

Theory

Let us consider the interaction of a Gunn diode with a low-frequency oscillatory circuit.
The form of the lf current–voltage characteristic is easily established by using Gunn's time-
dependent high-frequency current–voltage characteristic [6]. Figure 1 shows the form for the
Gunn characteristic and of the lf characteristic, and the approximation to the latter used in the
present calculation. The line AM represents the diode current when the domain is absent, and
PQ that when it is present. The voltage across the diode is limited to U_{lim}, at which the cur-
rent begins to increase rapidly with the voltage, usually because of the multiplication of carriers.
The lf current–voltage characteristic occupies an intermediate position between the characteristics with domain absent and present, and depends on the microwave load of the diode. The
Gunn effect is the instability of a large signal: large-amplitude oscillations appear immediately. Accordingly, the mean current suddenly drops by an amount ΔI at the critical voltage U_{cr}.
In our approximation to the lf characteristic, the differential resistances of the diode at voltages above and below critical are taken to be independent of the bias, with values R and r respectively; in Fig. 1, $R = \cot \beta$, $r = \cot \alpha$. The resistance R may be either negative or positive.
For real diodes, we always have $|R| > r$. Using this approximation for the current–voltage
characteristic, the equivalent circuit of a diode connected to a parallel R_l LC circuit, with a
bias voltage source in the inductive branch, may be represented as in Fig. 2. The load resistance in the equivalent circuit is

$$r_l = \begin{cases} \dfrac{rR_l}{(r + R_l)} \equiv r_e & \text{for} \quad u < U_{cr}, \tag{1} \\[2ex] \dfrac{RR_l}{(R + R_l)} \equiv R_e & \text{for} \quad u > U_{cr}, \tag{2} \end{cases}$$

and the current source is described by

$$I = \begin{cases} I_0 & \text{for} \quad u < U_{cr}, \tag{3} \\[2ex] I_0 + \dfrac{U_{cr}}{r} \cdot \dfrac{R - r}{R} - \Delta I & \text{for} \quad u > U_{cr}. \tag{4} \end{cases}$$

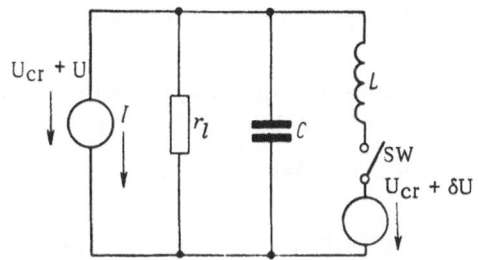

Fig. 2. Equivalent representation of a Gunn diode connected to an $R_l LC$ circuit.

In order to simplify the final expressions, we shall consider only the case of a critical bias, when the bias source voltage exceeds the critical value by a small amount δU. The case of large bias values will then be discussed qualitatively.

The voltage is taken to vary in the following manner. In the linear part of the current−voltage characteristic, where $U < U_{cr}$, the voltage across the diode and the circuit after applying the bias source will approach quasi-exponentially the source voltage. If $\delta U \ll U_{cr}$, the time derivative of the voltage and the capacitative current (which is proportional to that derivative) will be small at the critical voltage; the inductive current is given by

$$I_L^{(1)}\big|_{u=U_{cr}} = -I_0 - \frac{U_{cr}}{r_e}. \tag{5}$$

When the critical voltage is reached, the current through the diode decreases; since the voltage across it is stabilized by the capacitance of the circuit, the current will change by ΔI without a change in the voltage (AB in Fig. 1). The bias source current is stabilized by the inductance L, and the decrease of the diode current is compensated by the rise of the capacitative current. Accordingly, the circuit voltage increases above U_{cr}. An analytical expression for the time dependence of the circuit voltage may be obtained by solving the differential equation for the equivalent circuit shown in Fig. 2, with the circuit parameters given by Eqs. (2) and (4), which correspond to voltages above the critical value:

$$\frac{1}{L}\int U\,dt + \frac{U_{cr}+U}{R_e} + C\frac{d(U_{cr}+U)}{dt} + I_0 + \frac{U_{cr}}{r}\cdot\frac{R-r}{R} - \Delta I = 0. \tag{6}$$

Here, $U = u - U_{cr}$. The initial condition is given by Eq. (5). The solutions of Eq. (6) are easily found by means of the Laplace transformation, and have the form

$$U = \Delta I R_e f(t, R_e), \tag{7}$$

where

$$f(t, R_e) = \begin{cases} \dfrac{2\exp\left(-\dfrac{t}{2R_eC}\right)}{\sqrt{\dfrac{4R_e^2}{L}-1}} \sin\left[\dfrac{t}{2R_eC}\sqrt{\dfrac{4R_e^2}{L}-1}\right] & \text{for} \quad \dfrac{2|R_e|}{\sqrt{\dfrac{L}{C}}} > 1, \tag{8} \\[3em] \dfrac{t}{R_eC}\exp\left(-\dfrac{t}{2R_eC}\right) & \text{for} \quad \dfrac{2|R_e|}{\sqrt{\dfrac{L}{C}}} = 1, \tag{9} \\[3em] \dfrac{2\exp\left(-\dfrac{t}{2R_eC}\right)}{\sqrt{1-\dfrac{4R_e^2}{L}}} \operatorname{sh}\left[\dfrac{t}{2R_eC}\sqrt{1-\dfrac{4R_e^2}{L}}\right] & \text{for} \quad \dfrac{2|R_e|}{\sqrt{\dfrac{L}{C}}} < 1. \tag{10} \end{cases}$$

When $\dfrac{2|R_e|}{\sqrt{\dfrac{L}{C}}} \leq 1$, which corresponds to the solutions (9) and (10), periodic oscillations
do not occur in the circuit: if $R_e < 0$, the voltage increases aperiodically with time,* and if $R_e > 0$, it first increases and then tends aperiodically to the source voltage $U_{cr} + \delta U$. The solution Eq. (8), which corresponds to diode and circuit parameters such that $\dfrac{2|R_e|}{\sqrt{\dfrac{L}{C}}} > 1$, is
either an exponentially decreasing sinusoid (if $R_e > 0$) or an exponentially increasing sinusoid (if $R_e < 0$). At the instant

$$t_p = 2\pi |R_e| C \left(\frac{4R_e^2}{\frac{L}{C}} - 1 \right)^{-\frac{1}{2}} \tag{11}$$

the voltage $U = u - U_{cr}$ passes through zero, and the voltage u across the circuit passes through the critical value. The reverse of the above-mentioned jump takes place: the diode current suddenly increases by ΔI, the capacitative current decreases by ΔI, and the voltage remains unchanged. Because of the decrease in the capacitative current, the voltage across the circuit also decreases. An analytical expression for the time dependence of the circuit voltage can be derived in the same way as above by solving the differential equation for the equivalent circuit shown in Fig. 2, with the circuit parameters given by Eqs. (1) and (3), which correspond to subcritical voltages. The initial condition is found by substituting Eqs. (5), (7), (8), and (11) in the expression for the inductive current [the first term in Eq. (6)]. The form of the solutions is

$$U = -\Delta I r_e \left\{ 1 + \exp \left[-\pi \left(\frac{4R_e^2}{\frac{L}{C}} - 1 \right)^{-\frac{1}{2}} \right] \right\} f(t, r_e), \tag{12}$$

where $f(t, r_e)$ is the same function as in Eqs. (8)-(10) except that R_e is now replaced by r_e. The quantity r_e is always positive. All the solutions similar to Eqs. (8)-(10) for $f(t, r_e)$ correspond to periodic oscillations; in all three cases, the circuit voltage first decreases and then increases to a value not less than $U_{cr} + \delta U$. When it passes through U_{cr}, there is a transient process of the type (8) [possibly with an initial condition somewhat different from Eq. (5)], because of the discontinuity in the diode current. The processes of the form (7)-(8) and (12) are repeated periodically.

Using Eqs. (7) and (12), we can calculate the maximum positive and negative values of $U = u - U_{cr}$:

$$U_{max(+)} = \Delta I |R_e| F \left(\frac{2R_e}{\sqrt{\frac{L}{C}}} \right), \tag{13}$$

$$U_{max(-)} = -\Delta I r_e \left\{ 1 + \exp \left[-\pi \left(\frac{4R_e^2}{\frac{L}{C}} - 1 \right)^{-\frac{1}{2}} \right] \right\} F \left(\frac{2r_e}{\sqrt{\frac{L}{C}}} \right), \tag{14}$$

---

* Actually, periodic oscillations can occur even if $R_e < 0$. To discuss this case, we must consider the nonlinearity of the current-voltage characteristic in the range $u > U_{cr}$ (increase of differential resistance at $u = U_{lim}$; see Fig. 1).

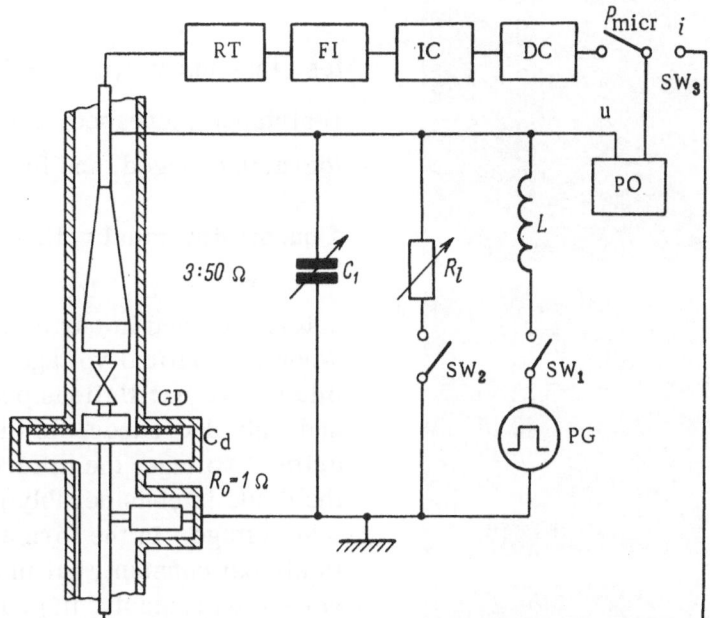

Fig. 3. Layout of apparatus used in measurements. GD) Gunn diode; RT) coaxial resistance transformer; FI) ferrite isolator; IC) isolating capacitor; DC) detector chamber; PO) pulse oscilloscope; C_d) disk capacitor (265 pF); i, u, P_{micr}) current, voltage, and detected microwave signal pulses; SW_1-SW_3) switches.

where

$$F(x) = \begin{cases} \dfrac{2\sin(\text{arc tg}\sqrt{x^2-1})}{\sqrt{x^2-1}} \exp\left[-\dfrac{\text{arc tg}\sqrt{x^2-1}}{\sqrt{x^2-1}}\right] & \text{for } |x| > 1, \\[2ex] \dfrac{2}{e} & \text{for } |x| = 1, \\[2ex] \dfrac{2\,\text{sh}(\text{arc th}\sqrt{1-x^2})}{\sqrt{1-x^2}} \exp\left[-\dfrac{\text{arc th}\sqrt{1-x^2}}{\sqrt{1-x^2}}\right] & \text{for } |x| < 1. \end{cases} \qquad (15)$$

As $|x|$ varies from zero to infinity, $F(x)$ decreases steadily from unity to zero.

Experimental

The apparatus shown in Fig. 3 was used in order to test the above theory. We also used another arrangement having the same low-frequency section but with the diode placed in a coaxial microwave resonator (no measurement was made of the current pulses). Similar results were obtained with both arrangements. Using the layout shown in Fig. 3, the Gunn diode GD under test had microwave load of about 3 Ω and an lf load consisting of the parallel $R_l LC$ circuit. At microwave frequencies the diode was grounded through a disk capacitor $C_d = 265$ pF. A current pulse i was taken from a resistor $R_0 = 1$ Ω. A pulsed voltage u and either a pulse of the detected microwave signal pulse $\overline{P}$ or the current pulse i were observed on the two-beam pulse oscilloscope PO. Figure 4 shows typical oscillograms of the voltage, current, and detected microwave signal obtained using this arrangement. Similar oscillograms were observed for all the diodes which had good or moderate microwave parameters (output power and efficiency). The oscillograms in Fig. 4a,b were observed in measurements on low-resistance

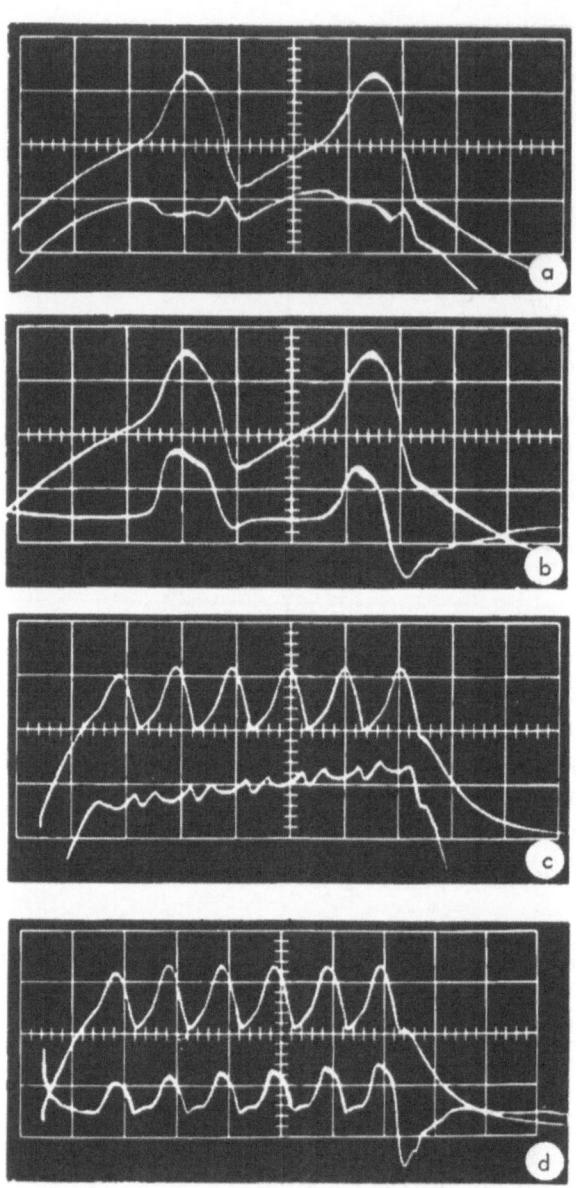

Fig. 4. Voltage, current, and detected microwave signal oscillograms. Low-resistance diodes (a, b) and high-resistance diodes (c, d). a, c) Peaks of voltage (upper trace) and current pulses; b, d) peak of voltage pulse and detected microwave signal pulse. a, b) Abscissa scale 0.2 μsec/cm, voltage scale 5 V/cm, current scale 0.5 A/cm; c, d) abscissa scale 0.5 μsec/cm, voltage scale 10 V/cm, current scale 0.2 A/cm.

diodes, for which the parameter $\dfrac{2r_e}{\sqrt{\frac{L}{C}}}$ was less than unity $\left(r_e = \dfrac{rR_l}{r + R_l}\right)$. As the diode resistance increased, the form of the oscillograms changed, and for $\dfrac{2r_e}{\sqrt{\frac{L}{C}}} > 1$, they had a quasi-sinusoidal voltage waveform (Fig. 4c).

The form of the current−voltage characteristic used in our calculation is realistic. When the critical voltage is exceeded (and microwave oscillations occur), the current suddenly decreases. In the range above the critical voltage, the differential resistance of the diode is considerably greater than below that voltage. In the former range, the current is almost constant, but in the latter range it varies appreciably, in proportion to the change in voltage.

The observed oscillograms are also in agreement with the theoretical results. For voltages above critical, Eqs. (7) and (8) show that the time dependence of the voltage is a quasi-sinusoidal half-period wave. For voltages below critical, depending on whether the parameter $\dfrac{2r_e}{\sqrt{\frac{L}{C}}}$, is greater or less than unity, the voltage waveform either is quasi-sinusoidal [as in Eqs. (12), (8), and Fig. 4c] or has a sharp drop followed by a slower quasi-exponential rise [as in Eqs. (12), (9), (10), and Fig. 4a].

According to Eqs. (7) and (12), when r_e and R_e increase above the values corresponding to $\dfrac{2r_e}{\sqrt{\frac{L}{C}}} = 1$ and $\dfrac{2|R_e|}{\sqrt{\frac{L}{C}}} = 1$ the oscillations tend to a sinusoidal form with period $T = 2\pi\sqrt{\dfrac{L}{C}}$. With diodes having sufficiently high resistance, sinusoidal voltage self-modulation is observed, with a period equal to $2\pi\sqrt{\dfrac{L}{C}}$ within the experimental error ($\approx 10\%$); the magnitude of the period does not depend appreciably on the diode bias.

According to the calculations, self-modulation can take place only when $\dfrac{2|R_e|}{\sqrt{\frac{L}{C}}} > 1$. The validity of this is confirmed by measurements to within the accuracy of determination of R_e (≈ 20-30%).

Again, within the accuracy of determination of R_e, agreement is found between the amplitudes of the low-frequency oscillations and their theoretical values, Eqs. (13) and (14); agreement extends to the region where the diode voltage becomes equal to U_{lim} (Fig. 1), which limits the amplitude of the lf oscillations. The voltage U_{lim} at which the diode current began to increase was marked by a blip in the current pulse at the maximum diode voltage.

Deviations from the above time dependences of the diode voltage and current are found for diodes having high resistance in a weak field and a small discontinuity of current ΔI when loaded at low frequencies with a parallel circuit having a large resistance R_l. Damped oscillations at the circuit resonance frequency are observed, apparently due to decreased losses in the circuit because of the greater differential resistance of the diode at the critical voltage. Oscillations, like those of a negative resistance with amplitude limitation at the critical voltage or at U_{lim} (Fig. 1), are also observed. The oscillations are sinusoidal, and the amplitude (up to the limit point) is close to the difference of constant-bias and critical voltages.

The following observations were made when the diode bias was increased above the critical value: (1) the width of the region where the voltage is below critical decreases in the oscillograms shown in Fig. 4a,b; (2) above a certain bias, there is a decrease of the amplitude, and subsequently a complete disappearance of self-modulation. The range of bias in which self-modulation is observed decreases with decreasing circuit load resistance R_l and diode current jump ΔI at the critical voltage. Both these relationships are explained by our model. The calculation given above shows that, for diodes having a parameter $\dfrac{2r_e}{\sqrt{\frac{L}{C}}} < 1$, the voltage in the

range $u < U_{cr}$ drops rapidly and then approaches the bias value in a quasi-exponential manner. The critical voltage will evidently be reached sooner when it differs considerably from the bias. Moreover, in the case of self-modulation, the voltage waveform above the critical value is sinusoidal with an exponential coefficient. Evidently, the self-modulation oscillations are possible only if the minimum of the sinusoid lies below the level of the critical voltage. The amplitude of this minimum is, according to Eq. (7), proportional to R_e and ΔI. Accordingly, the range of bias values in which self-modulation is possible must become narrower with decreasing circuit load resistance R_e and diode current jump ΔI.

Conclusions

1. An explanation is proposed for the generation by a Gunn diode of modulated microwave oscillations as a consequence of a discontinuous decrease of the mean diode current when microwave generation occurs, and of the interaction of the diode with a low-frequency oscillatory circuit. It is shown that the depth of modulation is always 100%, and in accordance with the accepted terminology, the phenomenon is called self-modulation.

2. In the case of self-modulation there is an ac voltage across the diode, which modulates the microwave signal. The amplitude, waveform, and frequency of the self-modulation voltage are calculated as functions of the diode and circuit parameters. The waveform of the voltage across the diode at a bias above the critical value (where modulation of the microwave signal occurs) is sinusoidal with an exponential coefficient; at a bias below the critical value, it is either again sinusoidal with an exponential coefficient if $\dfrac{2r_e}{\sqrt{\frac{L}{C}}} > 1$, or a fall is followed by a

quasi-exponential rise if $\dfrac{2r_e}{\sqrt{\frac{L}{C}}} \leqslant 1$. The oscillation period in the quasi-sinusoidal modulation

case is independent of the bias, and is approximately $2\pi\sqrt{\frac{L}{C}}$. If, however, $\dfrac{2r_e}{\sqrt{\frac{L}{C}}} < 1$, the period decreases with increasing bias. Relative to the applied bias, the voltage oscillations are asymmetric not only in shape but also in amplitude, the amplitudes of oscillation towards higher voltages being

$$\Delta I \, |R_e| \, F \frac{2R_e}{\sqrt{\dfrac{L}{C}}}$$

and toward lower voltages being

$$-\Delta I r_e \left\{ 1 + \exp\left[-\frac{\pi}{\sqrt{\left[\dfrac{4R_e^2}{\left(\dfrac{L}{C}\right)} - 1\right]}} \right] \right\} F\left(\frac{2r_e}{\sqrt{\dfrac{L}{C}}} \right),$$

where $0 < F(x) < 1$ and $F(x)$ is given by Eq. (15). Self-modulation is possible in a certain restricted range of bias values. This range becomes wider with increasing current discontinuity ΔI in the diode and increasing circuit load resistance R_l. For diodes having good or moderate microwave parameters (output power and efficiency), all these relations are experimentally confirmed.

3. The phenomenon of self-modulation can be used to obtain a modulated microwave signal from a Gunn diode. The modulation frequency is variable from zero to a value close to the microwave oscillation frequency, and the duty factor ranges from $\lessgtr 0.1$ to $\gtrless 0.5$. The frequency is varied by tuning the circuit and partly by changing the bias. The duty factor can be varied by changing the bias. Under self-modulation conditions, the diode voltage may be several times the applied bias. Hence, self-modulation can give higher output powers and efficiencies (this is especially true of the peak power and efficiency) than the same bias but without self-modulation. The latter has been experimentally confirmed in [2]. To avoid diode breakdown under self-modulation conditions, the value of R_e must be restricted.

Literature Cited

1.   S. V. Jaskolski and T. K. Ishii, Electronics Letters, 3:12 (1967).
2.   P. L. Fleming, Proc. IEEE, 54:799 (1966).
3.   J. E. Carroll, Radio and Electronic Engineer, 34:17 (1967).
4.   H. Hayakawa, M. Kikuchi, and Y. Abe, Japan. J. Appl. Phys., 5:734 (1966).
5.   M. Kikuchi, H. Hayakawa, and Y. Abe, Japan. J. Appl. Phys., 5:735 (1966).
6.   J. B. Gunn, IBM J. Res. Develop., 10:310 (1966).

THRESHOLD ENERGY FOR ELECTRON—HOLE
PAIR PRODUCTION BY HOT ELECTRONS IN GaAs

A. A. Gutkin, É. M. Magerramov,
D. N. Nasledov, and V. E. Sedov

A. F. Ioffe Physicotechnical Institute
Academy of Sciences of the USSR, Leningrad

Electron multiplication in GaAs p—n junctions with various impurity concentration gradients was measured
in the range of low multiplication factors, $M \leq 1.1$. The p—n junctions were formed by diffusion of zinc
into n-type material. The method of Chynoweth and McKay was applied to the experimental results to
obtain the threshold energy for impact ionization by electrons, $\varepsilon_{in} = 1.8 \pm 0.1$ eV, and the mean free path
of hot electrons ($\lambda = 50$ Å) in the case of scattering by optical phonons. By considering zero-phonon electron transitions accompanying impact ionization, it is shown that the experimental value of ε_{in} corresponds to that expected from the approximate band structure of GaAs.

Introduction

The threshold energy ε_i for electron—hole pair production is an important parameter
in the theory of avalanche multiplication in semiconductors. It has been determined experimentally for GaAs [1, 2] and for various other semiconductors [2-4] by comparing the dependence
of the rate of impact ionization on the electric field with Baraff's theoretical curves [5]. The
ionization rates were calculated from the measured multiplication of carriers in p—n junctions
with a linear or abrupt distribution of impurity concentration. This determination of ε_i may,
however, be unsatisfactory, for a number of reasons. Firstly, in [5], the carrier energy dependence of the ratio (r) of the impact ionization cross section to the total scattering cross
section was approximated by a step function: $r = 0$ for $\varepsilon < \varepsilon_i$, and $r = 0.5$ for $\varepsilon \geq \varepsilon_i$. The
energy ε_i thus found may therefore differ from the threshold ionization energy given by the band
structure of the material. Secondly, the theoretical curves are made to coincide with the experimental points by the choice of two parameters: the threshold ionization energy ε_i and the
mean free path λ for scattering by optical phonons. A slight change of λ may give agreement
with the theoretical curve (within the experimental error) for a fairly wide range of values of
ε_i. This causes an uncertainty in the value of ε_i exceeding that due to the first factor mentioned
above. For example, it has been found in [2, 3] that the threshold energies for electrons and

holes are indistinguishable in Si, Ge, and GaAs, and lie in the ranges $\varepsilon_g \leq \varepsilon_i \leq 1.5\,\varepsilon_g$ for Si and $\varepsilon_i = (1.5 \pm 0.5)\,\varepsilon_g$ for Ge and GaAs.

However, in order to relate this parameter to the band structure and decide the role of phonons in an impact ionization event during avalanche multiplication, the quantity ε_i must be more accurately determined, and the method of doing so must yield a value that corresponds to the onset of avalanche multiplication.

In [6], a method has been proposed for determining the threshold ionization directly from the curves giving the dependence of the multiplication factor on the voltage across the p−n junction. The values thus obtained for silicon in [6] and [7] are, however, much greater than those derived in [3] by the use of Baraff's theory [5], and there is considerable discrepancy between the results of [6] and [7]. These can be reconciled [8] taking account of the role of impact ionization by holes in [6], but the disagreement with [3] remains unexplained.

In other semiconductors, the threshold ionization energy has been determined only by comparison with Baraff's theory [5].

In the present paper, the threshold energy for impact ionization by electrons in GaAs is determined by the method proposed in [6] and the influence of impact ionization by holes is eliminated. The result is compared with the value predicted from the structure of the energy bands in k space.

Experimental Technique

The p−n junctions used in the measurements were prepared by diffusion of zinc into n-type GaAs having a carrier density of $3 \cdot 10^{16}$-$5 \cdot 10^{17}$ cm^{-3} at room temperature and a mobility of 3000-4000 cm^2/V · sec. The diffusion temperatures were 680 and 800°C. The diffusion times ranged from 3 to 15 min. After diffusion, the samples were etched. The final thickness of the p-type layer was at least $1.5\,\mu$. The dependence of the differential capacitance on the voltage, for all the p−n junctions, was of the form $C \propto (U_c - U)^{-1/n}$ with $2 < n < 3$.

As in [1-4, 6, 7], the photocurrent at various voltages was measured in order to study the carrier multiplication. The sample was illuminated on the p-type side with light from an incandescent lamp through water and SZS-8 glass filters. These filters transmitted light having photon energy $\hbar\omega \gtrsim 2$ eV, which (according to [9]) was entirely absorbed in the p-type layer, so that only electrons were injected into the p−n junction; they were then accelerated in the space-charge layer and became capable of ionization. The filters also prevented heating of the sample (which was placed in a current of air to improve the heat removal). In order to exclude the effect of field inhomogeneities and resulting microplasmas, the multiplication factors M were measured only for $M \leq 1.1$, although it is usually considered possible to make measurements of this type up to $M = 2$ [2]. Since the electron and hole ionization rates in GaAs are approximately equal [2], the low multiplication factors in our case allow impact ionization by holes to be neglected.

The photocurrent I_p was determined by measuring the voltage drop across a load resistance (6-150 Ω) connected in series in the p−n junction circuit. The measurements were made using dc and a PPTN-1 potentiometer. The voltage drop caused by the dark current was subtracted from the total voltage drop across the load during illumination. The dark current of some samples varied with time at constant voltage. The reason for this was not known. Such samples were not used subsequently. The change in carrier collection by the p−n junction caused by broadening of the space-charge layer with increasing voltage was, as usual [7], taken to be proportional to the change of width W of the space-charge layer, as a first approximation. In order to take account of this in determining the multiplication factor at a voltage U,

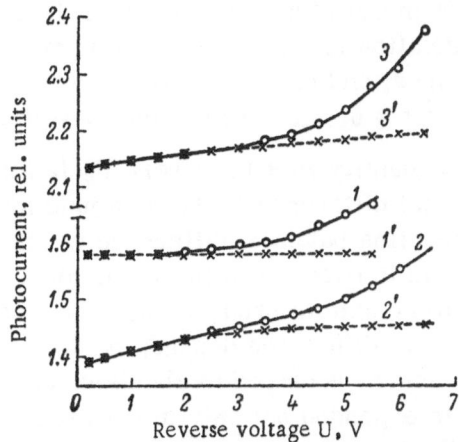

Fig. 1. Dependence of photocurrent on reverse voltage. Numbers along-side the curves are sample numbers. Dashed curves show the behavior of $I_p(0) + A[W(U) - W(0)]$.

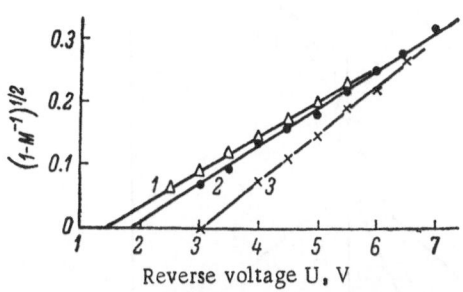

Fig. 2. Dependence of the function $(1 - M^{-1})^{\frac{1}{2}}$ on the reverse voltage. Numbers alongside the curves are sample numbers.

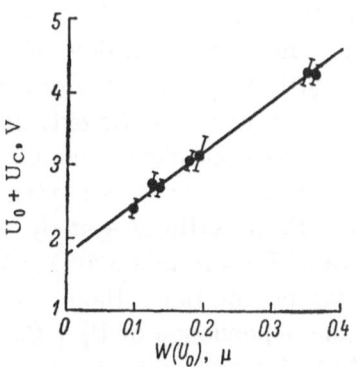

Fig. 3. Dependence of potential difference, to be experienced by an electron in order to cause multiplication, on the width of the space-charge layer.

the photocurrent corresponding to zero carrier multiplication was taken to be the sum of the photocurrent for $U = 0$ and a term proportional to the change of width of the space-charge layer, $W(U) - W(0)$. The multiplication factor was defined as

$$M(U) = \frac{I_p(U)}{I_p(0) + A[W(U) - W(0)]}.$$

The proportionality coefficient A was found by assuming that, at low voltages, the change in the photocurrent was entirely due to the change in the quantum efficiency.

The width of the space-charge layer was found from measurements of the differential capacitance of the p−n junctions [10].

All measurements were made at room temperature.

Results

Figure 1 shows the experimental dependence of the photocurrent on the reverse voltage across a p−n junction in three samples, and also, for each sample, the voltage dependence of the photocurrent in the absence of carrier multiplication, $I_p(0) + A[W(U) - W(0)]$. Initially, the curves of $I_p(U)$ and $I_p(0) + A[W(U) - W(0)]$ coincide for each sample, over a range of U greater than that used for the determination of A. This confirms the supposition that the initial change of the photocurrent is governed by the change of width of the space-charge layer and is proportional to that change.

In order to determine the threshold ionization energy ε_i, we have to find the maximum external voltage U_0 across the p−n junction for which the multiplication factor $M = 1$. To do this, it is convenient to use the function $(1 - M^{-1})^{\frac{1}{2}} = f(U)$, which is linear for small M ac-

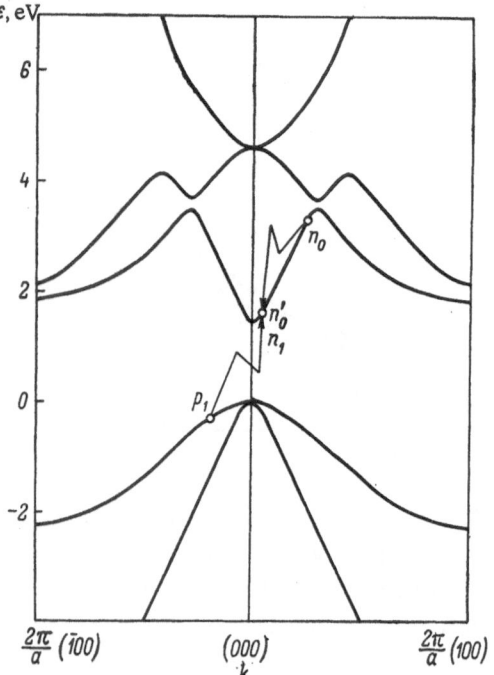

Fig. 4. Diagram of electron transitions for impact ionization in the band structure of gallium arsenide. The primary electron energy is close to the threshold value.

cording to [6]. Figure 2 shows such functions derived from the results in Fig. 1. It is seen that the relation is closely approximated by straight lines, and can therefore be extrapolated to $(1 - M^{-1})^{\frac{1}{2}} = 0$, which thus yields the voltage U_0.

The quantity $U_0 + U_c$ (where U_c is the contact potential difference in the p−n junction) is the accelerating potential difference which an electron has to experience in the distance $W(U_0)$ in order to be able to produce secondary pairs. This potential difference depends on $W(U_0)$, since the energy acquired by the electron in this distance is expended not only in ionization but also in other scattering processes. This leads to the difference between the values of U_0 for the various samples represented in Fig. 2. The energy losses in other scattering processes may be neglected if the path traversed by the electrons is small compared with the mean free path for these scattering mechanisms. The quantity $q(U_0 + U_c)$ for this case gives the threshold ionization energy ε_i. In a practical determination, the dependence of $q(U_0 + U_c)$ on the width $W(U_0)$ of the space-charge layer must be extrapolated to $W(U_0) = 0$ [6]. This dependence can be plotted from measurements of samples having various impurity concentration gradients in the space-charge layer. The results of a study of seven samples are shown in Fig. 3. For all the samples, the value of U_c was taken as 1.3 V, in accordance with the relations between the p−n junction capacitance and the voltage, found in [10]. It is seen from Fig. 3 that, within the experimental error, the dependence of $U_0 + U_c$ on $W(U_0)$ is closely approximated by a straight line. Extrapolation to $W(U_0) = 0$ gives, in accordance with the experimental conditions, a threshold energy $\varepsilon_i = 1.8 \pm 0.1$ eV for impact ionization by electrons in GaAs.

The energy lost by the electron per unit path is given by the slope of the line $U_0 + U_c = f[W(U_0)]$ as $7.1 \cdot 10^4$ eV/cm. If we assume, as in recent theories of avalanche multiplication [5, 11, 12], that the loss is due to the emission of optical phonons, whose energy in GaAs is 0.036 eV [13], then the mean free path of hot electrons for scattering by optical phonons is 50 Å.

Discussion

Our values of the parameters of avalanche multiplication of electrons in GaAs ($\varepsilon_i = 1.8 \pm 0.1$ eV, $\lambda = 50$ Å) are consistent with the results [2]* of a comparison between the experimental dependence of the ionization rate on the electric field and Baraff's theory: $\varepsilon_i = (1.5 \pm 0.5)\varepsilon_g$, $\lambda = 42-51$ Å.

We may compare the experimental value of ε_{in} with that predicted from the band structure of GaAs, assuming, as in [14-17], that the electron transitions accompanying impact ioniza-

---

* The value $\varepsilon_i = 1.7 \pm 0.3$ eV previously obtained [1] by comparison with Baraff's theory is shown in [2] not to be attributed to electrons alone, since the geometrical structure of the p−n junctions studied did not allow a distinction between the ionization properties of electrons and holes.

tion do not involve phonons. Then, the minimum energy at which a carrier can participate in impact ionization is determined by the laws of conservation of energy and crystal momentum. For impact ionization by electrons, these laws are

$$\varepsilon_{n0}(\mathbf{k}_0) - \varepsilon'_{n0}(\mathbf{k}'_0) = \varepsilon_g + \varepsilon_{n1}(\mathbf{k}_{n1}) + \varepsilon_{p1}(\mathbf{k}_{p1}),$$
$$\mathbf{k}_0 - \mathbf{k}'_0 = \mathbf{k}_{n1} - \mathbf{k}_{p1}.$$

Here, ε_g is the forbidden band width; ε_{n0} and $\mathbf{k}_0$ the energy and wave vector of the primary electron before ionization; ε_{n1}, $\mathbf{k}_{n1}$, ε_{p1}, $\mathbf{k}_{p1}$ the energies and wave vectors of the secondary electron and hole; ε'_{n0} and $\mathbf{k}'_0$ the energy and wave vector of the primary electron after ionization. Since the analytical form of the dependence of the energy on the crystal momentum at high carrier energies is unknown, the threshold energy for impact ionization must be determined graphically, using the results of numerical calculations of the band structure [18] in the form of graphs of $\varepsilon(\mathbf{k})$. The energy thus found is about 1.8 eV, in good agreement with our experimental value from the multiplication curves. Figure 4 shows the position of the primary electron and the transitions in the Brillouin zone for one such impact ionization event. Impact ionization with approximately the same primary electron energy is also possible for electrons whose crystal momentum is directed along the $\langle 110 \rangle$ direction. It should also be noted that approximately the same value of the threshold energy is obtained by a similar treatment of electron transitions in the case of impact ionization by holes belonging to a valence band split off by the spin-orbit interaction.

Literature Cited

1. R. A. Logan, A. G. Chynoweth, and B. G. Cohen, Phys. Rev., 128:2518 (1962).
2. R. A. Logan and S. M. Sze, Proc. Eighth Intern. Conf. on Physics of Semiconductors, Kyoto, 1966, in: J. Phys. Soc. Japan, 21:Supplement, 434 (1966).
3. C. A. Lee, R. A. Logan, R. L. Batdorf, J. J. Kleimack, and W. Wiegmann, Phys. Rev., 134:A761 (1964).
4. R. A. Logan and H. G. White, J. Appl. Phys., 36:3945 (1965).
5. G. A. Baraff, Phys. Rev., 128:2507 (1962).
6. A. G. Chynoweth and K. G. McKay, Phys. Rev., 108:29 (1957).
7. J. L. Moll and R. van Overstraeten, Solid State Electron., 6:147 (1963).
8. T. Ya. Puritis, Fiz. Tekh. Poluprov., 1:599 (1967).
9. M. D. Sturge, Phys. Rev., 127:768 (1962).
10. L. W. Aukerman, D. F. Kyser, and M. F. Millea, Solid State Electron., 8:119 (1965).
11. P. A. Wolff, Phys. Rev., 95:1415 (1954).
12. W. Shockley, Solid State Electron., 2:35 (1961).
13. W. Cochran, S. J. Fray, F. A. Johnson, J. E. Quarrington, and N. Williams, J. Appl. Phys., 32:2102 (1961).
14. A. R. Beattie, J. Phys. Chem. Solids, 23:1049 (1962).
15. R. J. Hodgkinson, Proc. Phys. Soc. (London), 82:1010 (1963).
16. A. A. Gutkin, É. M. Magerramov, D. N. Nasledov, and V. E. Sedov, Fiz. Tverd. Tela, 8:712 (1966).
17. J. R. Hauser, J. Appl. Phys., 37:507 (1966).
18. W. Saslow, T. K. Bergstresser, C. Y. Fong, M. L. Cohen, and D. Brust, Solid State Commun., 5:667 (1967).

ELECTRICAL FLUCTUATIONS
IN SILICON CARBIDE JUNCTIONS

Yu. S. Karpov, V. S. Galushko, and V. Mertins

Novgorod Branch
V. I. Ul'yanov-Lenin Leningrad Institute of Electrical Engineering

The parameters describing electrical fluctuations in junctions were taken to be the relative spectral density of nominal noise power (noise coefficient) and the spectral density of the short-circuit noise current. Measurements of the spectral density of fluctuations were made in the audiofrequency range by comparison with a noise generator. An increase of the forward voltage was accompanied by a nonmonotonic variation of the noise current and noise coefficient of a diode. The minimum noise intensity occurred at forward voltages between 1.9 and 3 V. This effect could be explained qualitatively by a decrease of the differential resistance when the voltage increased. When a junction was reverse-biased, the spectral density of the noise current varied very rapidly with the reverse current. This behavior was typical of the avalanche breakdown region of the current—voltage characteristic. The observed current—voltage characteristic was typical of surface breakdown. The occurrence of low-frequency noise in surface breakdown of silicon carbide junctions was in agreement with corresponding results for germanium diodes. When the reverse voltage was increased, the noise coefficient rose much more slowly than the spectral density of the noise current; this was due to the decrease of the differential resistance of the junction when the reverse voltage was increased. A study of the frequency dependence of the noise coefficient showed that, in the forward-bias case, there was either a typical low-frequency junction noise or a generation—recombination noise of the spreading resistance. In the reverse connection, some samples gave typical low-frequency fluctuations. In the majority of diodes, a generation—recombination noise component appeared against the low-frequency noise background. With both forward and reverse connection, the time constant of the generation—recombination noise was about 10^{-5} sec. The nature of the fluctuations having this time constant is not yet known.

The parameters describing electrical fluctuations in junctions are taken to be the relative spectral density of nominal noise power (noise coefficient) γ and the spectral density of the short-circuit noise current $\overline{i^2}/\Delta f$. Measurements of the spectral density of fluctuations were made by comparison with a noise generator. The experimental error did not exceed 15-20%.

Forward Connection. When the external voltage was increased, the short-circuit noise current and the noise coefficient of the diode varied nonmonotonically. Figure 1a shows typical curves. The minimum values of the noise current and noise coefficient occurred at

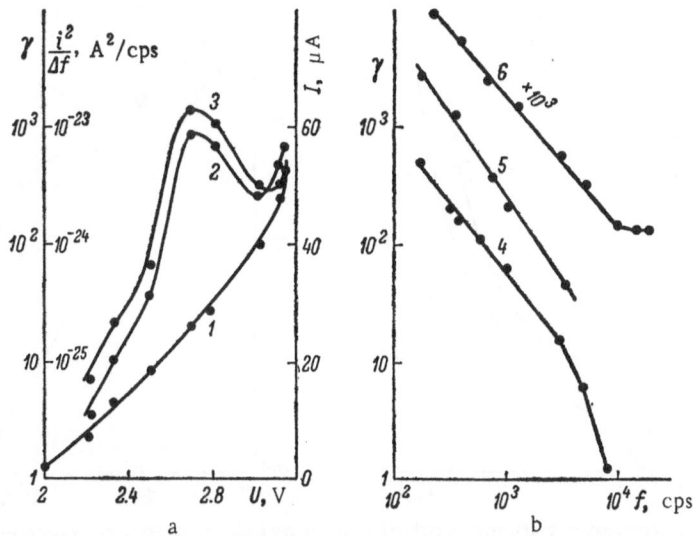

Fig. 1. Current−voltage and noise characteristics of forward-biased silicon carbide junctions. 1) Current−voltage characteristic; 2) mean square of noise current; 3) noise coefficient of one diode; 4-6) frequency dependences of noise coefficient for various diodes.

forward voltages between 1.9 and 3 V in different samples. This effect is similar to that found with germanium diodes [1], and can be qualitatively explained by a decrease of the differential resistance of the junction as the voltage increases.

A study of the frequency dependence of the noise coefficient gave two types of relation (Fig. 1b). Curves 5 and 6 show that, in these samples, typical low-frequency fluctuations predominated at frequencies below 10^4 cps (noise coefficient proportional to $f^{-\alpha}$). The values of α ranged from 1.1 to 1.4 in different samples. Curve 6 shows that the noise coefficient was independent of frequency above 10^4 cps. A comparison of the observed value of γ with the shot noise of the junction indicates that the latter gives values three orders of magnitude less than those observed. In this case, the generation−recombination noise of the spreading resistance appears to be predominant; it may become considerable in silicon carbide diodes, because of the incomplete ionization of impurity levels [2]. For sample 4, $\alpha \to 3$ with increasing frequency. This can be explained either by the effect of the junction capacitance, or by the presence of generation−recombination noise with a time constant of the order of 10^{-5} sec. An estimate shows that the capacitance effect becomes appreciable for capacitances above 0.2 μF. This value seems improbably high. The nature of generation−recombination noise having the time constant mentioned is not yet known.

<u>Reverse Connection</u>. Figure 2a shows typical dependences of the noise characteristics on the reverse voltage. The curve of the spectral density of the noise current resembles the current−voltage curve. We may therefore suppose that

$$\frac{\overline{i^2}}{\Delta f} = \exp (AI). \tag{1}$$

Here, A is a constant and I the reverse current of the diode. This rapid variation of the noise current spectral density with the reverse current was typical of the avalanche breakdown region of the current−voltage characteristic [3]. The observed current−voltage characteristic (Fig. 2a) was typical of surface breakdown. Thus, the occurrence of the low-frequency noise in surface breakdown of silicon carbide junctions was in agreement with corresponding results for

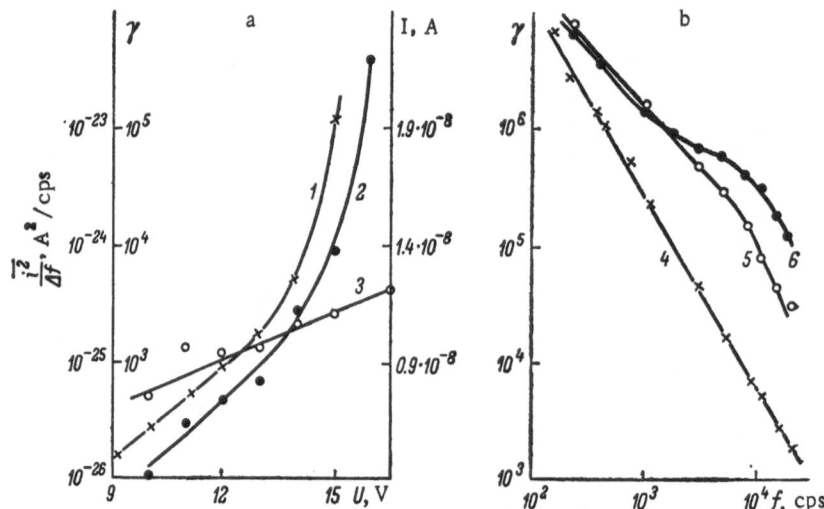

Fig. 2. Current—voltage and noise characteristics of reverse-
biased silicon carbide junctions. 1) Current—voltage characteristic; 2) mean square of noise current; 3) noise coefficient of one
diode; 4-6) frequency dependences of noise coefficient for various
diodes.

germanium diodes. When the reverse voltage was increased, the noise coefficient (and therefore the nominal noise power) rose much more slowly than the spectral density of the noise
current; this was due to the decrease of the differential resistance of the blocked junction when
the reverse voltage increased.

A study of the frequency dependence of the noise coefficient showed that, for some samples,
typical low-frequency fluctuations occurred, with α ranging from 1.1 to 1.6 (Fig. 2b, curve 4).
For most diodes, α increased from about unity to 3 with increasing frequency (Fig. 2b, curves
5 and 6). This indicated the same generation—recombination noise component as in the forward
connection.

A more detailed interpretation of the results will be possible following an investigation of
equivalent circuits simulating silicon carbide diodes.

We thank G. F. Kholuyanov and G. N. Violina for providing samples of junctions and for
discussing the results.

Literature Cited

1.   W. H. Fonger, Transistors, Vol. 1, RCA Laboratories, Princeton (1956), p. 239.
2.   A. van der Ziel, Fluctuation Phenomena in Semiconductors, Butterworths, London (1959).
3.   A. S. Tager, Fiz. Tverd. Tela, 6:2418 (1964).

STRAIN EFFECT IN POLYCRYSTALLINE FILMS
OF INDIUM ANTIMONIDE AND GALLIUM ANTIMONIDE

I. I. Fal'ko

Information is given on the technique of fabricating semiconducting films of indium antimonide and gallium antimonide having p-type and n-type conduction. Studies of the strain effect and of the temperature dependence of the resistance of prototype strain gauges are reported.

Miniature thin-film strain gauges, formed by vapor deposition in vacuum, have recently come into use in investigations of the strength of materials and in dynamometry. Soviet authors have described gauges based on thin films of bismuth and germanium, with strain sensitivities of the order of 20 and 30-70 respectively [1, 2].

In the measurement of very small strains ($\sim 10^{-5}$ to 10^{-6}), the use of semiconductor gauges with higher strain sensitivities is very attractive. It is known [3] that single crystals of certain semiconductors (InSb, GaSb, CdS, and others) have higher strain sensitivities than any other semiconductors so far studied, and they were therefore chosen for investigation.

InSb and GaSb films were prepared by thermal evaporation. They were vapor-deposited by a batch method in a vacuum of 10^{-4}-10^{-5} torr. In this method finely powdered material was supplied continuously to the evaporator, using a special feed device described in [4].

The substrate was type A mica, 0.01-0.03 mm thick, of the grade used in television sets. The evaporator was heated to 1100-1200°C. InSb and GaSb were deposited on substrates kept at temperatures of 300-470°C and 200-680°C, respectively. The n-type InSb films were made from n-type InSb powder; their electron mobility was 3000 $cm^2 \cdot V^{-1} \cdot sec^{-1}$ and the electron density was 10^{16} cm^{-3} at 300°K. To obtain p-type films, 2% copper was added to the original powder.

To prepare p-type GaSb films, n-type and p-type powders were used. When 1 to 4.5% by weight of Ga_2Te_3 was added to these, n-type GaSb films were obtained.

The resistivity, the strain sensitivity, and the temperature dependence of the resistivity were measured for the films obtained.

The resistivity measurements showed that there is a considerable dependence on the substrate temperature. For example, n-type GaSb films were blue and had a resistivity exceeding 600 $\Omega \cdot$ cm when condensed on a substrate kept at less than 550°C, whereas at higher temper-

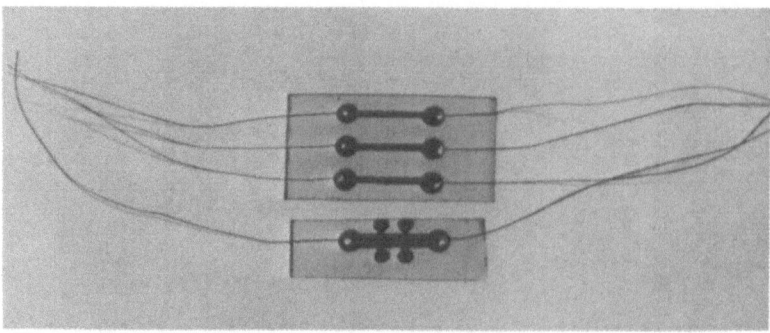

Fig. 1. Photograph of prototype gauges.

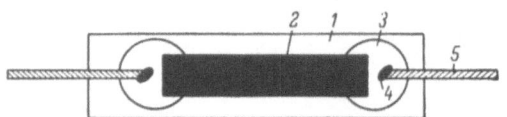

Fig. 2. Diagram of a prototype gauge.

atures the condensate formed a gray polycrystalline film having a resistivity 1-7 $\Omega \cdot$ cm. The low-resistivity gray samples were used for studying the strain effect in GaSb films.

The evaporation apparatus used was such as to give from 1 to 7 samples in each run. All samples were used to make strain gauges; some were used for the Hall effect studies (mobility and carrier density).

Figure 1 is a photograph and Fig. 2 a diagram of a prototype gauge. A substrate 1 carries a sensitive element 2, dimensions $1.5 \times 1.5 \times 0.004$ mm. Silver electrodes 3, also formed by vapor deposition, partly overlap the sensitive element. Copper leads 5 are attached to the silver electrodes with Wood's metal 4. The gauge is covered with thin capacitor paper or a layer of VL-5 lacquer or BF-2 cement. The gauge is attached to the part under test by means of 192-T cement with finely ground mica filler. After attachment, the gauge is kept at 60-70°C for 6-8 h.

The gauges were given a strain of the order of $0.2 \cdot 10^{-3}$ by stepwise loading and unloading.

As already mentioned, samples were prepared at various substrate temperatures. It was found that the strain sensitivity of p-type samples was not greatly affected by the substrate temperature. The values were from 4 to 16 for p-type InSb and 15 to 25 for p-type GaSb.

The strain sensitivity of n-type samples depended on the substrate temperature. At 300-400°C, InSb samples having a strain sensitivity of 10-26 were obtained; at 400-450°C, the range was 33-56.

For n-type GaSb samples, the largest strain sensitivity (71-74) occurred for substrate temperatures of 560-600°C. It was found that the strain sensitivity was positive for the p-type materials and negative for the n-type materials.

Figures 3-6 show the change in the resistance as a function of the strain, $\Delta R / R = f(\varepsilon)$, for gauges made from these materials. These graphs give tension and compression curves obtained by stepwise loading and unloading.

Figure 4 shows the dependence of the change in the resistance on the strain for a gauge whose sensitive element consisted of an n-type InSb film obtained at a substrate temperature of 450°C, and Fig. 6 for a gauge whose sensitive element was an n-type GaSb film obtained at a substrate temperature of 590°C. These graphs show that the relative change in the resistance varies almost linearly with the strain in the range investigated, for all types of gauge. The

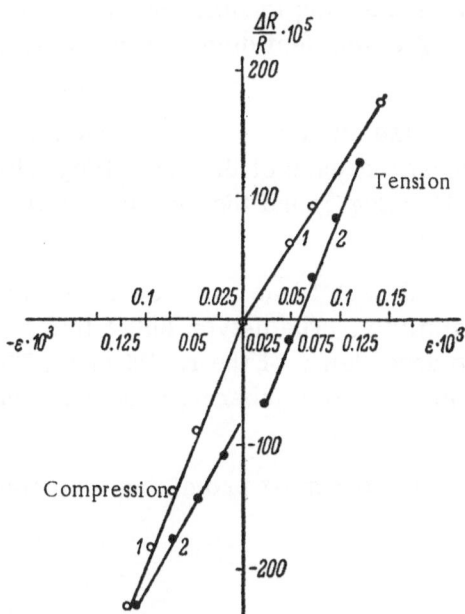

Fig. 3. Dependence of the change in the resistance on the strain, $\Delta R/R = f(\varepsilon)$, for p-type indium antimonide gauges. 1) Loading; 2) unloading.

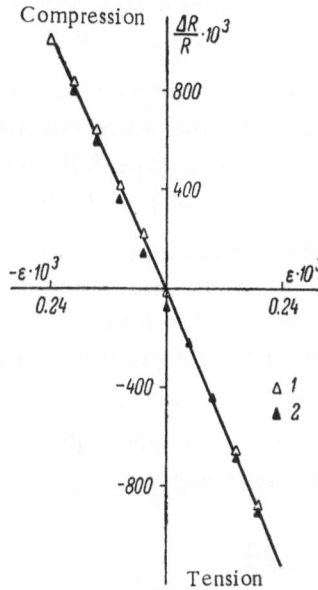

Fig. 4. Dependence of the change in the resistance on the strain, $\Delta R/R = f(\varepsilon)$, for n-type indium antimonide gauges. 1) Loading; 2) unloading.

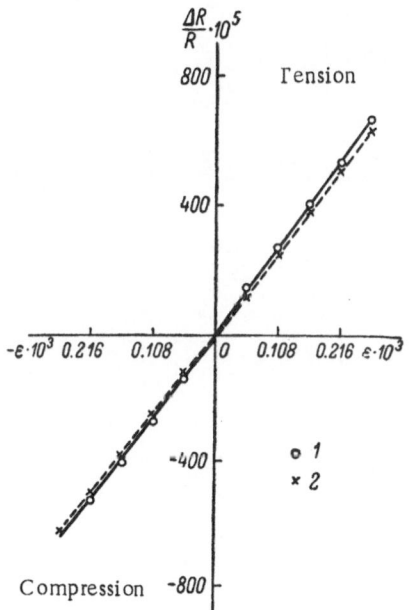

Fig. 5. Dependence of the change in the resistance on the strain, $\Delta R/R = f(\varepsilon)$, for p-type gallium antimonide gauges. 1) Loading; 2) unloading.

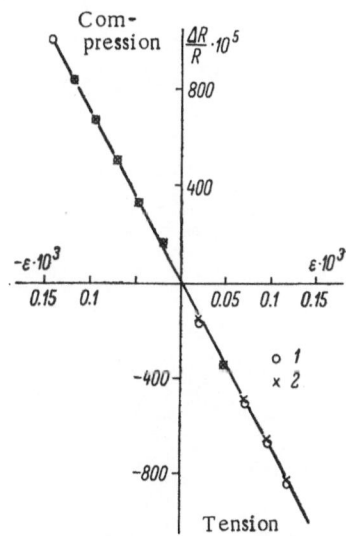

Fig. 6. Dependence of the change in the resistance on the strain, $\Delta R/R = f(\varepsilon)$, for n-type gallium antimonide gauges. 1) Loading; 2) unloading.

strain sensitivity values are found to differ for tension and compression. Some types of gauge show a considerable divergence between the curves obtained during loading and unloading; this is seen most clearly in the p-type InSb gauges (Fig. 3). The smallest divergence occurs for the low-resistance n-type GaSb gauges (Fig. 6).

The temperature dependence of the resistivity was also studied. All the gauges were found to have a large temperature coefficient. The largest increase of the resistivity with temperature occurred in the n-type InSb gauges, $(10-30) \cdot 10^{-3}$ deg^{-1}, and the smallest in the n-type GaSb gauges, $(0.02-0.06) \cdot 10^{-3}$ deg^{-1}.

The results obtained indicate that the p-type InSb and GaSb films have low strain sensitivity and a strong temperature dependence of the resistivity. The n-type InSb films have a higher strain sensitivity and a very strong temperature dependence of the resistivity. The n-type GaSb films show the greatest strain effect and only a weak temperature dependence of the resistivity.

From this, we can conclude that n-type GaSb films are the most promising materials for the production of strain gauges.

Literature Cited

1. G. N. Guk, Bismuth Film Resistance Strain Gauge, Abstract of Thesis, Novosibirsk (1962).
2. L. V. Novikov and R. S. Smirnova, Izmeritel'naya Tekhnika, No. 1, p. 33 (1967).
3. Yu. V. Ilisavskii, Semiconductor Strain Gauges, No. 6, Leningrad (1963).
4. P. S. Agalarzade and S. A. Semiletov, Kristallografiya, 8:298 (1963).

FLUX-METHOD CALCULATIONS OF
THE CHARACTERISTICS OF SEMICONDUCTOR DEVICES
HAVING AN ELECTRIC FIELD IN THE BASE REGION

I. M. Beskrovnyi

A method is proposed for using the flux equations to calculate the characteristics of drift transistors. The transport factor of a drift-transistor base is calculated for exponential and actual impurity distributions in the base. It is shown that the allowance for the actual field distribution in the base, especially the retarding section, has a considerable effect on the results. For sufficiently long lifetimes, the results given by the flux method agree exactly with those obtained by solution of the diffusion equations, but the flux-method results are more rigorous for short lifetimes.

The flux equations were first formulated by Shockley [1]. In [2], their relation to the P_1 approximation of transport theory was analyzed, and a modification of these equations was proposed in order to extend their range of applicability to very short lifetimes ($\tau \rightarrow 0$). In [3], flux equations valid in the presence of a constant electric field were derived. Their general form is

$$\frac{dr(x)}{dx} = -[k_0(1 \mp 2\delta) + m_0(1 \mp \delta)] r(x) + k_0(1 \pm 2\delta) l(x) + \frac{g(x)}{2}(1 \pm 2\delta), \tag{1}$$

$$\frac{dl(x)}{dx} = [k_0(1 \pm 2\delta) + m_0(1 \pm \delta)] l(x) - k_0(1 \mp 2\delta) r(x) - \frac{g(x)}{2}(1 \mp 2\delta), \tag{2}$$

where $r(x)$ and $l(x)$ are the carrier fluxes to the right and left respectively;

$$k_0 = \frac{\bar{v}}{4D}, \ m_0 = \frac{2}{\bar{v}\tau}, \ \delta = \frac{\mu E}{\bar{v}}, \ g(x) = \frac{P_n(x)}{\tau}; \tag{3}$$

$\bar{v}$, τ, D, and μ are, respectively, the mean thermal velocity, the lifetime, the diffusion coefficient, and the mobility of the minority carriers. The upper signs in the parentheses are taken if the field is in the direction of the flux $r(x)$, and the lower signs if it is in the direction of $l(x)$. Differentiating Eqs. (1) and (2) and solving the resulting simultaneous equations, we obtained a pair of separate second-order differential equations for $r(x)$ and $l(x)$.

The general solutions of these equations may be written, if the field is in the direction of $l(x)$:

$$r(x) = AR_{\infty+}e^{q+x} + Be^{-q-x} + f_-(x),$$ (4)

$$l(x) = Ae^{q+x} + BR_{\infty-}e^{-q-x} + f_+(x),$$ (5)

where

$$q_\pm = q' \mp \delta(2k_0 + m_0), \quad q' = \sqrt{2k_0 m_0 + m_0^2 + 4\delta^2 k_0^2},$$ (6)

$$R_{\infty+} = \frac{q' - m_0 \mp 2\delta k_0}{q' + m_0 \mp 2\delta k_0}.$$ (7)

A and B are constants determined by the boundary conditions; $f_\pm(x)$ are functions depending on the distribution of impurities.

Let us first consider the calculation of the transport factor β^* for a drift transistor, using the customary assumption that the distribution of impurities in the base is exponential [4, 5]. It is convenient to assume also that the collector junction is at x = 0 and the emitter junction at x = W, where W is the width of the base. Then, if

$$N_d(x) = N_C e^{ax},$$ (8)

we have

$$N_e = N_C e^{aW},$$ (9)

where N_e and N_c are the concentrations of donor impurities at the boundaries of the emitter and collector junctions. From Eqs. (8) and (9), it follows that

$$a = \frac{1}{W} \ln \frac{N_e}{N_C}.$$ (10)

It is also known [5] that

$$E(x) = \frac{kT}{e} \frac{1}{N_d(x)} \frac{dN_d(x)}{dx};$$ (11)

then, from Eqs. (8) and (11),

$$a = \frac{eE}{kT}.$$ (12)

Usually, we have

$$P_n(x) = \frac{1}{N_d(x)},$$ (13)

that is,

$$P_{nc} = P_{ne}e^{aW}$$ (14)

and so

$$P_n(x) = P_{ne}e^{aW}e^{-ax}.$$ (15)

Using Eqs. (3) and (12), we find that

$$a = 4\delta k_0,$$ (16)

and hence

$$g(x) = g_0 e^{4\delta k_0(W-x)},$$ (17)

$$\frac{dg(x)}{dx} = -4\delta k_0 g_0 e^{4\delta k_0(W-x)},$$ (18)

where

$$g_0 = \frac{P_{ne}}{\tau_p}.$$ (19)

We can now find the values of the functions $f_\pm(x)$. After double differentiation of Eqs. (4) and (5) and combination with Eqs. (1) and (2), we can derive the following equations for the functions $f_\pm(x)$:

$$\frac{d^2 f_\pm(x)}{dx^2} \mp 2\delta\,(2k_0 + m_0)\,\frac{df_\pm(x)}{dx} - [2k_0 m_0\,(1 - 2\delta^2) + m_0^2\,(1 - \delta^2)]\,f_\pm(x) =$$

$$= -\frac{g(x)}{2}\,[2k_0 + m_0\,(1 \pm \delta)]\,(1 \pm 2\delta) + \frac{1}{2}\,\frac{dg(x)}{dx}\,(1 \pm 2\delta). \tag{20}$$

Substitution of Eqs. (17) and (18) in Eq. (20) gives

$$f_\pm(x) = \frac{g_0}{2m_0}\,\frac{(1 \pm 2\delta)\,[2k_0\,(1 \mp 2\delta) + m_0\,(1 \pm \delta)]}{2k_0\,(1 + 4\delta^2) + m_0\,(1 - \delta^2)}\,e^{4\delta k_0(W-x)}. \tag{21}$$

If the small quadratic terms are neglected, then

$$f_+(x) = f_-(x) = \frac{g_0}{2m_0}\,e^{4\delta k_0(W-x)}. \tag{22}$$

To determine the constants A and B, we can use the following boundary conditions: at the collector junction,*

$$r(0) = 0, \tag{23}$$

and at the emitter junction,

$$\frac{2}{\bar{v}}\,[l(W) + r(W)] = P_{\mathrm{ne}}\exp\left(\frac{eU}{kT}\right), \tag{24}$$

where U is the voltage across the emitter junction.

Using Eqs. (23) and (24) with Eqs. (4), (5), and (22), we find

$$B = -AR_{\infty_+} \frac{g_0}{2m_0}\,e^{4\delta k_0 W}, \tag{25}$$

$$A = \frac{P_{\mathrm{ne}}\bar{v}\left[\exp\left(\dfrac{eU}{kT}\right) - 1\right] e^{\delta(2k_0 + m_0)W} + \dfrac{P_{\mathrm{ne}}\bar{v}}{4}\,(1 + R_{\infty_-})\,e^{4\delta k_0 W} e^{-q'W}}{(1 + R_{\infty_+})\,e^{q'W} - R_{\infty_+}\,(1 + R_{\infty_-})\,e^{-q'W}}. \tag{26}$$

Since the sum of the fluxes is proportional to the carrier density, and the difference of the fluxes is proportional to the current density [3], the values of p(x) and J(x) at any cross section of the device can be found immediately from the constants A and B. By definition,

$$\beta^* = \frac{J_{\mathrm{pc}}}{J_{\mathrm{pe}}} = \frac{l(0) - r(0)}{l(W) - r(W)}. \tag{27}$$

Substitution of Eqs. (22), (25), and (26) in Eqs. (4) and (5), and substitution of the resultant expressions in Eq. (27), followed by transformations, gives

$$\beta^* =$$

$$\frac{\left\{\left[\exp\left(\dfrac{eU}{kT}\right) - 1\right] e^\gamma + \dfrac{1}{2}\,(1 + R_{\infty_-})\,e^\alpha e^{-q'W}\right\} \dfrac{1 - R_\infty^{*2}}{1 - R_{\infty_-}} + \dfrac{1}{2}\,[(1 + R_{\infty_+})\,e^{q'W} - R_{\infty_+}\,(1 + R_{\infty_-})\,e^{-q'W}]\,e^\alpha}{\left\{\left[\exp\left(\dfrac{eU}{kT}\right) - 1\right] + \dfrac{1}{2}\,(1 + R_{\infty_-})\,e^{\alpha - \gamma} e^{-q'W}\right\}\left[\dfrac{1 - R_{\infty_+}}{1 - R_{\infty_-}}\,e^{q'W} + R_{\infty_+}e^{q'W}\right] + \dfrac{1}{2}\,[(1 + R_{\infty_+})\,e^{q'W} - R_{\infty_+}\,(1 + R_{\infty_-})\,e^{-q'W}]\,e^{\alpha - \gamma - q'W}},$$

$$\tag{28}$$

where

$$\gamma = \delta\,(2k_0 + m_0)\,W, \quad \alpha = 4\delta k_0 W, \quad R_\infty^{*2} = R_{\infty_+}R_{\infty_-}. \tag{29}$$

---

* It can be shown that this boundary condition is more rigorous than the condition p(0) = 0 generally used in solving the diffusion equations.

Using Eqs. (6), (7), and (29), we can bring Eq. (28) to the clearer form

$$\beta^* = \frac{\left[\exp\left(\frac{eU}{kT}\right) - 1\right] e^{\delta(2k_0 + m_0)W} + \left[\text{ch } q'W - \frac{2\delta k_0}{q'} \text{ sh } q'W\right] e^{4\delta k_0 W}}{\left[\exp\left(\frac{eU}{kT}\right) - 1\right]\left[\text{ch } q'W + \frac{2\delta k_0 + m_0}{q'} \text{ sh } q'W\right] + e^{\delta(2k_0 - m_0)W}}. \tag{30}$$

If the lifetime is sufficiently long, then

$$q' \approx \frac{1}{L_p}, \; k_0 \gg m_0, \tag{31}$$

and Eq. (30) agrees exactly with the result obtained from the solution of the diffusion equations [4, 5]. For short lifetimes, however, especially when $\tau \to 0$, Eq. (30) is to be preferred, as it is more rigorous.

The concept of the rigor of a particular expression derived by assuming an exponential distribution of impurities in the base region is itself arbitrary. The reason is that this assumption of an exponential impurity distribution may lead to much larger errors than is commonly supposed in the analysis of characteristics of drift transistors. Let us consider this further.

It is known [4] that the true distribution of impurities in the base cannot be purely exponential, but approximates to one of two idealized cases, depending on the fabrication technique used. In the ideal case of diffusion from a constant source, the distribution function is

$$N(x) = Q(\pi D t)^{\frac{1}{2}} \exp\left(-\frac{x^2}{4Dt}\right), \tag{32}$$

where $Q(\pi D t)^{\frac{1}{2}}$ is the volume concentration of an impurity at the surface of a semiconductor, and t is the diffusion duration.

In the other ideal case, that of a constant surface concentration, we have

$$N(x) = N_0 \left\{1 - \text{erf}\left[\frac{x}{2}(Dt)^{\frac{1}{2}}\right]\right\}, \tag{33}$$

where N_0 is the volume concentration of an impurity at the surface of a semiconductor.

The two distributions given by Eqs. (32) and (33) are similar to each other, but they differ noticeably from the exponential case. It has been shown in [6] that an allowance for this fact leads to a considerable change of the diffusion equation.

The usual form of the diffusion equation for an arbitrary impurity distribution function $f(x)$ is

$$\frac{d^2P}{dx^2} - \frac{d}{dx}\left[\frac{P}{f(x)}\frac{df(x)}{dx}\right] - \frac{P - P_n(x)}{L_p^2} = 0. \tag{34}$$

Using the normalized coordinate y = x/W, we can write Eq. (34) as

$$\frac{d^2P}{dy^2} - \frac{1}{F(y)}\frac{dF(y)}{dy}\frac{dP}{dy} - \left\{\frac{W^2}{L_p^2} - \frac{d}{dy}\left[\frac{1}{F(y)}\frac{dF(y)}{dy}\right]\right\} P = -W^2\frac{P_n}{D_p}. \tag{35}$$

If $F(y)$ is exponential, then

$$\frac{d}{dy}\left[\frac{1}{F(y)}\frac{dF(y)}{dy}\right] = 0, \tag{36}$$

TABLE 1

$\frac{N(0)}{N(W)}$	$2C^2$
10	4.6
10^2	9.2
10^3	13.8
10^4	18.4

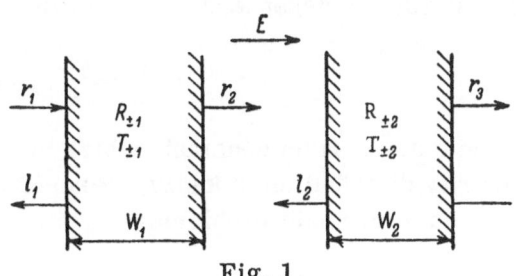

Fig. 1

and we get the customary equation

$$\frac{d^2P}{dx^2} - a\frac{dP}{dx} - \frac{P - P_{\text{ne}}e^{ax}}{L_p^2} = 0, \tag{37}$$

the solution of which is well known. But if

$$f(x) = \exp\left(-\frac{x^2}{4Dt}\right), \tag{38}$$

then

$$F(y) = \exp\left(-C^2y^2\right), \tag{39}$$

where

$$C = \frac{W}{4Dt}. \tag{40}$$

Then,

$$\frac{d}{dy}\left[\frac{1}{F(y)}\frac{dF(y)}{dy}\right] = -2C^2, \tag{41}$$

and Eq. (35) becomes

$$\frac{d^2P}{dy^2} - \frac{1}{F(y)}\frac{dF(y)}{dy}\frac{dP}{dy} - \left(\frac{W^2}{L_p^2} + 2C^2\right)P = -W^2\frac{P_n}{D_p}. \tag{42}$$

In any diffusion-drift device, $W^2/L_p^2 \ll 1$, whereas C^2, which can be found from the expression

$$C^2 = \ln\frac{N(0)}{N(W)}, \tag{43}$$

considerably exceeds unity, as may be seen from Table 1.

Thus, the coefficient of P in Eq. (42) is entirely independent of the diffusion length, at least until the latter decreases by a factor of hundreds; it depends only on the distribution function, whereas, in Eq. (37), the corresponding coefficient is uniquely determined by the value of the diffusion length.

In addition to the need to take into account the actual law of distribution of impurities, it is also necessary to consider the way in which the minority carrier mobility depends on the impurity concentration [5]. It has been shown in [5] that the dependence of the mobility μ_p on the impurity concentration N may be approximated by

$$\mu_p = 1800 \ \text{cm}^2 \cdot \text{V}^{-1} \cdot \text{sec}^{-1} \ (N < 10^{15} \ \text{cm}^{-3}), \tag{44}$$

$$\mu_p = 1800 - 500 \ \lg\frac{N}{10^{15}} \ \text{cm}^2 \cdot \text{V}^{-1} \cdot \text{sec}^{-1} \ (10^{15} < N < 10^{18}). \tag{45}$$

The coordinate dependence of the diffusion coefficient $D_p(x)$ can then be written as

$$D_p(x) = \frac{kT}{e}\,\mu_p(x) = \frac{kT}{e}\,(b + cx^2),\tag{46}$$

where b and c are certain constants which may be determined from the relations (44) and (45) if the impurity distribution is known. Since $D_p(x) \neq$ const, Eq. (42) becomes still more complex in the solution of dynamic problems:

$$\frac{\partial^2 p}{\partial x^2} - f_1(x)\,\frac{\partial p}{\partial x} - f_2(x)\,p - f_3(x)\,\frac{\partial p}{\partial t} = -f_4(x),\tag{47}$$

where

$$f_1(x) = 2ax - \frac{2cx}{b + cx^2},\tag{48}$$

$$f_2(x) = 2a\left(\frac{2cx}{b + cx^2} + 1\right) + \frac{e}{kT}\,\frac{1}{(b + cx^2)\,\tau_p},\tag{49}$$

$$f_3(x) = \frac{e}{kT}\cdot\frac{1}{b + cx^2},\quad f_4(x) = f_3(x)\,\frac{e^{ax^2}}{\tau_p}\,P_{\text{ne}}.\tag{50}$$

It is evident that the solution of such an equation is quite a difficult matter.

The most effective way of solving such problems seems to be by iteration of the reflection and transmission coefficients, as suggested in [7]. Figure 1 shows two layers of finite thicknesses W_1 and W_2, having the coefficients $R_{\pm 1}$, $T_{\pm 1}$, $R_{\pm 2}$, and $T_{\pm 2}$. The latter are the reflection and transmission coefficients, whose values will be derived below. If the field is in the direction of the flux r, as shown in Fig. 1, we can write the following equations (for $l_3 = 0$):

$$r_2 = r_1 T_{+1} + l_2 R_{-1},\tag{51}$$

$$l_2 = r_2 R_{+2},\tag{52}$$

$$r_3 = r_2 T_{+2},\tag{53}$$

$$l_1 = r_1 R_{+1} + l_2 T_{-1}.\tag{54}$$

Solving these for the fluxes l_1 and r_3, we can now find the coefficients $R_+^{(2)}$ and $T_+^{(2)}$ for the combined layer of thickness $W_1 + W_2$:

$$R_+^{(2)} = \frac{l_1}{r_1} = R_{+1}\,\frac{T_{+1} T_{-1}}{1 - R_{-1} R_{+2}},\tag{55}$$

$$T_+^{(2)} = \frac{r_3}{r_1} = \frac{T_{+1} T_{+2}}{1 - R_{-1} R_{+2}}.\tag{56}$$

Reversing the direction of the field, we can similarly obtain the values of $R_-^{(2)}$ and $T_-^{(2)}$.

In the general case, we can evidently divide any layer whose parameters vary with x into an arbitrary number n of sublayers, in each of which the parameters (and, in particular, the field) may be taken as constant. We can then calculate the reflection and transmission coefficients for the entire layer by the recurrence equations

$$R_\pm^{(n)} = R_\pm^{(n-1)} + R_{\pm n}\,\frac{T_\pm^{(n-1)} T_\mp^{(n-1)}}{1 - R_\mp^{(n-2)} R_{\pm n}},\tag{57}$$

$$T_\pm^{(n)} = \frac{T_\pm^{(n-2)} T_{\pm n}}{1 + R_\pm^{(n-1)} R_{\mp n}}.\tag{58}$$

TABLE 2

Sec-tion	Parameters			
	E_i, V/cm	μ_i, cm^2 · V^{-1}·sec^{-1}	D_i, cm^2/sec	W_i, μ
I	600	1000	25	0.8
II	420	700	18	0.6
III	280	450	12	1.0
IV	4000	465	11.5	0.13

Let us now determine the values of the coefficients $R_{\pm}(W)$, $T_{\pm}(W)$ for a uniform layer with a constant field. To do so, we use Eqs. (4) and (5). The foregoing recurrence equations can be used only if the signal is not small so that the equilibrium carrier density is negligible compared with the excess density. Then, Eqs. (4) and (5) become (with the flux r in the direction of the field)

$$r = AR_{\infty-}e^{q-x} + Be^{-q+x}, \qquad (59)$$

$$l = Ae^{q-x} + BR_{\infty+}e^{-q+x}. \qquad (60)$$

If r(0) = 0, then, from Eq. (59),

$$B = -AR_{\infty-}, \qquad (61)$$

and we have, for a layer of thickness W,

$$R_{\mp}(W) = \frac{r(W)}{l(W)} = \frac{\dfrac{k_0(1 \pm 2\delta)}{q'}\,\text{sh}\,q'W}{\text{ch}\,q'W + \dfrac{k_0 + m_0}{q'}\,\text{sh}\,q'W}, \qquad (62)$$

$$T_{\mp}(W) = \frac{r(0)}{l(W)} = \frac{e^{\mp\delta(2k_0+m_0)}}{\text{ch}\,q'W + \dfrac{k_0+m_0}{q'}\,\text{sh}\,q'W}. \qquad (63)$$

(The lower sign corresponds to the opposite direction of the field.)

It is very easy to calculate the base transport factor β^*, using the coefficients R and T. Figure 2 represents the base region of thickness W, the collector junction being taken, for convenience, at the origin. The direction of the field is, of course, from the emitter to the collector. The boundary condition at the collector is r(0) = 0. For the fluxes shown in Fig. 2, the following relations apply:

$$l(0) = l(W)\,T_+(W), \qquad (64)$$

$$r(W) = l(W)\,R_+(W). \qquad (65)$$

The transport factor β^* can be determined from the relation

$$\beta^* = \frac{J_c}{J_e} = \frac{l(0) - r(0)}{l(W) - l(W)} = \frac{T_+(W)}{1 - R_+(W)}. \qquad (66)$$

Thus, the base transport factor for a nonuniform distribution of the field and of the semiconductor parameters in the base region can be calculated by dividing that region into n sublayers each having a uniform field E_i, calculating the coefficients $R_{\pm i}$, $T_{\pm i}$ for each sublayer, calculating the total coefficients $R_{+(W)}$, $T_{+(W)}$ for the whole base region by means of Eqs. (57) and (58), and calculating β^* from Eq. (66). If the base layer may be regarded as uniform (i.e., a single sublayer), then substitution of Eqs. (62) and (63) in Eq. (66) gives

$$\beta^* = \frac{e^{\delta(2k_0+m_0)W}}{\text{ch}\,q'w + \dfrac{2k_0+m_0}{q'}\,\text{sh}\,q'W}, \qquad (67)$$

which is exactly the same as the expression obtained from Eq. (30) when U is sufficiently large. When τ is so large that we can take $k_0 \gg m_0$ and q' = 1/L_p, Eq. (67) agrees exactly with the result obtained from the solution of the diffusion equations [4, 5]:

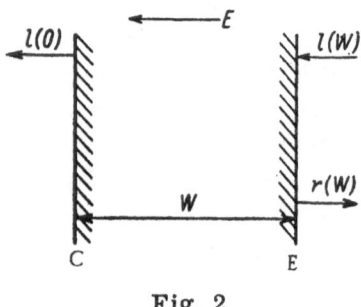

Fig. 2

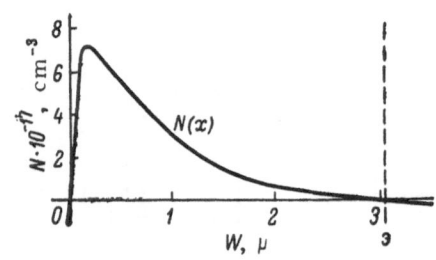

Fig. 3. Distribution of the excess density in the base of the drift transistor IT308.

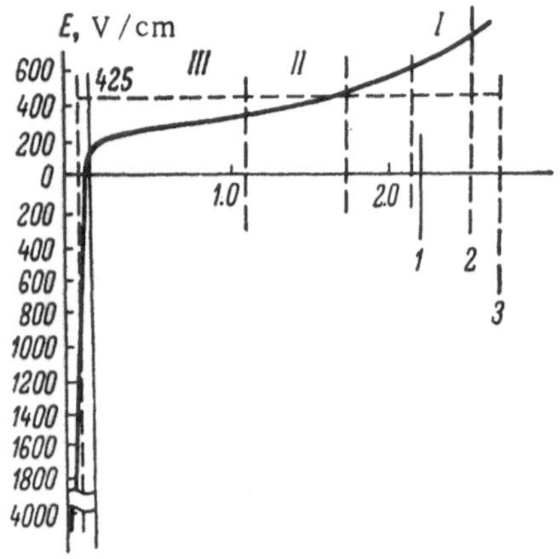

$$\beta^* = \frac{e^{2\delta k_0 W}}{\operatorname{ch} \dfrac{W}{L_p} + 2k_0 L_p \operatorname{sh} \dfrac{W}{L_p}} \cdot \qquad (68)$$

The iteration calculation has been carried out for the transistor IT308. Figure 3 shows the actual distribution of the excess density in the base of this drift transistor; Fig. 4 shows the corresponding field distribution [8]. The base region was divided into four sections as shown in Fig. 4, and the parameters for each section were taken as in Table 2.

β^* was calculated for three cases:

1) using Eq. (68), derived from the diffusion approximation: $\beta^* = 0.99636$.

2) assuming the base layer to be a single uniform sublayer, i.e., using Eq. (67): $\beta^* = 0.99578$.

Fig. 4. Field distribution in the base of the transistor IT308. 1) Junction boundary at U_{br}; 2) collector junction boundary at rated voltage of collector; 3) collector junction boundary at contact potential.

3) dividing the base layer into four sublayers with parameters as shown in Table 2: $\beta^* = 0.98105$.

It is interesting to note that the value of β^* obtained when the base is divided into four sublayers is the same as the experimentally determined value of α. This, however, does not definitely prove that the calculations are accurate, since it is not possible to make a separate experimental measurement of the emitter efficiency and the transport factor.

The following conclusions may be drawn from these results.

The flux method is adequate for the analysis of the characteristics of drift semiconductor devices with either an exponential or any other distribution of impurities.

When the actual impurity distribution is taken into account, the results are considerably affected.

Literature Cited

1.   W. Shockley, Phys. Rev., 125:1570 (1962).
2.   I. M. Beskrovnyi and A. A. Kostritsa, Izv. Akad. Nauk Beloruss. SSR, Ser. Fiz.-Mat. Nauk (1969).

3.   I. M. Beskrovnyi and V. A. Botvin, Fiz. Tekh. Poluprov., 2:1391 (1968).

4.   A. V. Krasilov and A. F. Trutko, Calculation Methods for Transistors, Energiya, Moscow (1964).

5.   N. S. Spiridonov and V. I. Vertogradov, Drift Transistors, Sov. Radio, Moscow (1964).

6.   G. C. Jain and R. M. S. Al-Rifai, J. Appl. Phys., 37:2401 (1966).

7.   E. F. Pulver and J. P. McKelvey, Phys. Rev., 149:617 (1966).

8.   D. I. Smetanina and Yu. A. Sher, Voprosy Radioélektroniki, Ser. Poluprovodnikovye Pribory, 129 (1964).

THERMALLY STIMULATED CURRENTS IN
p−n JUNCTIONS IN GALLIUM PHOSPHIDE

N. M. Kolchanova, R. F. Mamedova,
M. A. Mirdzhalilova, and D. N. Nasledov

A. F. Ioffe Physicotechnical Institute
Academy of Sciences of the USSR, Leningrad

A study was made of the energy positions of impurity levels in gallium phosphide p−n structures, using thermally stimulated currents. The thermally stimulated conductivity was investigated in the range 80-300°K at various heating rates. Several methods were used to calculate the energy positions of the impurity levels and they gave consistent results. Levels at 0.27-0.3, 0.06, and 0.04 eV were found. The same levels appeared in a study of recombination radiation spectra and of the temperature dependence of the Hall coefficient. The densities and capture cross sections of these levels were calculated.

One method of determining the energy spectrum of impurity levels lying in the forbidden band of semiconductors is to investigate the curves of the thermally stimulated current. This has been successfully used to study the properties of CdS, ZnS, etc. It was recently shown that the method of thermally stimulated currents can be applied to estimate the positions of impurity levels in p−n structures. We have used it to derive further information about the impurity spectrum in diodes based on gallium phosphide.

Results and Discussion

We examined alloyed diodes having hole densities from 10^{14} to 10^{16} cm^{-3} and mobilities about 80 cm$^2 \cdot$ V$^{-1} \cdot$ sec^{-1} at room temperature. The photodiodes were cooled to liquid-nitrogen temperature and illuminated from the p-type side for 10 min, and then uniformly heated to 300°K.

The photodiodes were examined with 0.5 to 5 V applied in the reverse direction, using various heating rates. When the diodes were not previously illuminated, the temperature dependence of the dark current had the usual form shown in Fig. 1, curve 2. Curves denoted by 1 in Fig. 1 show the current as a function of the temperature for the same sample previously illuminated at 80°K; they have three peaks. The thermally stimulated current curve for some diodes showed only two peaks (Fig. 2). Varying the reverse voltage did not appreciably alter the form of these curves.

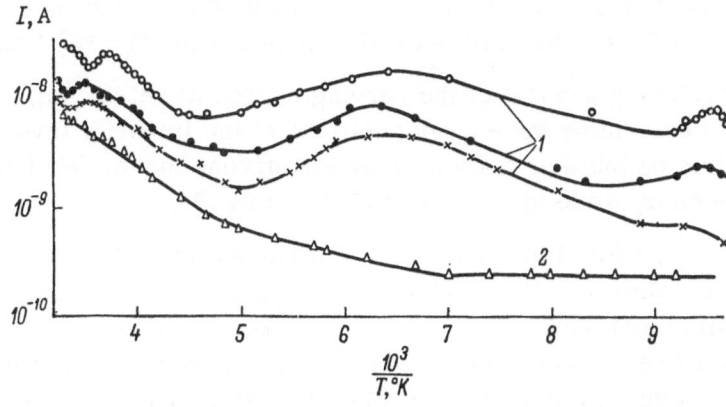

Fig. 1. Thermally stimulated currents in gallium phosphide diode 8A. 1) After illumination, at various reverse voltages; 2) dark current.

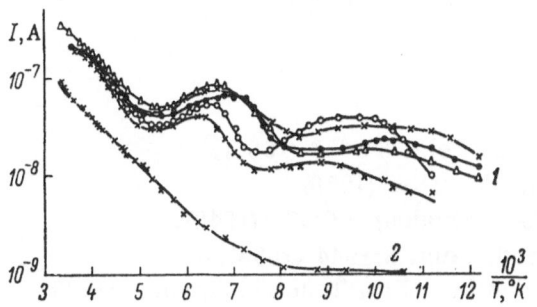

Fig. 2. Thermally stimulated currents in gallium phosphide diode 20. 1, 2) As in Fig. 1.

Let us first determine the energy position of the impurity level responsible for the high-temperature maximum of curves 1 in Fig. 1. It has been shown in [1] that the excess current at the maximum of the thermally stimulated curve for a reverse-biased diode with a low-resistance n-type region is given by Bube's method (which assumes that the maximum current corresponds to the coincidence of the Fermi level with the trap level) as

$$i_{max} = q \frac{L_n}{\tau_n} N_c e^{-\frac{\Delta E_m}{k T_{max}}} S \delta,$$

where i_{max} is the maximum current, T_{max} the temperature of the maximum, L_n and τ_n the electron diffusion length and lifetime at temperature T_{max}, S the area of the junction, N_c the density of states in the conduction band, δ the quantum efficiency, and ΔE_m the trap energy. To estimate ΔE_m, we must know the ratio L_n/τ_n and δ. The latter was taken to be unity. Since the high-temperature maximum was located almost at room temperature, the values of τ_n and L_n at 300 and 270°K were almost the same. In the calculation, they were taken to have their values at 300°K ($L_n = 0.1 \mu$, $\tau_n = 10^{-7}$ sec). The energy position of the level, found in this way for various samples and heating rates, varied only slightly in the range 0.28-0.3 V.

The position of the impurity level was also determined from the change of the temperature of the maximum of the curve with varying rate of heating of the sample [2]. This method indicated a trap level having an activation energy of 0.24-0.27 eV.

A similar value for the position of the level, 0.3 eV, was obtained by the initial-rise method, first proposed by Garlick and Gibson [3]. The position of the level in the forbidden band of gallium phosphide was also estimated by the method described in [4]. This indicated an impurity level at 0.24-0.27 eV. The position determined by Bube's method, with allowance for strong retrapping, was in fair agreement with the value found by the other methods, which do not take account of the recombination kinetics. We may suppose that rapid retrapping occurs at the 0.27 eV level. This is confirmed by the absence of any slow decay of the photoconductivity when the action of radiation ceases.

The concentration of impurity centers was determined from the thermally stimulated current curve, and hence the capture cross section of this level ($2 \cdot 10^{-17}$ cm^2) was deduced.

Similar methods, taking account of the varying degree of retrapping of nonequilibrium carriers, were used to determine the energy positions of the impurity levels corresponding to the other two maxima on the thermally stimulated current curve. We found two shallow levels, with activation energies of about 0.06-0.07 and 0.04 eV.

The energies for the three impurity levels determined from the curves were in good agreement with those obtained by other methods. The temperature dependence of the Hall coefficient in the original material doped with zinc and tin showed the presence of acceptor levels having activation energies of 0.04 and 0.32 eV, and a donor level at 0.06 eV. The same levels were found in a study of recombination radiation spectra [5]. These values agree well with the impurity level energies estimated from the thermally stimulated current curves. The level at 0.27-0.3 eV may be the deep acceptor level of silicon; the second level, at 0.06 eV, may be the silicon donor level. The value 0.04 eV corresponds to the ionization energy of zinc.

Thus, observations of the thermally stimulated current through a p−n junction can be utilized for the detection and investigation of local levels in gallium phosphide semiconductor devices.

Literature Cited

1.   M. A. Mirdzhalilova and L. G. Paritskii, Fiz. Tverd. Tela, 8:3090 (1966).
2.   K. W. Böer, S. Oberländer, and J. Voigt, Ann. Physik, 2:130 (1958).
3.   G. F. J. Garlick and A. F. Gibson, Proc. Phys. Soc. (London), 60:574 (1948).
4.   A. P. Kulshreshtha and V. A. Goryunov, Fiz. Tverd. Tela, 8:1944 (1966).
5.   R. F. Mamedova, D. N. Nasledov, and S. V. Slobodchikov, Fiz. Tekh. Poluprov., 1:509 (1967).

$Al_X Ga_{1-X} As$—GaAs HETEROJUNCTIONS

Zh. I. Alferov, V. M. Andreev, V. I. Korol'kov, E. L. Portnoi, and D. N. Tret'yakov

A. F. Ioffe Physicotechnical Institute
Academy of Sciences of the USSR, Leningrad

Electrical properties and injection luminescence were studied in $Al_xGa_{1-x}As$—GaAs n—p and p—n heterojunctions containing various amounts of aluminum (E_g = 1.52-1.9 eV at 300°K). A band model of the heterojunctions was established and its principal parameters were determined. It is shown that the main discontinuity occurs in the conduction band. The size ΔE_C of this discontinuity varies linearly with the aluminum content in $Al_xGa_{1-x}As$; the discontinuity ΔE_V in the valence band is almost zero. The injection luminescence spectra agree with the proposed model. The mechanism of current flow is compared with features of the electroluminescence spectra. At room temperature and above, the initial (low-current) section, the forward branch of the current—voltage characteristic for n—p heterojunctions is determined by recombination in the space-charge layer, and at current densities above 1-5 A/cm², by recombination in the bulk of the diode. At low temperatures, the forward branch is given by the theory of Shockley, Noyce, and Sah. For p—n heterojunctions, the initial section of the current—voltage characteristic is of the tunnel type; at about 5 A/cm² and above, the mechanism of current flow is similar to that in n—p heterojunctions.

AlAs—GaAs heterojunctions attracted our attention because of the very similar values of the lattice constants of the two materials (5.65 Å for GaAs, 5.66 Å for AlAs) [1]. In epitaxial single-crystal heterojunctions of this system, we may expect to obtain structures having no surface states at the interface, and this pair appears to be one of the most promising for the fabrication of heterojunction-based injection devices. A band model of a heterojunction [2] has been derived from studies of GaAs—Ge heterojunctions, which were the first to be systematically studied. This model has been found applicable to GaP—GaAs heterojunctions also [3]. In GaAs—Ge heterojunctions, despite the similarity of the lattice constants, it has not been possible to achieve efficient injection, and many properties of the system are not in satisfactory agreement with the theoretical model. The reason is probably related to the formation of new interfacial phases, whose nature and properties are unknown. In a heterojunction, especially one consisting of a semiconducting element and a compound, when one material forms impurity centers in the other material, intermediate phases may form at the interface because of mutual solubility, diffusion, and the possible occurrence of new chemical compounds. In this respect, heterojunctions between isotypic semiconductors (e.g., $A^{III}B^V$—$A^{III}B^V$, $A^{II}B^{VI}$—$A^{II}B^{VI}$) are much to be preferred, since the mutual solubility and diffusion of the components affect only the abruptness of the junction. AlAs and GaAs form a continuous series of solid solutions, whose electrical properties and band structure are similar to those of the extensively studied system $GaAs_xP_{1-x}$. AlAs is a semiconductor having an "indirect" band structure and forbidden band width 2.16 ± 0.1 eV [4]; in $Al_xGa_{1-x}As$ solid solutions, the change from the band structure of GaAs (a "direct" semiconductor) to that of AlAs occurs at an aluminum concentration of about 50% [5]. The properties of AlAs have not been much studied, partly on account of the chemical instability of this compound.

Experimental Samples and Their Principal Parameters

The $Al_xGa_{1-x}As - GaAs$ heterojunctions were prepared by epitaxial growth of $Al_xGa_{1-x}As$ solid solutions on single-crystal substrates of GaAs. Studies of the electroluminescence spectra of n-type $Al_xGa_{1-x}As - $p-type GaAs heterojunctions and the photoluminescence of epitaxial layers, and direct determinations of the composition by means of the X-ray microanalyzer, showed that the solid solutions had variable compositions, and the forbidden band width was greatest at the interface, decreasing away from the interface to a value close to that for GaAs. No decomposition or instability of the solid-solution layers was observed. We studied p−n and n−p heterojunctions (i.e., respectively p-type and n-type $Al_xGa_{1-x}As$ and n-type and p-type GaAs) whose principal parameters were as shown in Table 1.

Recombination Radiation Spectra

The luminescence spectra of n-type $Al_xGa_{1-x}As - $p-type GaAs and p-type $Al_xGa_{1-x}As - $n-type GaAs heterojunctions were measured at current densities ranging from 10^{-1} to $5 \cdot 10^3$ A/cm^2 and at 300 and 77°K.

The injection luminescence spectra of the p−n heterojunctions had luminescence bands due to n-type GaAs but no shortwave bands due to radiative recombination in the wide-gap material. This result appears to be due to the unipolar injection first experimentally demonstrated for heterojunctions in a study of radiative recombination in p-type GaP−n-type GaAs heterojunctions [6].

The injection luminescence spectra are particularly interesting for n-type $Al_xGa_{1-x}As - $p-type GaAs heterojunctions, because they include radiative recombination bands of the narrow- and wide-gap materials. Figure 1 shows the luminescence spectrum of one sample for two values of the current in the forward direction, corresponding to different parts of the current− voltage characteristic (see below). The position and shape of the luminescence band at $h\nu = 1.38$ eV ("edge" luminescence in p-type GaAs) did not vary with the current, but the shape of the shortwave radiation spectrum resulting from recombination in the solid solution having a variable energy gap (forbidden-band width) depended considerably on the forward current. It seems that, for a small forward bias, recombination takes place in a narrow region of the wide-gap material near the heterojunction interface, but as the current increases, the recombination region becomes broader. From these results, and the knowledge how the composition varies through the epitaxial layers of the solid solution, we can find the diffusion length L_d of minority carriers in the $Al_xGa_{1-x}As$ epitaxial layer. At 77°K (Fig. 2), at both small and large currents, recombination in the solid solution takes place in a narrow region near the interface. As the current through the heterojunction increases, although the wide-gap part of the junction is lightly doped ($N \approx 10^{15}$ cm^{-3}), the "edge" luminescence band is found to become narrower because of an inversion of population in GaAs, due to injection by the heterojunction of an electron density exceeding the equilibrium density in the wide-gap emitter. This "super-injection" effect has been theoretically predicted [7]; experimentally, it was first discovered by ourselves in these same n-type $Al_xGa_{1-x}As - $p-type GaAs heterojunctions [8]. The absence

TABLE 1

Batch No.	Type of hetero-junction	Substrate and density, cm⁻³	Epitaxial layer	
			forbidden band width E_g, eV	density, cm⁻³
114			1.74	$p = 2.5 \cdot 10^{17}$
159	$p-n$	$n = 2 \cdot 10^{18}$	1.85	$p = 1.5 \cdot 10^{17}$
113			1.85	$p = 3.7 \cdot 10^{17}$
112			1.9	$p = 4 \cdot 10^{17}$
133			1.66	$n = 6 \cdot 10^{13}$
149			1.80	$n = 2.6 \cdot 10^{15}$
164	$n-p$	$p = 1.5 \cdot 10^{19}$	1.87	$n = 10^{17}$
144c			1.91	$n = 5.4 \cdot 10^{15}$
147			1.71	$n = 9 \cdot 10^{14}$
264			1.52	$n = 5 \cdot 10^{14}$
150			1.88	$n = 10^{16}$

N o t e .   The value of E_g refers to the substrate-layer interface. All values are given for 300°K.

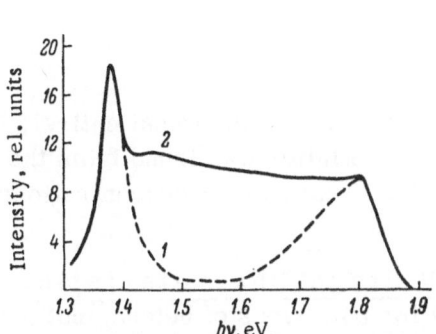

Fig. 1. Electroluminescence spectra of p-type GaAs−n-type $Al_xGa_{1-x}As$ heterojunction at 300°K. Junction area ≈5 · 10⁻³ cm². Current, mA: 1) 1; 2) 100.

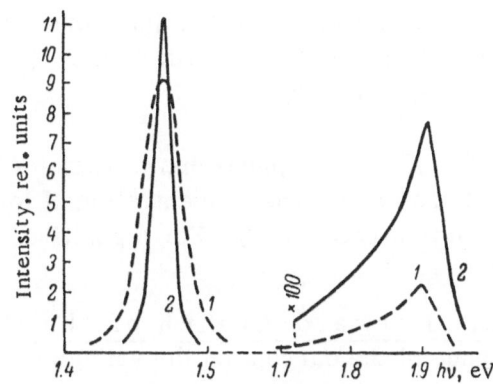

Fig. 2. Electroluminescence spectra of p-type GaAs−n-type $Al_xGa_{1-x}As$ hetero-junction at 77°K. 1) 10 mA dc; 2) 40 A in pulses 1 μsec long, repetition frequency 40 cps.

of radiation bands due to n-type GaAs in the injection luminescence of these heterojunctions, and the "superinjection" effect, in our view exclude the presence of a "spurious" p−n junction in either the GaAs or the $Al_xGa_{1-x}As$.

Electrical Properties and Band Model of Heterojunctions

1. Capacitance − Voltage Characteristic.   For all the samples studied, the voltage dependence of the capacitance is described by the expression $C \propto (V_0^C - V)^{-\frac{1}{2}}$ (an abrupt junction). For p-type $Al_xGa_{1-x}As$ −n-type GaAs heterojunctions, no deviation from the straight line $1/C^2 = f(V)$ was observed in the voltage dependence of the capacitance in the range of positive bias 0–0.8 V (i.e., the impurity distribution in the space-charge layer was uniform). The capacitive voltage intercepts V_0^C for various compositions of p-type $Al_xGa_{1-x}As$ were in the range from 1.18 to 1.4 V, and agreed with the current−voltage intercepts V_0^I. The thickness of the space-charge layer for samples in this group at zero bias was $W_0 = (8-14) \cdot 10^{-6}$ cm. In n-type $Al_xGa_{1-x}As$ −p-type GaAs heterojunctions, the GaAs substrate had a hole density

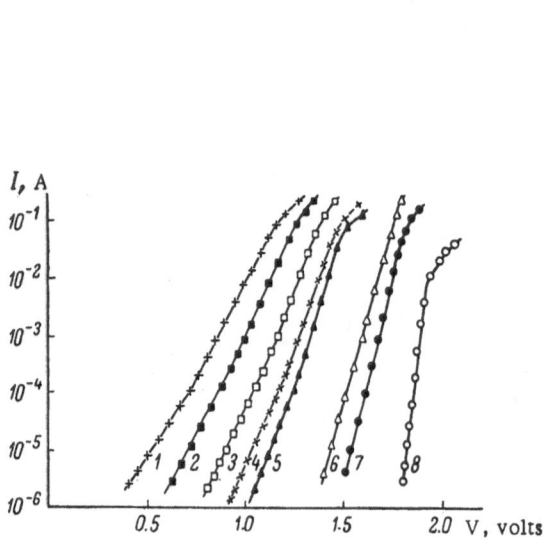

Fig. 3. Forward branch of the current−voltage characteristic for p-type GaAs−n-type Al_xGa_{1-x} As heterojunctions. T, °K: 1) 493; 2) 433; 3) 373; 4) 333; 5) 295; 6) 180; 7) 140; 8) 77.

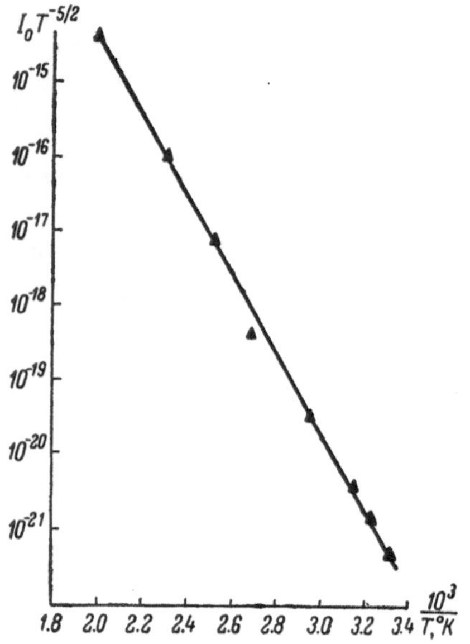

Fig. 4. Temperature dependence of $I_{01}T^{-5/2}$.

$1.5 \cdot 10^{19}$ cm^{-3}. The space-charge region (thickness $W_0 \sim 10^{-4}$ cm) was almost entirely in the wide-gap material. The concentration of ionized donors in the latter was found from the slope of the relation $1/C^2 = f(V)$. For a positive bias, a deflection toward lower concentrations appeared.

2. Forward Branch of the Current − Voltage Characteristic. The dependence of the current on the voltage was exponential for both types of heterojunction:

$$I = I_0 e^{\frac{qV}{\eta kT}}. \tag{1}$$

However, the mechanisms of current flow in these heterojunctions were probably different.

(a) n-Type Al_xGa_{1-x} As − p-Type GaAs Heterojunctions. In the temperature range 300-500°K, the forward branch consists of two exponential parts:

$$I = I_{01} e^{\frac{qV}{\eta_1 kT}} + I_{02} e^{\frac{qV}{\eta_2 kT}}. \tag{2}$$

In the first part, $\eta_1 = 2$; in the second part, $\eta_2 = 1.2$-1.4. Below room temperature, the dependence of the current on the voltage has only one section, with $\eta_1 = 2$ (Fig. 3). The first part of the current−voltage characteristic for small currents has the form typical of a current due to recombination in the space-charge layer. According to the model of Shockley, Noyce, and Sah [9], the pre-exponential factor is

$$I_{01} = \frac{kTn_i}{\tau_0 E_m} \propto \frac{kT^{5/2}}{\tau_0 E_m} e^{-\frac{E_g}{2kT}}, \tag{3}$$

where E_m is the maximum field in the space-charge layer, τ_0 the effective lifetime, and n_i the intrinsic carrier density in the material where the recombination occurs. It is seen from

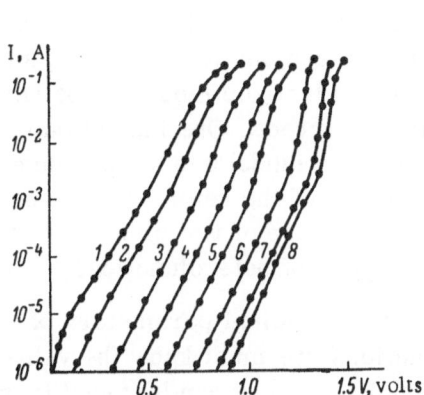

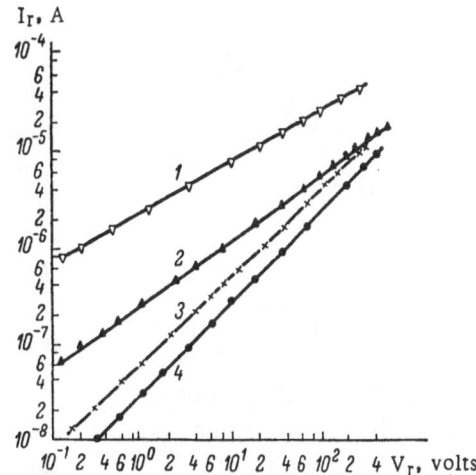

Fig. 5. Forward branch of the current−voltage characteristic for n-type GaAs−p-type $Al_xGa_{1-x}As$ heterojunctions. T, °K: 1) 535; 2) 525; 3) 333; 4, 5) 300; 6) 275; 7) 110; 8) 77.

Fig. 6. Reverse current as a function of the voltage at various temperatures. T, °K: 1) 583; 2) 450; 3) 420; 4) 300–400.

Fig. 4 that an exponential temperature dependence of $I_{01}T^{-5/2}$ is observed. The found value $E_g = 1.96$ eV is close to the forbidden band width for $Al_xGa_{1-x}As$ of this composition at 0°K ($E_g = 1.80$ eV for 300°K; see Table 1). The effective lifetime in $Al_xGa_{1-x}As$, determined from Eq. (3), is $\tau_0 \approx 3 \cdot 10^{-8}$ sec.

(b) p-Type $Al_xGa_{1-x}As$−n-Type GaAs Heterojunctions. The current−voltage characteristic (Fig. 5) again shows two exponential parts. In the first part, the slope of log I = f (V) (77-535°K) remains constant with varying temperature; this is typical of carrier tunneling. The tunnel mechanism of current flow in this range is also indicated by the relatively slight temperature dependence of the pre-exponential factor I_{01}.

The second part is described by $I = I_{02}e^{\frac{qV}{\eta_2 kT}}$, where $\eta_2 = 1.15$-1.3 for T > 300°K; $\eta_2 = 2$ for T < 300°K. This relationship indicates that the current is mainly due to recombination in the bulk of GaAs at high temperatures and due to recombination in the space-charge layer at low temperatures. The voltage intercepts V_0^I are 1.2-1.4 V for p-type $Al_xGa_{1-x}As$−n-type GaAs heterojunctions and 1.4-1.7 V for n-type $Al_xGa_{1-x}As$−p-type GaAs junctions, moving to larger values with increasing aluminum content; for V > V_0^I, the forward current varies linearly with the applied voltage for both types of heterojunction.

3. Reverse Branch. For p-type $Al_xGa_{1-x}As$−n-type GaAs heterojunctions, the reverse breakdown voltages were 10-15 V, decreasing with decreasing temperature. The calculated value of the critical field at breakdown was ~10^5 V/cm. For n-type $Al_xGa_{1-x}As$−p-type GaAs heterojunctions, the breakdown voltage was usually 300-400 V, and for some samples, 800-900 V. Figure 6 shows the function log I_r = f (log V_r) in the temperature range 300-590°K. Throughout this temperature range, we have $I_r \propto V^\gamma$. In the range 300-400°K, the reverse current is independent of the temperature, and $\gamma = 1$; this is probably a result of surface "leaks." As the temperature increases, γ decreases to 0.5 at 590°K, so that thermal generation in the space-charge layer appears to become the predominant mechanism.

4. The Band Model. The principal parameters of the band model of the heterojunction are the conduction band discontinuity ΔE_c and the valence band discontinuity ΔE_v:

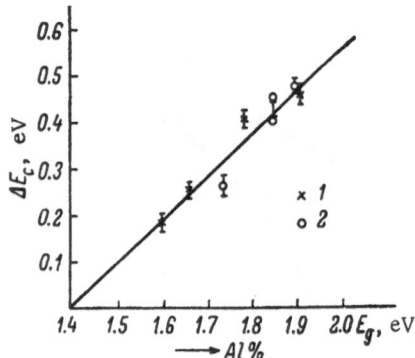

Fig. 7. Dependence of the conduction band discontinuity on the forbidden band width of $Al_xGa_{1-x}As$ in $GaAs-Al_xGa_{1-x}As$ heterojunctions. 1) n-type $Al_xGa_{1-x}As$ −p-type GaAs; 2) p-type $Al_xGa_{1-x}As$ −n-type GaAs.

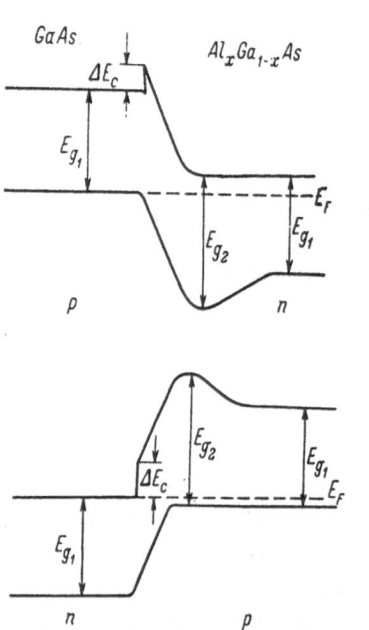

Fig. 8. Band model of p−n and n−p heterojunctions in the GaAs −AlAs system.

$$\left. \begin{array}{l} \Delta E_c = \theta_1 - \theta_2, \\ \Delta E_v = (E_{g_2} + \theta_2) - (E_{g_1} + \theta_1) = \Delta E_g - \Delta\theta. \end{array} \right\} \quad (4)$$

The expressions (4) are valid when there are no charged states at the heterojunction interface, and they can be used only if this idealized model of the heterojunction is found to be correct. One indication of the correctness of the chosen model can be the adequacy of the principal parameters of the heterojunction as determined from the luminescence of two or more different (e.g. p−n and n−p) types of heterojunction.

To determine the principal parameters of the band model of heterojunctions, we must know the values of the contact potential V_D. These can be found from measurements of the current and capacitative voltage intercepts. We have studied p−n and n−p heterojunctions in the AlAs−GaAs system. In both cases, the GaAs was heavily doped, and the Fermi level was above the bottom of the conduction band in n-type GaAs, and below the top of the valence band in p-type GaAs. Then,

$$\Delta E_c = V_D^{n-p} - E_{g_1} + \mu_1 - \mu_2 = E_{g_2} - V_D^{p-n} - \mu_2 + \mu_1. \quad (5)$$

Knowing the positions of the Fermi level in the two semiconductors, their forbidden band widths and the values of the contact potentials, we can easily find ΔE_c and ΔE_v. Figure 7 shows the values of ΔE_c thus determined for various aluminum contents in the $Al_xGa_{1-x}As$ solid solution near the interface. It is seen that ΔE_c increases linearly with increasing width of the forbidden band in the wide-gap part of the heterojunction, while ΔE_v is negligible (of the order of kT). Thus, for all the $Al_xGa_{1-x}As$ compositions studied, the band model of heterojunctions is as shown in Fig. 8.

Discussion of Results

Information about the processes which occur in $Al_xGa_{1-x}As$−GaAs heterojunctions has been obtained by two independent routes: measurements of electrical properties and injection luminescence. It is naturally interesting to compare the results.

1. The electroluminescence spectra of p−n and n−p heterojunctions are in qualitative agreement with the heterojunction band model derived from electrical measurements. Because of the small value of ΔE_v, the holes from p-type GaAs having a dopant concentration several orders of magnitude higher than that in n-type $Al_xGa_{1-x}As$ are injected into the wide-gap material; the injection luminescence spectra of n-type $Al_xGa_{1-x}As$−p-type GaAs heterojunctions therefore show a radiation band due to the solid solution. For p-type $Al_xGa_{1-x}As$−n-type GaAs heterojunctions, the conduction band discontinuity ΔE_c prevents injection into the wide-gap material, and there are therefore no bands due to radiation in that material.

2. The variable forbidden band width of the $Al_xGa_{1-x}As$ solid solutions indicates a connection between the luminescence spectrum of these solid solutions and the current—voltage relations for n-type $Al_xGa_{1-x}As$—p-type GaAs heterojunctions. The change in the spectral distribution of the luminescence of the solid solutions, observed at 300°K for varying forward current, is in good agreement with the existence of two parts of the current—voltage characteristic of the n—p heterojunction. At a small positive bias, when the current is due to recombination in the space-charge layer, the spectral distribution of luminescence from a solid solution with variable forbidden band width shows that recombination takes place in a narrow layer near the interface. As the current becomes larger, recombination within the material predominates and radiative recombination in a wide layer is observed for solid solutions whose forbidden band width varies from a maximum at the interface to a value close to that of GaAs. At 77°K, where the current—voltage relation is described by $I = I_0 \exp(qV/2kT)$, the luminescence spectrum of n-type $Al_xGa_{1-x}As$ over a wide range of currents corresponds to recombination in a narrow layer near the interface.

Thus, in the n—p heterojunctions studied, when the bias is fairly small, there appears to be a predominant injection into the wide-gap material, and the current—voltage relation is governed by recombination processes occurring in that material. This can be explained bearing in mind that the valence band discontinuity is negligible, and the hole density in p-type GaAs is 3 to 4 orders of magnitude greater than the donor concentration in n-type $Al_xGa_{1-x}As$. However, at high currents, there is a redistribution of the current due to the special nature of injection in heterojunctions [6, 7].

These results give reason to suppose that the defect density at the interface in $Al_xGa_{1-x}As$ — GaAs heterojunctions is small and does not appreciably affect their injection properties. The absence of charged states at the interface enables us to determine the principal parameters ΔE_c and ΔE_v of the heterojunction.

The authors thank V. M. Tuchkevich for his interest in this work, and E. A. Gamilko and A. P. Ermakov for their help in preparing the samples.

Literature Cited

1. N. A. Goryunova, The Chemistry of Diamond-like Semiconductors, Chapman and Hall, London (1965).
2. R. L. Anderson, Solid State Electron., 5:341 (1962).
3. Zh. I. Alferov, V. I. Korol'kov, and M. K. Trukan, Fiz. Tverd. Tela, 8:3513 (1966).
4. F. Herman, J. Electronics, 1:103 (1955).
5. J. F. Black and S. M. Ku, J. Electrochem. Soc., 113:249 (1966).
6. Zh. I. Alferov and D. Z. Garbuzov, Fiz. Tverd. Tela, 7:2375 (1965).
7. Zh. I. Alferov, V. B. Khalfin, and R. F. Kazarinov, Fiz. Tverd. Tela, 8:3102 (1966).
8. Zh. I. Alferov, V. M. Andreev, V. I. Korol'kov, E. L. Portnoi, and D. N. Tret'yakov, Fiz. Tekh. Poluprov., 2:1016 (1968).
9. C. T. Sah, R. N. Noyce, and W. Shockley, Proc. IRE, 45:1228 (1957).

RECOMBINATION MECHANISMS IN PHOTOSENSITIVE FILMS OF THE PbS GROUP

B. V. Izvozchikov and I. A. Taksami

A. F. Ioffe Physicotechnical Institute
Academy of Sciences of the USSR, Leningrad

Electrical and photoelectric properties of films of the PbS group were examined as functions of temperature and pressure. The results are explained by assuming that the forbidden band contains one deep level and some shallow levels that are completely ionized in the temperature range from 300 to 90°K. At "low" temperatures (150 to 90°K), the trapping levels become recombination levels.

There have been many papers on the mechanism of photoconduction and carrier recombination in films of the PbS group. Nevertheless, there is still no single theory which satisfactorily accounts for the variation of the electrical and photoelectric properties of lead chalcogenide films over a wide range of temperatures.

A widely used model due to Petritz [1] gives a good description of the properties of films of the PbS group in the temperature range from 300 to 200°K. According to this model, the films consist of crystallites separated by oxide barriers. The barrier height E_b is related to the thickness of the oxide film and to the crystallite packing density. The dark conductivity of a film is given by

$$\sigma_d = e n_p \mu^*, \tag{1}$$

where $\mu^* = \mu \exp(-E_b/kT)$.

The acceptor levels of oxygen, which have a high probability of minority-carrier (electron) trapping and a low probability of subsequent hole trapping, are located at a distance E_{tv} above the edge of the valence band and E_{tc} below the bottom of the conduction band.

This model has been experimentally confirmed by measurements of the temperature dependences of various parameters of photosensitive films of the PbS group, such as the dark conductivity and the carrier density, mobility, and lifetime* in the temperature range from

---

* It has been shown [2] that, in this temperature range, the photoconductivity time constant is equal to the majority carrier lifetime.

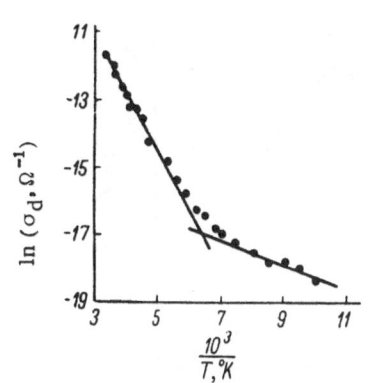

Fig. 1. Temperature dependence of the dark conductivity.

300 to 200°K [1, 3, 4]. At lower temperatures, however, these dependences, apart from the mobility,[†] are quite different [4, 5] from those in the above-mentioned range, and are not accounted for by Petritz's model. In some papers [4] the change in the temperature dependences of the parameters of films of the PbS group has been explained by the influence of background radiation; in others [5], the interpretation of the results has been based on a model having two deep levels in the forbidden band of PbS, from which it follows that the majority carrier lifetime and the photoconductivity time constant are equal only at room temperature, and differ considerably at lower temperatures.

To elucidate the true mechanisms of conduction and recombination of carriers in films of the PbS group at "low" temperatures, it is necessary to analyze a series of parameters which determine these mechanisms.

In the present work, the aim was to study the electrical and photoelectric properties of polycrystalline films of PbS and PbSe over a wide range of temperatures at atmospheric pressure, and their pressure dependence at 93°K.

The measurements were made using a large number of high-sensitivity PbS and PbSe films prepared at our Institute.

The techniques of producing high pressures and carrying out the measurements have been described in [6, 7]. The effect of background radiation was eliminated by the design of the instrument used for measuring the temperature dependences of the parameters of the films.

The results of the experiments using PbS and PbSe films were basically similar, and therefore, we shall give below only the results for lead sulfide films.

Figures 1-3 show, on a semilogarithmic scale, typical temperature dependences of the dark conductivity σ_d, the carrier density n_p, the carrier mobility μ^*, the photoconductivity time constant τ, and the photoconductivity $\Delta\sigma$. In the temperature range from 293 to 170°K, the curves in these diagrams are in good agreement, both qualitative and quantitative, with the majority-carrier model [1]. However, at "low" temperatures (150 to 93°K), Figs. 1 and 3 show that σ_d and τ vary with temperature much less rapidly than at "high" temperatures.[‡] It is interesting to note that the curves of $\ln \sigma_d = f(T)$ at "low" temperatures and $\ln \mu^* = f(T)$ have almost the same slope.

The temperature dependence of the photoconductivity (Fig. 3) is consistent with the temperature dependences of μ^* and τ (Figs. 2 and 3); this indicates that the majority carrier lifetime is the same as the photoconductivity time constant throughout the temperature range studied.

The similarity of the temperature dependences of σ_d and τ can be explained by assuming that the forbidden band of PbS contains one deep level (E = 0.20 eV) and some shallow levels that are completely ionized in the temperature range studied. In the range from 293 to 170°K,

---

[†] It has been found experimentally [5] that, in the temperature range from 300 to 120°K, the variation of the carrier mobility in PbS layers is accurately represented by $\exp(-E_b/kT)$.

[‡] The temperature dependences reported here for the parameters of PbS films are basically similar to those published elsewhere [4, 5].

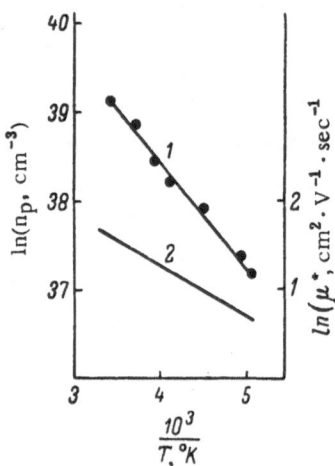

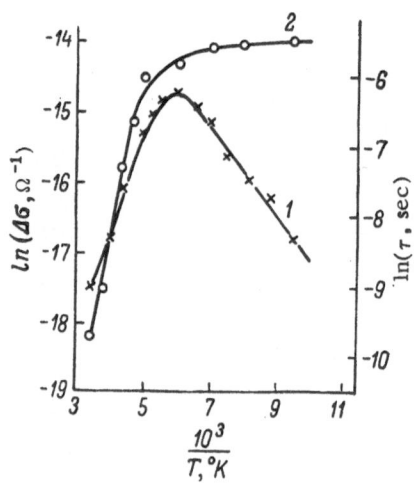

Fig. 2. Temperature dependences of the carrier density (1) and mobility (2).

Fig. 3. Temperature dependences of the photoconductivity (1) and the photoconductivity time constant (2).

the carrier density is determined by the deep acceptor levels of oxygen and varies rapidly with temperature; in the range from 150 to 93°K, it is determined by the shallow, completely ionized. acceptor levels, and is therefore independent of temperature.

The temperature dependence of τ in the range from 150 to 93°K can be explained by assuming that, at sufficiently low temperatures, the thermal transfer of electrons from the trapping levels to the conduction band becomes improbable, and these levels become recombination centers. Since the carrier density is independent of temperature in this range, we find that the lifetime is only slightly temperature-dependent, in accordance with the variation of the thermal velocity of the carriers and the capture cross section.

The foregoing interpretation of the temperature dependences of the parameters of the films of the PbS group is confirmed by studies of the influence of pressure on their "low-temperature" properties. Increasing the pressure on PbS or PbSe films is to some extent equivalent to raising the temperature (the density and mobility of the majority carriers increase, and their lifetime decreases) [7]. Calculation and experiment show that pressures above 6000-8000 atm acting on a PbS film at 93°K cause the "apparent" temperature of the layer to be raised above the "low-temperature" range. Consequently, we shall here discuss the results of experiments in the range from 1 to 6000 atm.

Figure 4 shows, on a semilogarithmic scale, the pressure dependences of the dark conductivity for several PbS films at 93°K. The pressure dependence may be written

$$\frac{\partial \ln \sigma_d^P / \sigma_d^0}{\partial P} = \frac{\partial \ln n_p^P / n_p^0}{\partial P} + \frac{\partial \ln \mu_p^* / \mu_0^*}{\partial P}. \tag{2}$$

It follows from Fig. 4 that

$$\frac{\partial \ln \sigma_d^P / \sigma_d^0}{\partial P} = 3.9 \cdot 10^{-4} \, \text{atm}^{-1}. \tag{3}$$

As already mentioned, the scattering of carriers in lead chalcogenide films takes place mainly at intercrystallite barriers. A decrease in the barrier height, due to a change in the spacing between the crystallites when pressure is applied, depends on the compressibility of

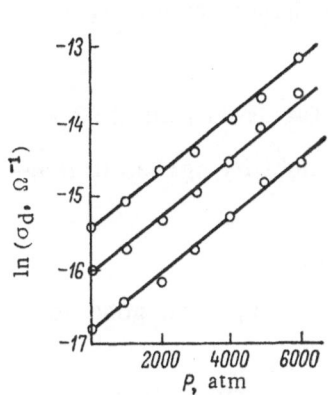

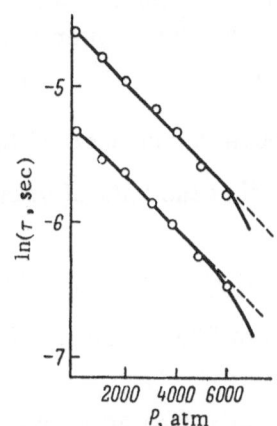

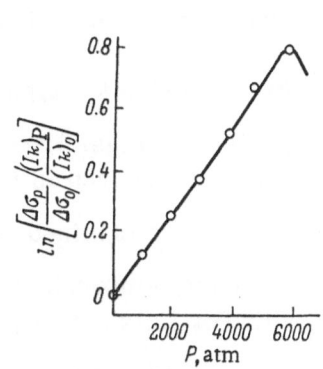

Fig. 4. Pressure dependence of the dark conductivity for several PbS films. T = 93°K.

Fig. 5. Pressure dependence of the hole lifetime for PbS films. T = 93°K.

Fig. 6. Typical pressure of $\ln\left[\dfrac{\Delta\sigma_p}{\Delta\sigma_0}\Big/\dfrac{(Ik)_P}{(Ik)_0}\right]$ for PbS film. T = 93°K. I is the incident radiation intensity and k the absorption coefficient.

the substrate.* It may be supposed that this compressibility does not alter greatly over the pressure and temperature ranges studied [8]. Then, the rate of change of the height of the intercrystallite barriers with pressure at 93°K must be approximately equal to the rate of change at room temperature. According to [7], we have

$$\frac{\partial \ln \mu_P^*/\mu_0^*}{\partial P}\bigg|_{93°\,K} \approx 4 \cdot 10^{-4}\ \text{atm}^{-1}. \tag{4}$$

The agreement of the numerical coefficients in Eqs. (3) and (4) shows that at 93°K the pressure dependence of the conductivity (for $p \le 6000$ atm) is mainly due to a change in the mobility. The carrier density is independent of pressure in the range concerned, just as it is independent of temperature between 150 and 93°K.

Figure 5 shows the pressure dependence of the carrier lifetime for several PbS film at 93°K. The rate of change of τ is approximately the same for all the samples studied:

$$\frac{\partial \ln \tau^P/\tau^0}{\partial P} - -1.9 \cdot 10^{-4}\ \text{atm}^{-1}. \tag{5}$$

The value obtained for $\partial \ln \dfrac{\tau^P}{\tau^0}\Big/\partial P$ is considerably less than that expected if the oxygen acceptor levels continue to act as α-type trapping levels [9] in the same way as they do near room temperature. Since the forbidden band width of films in the PbS group varies with pressure at a uniform rate throughout the temperature range from 300 to 90°K [6], we may reasonably suppose that the energy gap between the bottom of the conduction band and the oxygen acceptor level likewise varies with pressure at a constant rate in the same temperature range. A simple calculation shows that the rate of change of the hole lifetime with pressure at 93°K must then be $-8.5 \cdot 10^{-4}$ atm^{-1} [7]. Consequently, from the results obtained, we may conclude that the hole lifetime at 93°K is independent of the gap E_{tc} between the bottom of the conduction

---

*We have experimentally confirmed this in a study of the pressure dependence of the conductivity of PbS and PbSe films on substrates having various compressibilities.

band and the acceptor level. This confirms the hypothesis stated above (in order to explain the "low-temperature" behavior of τ) that the oxygen acceptor levels change from α-type trapping levels to recombination levels.

Figure 6 shows a typical pressure dependence of $\ln\left[\frac{\Delta\sigma_P}{\Delta\sigma_0}\Big/\frac{(Ik)_P}{(Ik)_0}\right]$ for PbS films at 93°K. A comparison of Figs. 5 and 6 shows that the rate of variation of the mobility agrees in magnitude with that derived in Eq. (4).

The results obtained in this work yield the following conclusions.

1. In the temperature range from 300 to 170°K, films of the PbS group show good agreement with Petritz's model [1].

2. In the temperature range from 150 to 90°K, the carrier density in films of the PbS group is determined by shallow, completely ionized levels, and the deep acceptor levels of oxygen are transformed from α-type trapping levels into recombination levels; this is responsible for the slight temperature dependence of the dark conductivity and majority carrier lifetime in this temperature range.

Literature Cited

1. R. L. Petritz, F. L. Lummis, H. E. Sorrows, and J. F. Woods, in: R. H. Kingston (ed.), Semiconductor Surface Physics, University of Pennsylvania Press, Philadelphia (1957), p. 229.
2. F. M. Klaassen and J. Blok, Physica, 24:975 (1958).
3. F. M. Klaassen, J. Blok, H. C. Booy, and F. J. de Hoog, Physica, 26:623 (1960).
4. J. N. Humphrey, Appl. Optics, 4:665 (1965).
5. G. Giroux, Can. J. Phys., 41:1840, 1856 (1963).
6. B. V. Izvozchikov and I. A. Taksami, Fiz. Tekh. Poluprov., 1:152 (1967).
7. B. V. Izvozchikov and I. A. Taksami, Fiz. Tekh. Poluprov., 1:1178 (1967).
8. P. W. Bridgman, The Physics of High Pressure, Bell, London (1949).
9. S. M. Ryvkin, Photoelectric Effects in Semiconductors, Consultants Bureau, New York (1964).

AN INVESTIGATION OF
A PRESSURE-SENSITIVE TRANSISTOR

L. N. Syrkin and N. N. Feoktistova

A theoretical analysis is given of the dependence of the parameters of a transistor on a force F applied to the emitter and distributed nonuniformly across its surface. The analysis is carried out on the assumption that the current is due to diffusion of the minority carriers. The changes in the transistor parameters are due to the influence of mechanical stress σ on the effective width of the forbidden band. The dependence of the minority carrier lifetime on σ is ignored. The transistor is represented by two parallel transistors with equivalent areas S_F and $S_0 - S_F$, the first of which is assumed to be subjected to a uniaxial compression and the second is taken to be free of mechanical stresses. The two limiting cases of isotropic and anisotropic distribution of the normal stresses σ_z are considered. In the first case, the gradient $\nabla_z \sigma_z$ in the region of the p−n junction is practically equal to zero whereas in the second case the gradient is very large. It is shown that in the isotropic case the static current gain $B = I_C/I_b$ is independent of the force F. However, the collector (I_C) and base (I_b) currents should depend most strongly on F when the input is short-circuited. In the anisotropic case, the factor B decreases as F increases, and the pressure sensitivity $\Gamma_C = \partial I_C/\partial F$ passes through a maximum; the dependence of I_C on F is strongest under the open-circuit conditions. The main conclusions of the theory are compared with the results of an experimental investigation of germanium planar mesa transistors subjected to pressure concentrated at a point. The principal experimental and theoretical results are in good agreement. However, the presence of a wide zero-sensitivity region and of hysteresis of the dependence of I_C on F are not explained by the theory. It is suggested that these effects may be due to plastic deformation of a crystal near the tip of the needle used to apply the pressure to the transistor.

A theoretical analysis is given of the dependences of the collector and base currents on the mechanical stresses applied in the region of the emitter p−n junction by means of a needle. It is assumed that the current is due to the minority carrier diffusion. Two limiting cases of the elastic stress distribution are considered: isotropic and anisotropic pressure. Expressions are obtained for the alternating components of the collector and base currents in the case of the simultaneous application of static and alternating mechanical stresses. It is shown that the dependence of the pressure sensitivity of the collector current on the applied force passes through a maximum. The principal results of an experimental investigation of pressure− sensitive transistors are reported. These results are compared with the theoretical conclusions.

Introduction

There have been many investigations of the dependences of diode and transistor characteristics on nonuniform mechanical stresses but there is still no agreement about the mechanism of the effects of such stresses.

According to some investigators [1-6], changes in the electrical characteristics of a p−n junction subjected to pressure at a point, applied by means of a needle, are primarily due to a change in the forbidden band width.

According to other investigators, such changes in the electrical characteristics of a p−n junction are principally due to a reduction in the minority carrier lifetime τ caused by the appearance of generation−recombination centers (it is assumed that the concentration of such centers is high) [7-9].

An analysis of the available experimental data shows that, although a fall in τ does indeed take place [9-11], the observed effect can be explained by a change in the lifetime only if it is postulated that τ decreases by a factor of at least 10^8-10^9 under the action of pressure of $\sigma = 10^9$-10^{10} dyn/cm^2 concentrated at a point. This is highly unlikely. Direct measurements of τ in germanium p−n junctions have demonstrated that compressive stresses up to $4 \cdot 10^9$ dyn · cm^{-2} alter τ by a factor not greater than 2-2.5 [10].

The present paper develops further a model according to which the influence of pressure concentrated at a point on the characteristics of a p−n junction is due to a change in the forbidden band width; this model is applied to planar transistors.

In the case of transistors, the effect is mainly due to mechanical stresses applied to the emitter junction: the hole component of the emitter current I_{ep} governs, in the first approximation, the collector current I_c and the electron component of the emitter current I_{en} determines the base current I_b.* The change in I_{ep} under the action of pressure is governed by the change in the minority carrier (hole) density near the emitter junction on the base side; and the change in I_{en} is determined by the change in the electron density on the emitter side of the same junction. Therefore, an investigation of the influence of pressure on the transistor characteristics can be used (in contrast to similar investigations of diode structures) to consider separately changes in the hole and electron components of the emitter current due to various types of stressed state in the region of the p−n junction. There are potentially many practical applications of pressure-sensitive transistors ("piezotransistors"), which combine the functions of an electromechanical transducer and a current amplifier with low input and high output impedances.

Theory

It is interesting to consider thin-base transistors whose collector current is not very sensitive to recombination losses in the base and whose base current is governed primarily by the electron current of the emitter. This allows us to ignore the recombination processes in the base. We shall also ignore the recombination processes on the surface and in the region of the p−n junction, as well as the edge effects and the influence of contacts.

Let us assume that the emitter and collector junctions are parallel to the emitter surface and that a force F, applied through the tip of a needle, is perpendicular to these junctions. We shall ignore the influence of tangential mechanical stresses on the transistor parameters. This does not give rise to appreciable errors because, although the tangential stresses extend

---

*We shall consider specifically a p−n−p transistor.

over a considerable region around the tip of a needle, these stresses are relatively small and they do not give rise to appreciable stress gradients near the point of application of the needle [9, 12].

The influence of the normal stresses σ_z can be described by representing a transistor (subjected to pressure at a point) by two transistors connected in parallel: one of them (the "stressed" transistor) has an equivalent p−n junction area s_F and is subjected to a uniaxial pressure; the other (the "unstressed" transistor) has an equivalent junction area $s_0 - s_F$ and is free of mechanical stresses (s_0 is the actual emitter junction area).

The following inequalities are satisfied under normal working conditions of a transistor:

$$-qU_{be} \gg kT, \quad -q\,(U_{ce} - U_{be}) \gg kT, \tag{1}$$

where U_{be} is the base-emitter voltage; U_{ce} is the collector−emitter voltage; T is the absolute temperature; q is the electronic charge; k is the Boltzmann constant.

Using the conditions given in Eq. (1) and the assumptions listed in the preceding paragraphs, the expressions for the collector I_c^F and the base I_b^F currents of a transistor subjected to pressure can be written in the following form:

$$
\left.
\begin{aligned}
I_c^F &= \frac{s_F}{s_0}\left[-I_{C1}^F \exp\left(-\frac{qU_{be}^F}{kT}\right) + I_{C2}^F\right] + \frac{s_0 - s_F}{s_0}\left[-I_{C1}^0 \exp\left(-\frac{qU_{be}^F}{kT}\right) + I_{C2}^F\right] = \\
&= -I_{C1}^0\left(1 - \varkappa + \varkappa\frac{I_{C1}^F}{I_{C1}^0}\right)\exp\left(-\frac{qU_{be}^F}{kT}\right) + I_{C2}^0\left(1 - \varkappa + \varkappa\frac{I_{C2}^F}{I_{C2}^0}\right), \\
I_b^F &= -I_{b1}^0\left(1 - \varkappa + \varkappa\frac{I_{b1}^F}{I_{b1}^0}\right)\exp\left(-\frac{qU_{be}^F}{kT}\right) + I_{b2}^0\left(1 - \varkappa + \varkappa\frac{I_{b2}^F}{I_{b2}^0}\right).
\end{aligned}
\right\} \tag{2}
$$

These expressions are deduced from the formula of Moll and Ebers [13]. Here, $\varkappa = s_F/s_0$, and the values of I_{c1}, I_{c2}, I_{b1}, I_{b2} are expressed in terms of the reverse saturation currents of the emitter and collector junctions I_{e0}, I_{c0}, and the normal and inverse current gains α_N, α_I in the common-base configuration

$$I_{C1} = \frac{\alpha_N I_{e0}}{1 - \alpha_N\alpha_I}, \quad I_{b1} = \frac{(1 - \alpha_N)\,I_{e0}}{1 - \alpha_N\alpha_I}, \quad I_{c2} = -\frac{I_{c0}}{1 - \alpha_N\alpha_I}, \quad I_{b2} = \frac{(1 - \alpha_I)\,I_{c0}}{1 - \alpha_N\alpha_I}. \tag{3}$$

These quantities have the following physical meaning: I_{c1}, I_{b1} are the collector and base currents in the case of inverse operation provided that $|U_{be}| = |U_{ce}| \gg kT/q$; I_{c2}, I_{b2} are the same currents in the case of normal operation of the transistor whose input is short-circuited ($U_{be} = 0$), provided that $|U_{ce}| \gg kT/q$. In the formulas of Eq. (2), the superscript F means that a force F is applied to the transistor, and the superscript 0 that no force is applied. The currents I_{c1}, I_{c2}, I_{b1}, I_{b2} and the current gains α_N, α_I in Eq. (3) can also have the superscripts F or 0. The pressure dependences of these quantities are governed by the pressure dependences of the electron and hole components of the currents I_{e0} and I_{c0} in the "stressed" transistor:

$$I_{e0} = I_{en0} + I_{ep0} = qs_0\left(n_{ep0}\sqrt{\frac{D_{en}}{\tau_{en}}} + p_{en0}\sqrt{\frac{D_{ep}}{\tau_{ep}}}\right), \tag{4}$$

$$\alpha_N = \frac{I_{ep0}}{I_{e0}} = \frac{I_{ep0}}{I_{ep0} + I_{en0}}. \tag{5}$$

(The values of I_{c0} and α_I are given by similar expressions.)

In Eqs. (4) and (5) the symbols I_{ep0}, I_{en0} represent the hole and electron components of I_{e0}; p_{en0}, D_{ep}, τ_{ep} are, respectively, the equilibrium density, the diffusion coefficient, and the lifetime of holes in the n-type region (i.e., in the base) near the emitter junction; n_{ep0}, D_{en}, τ_{en} are the corresponding parameters of the electrons in the emitter near the junction. The values of D and τ are assumed to be independent of pressure.

Uniaxial pressure is known to reduce the effective forbidden band width E_g of germanium and silicon; this causes p_{n0} and n_{p0} to rise exponentially. In particular,

$$p_{n0}^F = p_{n0}^0 \exp\left(-\frac{\Delta E_g}{kT}\right) = p_{n0}^0 \exp\left(-\frac{\gamma F}{s_F kT}\right) = p_{n0}^0 \exp\left(-\frac{\gamma \sigma_z}{kT}\right), \tag{6}$$

where γ is a coefficient which depends on the direction of the applied force relative to the crystallographic axes and is related by simple expressions to the deformation potential constants.*

In order to develop an approximate theory of a transistor subjected to pressure at a point, it is necessary to assume some simplified model of the distribution of the mechanical stresses under the pressure-transmitting needle. Such a model is needed because the actual distribution of stresses is very complex and its details are unimportant in the derivation of the basic relationships governing the behavior of a pressure-sensitive transistor.

We shall assume that the stressed region of our transistor is in the form of a cylinder with an equivalent cross sectional area s_F. The axis of this cylinder is perpendicular to the p−n junction planes and the normal component of the stresses σ_z differs from zero only within this cylindrical volume. Moreover, in the first approximation, the normal component of the stresses depends only on the depth of a given layer.† Thus, the quantity ΔE_g in Eq. (6) is a function of z. Irrespective of the nature of this function, we can introduce the following quantities: ΔE_{g1} is the value of ΔE_g near the emitter junction on the emitter side; $\Delta E_{g1} + \delta E_{g1}$ is the value of ΔE_g near the emitter junction on the base side; ΔE_{g2} is the value of ΔE_g near the collector junction on the base side; $\Delta E_{g2} + \delta E_{g2}$ is the value of ΔE_g near the collector junction on the collector side. It is evident that δE_{g1} and δE_{g2} are governed by the average gradients of σ_z near the emitter and collector p−n junctions whose equivalent thicknesses are δ_e and δ_c:

$$\delta E_{g1} = \gamma \delta_e \nabla_z \sigma_{z\,e} \quad \delta E_{g2} = \gamma \delta_c \nabla_z \sigma_{z\,C}. \tag{7}$$

It follows from Eqs. (4)–(7) that, in the case of the "stressed" transistor, we have

$$I_{e0}^F = I_{en0}^F + I_{ep0}^F = I_{en0}^0 \exp\left(-\frac{\Delta E_{g1}}{kT}\right) + I_{ep0}^0 \exp\left(-\frac{\Delta E_{g1} + \delta E_{g1}}{kT}\right), \quad \Big\}$$
$$I_{C0}^F = I_{Cn0}^F + I_{Cp0}^F = I_{Cn0}^0 \exp\left(-\frac{\Delta E_{g2} + \delta E_{g2}}{kT}\right) + I_{Cp0}^0 \exp\left(-\frac{\Delta E_{g2}}{kT}\right), \quad \Big\} \tag{8}$$

$$\alpha_N^F = \left[1 + \frac{1}{\beta_N^0} \exp\left(\frac{\delta E_{g1}}{kT}\right)\right]^{-1}; \quad \alpha_I^F = \left[1 + \frac{1}{\beta_I^0} \exp\left(-\frac{\delta E_{g2}}{kT}\right)\right]^{-1}, \tag{9}$$

where β_N and β_I are, respectively, the normal and inverse current gains in the common-emitter configuration $\left(\beta = \frac{\alpha}{1 - \alpha}\right)$. It follows from Eqs. (9) and (7) that the dependences of $\alpha_{N;I}$ and $\beta_{N;I}$ on F can be observed only for nonzero stress gradients $\nabla_z \sigma_z$ near the p−n junctions.

---

* The coefficient γ for germanium and silicon subjected to uniaxial compression along the [111] direction is negative and its values are, respectively, $-19.6 \cdot 10^{-12}$ and $-10.9 \cdot 10^{-12}$ eV · cm² · dyn⁻¹ [1].

† This approximation is permissible although it does not include an explicit dependence of s_F on z. In fact, the current I_C^F depends exponentially on σ_z and is related to s_F and $\varkappa$ by a relatively weak linear dependence.

The substitution of Eqs. (8) and (9) into Eqs. (3)-(5) gives

$$y_1 = \frac{I_{b1}^F}{I_{b1}^0} = \Phi \exp\left(-\frac{\Delta E_{g1}}{kT}\right), \quad y_2 = \frac{I_{b2}^F}{I_{b2}^0} = \Phi \exp\left(-\frac{\Delta E_{g2} + \delta E_{g2}}{kT}\right),$$

$$y_3 = \frac{I_{C1}^F}{I_{C1}^0} = \Phi \exp\left(-\frac{\Delta E_{g1} + \delta E_{g1}}{kT}\right),$$

$$y_4 = \frac{I_{C2}^F}{I_{C2}^0} = \Phi\left[1 + \frac{1}{\beta_I^0}\exp\left(-\frac{\delta E_{g2}}{kT}\right)\right]\exp\left(-\frac{\Delta E_{g2}}{kT}\right),$$

(10)

where

$$\Phi = (1 - a_I^0 a_N^0) \frac{\left[1 + \frac{1}{\beta_N^0}\exp\left(\frac{\delta E_{g1}}{kT}\right)\right]\left[1 + \frac{1}{\beta_I^0}\exp\left(-\frac{\delta E_{g2}}{kT}\right)\right]}{\frac{1}{\beta_N^0}\exp\left(\frac{\delta E_{g1}}{kT}\right) + \frac{1}{\beta_I^0}\exp\left(-\frac{\delta E_{g2}}{kT}\right) + \frac{1}{\beta_N^0 \beta_I^0}\exp\left(\frac{\delta E_{g1} - \delta E_{g2}}{kT}\right)}.$$

Equations (10) and (2) give the dependences of I_c and I_b on F for an arbitrary distribution of the stresses σ_z along the z axis.

Since $U_{be}^F = E_{be} - Z_\Gamma I_b^F$, where E_{be} is the emf of the external source and Z_Γ is the input resistance of the transistor, it follows that the expressions in Eqs. (2) and (10) can be regarded as a system of transcendental equations for I_c^F and I_b^F. Differentiating these equations with respect to F and assuming that — in a certain range of values of F — the quantities s_F and $\varkappa$ are independent of F, we obtain a system of linear equations for the alternating components of the collector and base currents (i_c, i_b), which appear in the presence of an alternating component of the force $f \ll F$. This system can be solved for i_b and i_c:

$$i_b = \frac{\partial I_b^F}{\partial F}f = \frac{\varkappa f\left[I_{b2}^0\frac{dy_2}{dF} - I_{b1}^0\frac{dy_1}{dF}\exp\left(-\frac{qU_{be}^F}{kT}\right)\right]}{1 + I_{b1}^0(1 - \varkappa + \varkappa y_1)\frac{qZ_\Gamma}{kT}\exp\left(-\frac{qU_{be}^F}{kT}\right)},$$

$$i_c = \frac{\partial I_c^F}{\partial F}f = -i_b\frac{qZ_\Gamma I_{C1}^0}{kT}(1 - \varkappa + \varkappa y_3)\exp\left(-\frac{qU_{be}^F}{kT}\right) -$$
$$- I_{C1}^0\varkappa f\frac{dy_3}{dF}\exp\left(-\frac{qU_{be}^F}{kT}\right) + I_{C2}^0\varkappa f\frac{dy_4}{dF}.$$

(11)

Before comparing the theoretical conclusions with the experimental data, we shall consider two idealized special cases.

A. I s o t r o p i c P r e s s u r e .   In this case, the dependence of σ_z on z is weak ($\nabla_z\sigma_z \to 0$) so that we may assume that σ_z is the same on both sides of both p—n junctions. It follows directly that

$$\Delta E_{g1} = \Delta E_{g2} = \Delta E_g, \quad \delta E_{g1} = \delta E_{g2} = 0, \quad \Phi = 1,$$
$$y_1 = y_2 = y_3 = y_4 = \exp\left(-\frac{\Delta E_g}{kT}\right) = \exp\left(-\frac{\gamma F}{s_F kT}\right).$$

(12)

The forbidden band width throughout the stressed region of the transistor decreases by the same amount and, therefore, the relative increase is the same for the electron and hole components of the emitter and collector currents.

The alternating components of the currents can be deduced by substituting Eq. (12) into Eq. (11).

The collector I_c^F and base I_b^F currents are then given by the formulas

$$I_c^F = \left[-I_{C1}^0 \exp\left(-\frac{qU_{be}^F}{kT} \right) + I_{C2}^0 \right] \left[1 - \varkappa + \varkappa \exp\left(-\frac{\gamma F}{s_F kT} \right) \right], \left. \vphantom{\begin{array}{c} 1 \\ 1 \end{array}} \right\}$$

$$I_b^F = \left[-I_{b1}^0 \exp\left(-\frac{qU_{be}^F}{kT} \right) + I_{b2}^0 \right] \left[1 - \varkappa + \varkappa \exp\left(-\frac{\gamma F}{s_F kT} \right) \right], \left. \vphantom{\begin{array}{c} 1 \\ 1 \end{array}} \right\} \tag{13}$$

$$i_b = I_b^F A f \left[1 - \frac{qZ_\Gamma (I_b^F - I_{b2}^F)}{kT} \right]^{-1}, \quad i_c = I_c^F A f \left[1 + \frac{\dfrac{qZ_\Gamma I_b^F}{kT}\left(1 - \dfrac{I_{C2}^F}{I_c^F} \right)}{1 - \dfrac{qZ_\Gamma I_b^F}{kT}\left(1 - \dfrac{I_{b2}^F}{I_b^F} \right)} \right]^{-1}, \tag{14}$$

where

$$A = -\frac{\gamma}{s_0 kT} \exp\left(-\frac{\gamma F}{\varkappa s_0 kT} \right) \left[1 - \varkappa + \varkappa \exp\left(-\frac{\gamma F}{\varkappa s_0 kT} \right) \right]^{-1}. \tag{14a}$$

The pressure sensitivity of the transistor $\Gamma_{c;b}$ is equal to $(1/f)i_{c;b}$ and, in the isotropic case, it has a maximum under constant-voltage conditions at the input ($U_{be}^F = U_{be}^0 = U_{be}$), i.e., when the stronger of the following two inequalities is satisfied:

$$\left| \frac{qZ_\Gamma (I_b^F - I_{b2}^F)}{kT} \right| \ll 1, \quad \left| \frac{Z_\Gamma I_b^F}{U_{be}} \right| \ll 1. \tag{15}$$

In this case, it follows from Eqs. (13) and (15) that

$$\Gamma_c = \Gamma_b B \simeq -\frac{\gamma I_c^0}{s_0 kT} \exp\left(-\frac{\gamma F}{s_F kT} \right), \tag{16}$$

where $B = I_c/I_b$ (in the isotopic case $B^F = B^0 = B$). It is evident from Eqs. (16) and (14a) that when the force F is increased the absolute values of the currents I_b^F, I_c^F and of the sensitivity Γ_c all increase (we must remember that $I_c < 0$, $I_b < 0$, $\gamma < 0$).

Under constant-current input conditions $\left(I_b^F \simeq I_b^0 = I_b; \ \left| \dfrac{qZ_\Gamma (I_b^F - I_{b2}^F)}{kT} \right| \gg 1 \right)$ an appropriate transformation of the expression in Eq. (14) gives

$$\Gamma_c \simeq -AB \frac{kT}{qZ_\Gamma} + \frac{\gamma I_{C0}}{\alpha_N s_0 kT} \exp\left(-\frac{\gamma F}{s_F kT} \right). \tag{16a}$$

Since $Z_\Gamma \to \infty$ and $I_{c0} \ll I_c^0$, we find that, in practice, $\Gamma_c \to 0$ but it still remains negative (< 0), i.e., when F is increased we should observe a slight increase in the absolute value of I_c^F.

B. Anisotropic Pressure. In this case, the stress gradient near the emitter p−n junction is $\nabla_z \sigma_z \neq 0$. The value of σ_z decreases rapidly when z is increased and the mechanical stresses do not strongly influence the minority carrier density at the boundary of the space-charge layer on the base side of the emitter p−n junction (this applies even more so near the collector junction):

$$\Delta E_{g1} \simeq -\delta E_{g1} = \Delta E_g, \quad \Delta E_{g2} \simeq \delta E_{g2} \simeq 0. \tag{17}$$

Under these conditions, the electron component of the saturation current of the emitter junction, I_{en0}, increases with increasing F but the hole component, I_{ep0}, and I_{c0}, are independent of F.

Substitution of Eq. (17) into Eq. (10) gives

$$y_2 = y_3 = y_4 = y_1 \exp\left(\frac{\Delta E_g}{kT}\right) \simeq (1 - \alpha_I^0 \alpha_N^0)\left[1 + \frac{\alpha_I^0}{1 - \alpha_I^0 + \frac{1}{\beta_N^0}\exp\left(-\frac{\Delta E_g}{kT}\right)}\right]. \tag{18}$$

(It is assumed that in the case of germanium and silicon planar transistors α_N = 0.9-0.98, α_I = 0.2-0.5.)

Further transformation of the formulas in Eq. (11), carried out using Esing Eqs. (17) and (18), gives expressions for i_c and i_b in the anisotropic case:

$$i_c = i_b \frac{qZ_\Gamma(I_c^F - I_{C2}^0)}{kT} = Af(I_b^F - I_{b2}^0)(I_c^F - I_{C2}^0)\frac{qZ_\Gamma}{kT}\left[1 - \frac{qZ_\Gamma(I_b^F - I_{b2}^0)}{kT}\right]^{-1}, \tag{19}$$

where A is defined by Eq. (14a).[†]

In contrast to the isotropic case, the strongest pressure sensitivity of the collector current Γ_c is observed under stable base current (I_b) conditions, i.e., under constant-current source conditions: $I_b^F \approx I_b^0$.

The value of I_c^F decreases rapidly with increasing F. In the first approximation, Γ_c is proportional to I_c^0 and $\Gamma_b \to 0$:

$$\left.\begin{array}{l} I_c^F \simeq \dfrac{I_c^0}{1 - \varkappa + \varkappa \exp\left(-\dfrac{\gamma F}{s_0 \varkappa kT}\right)}, \\[3ex] \Gamma_c \simeq -A(I_c^F - I_{C2}^0) \simeq \dfrac{\gamma}{s_0 kT}\exp\left(-\dfrac{\gamma F}{s_0 \varkappa kT}\right)\dfrac{I_c^0 - I_{C2}^0}{\left[1 - \varkappa + \varkappa \exp\left(-\dfrac{\gamma F}{s_0 \varkappa kT}\right)\right]^2}. \end{array}\right\} \tag{20}$$

Under constant-voltage source conditions, the dependence of I_c^F and I_b^F on F is found by substituting Eq. (18) into Eq. (2) and using the conditions $U_{be}^F = U_{be}^0$; the expressions for Γ_c and Γ_b are obtained from Eqs. (19) and (15):

$$\left.\begin{array}{l} I_b^F \simeq (I_b^0 - I_{b2}^0)\left[1 - \varkappa + \varkappa \exp\left(-\dfrac{\gamma F}{s_0 \varkappa kT}\right)\right], \\[2ex] I_c^F \simeq I_c^0[1 - \varkappa(1 - y_3)], \quad \Gamma_c = \Gamma_b \dfrac{(I_c^F - I_{C2}^0)qZ_\Gamma}{kT} \to 0. \end{array}\right\} \tag{21}$$

Thus, when F is increased the current I_b^F rises rapidly, whereas I_c^F decreases slowly. [In fact, it follows from Eq. (18) that when ΔE_g < 0 we have y_3 < 1; y_3 = 1 only when ΔE_g = 0.] In the anisotropic case, the static gain $B^F = I_c^F/I_b^F$ decreases with increasing F: under constant-voltage source conditions it decreases because of an increase in I_b^F; under constant-current source conditions, it decreases because of a decrease in I_c^F.

It is evident from Eqs. (20) and (6) that, in contrast to the isotropic case, the dependence of Γ_c on F has a maximum at F = F* (constant-current input):

$$\Gamma_{c\,max} = -\frac{\gamma(I_c^0 - I_{C2}^0)}{4\varkappa s_0 kT(1 - \varkappa)} \simeq \frac{\gamma j_0}{4kT\varkappa(1 - \varkappa)}, \tag{22}$$

[†] The expression $\left[1 - \dfrac{qZ_\Gamma(I_b^F - I_{b2}^{F;0})}{kT}\right]$ in Eqs. (14) and (19) is always larger than zero because I_b^F < 0 and I_{b2}^F > 0.

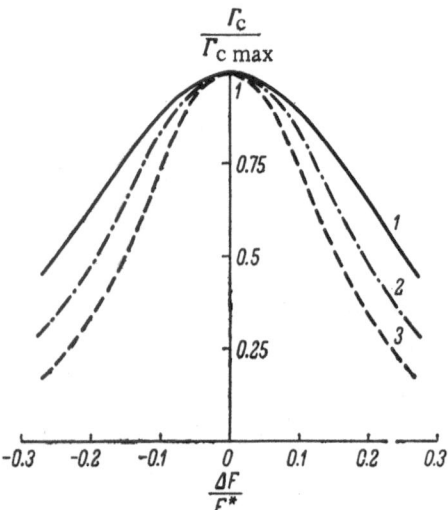

Fig. 1. Theoretical dependence of $\Gamma_c / \Gamma_{c\,max}$ on $\Delta F/F^*$ for different values of $\varkappa$: 1) $\varkappa = 10^{-3}$; 2) $\varkappa = 10^{-4}$; 3) $\varkappa = 10^{-5}$.

where j_0 is the current density in the absence of pressure;

$$F^* = -\frac{\varkappa s_0 kT}{\gamma} \ln \frac{1-\varkappa}{\varkappa}. \tag{23}$$

Equation (22) can be used to determine the values of $\varkappa$ and s_F. Since, usually, $\varkappa \ll 1$, it follows that

$$\varkappa \simeq \frac{\gamma j_0}{4kT\Gamma_{c\,max}}. \tag{24}$$

The dependence $\Gamma_c(F)$ can be calculated by substituting the value $\varkappa = s_F/s_0$ into the expression for Γ_c given in Eq. (20). However, in a comparison of the theory with the experimental data, it is more convenient to consider the function $\Gamma_c(F - F^*)$.

We can also introduce dimensionless variables: $\Gamma_c / \Gamma_{c\,max}$ and $\Delta F/F^* = (F - F^*)/F^*$. In this case, the dependence of the pressure sensitivity on the force is given by (Fig. 1):

$$\frac{\Gamma_c}{\Gamma_{c\,max}} = \left[\mathrm{ch}\,\frac{\gamma \Delta F}{2s_F kT} \right]^{-2} = \left[\mathrm{ch}\left(\frac{\Delta F}{2F^*} \ln \frac{1-\varkappa}{\varkappa} \right) \right]^{-2}, \tag{25}$$

where

$$\ln \frac{1-\varkappa}{\varkappa} = 4(1-\varkappa)\frac{F^*\Gamma_{c\,max}}{\Gamma_c^0}. \tag{26}$$

At low values of $\varkappa$, Eq. (26) simplifies to

$$\varkappa \simeq \exp\left(-\frac{4F^*\Gamma_{c\,max}}{\Gamma_c^0} \right). \tag{27}$$

Thus, using simple assumptions about the physical mechanism of the pressure sensitivity of transistors, based on the influence of elastic stresses on the effective forbidden band width, we can develop an approximate theory of a piezotransistor subjected to pressure concentrated at a point.

Experimental Results and Discussion

Germanium planar mesa transistors of the n−p−n type were investigated.[†] The space-charge layer thickness near the emitter junction was 0.2-0.3 μ, the junction depth was 1.5-2.5 μ, and the junction area $\sim 20 \times 50\,\mu$. The pressure was applied to the emitter via a corundum needle, whose tip was rounded to a radius $R \approx 20\,\mu$, or via a steel needle ($R \approx 60\,\mu$). The needle could be easily replaced.

The pressure applied to the needle was varied smoothly between 0 and 30 g, using a force-calibrated electrodynamic transducer (taken from an IGD-1 loudspeaker), a special micro-manipulator, and an MBS-2 microscope.

[†] Control measurements showed that there was no basic difference in the behavior of n−p−n and p−n−p transistors under pressure.

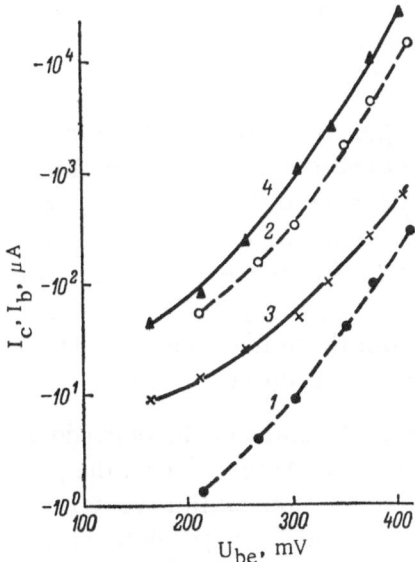 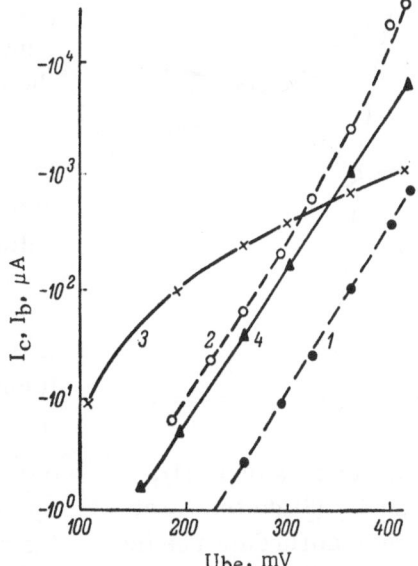

Fig. 2. Dependences of the currents I_c and I_b in a transistor on the voltage U_{be} in the absence of mechanical stresses and when a steel needle of R = 80 μ tip diameter was pressed against the emitter (F = 5.3 g). 1) I_b^0; 2) I_c^0; 3) I_b^F; 4) I_c^F.

Fig. 3. Dependences of the currents I_c and I_b in a transistor on the voltage U_{be} in the absence of mechanical stresses and when a corundum needle of R = 20 μ tip diameter was pressed against the emitter (F = 5.1 g). 1) I_b^0; 2) I_c^0; 3) I_b^F; 4) I_c^F.

The apparatus employed made it possible to transmit to the transistor, via the needle tip, a sinusoidal alternating mechanical signal of a given amplitude, which was additional to a static force. This made it possible to determine the pressure sensitivity of the transistor Γ_c as a function of the initial force F.

Figures 2 and 3 show the dependences of I_c and I_b on U_{be} for F = 0 and F = 5 g applied to the emitter p–n junction using two methods: a relatively blunt steel needle (Fig. 2) or a relatively sharp corundum needle (Fig. 3). In the first case, the currents I_c and I_b increased with pressure throughout the full range of U_{be} under the $U_{be}^F = U^0$ conditions (constant-voltage input). Under constant-current source conditions (i.e., $I_b^F = I_b^0$), the dependence of I_c on F (at least in a certain range of values of I_b) is quite weak (Fig. 2).

Thus, using a metal needle, we could, in the first approximation, establish an isotropic pressure [Eqs. (14)-(16)]. In agreement with the theoretical conclusions, it was found that $B^F \approx B^0$.

A completely different effect was obtained when a sharp corundum needle was pressed against the emitter. Under constant-voltage source conditions, in agreement with Eq. (21), I_c^F decreased slightly and I_b^F increased strongly with increasing force F. Under constant-current source conditions, the equality $I_b^F = I_b^0$ was satisfied only by a pronounced drop in U_{be}^F (Fig. 3), which gave rise to an even stronger decrease in I_c with increasing F (obviously, the current density changed not only in the stressed but also in the unstressed part of the transistor). When the base current was large, the dependence of I_c^F on F was much weaker, and this was due to the influence of the volume resistance of the base (R_b). When U_{be} increased, the resistance of the space-charge region decreased exponentially and the value of R_b, which was independent of F (at least in the first approximation), began to play the dominant role. Therefore, at sufficiently high values of U_{be}, the base current I_b ceased to depend on F, and I_c^F began to decrease ap-

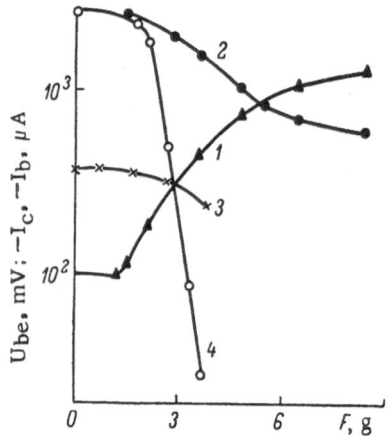

Fig. 4. Dependences of I_b (1), I_c (2, 4), and U_b (3) on the force F, applied by a corundum needle (R = 20 μ) pressed against the emitter. Input conditions: 1, 2) $U_{be}^F \approx U_{be}^0 = 0.36$ V; 3, 4) $I_b^F \approx I_b^0 = 100$ μA.

preciably with increasing F (i.e., constant-current source conditions at the input had been established in relation to the emitter p–n junction).

The current gain B^F decreased with increasing F, irrespective of the conditions at the input (Fig. 3). The experimental dependences of I_c, I_b, and U_{be} on F in the anisotropic pressure case are shown in Fig. 4. The rise of I_b and the fall of I_c with increasing force were exponential (for F > 1.5 g), in agreement with Eqs. (20) and (21). Deviations from these dependences, due to the influence of R_b, were observed at F > 5 g.

Thus, the general nature of the dependences of the currents in a planar mesa transistor on the pressure applied at a point are in good qualitative agreement with the theoretical predictions.[†] In order to carry out a quantitative comparison we shall consider theoretical and experimental dependences of I_c^F and Γ_c on F in the anisotropic case (Figs. 5 and 6). The curves denoted by 1 in these figures were determined experimentally, those denoted by 2 were calculated using Eq. (20), and those denoted by 3 were calculated using formulas similar to Eq. (20) but derived on the assumption that s_F depends on F (these formulas are not given in the present paper). The dependence of s_F on F was found approximately by solving the problem of a spherical die of radius R being forced into an elastic half-space [14]. As in [7, 12, 5], it was assumed that the area of the stressed part of the emitter junction $s_F = \pi r_F^2$ was equal to the contact area s_{con}, i.e.,

$$s_F = s_{con} = \pi r_{con}^2 = \pi K_1^2 (FR)^{\frac{2}{3}} \simeq \pi \left[\frac{3}{4} FR \left(\frac{1 - \sigma_1^2}{E_1} + \frac{1 - \sigma_2^2}{E_2} \right) \right]^{\frac{2}{3}}, \tag{28}$$

where K_1 is a coefficient which depends on the Young's moduli (E_1, E_2) and the Poisson ratios (σ_1, σ_2) of the materials of the two bodies in contact. The values of r_F (for the curves denoted by 2) and K_1 (for the curves denoted by 3), used in plotting the theoretical curves in Figs. 5 and 6, were determined by equating the theoretical and experimental values of $\Gamma_{c\,max}$.

Equation (28) predicts a value of K_1 which is 1.5-2 times higher than that employed in our calculations. If this value of K_1 is used, the values of $\Gamma_{c\,max}$ and of the calculated dependences $I_c^F(F)$ and $\Gamma_c(F)$ differ substantially from those obtained experimentally. The calculated curve $\Gamma_c(F - F^*)$, shown in Fig. 6, fits the experimental dependence only if it is assumed that $s_F =$ const. Curve 3, calculated on the assumption that the stressed area varies with the applied force, differs considerably from the experimental dependence.

It is obvious that the assumption that the stressed area s_F of the emitter junction at a depth ≤1.5-2 μ is equal to the area of the contact, s_{con}, between the needle and the crystal [Eq. (28)], is not really justified. This conclusion follows also from an analysis of the dependence of r_F on the tip radius R. We shall now compare the calculated dependence $\Gamma_{c\,max}(r_F)$, plotted on the basis of Eq. (22), with the experimentally determined dependence of $\Gamma_{c\,max}$ on R (curves 1 and 2 in Fig. 7a). We can then obtain r_F as a function of R (curve 1 in Fig. 7b). This func-

---

[†] In addition to those described, we also investigated experimentally the dependences of Γ_c on I_c^0 and on Z_Γ. Both dependences were in good agreement with the theoretical conclusions that follow from Eqs. (19) and (20).

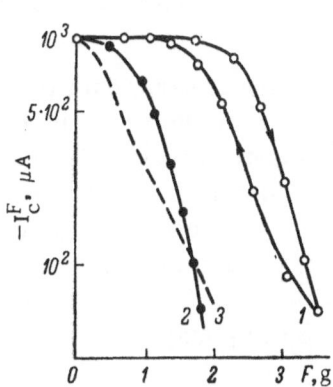 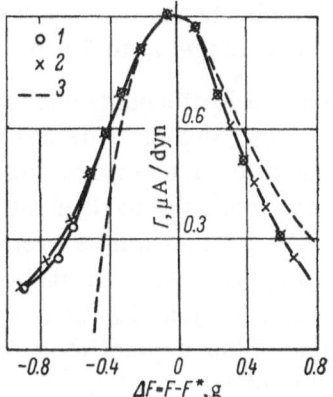

Fig. 5. Experimental (1) and theoretical (2 and 3) dependences of I_c^F on F for anisotropic pressure and various assumptions about the stressed emitter area s_F (R = 20 μ, $s_0 = 10^3$ μ^2); 2) s_F = 21.5 μ^2 = const; 3) $s_F = \pi K_1^2 (FR)^{2/3}$, $K_1 = 1.52 \cdot 10^{-4}$ cm$^{2/3}$ · dyn$^{-1/3}$.

Fig. 6. Experimental (1) and theoretical (2 and 3) dependences of the pressure sensitivity Γ_c on $\Delta F = F - F^*$ for various assumptions about the stressed emitter area s_F (R = 20 μ, $s_0 = 10^3$ μ^2); 2) s_F = 21.5 μ^2; 3) $s_F = \pi K_1^2 (FR)^{2/3}$, $K_1 = 1.52 \cdot 10^{-4}$ cm$^{2/3}$ · dyn$^{-1/3}$.

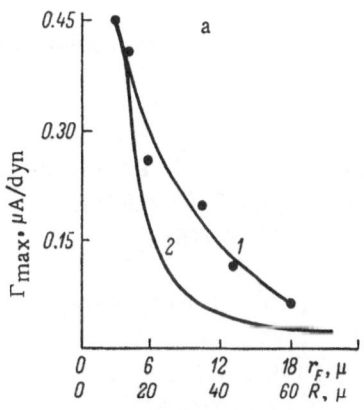 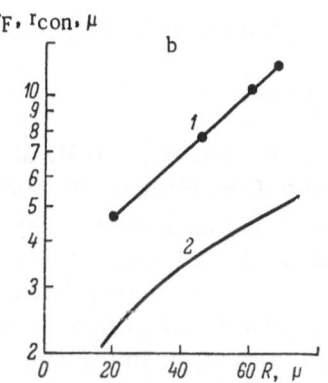

Fig. 7. a) Dependences of $\Gamma_{c\,max}$ on the radius of the needle tip R (1) and on the radius of the stressed region r_F (2); $s_0 = 10^3$ μ. b) Dependences of r_F (1) and of the radius of the contact area r_{con} (2) on R.

tion can be approximated satisfactorily by the empirical formula $r_F = r_0 \exp(aR)$, where $r_0 \approx$ 2.5 μ, $a \approx 2 \cdot 10^{-2}$ μ^{-1}. The values of r_0 and a can vary within wide limits for different transistors and different points on the surface of the same transistor, but the dependence is always exponential. This means that, as R → 0, the value of $s_F = \pi r_F^2$ (curve 1 in Fig. 7b) remains finite, which does not agree with the behavior of s_{con} given by Eq. (28). In general, there is a considerable difference between the dependences of s_{con} and s_F on R.

It must be stressed that some features of the observed effect cannot be explained within the framework of our theory. These features include the zero-sensitivity region in the $I_c^F(F)$ curves, observed for most samples.

It is evident from curve 1 in Fig. 5 that I_c^F is independent of the force when $F < 1.5$ g. In some cases (for example, in the case of transistors with deep emitter junctions), zero sensitivity is observed over a range up to 10-20 g.

An electromechanical hysteresis, characterizing the experimentally obtained dependence $I_c^F(F)$ represented by curve 1 in Fig. 5, is another feature which cannot be explained by our model. However, all such features can be interpreted qualitatively if it is assumed that, at pressures of 10^9-10^{10} dyn/cm^2, the part of the crystal lying immediately under the needle tip behaves as a viscoelastic medium whose properties are not described by the classical theory of elasticity [for example, Eq. (28)]. On the other hand, plastic deformation may not reach the depth at which the emitter junction is located, where the material is still in a uniaxially stressed state [12]. Under these conditions, the value of s_F may be almost independent of F in a wide range of values of F.

Plastic deformation in the surface layer under the needle is supported by observations of the dependence $I_c^F(F)$ obtained by cyclic variation of the force F. Each new loading operation reduces the zero-sensitivity region as well as the width of the hysteresis loop and it increases somewhat the value of Γ_c. This process, which ends after five to seven cycles, may reduce the zero-sensitivity region by a factor of 2-3 so that the experimental dependence coincides almost completely with the theoretical $I_c^F(F)$ curve. It is very likely that the cause of this phenomenon is plastic flow [15] and subsequent deformation hardening of a thin semiconductor layer between the needle tip and the emitter p−n junction. It is also possible that a more detailed investigation of the elastic and plastic properties of a crystal near the needle tip and in the region of the emitter junction will give a quantitative description of the electromechanical hysteresis observed in pressure-sensitive transistors and will explain the high gradient of the normal component of mechanical stresses in the thin space-charge layer.

The authors are grateful to G. E. Pikus for his valuable comments on this paper.

Literature Cited

1. J. J. Wortman, J. R. Hauser, and R. M. Burger, J. Appl. Phys., 35:2122 (1964).
2. J. J. Wortman and J. R. Hauser, J. Appl. Phys., 37:3527 (1966).
3. R. Edwards, IEEE Trans. Electron Dev., ED-11:286 (1964).
4. K. Bulthuis, Philips Res. Rep., 21:85 (1966).
5. A. L. Polyakova and V. V. Shklovskaya-Kordi, Fiz. Tverd. Tela, 8:208 (1966).
6. A. L. Polyakova, Akust. Zh., 13:256 (1967).
7. W. Rindner and I. Braun, J. Appl. Phys., 34:1958 (1963).
8. K. E. Preece and P. R. Selway, Proc. IEEE, 53:618 (1965).
9. Y. Matukura, Japan. J. Appl. Phys., 3:256, 304 (1964).
10. V. I. Gaman and V. F. Agafonnikov, Izv. Vuzov, Fizika, No. 6, p. 54 (1967); No. 12, p. 97 (1967).
11. V. S. Shadrin and V. A. Gridchin, Izv. LÉTI, No. 57, Pt. III, p. 73 (1966).
12. K. Bulthuis, J. Appl. Phys., 37:2066 (1966).
13. I. P. Stepanenko, Fundamentals of the Theory of Transistors and Transistor Circuits, Énergiya, Moscow (1967).
14. L. D. Landau and E. M. Lifshits, Theory of Elasticity, 2nd ed., Pergamon Press, Oxford (in press).
15. V. P. Alekhin, Author's Abstract of Dissertation for Candidate's Degree, Moscow (1967).

POWER OF THE RADIATION EMITTED DURING THE
TUNNEL BREAKDOWN OF SILICON p—n JUNCTIONS

V. G. Mel'nik

The results are reported of measurements of the absolute output power, external quantum efficiency, and device efficiency observed during the tunnel breakdown of silicon p—n junctions. The radiation power obeyed the law $6 \cdot 10^{-7} \exp(4.6 \Delta V)$ W/cm^2 for 4-5.5 V across the junction. The device efficiency was approximately $10^{-5}\%$ and its quantum efficiency was 10^{-6} photons/hole.

The present paper describes the results of measurements of the absolute power of the recombination radiation generated during the tunnel breakdown of alloyed silicon p—n junctions. A qualitative description of such radiation is given in [1] and its spectral characteristics are described in [2]. The radiation spectrum of a tunnel breakdown is concentrated primarily in the range 1-2 eV and has a maximum at 1.12 eV.

Investigations of the energy characteristics and the efficiency of such a source of radiation are of scientific and technical interest, for example, in connection with applications in semiconductor optoelectronics.

We investigated special p$^+$—n junctions with a guard ring, prepared from n-type silicon by the diffusion and melting method. This technique has been described in detail in the literature on the subject. The emitting surface of each device had a diameter of 0.5 mm. The current—voltage characteristics of the junctions were the usual "soft" tunnel characteristics with a current of 10 mA for a voltage of ~3.2 V. The breakdown voltage varied from sample to sample. The "softness" of the current—voltage characteristic made it possible to measure the radiation power as a function of the current through the junction and of the voltage across it.

The measurements were carried out using a calibrated optoacoustic detector with a quartz window. The junction was attached to a massive copper holder and placed in front of the detector window without any intermediate optical system. We ignored the absorption in the layer of silicon through which the radiation emerged because the thickness of this layer did not exceed 0.5 μ. We took into account the effects associated with the Joule heating. Measurements of the thermal resistance between the sample and the holder showed that the temperature of the sample could rise by 100 deg C when the power dissipated in it reached 3 W. Integration of the Planck distribution function in the range of frequencies corresponding to the transmission of the quartz window showed that the flux of radiation due to overheating of the

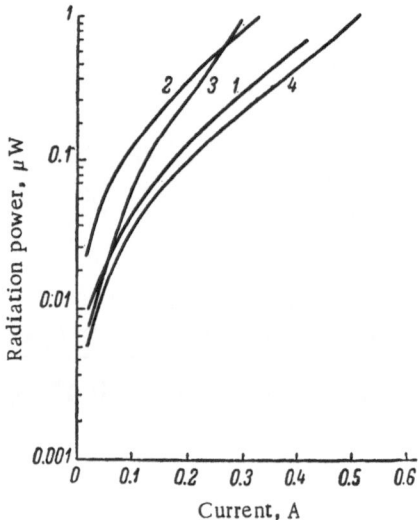

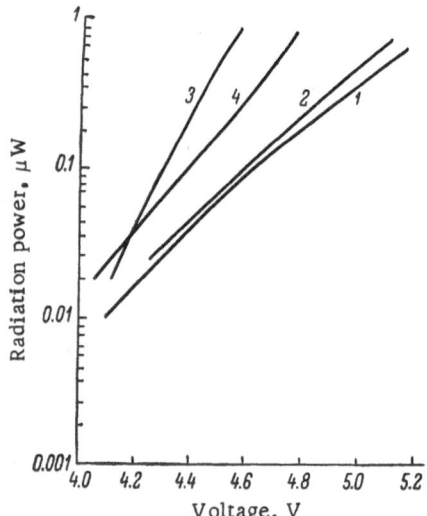

Fig. 1. Absolute integrated radiation power as a function of the current for a p—n junction of 0.5 mm diameter. The numbers alongside the curves are the sample numbers.

Fig. 2. Absolute integrated radiation power as a function of the voltage for a p—n junction of 0.5 mm diameter. The numbers alongside the curves are the sample numbers.

sample amounted to 10^{-10}-10^{-11} W, which was 2-3 orders of magnitude lower than the useful signals observed in our experiments.

Figure 1 shows some of the dependences of the radiation power on the current, recorded for different samples, and Fig. 2 shows the corresponding dependences on the voltage (for the same samples).

It is evident from Fig. 2 that, beginning from about 4 V, the dependence of the radiation power P on the voltage rise ΔV (expressed in volts) could be described by the formula

$$P = 10^{-3} \exp(4.6 \Delta V) \, \mu W.$$

and the specific power was given by $6 \cdot 10^{-7} \exp(4.6 \Delta V)$ W/cm^2.

The efficiency of the junctions, considered as radiation sources, increased with increasing electrical power and — in the range of the investigated currents and voltages — it rose from $6 \cdot 10^{-6}$ to $3 \cdot 10^{-5}\%$. We determined the spectral characteristic of the radiation emitted by sample 3 when carrying a current of 120 mA, and we plotted this characteristic as a dependence of the relative number of photons per 1 eV of the photon energy. (The technique used in such measurements has been described in detail in the literature.) The spectrum of the sample occupied the interval 1.1-2.2 eV and had a maximum at 1.13 eV. Graphical integration of the spectral characteristic was employed to determine the total number of photons emitted for the given value of the current. This number of photons was found from the expression

$$N = \frac{P}{\int\limits_{\varepsilon_0}^{\varepsilon_1} f(\varepsilon) \, \varepsilon \, d\varepsilon} \int\limits_{\varepsilon_0}^{\varepsilon_1} f(\varepsilon) \, d\varepsilon,$$

where P is the radiation power; $f(\varepsilon)$ is the number of photons of energy ε in an interval of 1 eV; ε_0 and ε_1 are the minimum and maximum energies of photons in the spectrum.

According to our measurements, $N = 5 \cdot 10^{11}$ photons/sec, which yielded an external quantum efficiency of 10^{-6} photons/hole for a current of 120 mA. The quantum efficiency decreased with increasing current.

This value of the quantum efficiency was two orders of magnitude higher than the avalanche breakdown efficiency reported by Chynoweth and McKay [3]. This difference could be due to two factors.

Firstly, Chynoweth and McKay [3] determined the quantum efficiency using very small currents so that they were able to determine the number of emitted photons with a counter; secondly, they took into account only the visible part of the radiation because their radiation detector was a photomultiplier with an S-4-type cathode.

Our efficiency was the external quantum efficiency. Assuming that the refractive index of silicon for the least energetic photons was 3.5, we found that integration of the Fresnel formulas over all angles of incidence yielded a value of the internal quantum efficiency which was about two orders of magnitude higher than the external efficiency. We ignored the multiple reflection in silicon in this calculation because of the high value of the absorption coefficient at all wavelengths in the radiation spectrum and because of the relatively large thickness of the samples (0.5-0.6 mm).

Literature Cited

1.   A. G. Chynoweth and K. G. McKay, Phys. Rev., 106:418 (1957).
2.   M. Migitaka, Solid State Electron., 8:295 (1965).
3.   A. G. Chynoweth and K. G. McKay, Phys. Rev., 102:369 (1956).

INFLUENCE OF X-RAY RADIATION ON
SOME PARAMETERS OF p—n JUNCTIONS IN SILICON

O. A. Klimkova and O. R. Niyazova

Nuclear Physics Institute
Academy of Sciences of the Uzbek SSR, Tashkent

The paper describes an investigation of the influence of irradiation with 50-keV x rays on the current—voltage I(V) and capacitance—voltage C(V) characteristics of silicon diodes. It was found that such irradiation produced permanent changes in these characteristics.

The effects of various radiations (γ rays, reactor neutrons, high-energy electrons, protons, x rays) on semiconductor p—n structures have been widely investigated [1, 2]. However, a possible influence of radiations of relatively low energies on the radiation stability of devices has not been studied in sufficient detail, in spite of the fact that many authors have pointed out the important role of "subthreshold" energies in the process of radiation-defect formation in semiconductors [3-5]. We investigated radiation-induced changes in p—n junctions in silicon irradiated with x rays for the purpose of detecting permanent changes in the current—voltage and capacitance—voltage characteristics.

We studied point-alloyed diodes, prepared by alloying aluminum to n-type silicon of 100 $\Omega \cdot$ cm resistivity. Phosphorus was used as the dopant. The junction area was $3 \cdot 10^{-3}$ cm^2. The base contacts were made by alloying with antimony-doped gold. The current—voltage (Fig. 1) and capacitance—voltage (Fig. 2) characteristics were determined. The capacitance was measured using an IIEV-1 meter. The error in the measurement of the capacitance and of the current—voltage characteristics was 6 and 5%, respectively. Irradiation was carried out using a unit of the URS-70 type with a BSV tube fitted with a nickel anticathode. This unit was capable of producing x rays up to 70 keV energy when the tube current was 30 mA. The dose rate, for a tube current of 20 mA and a voltage of 50 kV, was 100 rad/sec (measured using ferrosulfate and film dosimeters). The temperature during irradiation was 25 ± 3°C. Changes in the barrier capacitance and in the current through the junction were determined as a function of the reverse bias voltage V before and after irradiation. These parameters were determined at 25 ± 0.1°C after 24-h annealing at 25 ± 3°C. The results obtained are presented in Figs. 1 and 2.

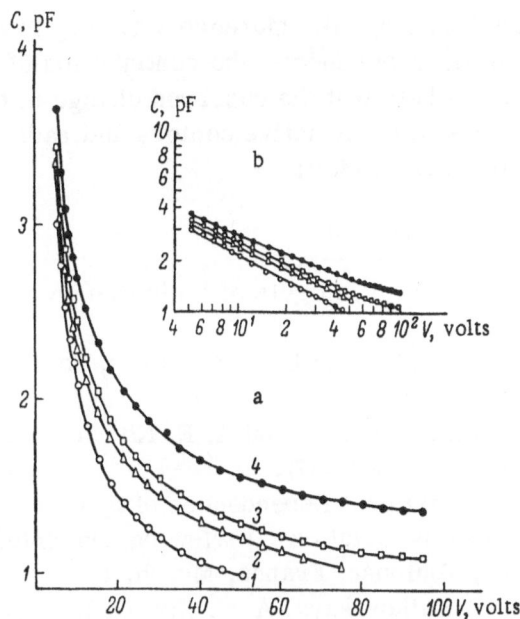

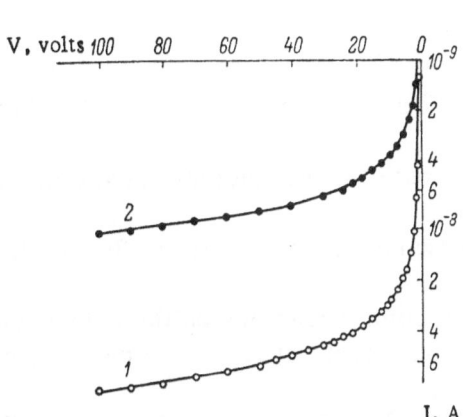

Fig. 1. Current—voltage characteristics of a reverse-biased silicon diode before (1) and after (2)
irradiation with an x-ray dose of
17.1 · 10⁶ rad.

Fig. 2. Capacitance—voltage characteristics of the barrier capacitance of a silicon diode: 1) before irradiation; 2-4) after
irradiation with x-ray doses (rad): 2) 4.7 ·
10⁶; 3) 10.6 · 10⁶; 4) 17.1 · 10⁶.

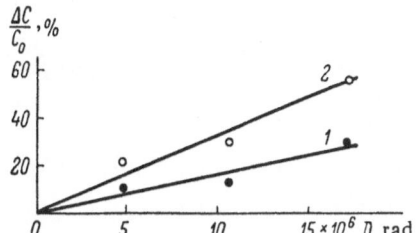

Fig. 3. Dose dependences of the
relative change in the barrier capacitance of a diode at V = 10 V
(1) and V = 50 V (2).

Curves 2, 3, and 4 in Fig. 2 represent the
results of irradiation with doses of $4.7 \cdot 10^6$,
$10.6 \cdot 10^6$, and $17.1 \cdot 10^6$ rad, respectively. The
barrier capacitance decreased with increasing
absorbed dose. Figure 3 shows the dose dependences of the relative change in the capacitance
for two values of the reverse bias voltage. The
capacitance increased by 30% at V = 10 V and by
56% at V = 50 V.

The capacitance of a reverse-bias junction
is a function of the voltage and current flowing
through the junction [6, 7]. The capacitance of
an abrupt reverse-biased junction, heavily doped in the p-type region, is given by the expressions

$$C(V) = S \sqrt{\frac{q \varepsilon N}{8\pi (V + V_c)}},$$

$$C(I) = \frac{I w_0}{D} \left[\frac{\varepsilon}{2qN(V + V_c)} \right]^{\frac{1}{2}},$$

where V is the potential across the junction; q is the electronic charge; ε is the permittivity
of silicon; N is the donor concentration in the n-type region; S is the junction area; I is the
current through the junction; w_0 is the depth of penetration in the n-type region of the injected
minority carriers; D is the diffusion coefficient of holes; V_c is the contact potential of the junction. The experimentally observed change in the capacitance (Fig. 2) could not be due to the
current component (the second equation) since the reverse current decreased with increasing

absorbed dose (Fig. 1). Moreover, at very low values of the reverse current (such as were observed in our experiments) the contribution of the current to the capacitance could be neglected. It was very likely that the observed change in the capacitance was due to the appearance of additionally electrically active centers and radiation-stimulated changes in the nature of the impurity distribution [8, 9].

Literature Cited

1. V. S. Vavilov, Effects of Radiation on Semiconductors, Consultants Bureau, New York (1965).
2. G. Dearnaley and D. C. Northrop, Semiconductor Counters for Nuclear Radiations, Spon, London (1963).
3. S. V. Starodubtsev and A. E. Kiv, in: Radiation Physics of Nonmetals [in Russian], Naukova Dumka, Kiev (1967).
4. V. S. Vavilov, "Dependences of radiation effects in semiconductors on the nature and energy of radiation" (review paper), Intern. Conf. on Effects of Radiation on Semiconductors, Toulouse, France, March, 1967.
5. M. A. Zaikovskaya, A. E. Kiv, O. R. Niyazova, and S. V. Starodubtsev, Fiz. Tekh. Poluprov., 1:1131 (1967).
6. L. S. Berman, Nonlinear Semiconductor Capacitors, Fizmatgiz, Moscow (1963).
7. J. R. Collard and F. Sterzer, Appl. Phys. Letters, 5:165 (1964).
8. S. V. Starodubtsev, O. R. Niyazova, and M. A. Kaneev, Fiz. Tverd. Tela, 9:872 (1967).
9. A. E. Kiv and F. T. Umarova, Fiz. Tverd. Tela, 9:2225 (1967).

AMPLIFICATION OF THE PHOTOCURRENT IN SEMICONDUCTING PHOTORESISTORS WITH NEUTRAL CONTACTS

G. A. Kazantsev and I. I. Taubkin

Calculations are reported of the steady-state distribution of photocarriers and of the photocurrent gain for various intensities of an electric field applied to a homogeneous, uniformly illuminated, semiconducting crystal which has neutral contacts, an ambipolar photoconductivity, and a linear law of trap filling. Calculations are also reported of the dependence of the gain on the ratio of the sample size to the diffusion length of the minority carriers. The maximum value of the voltage sensitivity is estimated for photoresistors made of p-type indium antimonide single crystals. The results are given of a calculation of the coordinate dependence of the photocurrent gain in the case of local illumination of the sensitive surface.

The possibility of amplifying the photocurrent in single-crystal semiconducting photoresistors and the study of the maximum value of the steady-state gain have been reported in many papers (see, for example, relevant chapters and bibliography in [1, 2] or [3]).

It is established in [3] that the photocurrent gain k_I of semiconducting crystals with neutral contacts and ambipolar photoconductivity becomes saturated at electric fields E in which the photocarrier extraction becomes important. These fields are $E > l/\mu\tau$, where l is the distance between the contacts; μ and τ are, respectively, the mobility and lifetime of the minority carriers.

It is also known [4] that the thickness of the "photoblocking" regions near neutral contacts with semiconducting crystals which exhibit ambipolar photoconductivity is of the order of the minority-carrier diffusion length l_d, even in very weak electric fields. This reduces k_I, particularly when $l \lesssim l_d$, which must be taken into account in investigations of microscopic single-crystal photoresistors.

In some applications, it is interesting to know the distribution of the sensitivity across the illuminated area, which is obviously nonuniform in the case of neutral contacts subjected to "strong" $(E > l/\mu\tau)$ electric fields and in the case of layer dimensions which are small compared with l_d.

The present paper reports the results of a calculation of the steady-state distribution of photocarriers and of the photocurrent gain of a homogeneous semiconductor crystal with neutral

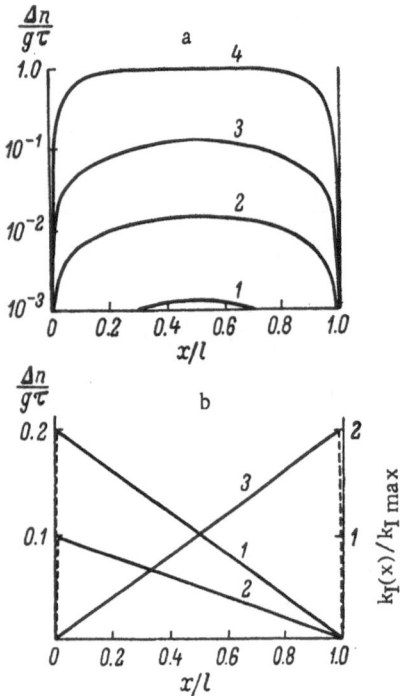

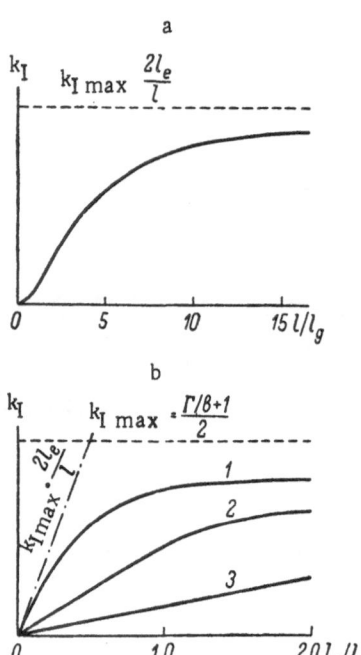

Fig. 1. Distribution of the photo-electron density along a uniformly illuminated crystal with neutral contacts. a) $l_e \ll l$, $l_e \ll l_d$: 1) $l = 0.1 l_d$; 2) $l = 0.33 l_d$; 3) $l = l_d$; 4) $l = 10 l_d$. b) $l_e > l$, $l_e > l_d$: 1) $l_e = 5l$; 2) $l_e = 10l$; 3) distribution of the photocurrent gain [Eq. (3)].

Fig. 2. Dependences of the photo-current gain k_I of a uniformly illuminated photoresistor with neutral contacts [Eq. (2)]. a) $k_I(l/l_d)$ for $l_e \ll l$, $l_e \ll l_d$; b) $k_I(l_e/l)$ for $l \gg l_d$ (1), $l = 2l_d$ (2), and $l = l_d$ (3).

contacts in the case of volume photogeneration of electron−hole pairs and an arbitrary value of E. It is assumed that the semiconductor is extrinsic (for example, $p_0 \gg n_0$, where p_0 and n_0 are equilibrium hole and electron densities) and that it contains (like crystals of p-type indium antimonide [5]) recombination levels whose electron and hole-capture cross sections are different and independent of E.

A system of basic equations, similar to that given in [6],* is solved making the usual assumption that the effects of space charge can be neglected (this assumption does not give rise to appreciable errors in the expressions for the photocarrier densities Δn and Δp, with the exception of very narrow regions near the contacts, whose contribution to the photosignal is negligibly small). It is also assumed that the rate of carrier photogeneration g is low so that the electron and hole lifetimes are constant. Then, taking into account the conditions at the neutral contacts, $\Delta n|_{x=0,l} = 0$, the value of Δn is given by (Fig. 1):

$$\Delta n = g\tau \left[1 - \frac{1 - e^{-\frac{l}{l_2}}}{e^{\frac{l}{l_1}} - e^{-\frac{l}{l_2}}} e^{\frac{x}{l_1}} - \frac{e^{\frac{l}{l_1}} - 1}{e^{\frac{l}{l_1}} - e^{-\frac{l}{l_2}}} e^{-\frac{x}{l_2}} \right], \tag{1}$$

where (by analogy with [4])

$$l_1 = \frac{2l_d^2}{\sqrt{l_e^2 + 4l_d^2} - l_e}; \quad l_2 = \frac{2l_d^2}{\sqrt{l_e^2 + 4l_d^2} + l_e}; \quad l_e = \mu\tau E.$$

_____

*The influence of surface recombination and conductivity of contact-free faces of a crystal is ignored.

The photocurrent gain k_I under short-circuit conditions (when the current sensitivity has its maximum value) is given by (Fig. 2):

$$k_I = \frac{\frac{\Gamma}{b}+1}{2}\frac{2l_e}{l}\left[1+\frac{l_1+l_2}{l}\frac{1+e^{\frac{l}{l_1}-\frac{l}{l_2}}-e^{\frac{l}{l_1}}-e^{-\frac{l}{l_2}}}{e^{\frac{l}{l_1}}-e^{\frac{l}{l_1}}}\right].\tag{2}$$

Here, Γ is the ratio of the hole and electron lifetimes; b is the ratio of the electron and hole mobilities.

It follows from Fig. 1a that the minority photocarrier density in a sample with an ambipolar photoconductivity and neutral contacts is considerably less than $g\tau$, even in weak fields (for $l \gg l_d$ near the contacts and $l \lesssim l_d$ throughout the crystal) and, consequently, the value of k_I is much lower for sensitive layers of dimensions $l/d < 3\text{-}5$ than the corresponding value for thicker layers (Fig. 2a).

Figure 2b shows the dependences of k_I on l_e/l, which are linear in the initial region; when $l_e/l > 1$, the gain tends to a saturation value $k_{I\,max} = [(\Gamma/b)+1]/2$, which is identical with that calculated in [7]. It must be pointed out that, when $l_e > l$, the photocarrier density in a crystal increases linearly on approach to the anode (Fig. 1b).

Figure 1b shows also the distribution (curve 3) of the photocurrent gain $k_I(x)$, calculated for the case of local illumination, which is restricted to a narrow strip of the sensitive area parallel to the contacts:

$$k_I(x) = k_{I\,max}\frac{2l_e}{l}\left[1-\frac{1-e^{-\frac{l}{l_2}}}{e^{\frac{l}{l_1}}-e^{-\frac{l}{l_2}}}e^{\frac{l-x}{l_1}}-\frac{e^{\frac{l}{l_1}}-1}{e^{\frac{l}{l_1}}-e^{-\frac{l}{l_2}}}e^{-\frac{l-x}{l_2}}\right].$$

When $l_e \ll l$, $l_e \ll l_d$, the distribution $k_I(x)$ is identical with the distribution $\Delta n(x)$ in the case of uniform illumination of the sensitive layer; when $l_e \gg l$, $k(x)$ is, in contrast to $\Delta n(x)$, independent of l_e and reaches its maximum value as $x \to l$.

By way of example, we shall apply the relationships obtained to determine the maximum voltage sensitivity $S_{\lambda\,max}$ of p-type InSb photoresistors with neutral contacts, calculated at the maximum of the spectral characteristic. When such a photoresistor is connected to a matched load, we have

$$S_{\lambda\,max} \approx \frac{(1-R)\,k_{I\,max}\,r}{2E_g},$$

where R is the reflection coefficient; r is the resistance of the photosensitive layer; E_g is the forbidden band width of the semiconducting layer (in the case of indium antimonide, $E_g \sim 0.23$ eV at 77°K). A square-shaped photoresistor, 20 μ thick, is found to have a sensitivity $S_{\lambda\,max} \approx 2 \cdot 10^6$ V/W for the following values of the parameters: $p_0 = 10^{13}$ cm^{-3}, $\mu_p = 7 \cdot 10^3$ cm$^2 \cdot$ V$^{-1} \cdot$ sec^{-1}, b = 30 and $\Gamma = 2 \cdot 10^3$. The critical field $E = l/\mu\tau$ for $l = 200\,\mu$ is 20-200 V/cm [$\mu = (2\text{-}5) \cdot 10^5$ cm$^2 \cdot$ V$^{-1} \cdot$ sec^{-1}, $\tau = (5\text{-}20) \cdot 10^{-10}$ sec]. Partial saturation of the electron current at these field intensities probably does not greatly alter $S_{\lambda\,max}$. For large values of l, the maximum gain of the photocurrent in a p-type indium antimonide photoresistor with an efficient heat sink is obviously determined by avalanche breakdown.*

---

*Our measurements on p-type InSb samples with $l_e < l$ showed that k_I increased linearly with E (i.e., the hole lifetime was constant) right up to the breakdown voltage.

Literature Cited

1. R. H. Bube, Photoconductivity of Solids, Wiley, New York (1960).
2. A. Rose, Concepts in Photoconductivity and Allied Problems, Interscience, New York (1963).
3. F. Stöckmann, Proc. Photoconductivity Conf., Atlantic City, 1954, publ. by Wiley, New York (1956), p. 269.
4. S. M. Ryvkin, Photoelectric Effects in Semiconductors, Consultants Bureau, New York (1964).
5. A. S. Volkov and V. V. Galavanov, Fiz. Tekh. Poluprov., 1:163 (1967).
6. E. S. Rittner, Proc. Photoconductivity Conf., Atlantic City, 1954, publ. by Wiley, New York (1956), p. 215.
7. F. Stöckmann, Phys. Status Solidi, 2:517 (1962).

ANALYSIS OF PROCESSES IN MULTILAYER SEMICONDUCTOR STRUCTURES OF THE n—p—n—p—n—p TYPE

A. A. Lebedev

A. F. Ioffe Physicotechnical Institute
Academy of Sciences of the USSR, Leningrad

The possibilities are considered of using six-layer semiconductor structures of the n—p—n—p—n—p type as control elements in switching devices with higher permissible forward and reverse voltages than those of the corresponding p—n—p—n structures. The power losses in devices of this type need not exceed the level of losses in ordinary thyristors. An analysis is made of the switching conditions in such structures and of their static characteristics. The characteristics of the turn-on and turn-off transient processes of these devices are discussed.

It is known that the maximum breakover voltage of controlled semiconductor rectifiers (thyristors), based on p—n—p—n structures, cannot exceed (under static or quasi-static conditions) the breakdown voltage of the central (collector) p—n junction. On the other hand, the value of the highest permissible voltage in the case of reverse polarity of the emf source in the anode circuit is limited by the breakdown voltage of the emitter p—n junction adjoining a wide lightly doped base. Thus, in order to increase the working voltage of a thyristor, it is necessary to increase the electric strength of the relevant p—n junctions. In order to avoid reduction in the breakover voltage because of overlap of the wide lightly doped base by the space charge layer of the collector or emitter p—n junctions, it is necessary to increase the width of this base. However, this increase in width either raises the forward voltage drop across the device in the on state (if the minority carrier lifetime in that base remains constant) or it raises the breakdown voltage if the lifetime is increased.

However, in principle, the working voltages of a thyristor can be increased by a method which avoids all these undesirable effects. This method, suggested by I. V. Grekhov and V. B. Shuman, involves the use of a larger number of p—n junctions. One of the simplest multilayer structures which can be used is a six-layer n—p—n—p—n—p structure (Fig. 1).

We shall consider processes in such a structure subjected to a forward voltage, i.e., we shall consider the case when a positive potential is appied to the p-type region on the extreme right in Fig. 1 and a negative potential to the n-type region on the extreme left. In this case, the j_1, j_3, j_5 p—n junctions are biased in the forward direction and j_2, j_4 junctions are

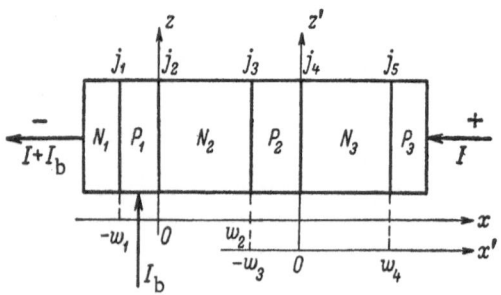

Fig. 1. Schematic representation of a six-layer structure of the n−p−n−p−n−p type.

biased in the reversed direction. Thus, unlike a p−n−p−n structure which has one collector junction, the six-layer structure has two collector junctions and when the avalanche breakdown voltages of the j_2 and j_4 p−n junctions are equal, the breakover voltage may be twice as high as that of a four-layer structure.

The current−voltage characteristic of a six-layer structure of this type has, like a p−n−p−n structure, two stable states provided the injection efficiency of at least one of the emitter p−n junctions or the transport factor of one of the inner base regions depends on the current.

The condition for the inversion of the sign of the voltage across both collector p−n junctions j_2 and j_4 can be found as follows. A six-layer structure can be considered as a combination of the p−n−p−n structures, $N_1 - P_1 - N_2 - P_2$ and $N_2 - P_2 - N_3 - P_3$, connected in series and sharing a common emitter p−n junction j_3. The condition for the inversion of the voltage across the collector p−n junction j_2 in the $N_1 - P_1 - N_2 - P_2$ structure is of the form

$$\gamma_{12}\beta_{12} + \gamma_{32}\beta_{32} = \alpha_{12} + \alpha_{32} \geqslant 1, \tag{1}$$

where $\gamma_{i,i\pm1}$ is the efficiency of injection from an i-th p−n junction into an (i ± 1)-th p−n junction; $\beta_{i,i\pm1}$ is the transport factor for the flow of carriers from an i-th to an (i ± 1)-th p−n junction. Obviously, $\gamma_{i,i+1} = 1 - \gamma_{i,i-1}$. Similarly, the condition for the inversion of the sign of the voltage across the collector p−n junction j_4 in the $N_2 - P_2 - N_3 - P_3$ structure is of the form

$$\gamma_{54}\beta_{54} + \gamma_{34}\beta_{34} = \alpha_{54} + \alpha_{34} \geqslant 1. \tag{2}$$

A six-layer structure is switched to the on state only if both conditions, i.e., Eqs. (1) and (2), are satisfied.

Since $\gamma_{32} = 1 - \gamma_{34}$, it is clear that the conditions (1) and (2) impose certain restrictions on the range of permissible values of the injection efficiency of emitter p−n junction j_3, namely

$$\frac{1 - \gamma_{12}\beta_{12}}{\beta_{32}} \leqslant \gamma_{32} \leqslant \frac{\beta_{34} + \gamma_{54}\beta_{54} - 1}{\beta_{34}}. \tag{3}$$

Thus, in order to facilitate the turn on of the n−p−n−p−n−p structure, we should ensure that the emitter p−n junction j_3 is not strongly asymmetric.

If one of the conditions (1) or (2) is not satisfied, the device cannot be turned on. In order to go over to the negative differential resistance region in the current−voltage characteristic of the six-layer structure, we must satisfy simultaneously the following conditions (which are deduced from the well-known results for p−n−p−n structures):

$$\alpha_{12}^* + \alpha_{32}^* \geqslant 1, \tag{4}$$

$$\alpha_{54}^* + \alpha_{34}^* \geqslant 1, \tag{5}$$

where $\alpha_{ik}^* = \alpha_{ik} + I \frac{d\alpha_{ik}}{dI}$ are the differential current gains of the corresponding three-layer structures.

Thus, when a forward voltage is applied, the current−voltage characteristic of a six-layer structure (Fig. 2) is similar to the characteristic of a p−n−p−n structure. The basic

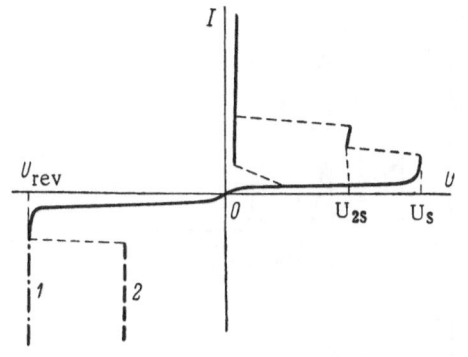

Fig. 2. Current–voltage characteristic of an n–p–n–p–n–p struc-
ture: 1) when $\alpha_{23}^* + \alpha_{43}^* \geq 1$; 2) when
$\alpha_{23}^* + \alpha_{43}^* \leq 1$.

difference is that the maximum switching voltage is now governed not by one but by two p–n junctions. In the on state when the collector p–n junctions are in a state of saturation, the polarity of the voltage across these junctions is opposite to the polarity of the external voltage. Consequently, the voltage drop across the whole six-layer structure is equal to the voltage drop across one p–n junction and the voltage drop across the base regions of the structure. It follows that the total voltage drop across a six-layer structure is less than that across two thyristors (p–n–p–n structures) connected in series. More-over, since the values of the current which satisfy the conditions (4) and (5) for a six-layer structure may differ in magnitude, the current–voltage charac-teristic of the whole structure has two negative re-sistance regions.

The breakover voltage of a six-layer structure can be controlled, as in a four-layer structure, by making a connection to one of the outer base regions. However, we must satisfy an additional condition. In order to alter the breakover voltage of the whole structure, we must be able to saturate not only the collector junction j_2 but also the other collector junction j_4, keeping the current through the structure at a constant value. We can see that the stronger the inequality $\gamma_{23} > \gamma_{21}$ and the higher the values of the current gains α_{12}, α_{23}, and α_{34}, the easier it is to satisfy this condition. The gate current flowing into the base P_1 should be due to the injection of holes by the junction j_2, which is saturated by the gate current I_b.

If the polarity of the external voltage is reversed, the p–n junctions j_2 and j_4 are for-ward biased and the other junctions (j_1, j_3, j_5) are reverse biased. Since the junction j_1 should have an artificial leakage, the reverse voltage of the structure is governed only by the junc-tions j_3 and j_5. If the parameters of the structure $P_1 - N_2 - P_2 - N_3$ are such that, even at low values of the current, the turn-on condition of this structure is satisfied, i.e.,

$$a_{23}^* + a_{43}^* \geqslant 1, \tag{6}$$

the reverse branch of the current–voltage characteristic of the n–p–n–p–n–p structure should have a negative resistance region (Fig. 2). However, the value of the holding voltage in that region is not less than the electrical breakdown voltage of the junction j_5. When

$$a_{23}^* + a_{43}^* \leqslant 1, \tag{7}$$

there is no negative differential resistance region in the reverse branch of the characteristic. It must be pointed out that the conditions (3), (7) and the control condition $\gamma_{23} > \gamma_{21}$ are not mu-tually exclusive and can be satisfied simultaneously. This happens when $\gamma_{45} \gg \gamma_{43}$, i.e., when the dopant concentration in the P_2 region is higher than that in the N_3 region.

The transient processes occurring during the switching of six-layer structures differ somewhat from the corresponding processes in p–n–p–n structures. The nature of the turn-on process in an n–p–n–p–n–p structure depends strongly on the asymmetry of injection of the central emitter junction.

We shall consider the turn-on process in an n–p–n–p–n–p structure by a gate current flowing into the base region P_1 (Fig. 1) in the case when $\gamma_{32} > \gamma_{34}$. The time dependences of the voltages V_2 and V_4 across the collector junctions j_2 and j_4 and of the current I flowing through the whole structure are all given in Fig. 3. Let us assume that, at a moment t_0, a cur-

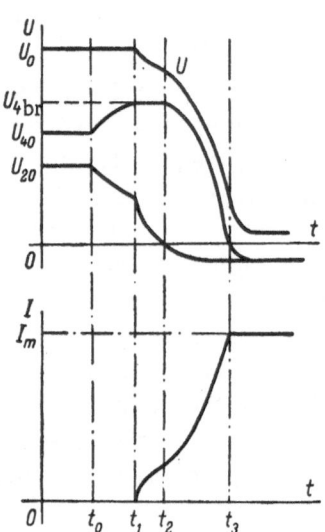

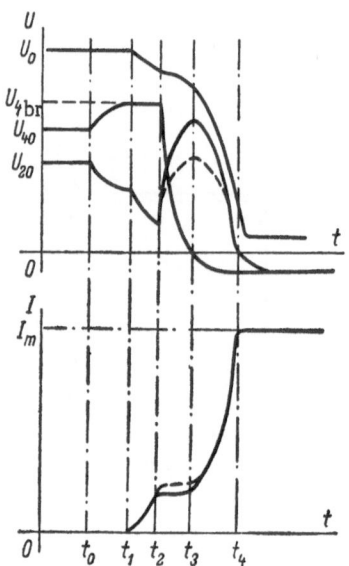

Fig. 3. Turn-on characteristics of an n−p−n−p−n−p
structure for $\gamma_{32} > \gamma_{34}$ when
the breakdown voltage of the
p−n junction j_4 is reached.

Fig. 4. Turn-on characteristics of an n−p−n−p−n−p
structure for $\gamma_{32} < \gamma_{34}$ when
the breakdown voltage of the
p−n junction j_4 is reached.

rent pulse in the form of a unit function flows into the base region P_1 through a gate circuit
connected to the emitter p−n junction j_1. After suffering an intrinsic time delay in the N_1-
P_1-N_2 structure, the nonequilibrium carriers injected by the emitter p−n junction j_1 appear
at the collector junction j_2 and the process of discharge of the barrier capacitance of this junction begins; this process reduces the voltage V_2. During this process, the current through the
junctions j_3, j_4, j_5, as well as the current through the load resistance, remains constant and equal
to the reverse current through the collector junction j_4. Therefore, a drop in V_2 is accompanied
by a rise of V_4, so that the total voltage across the structure remains constant. This fall of the
voltage V_2 continues until the voltage V_4 increases to the breakdown voltage of the junction j_4
(this happens after a time t_1). From this moment onward, the current through the whole structure begins to increase. The rate of rise of the current is governed by the $N_1-P_1-N_2$ transistor. The rise of the current through the whole structure, calculated ignoring carrier injection by the junction j_3, continues until the current reaches a value $I_b[\alpha_1/(1-\alpha_1)]$. The voltage
across the collector p−n junction j_4 remains equal to the breakdown voltage of this junction, and
the voltage across the collector junction j_2 decreases because of an increase in the voltage drop
across the load resistance. The flow of a current through the p−n junction j_3 results in hole
injection mainly into the base N_2. These holes reach the collector junction j_2 and give rise to a
further fall of the reverse voltage across this junction; eventually the sign of the voltage V_2 is
reversed (after a time t_3). From this moment onward, practically the whole external voltage is
applied to the collector junction j_4, and an avalanche-like rise of the current through the whole
structure begins. In this process, the electrons injected by the junction j_1 diffuse and drift
through the base P_1; they reach the base N_2 and are re-injected through the junction j_3, pass
through the base P_2, and reach the base N_3 as the majority carriers. These processes make
possible whole injection by the junction j_5 and the subsequent appearance of holes in the bases
P_2 and P_1. The rise of the current through the whole structure reduces the voltage drop across
the load resistance and the voltage V_4 across the collector junction j_4.

    When the current through the structure reaches the value $I_m = (E - \Delta V)/R_l$, where R_l
is the load resistance and ΔV is approximately equal to the voltage drop across the whole

structure in the on state, the subsequent rise of the current is limited by the external circuit parameters (after a time t_4). A steady state is established [1-2] in which the sign of the voltage across the collector junction j_4 is inverted and this junction becomes saturated.

We shall now consider the turn-on process in a six-layer structure when $\gamma_{32} < \gamma_{34}$ (Fig. 4). The first stage of the turn-on process from the moment t_0, when a gate current pulse is applied to the base P_1, to a moment t_1, when the breakdown voltage of the junction j_4 is reached, is the same as in the case just considered provided the ratio of γ_{21} and γ_{23} is the same. During the second stage of the turn-on process, we observe, as in the preceding case, a rise of the current through the structure and a fall of the voltage across the collector p−n junction j_2. The rise of the current continues until it reaches the value $I_b[\alpha_{12}/(1 - \alpha_{12})]$ at a moment t_2. Further rise of the current is limited because, in view of the assumed condition $\gamma_{32} < \gamma_{34}$, the emitter junction j_3 injects mainly electrons and, until a sufficiently large number of holes is accumulated in the base P_2, the injection of holes by the p−n junction j_3 does not alter greatly the gate current in the base P_1. Therefore, during the third stage of the turn-on process, from t_2 to t_3, the current through the structure remains approximately constant. Because of this restriction on the value of the current imposed by the $N_1 - P_1 - N_2$ transistor, the process which occurs in the $N_2 - P_2 - N_3 - P_3$ part of the structure is similar to the process of establishment of a steady state in a p−n−p−n structure, and the result of this process is a rapid drop in the voltage across the collector junction j_4. Since the voltage across the whole structure is not affected by this process, the voltage across the junction j_2 increases. After a time t_3, the voltage V_4 decreases to zero and its sign is inverted. At this moment, the p−n junction j_3 begins to inject holes in appreciable amounts and this injection increases the total number of holes reaching the base P_1, triggering off the avalanche-like rise of the current described in the preceding paragraphs. As in the preceding case, the rise of the current through the structure is accompanied by a fall of the voltage V_2 across the collector junction j_2, which is still biased in the reverse direction. From the moment t_4, the current of the structure is limited by the external circuit parameters and a forward bias is gradually established across the junction j_2. If the current is greatly restricted by the transistor $N_1 - P_1 - N_2$ during the time interval t_2-t_3, the avalanche breakdown voltage across the collector junction j_2 may be reached during this time interval. If this happens, the voltage rise across the junction j_2 during the time interval t_2-t_4 is much lower.

We have considered so far the turn-on process in an n−p−n−p−n−p structure under conditions such that the voltage applied to the structure is higher than the electric breakdown voltage of the p−n junction j_4. However, a six-layer structure can be turned on using a voltage lower than the electric strength of the junction j_4. In this case, the turn-on process develops as follows. First, as in the preceding two cases, the barrier capacitance of the collector p−n junction j_2 begins to discharge and this is accompanied by a fall of the voltage V_2 and a rise of the voltage V_4, which continue until practically the whole external voltage is applied to the p−n junction j_4 (at a moment t_1). Next, because the current through the $P_2 - N_3 - P_3$ structure remains equal to the reverse current through the collector junction j_4, the junction j_2 becomes saturated. The next stage of the turn-on process depends on the injection efficiency γ_{23} of the p−n junction j_2. If the impurity concentration N_2 in the second base region is considerably higher than the impurity concentration P_1 in the first p-type base, an increase in the nonequilibrium hole density in the base region N_2 may be insufficient for subsequent development of the turn-on process. However, if the value of the injection efficiency γ_{23} of the junction j_2 in the direction of the junction j_3 is sufficiently high, holes injected by j_2 pass through N_2 and reach P_2, where they are the majority carriers. This produces injection from the junction j_3. If $\gamma_{32} > \gamma_{34}$, the next stage is an avalanche-like rise of the current with the collector p−n junction j_2 still in the state of saturation (Fig. 5). If $\gamma_{32} < \gamma_{34}$, the electrons injected by j_3 reach, after an intrinsic delay time, the collector junction j_4 and current begins to rise through the whole six-layer structure (time t_2, Fig. 6). This causes the collector junction j_2 to depart from its state of saturation. Next, the process continues as in the case of $\gamma_{32} < \gamma_{34}$.

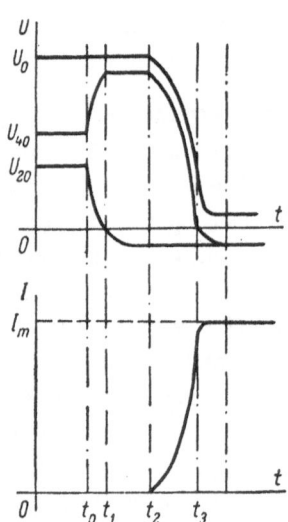

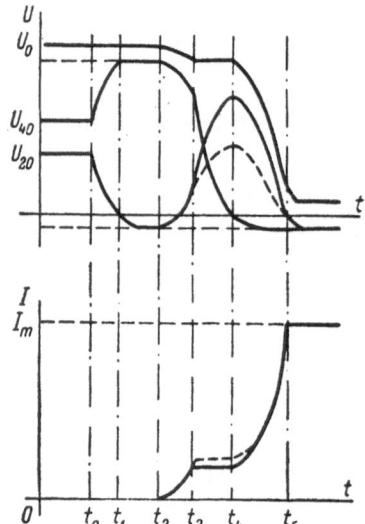

Fig. 5. Turn-on characteristics of an n−p−n−p−n−p
structure for $\gamma_{32} > \gamma_{34}$ when
the breakdown voltage of the
p−n junction j_4 is not reached.

Fig. 6. Turn-on characteristics of an n−p−n−p−n−p
structure for $\gamma_{32} < \gamma_{34}$ when
the breakdown voltage of the
p−n junction j_4 is not reached.

The turn-on process in a symmetric six-layer structure switched on by an anode current
or by a gate current (flowing into both outer base regions) is similar to the turn-on process in
a four-layer structure. The reverse voltage is divided equally between the two collector p−n
junctions.

It follows from our analysis that when a six-layer structure is triggered by passing a gate
current into the base P_1, the switching process is smoothest and fastest if $\gamma_{32} < \gamma_{34}$ and $\gamma_{21} < \gamma_{23}$
(Fig. 3). It is also obvious that the injection efficiencies of the outer emitter junctions, γ_{12} and
γ_{54}, should be close to unity at currents exceeding the turn-on value.

When the gate current flowing into the base P_1 is increased, the duration of the initial
stages of the turn-on process, up to the onset of the avalanche rise of the current, may be reduced to very small values. Thus, the greatest interest lies in the regenerative process of
the current rise, i.e., in the interval t_2-t_3 (Figs. 3 and 5). We shall assume that the injection
level in all base regions of the structure is low and there is no electric field in these regions.
Moreover, we shall postulate that the injection efficiencies of the outer emitter junctions are
unity and the intrinsic current of the reverse-biased collector junction j_4 is negligibly small.
We shall consider the one-dimensional model and employ the usual assumptions made in the
theory of transistors.

The influence of a high injection level and of the discharge of the barrier capacitance of
the junction j_4 on the turn-on process of a six-layer structure has not been considered in the
analysis given so far. This influence can be deduced by analogy with the role played by the
same two factors in a p−n−p−n structure [3].

In order to simplify the intermediate stages of the analysis, it is convenient to use two
coordinate axes along the normal to the p−n junction planes: one of these axes (x) is assumed
to have its origin in the plane of the collector junction j_2 and the other axis (x') has its origin
in the j_4 plane. The thickness of both collector junctions is assumed, as in [3], to be negligible.

In this approximation, the motion of the minority carriers in the i-th base region is described by the equations

$$\tau_i \frac{\partial n_i}{\partial t} = L_i^2 \frac{\partial^2 n_i}{\partial x^2} - n_i. \tag{8}$$

In view of our assumptions ($\gamma_{12} = \gamma_{54} = 1$), the outer emitter regions are excluded from the analysis and the base regions are numbered as follows: $i = 1$ for P_1, $i = 2$ for N_2, $i = 3$ for P_2, and $i = 4$ for N_3, so that odd subscripts in Eq. (8) refer to nonequilibrium electrons and even subscripts refer to nonequilibrium holes. We shall also use the fold symbols: $n_i(x, t)$, n_{0i} are the nonequilibrium and equilibrium densities of the minority carriers in the i-th base; τ_i is the minority carrier lifetime; $L_i = \sqrt{D_i \tau_i}$ is the diffusion length; w_i is the thickness of the i-th base; D_i is the diffusion coefficient of the minority carriers in the i-th base.

The conditions of quasi-neutrality in the base regions, which are the boundary conditions of the problem, are of the following form:

in base P_1,

$$-D_1 \frac{\partial n_1}{\partial x}\bigg|_{w_1} + D_1 \frac{\partial n_1}{\partial x}\bigg|_0 = \frac{I_b}{q} + D_2 \frac{\partial n_2}{\partial x}\bigg|_0; \tag{9}$$

in base N_2,

$$D_2 \frac{\partial n_2}{\partial x}\bigg|_{w_2} - D_2 \frac{\partial n_2}{\partial x}\bigg|_0 = -D_1 \frac{\partial n_1}{\partial x}\bigg|_0 + D_3 \frac{\partial n_3}{\partial x}\bigg|_{-w_3}; \tag{10}$$

in base P_2,

$$-D_3 \frac{\partial n_3}{\partial x}\bigg|_{-w_3} + D_3 \frac{\partial n_3}{\partial x}\bigg|_0 = D_4 \frac{\partial n_4}{\partial x}\bigg|_0 - D_2 \frac{\partial n_2}{\partial x}\bigg|_{w_2}; \tag{11}$$

in base N_3

$$D_4 \frac{\partial n_4}{\partial x}\bigg|_{w_4} - D_4 \frac{\partial n_4}{\partial x}\bigg|_0 = -D_3 \frac{\partial n_3}{\partial x}\bigg|_0. \tag{12}$$

Two additional boundary conditions can be obtained by writing down the relationships between the nonequilibrium minority carrier densities on both sides of the p-n junctions j_2 and j_3. We thus obtain

$$\frac{n_1}{n_{01}}\bigg|_0 = \frac{n_2}{n_{02}}\bigg|_0 \tag{13}$$

for the junction j_2 and

$$\frac{n_2}{n_{02}}\bigg|_{w_2} = \frac{n_3}{n_{03}}\bigg|_{-w_3} \tag{14}$$

for the junction j_3.

Since, throughout the avalanche process, the voltage across the collector junction j_4 remains reversed, we may assume that the densities of the nonequilibrium minority carriers near this junction are zero, i.e.,

$$n_3(0, t) = n_4(0, t) = 0. \tag{15}$$

Thus, we have four second-order nonstationary equations of continuity (8) and eight corresponding boundary conditions (9)-(15). Without committing appreciable errors, we can assume that the initial conditions are all zero.

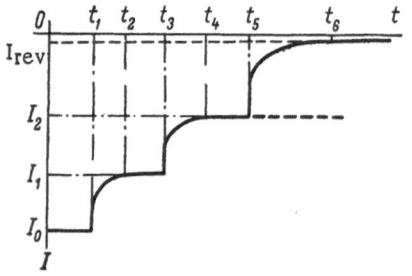

Fig. 7. Dependence $I = f(t)$ when an n−p−n−p−n−p structure is turned off by a reverse anode voltage.

The solution of this problem shows that the avalanche-like rise of the current is described, in the first approximation, by an exponential law similar to that which is obtained for a four-layer structure:

$$I(t) = I_0 \left(e^{\frac{t}{\tau_a}} - 1 \right),\qquad(16)$$

where I(t) is the current through the structure; I_0 is a quantity of the order of the turn-on current; τ_a is the time constant of the rise of the current to the turn-on value. The time constant τ_a is given by the equation

$$(\mathrm{ch}\,a_1 + \mathrm{ch}\,a_2 - \mathrm{ch}\,a_1\,\mathrm{ch}\,a_2)\left\{ \delta_{23}a_2\,\mathrm{sh}\,a_2\,\frac{\mathrm{sh}\,a_3}{a_3}(\mathrm{ch}\,a_4 - 1) + \mathrm{ch}\,a_2\,(\mathrm{ch}\,a_3\,\mathrm{ch}\,a_4 - \mathrm{ch}\,a_3 - \mathrm{ch}\,a_4) \right\} +$$

$$+\,[\delta_{12}a_1\,\mathrm{sh}\,a_1 + a_2\,\mathrm{sh}\,a_2\,(\mathrm{ch}\,a_1 - 1)]\left\{ \delta_{23}\,\mathrm{ch}\,a_2\,\frac{\mathrm{sh}\,a_3}{a_3}(\mathrm{ch}\,a_4 - 1) + \frac{\mathrm{sh}\,a_2}{a_2}(\mathrm{ch}\,a_3\,\mathrm{ch}\,a_4 - \mathrm{ch}\,a_3 - \mathrm{ch}\,a_4) \right\} +$$

$$+\,\mathrm{ch}\,a_3\,(\mathrm{ch}\,a_1 - 1)\,(\mathrm{ch}\,a_4 - 1) = 0,\qquad(17)$$

where

$$a_i = \frac{w_i}{L_i}\sqrt{1 + \frac{\tau_i}{\tau_a}},\quad \delta_{ik} = \frac{n_{0i}D_i w_k}{n_{0k}D_k w_i}.$$

For a symmetric structure $1/\tau_a$ is given by the same expression as for a p−n−p−n structure [4], namely $1/\tau_a = (0.84/\theta) - (1/\tau)$, where $\theta = w^2/2D$ is the transit time and τ is the lifetime of the minority carriers.

We shall now consider the turn-on process in an n−p−n−p−n−p structure, triggered by a reverse anode voltage. Here, in principle, we can have two different processes depending on the value of the sum of the steady-state current gains $\alpha_{23} + \alpha_{43}$, which has been mentioned before. The first case, when $1 < (\alpha_{23} + \alpha_{43}) < 2$, is observed when, for example, an n−p−n−p−n−p structure is prepared by triple diffusion into a material initially of, say, the p-type, so that $P_2 > N_3 > P_2 < N_2 < P_1 < N_1$. If a constant reverse voltage is applied to such a structure in the on state, the following process is observed (Fig. 7). First, the current through the structure remains constant and equal to $I_0 = V_0/R_l$, where V_0 is the voltage supplied by the external source and R_l is the load resistance (it is assumed that V_0 is considerably larger than the voltage drop across the structure in the on state). After a time t_1, when the excess density of the minority carriers at the junction j_1 decays to zero, the current decreases rapidly because of an increase of the voltage V_0 across this junction. This continues until the breakdown voltage V_{br1} is reached at a moment t_2. Then, the current again remains constant and its value is now

$$I_1 = \frac{V_0 - V_{br1}}{R_l}.\qquad(18)$$

At a moment t_3, the densities of the minority carriers near the junction j_5 decreases to zero and the consequent rise of the reverse voltage across this junction is responsible for a rapid fall of the current through the whole structure to the value

$$I_2 = \frac{V_0 - V_{br1} - V_{br5}}{R_l},\qquad(19)$$

which is reached at a moment t_4 when the junction j_5 begins avalanching. The remaining $P_1−N_2−P_2−N_3$ structure is still turned on in the forward direction and, if the value of I_2 is sufficiently high, this produces an irreversible breakdown in the junction j_5 because the junction j_1 is very heavily doped and has artificial leakage.

A different variant of the turn-off process is observed when $\alpha_{23} + \alpha_{43} < 1$. This case is observed, for example, when the impurity concentrations in the base regions obey the following inequalities: $N_1 > P_1 > N_2 < P_2$, $P_2 > N_3 < P_3$. The dependence of the current on time (Fig. 7) up to a moment t_4 is the same as in the case just discussed. The only difference is that the second junction (after j_1) to reach the avalanche breakdown conditions is the junction j_3. Consequently, the current is

$$I_2 = \frac{V_0 - V_{br\,1} - V_{br\,3}}{R_1}.$$ (20)

However, in this case, in addition to the concentrations specified by the inequalities just quoted, we must satisfy the condition $\tau_2 < \tau_4$. Then, the junction j_3 breaks down and carries a sufficiently high current only during the period t_4-t_5. At a moment t_5, the density of the excess nonequilibrium carriers at the junction j_5 reaches zero and the current begins to drop rapidly, corresponding to a rise of the reverse voltage across this junction; this is followed by a slow decay of the current, corresponding to charge dispersal process with a time constant discussed in [5]. In the $P_2 \gg N_3$ case this time constant is equal to t_4. It follows that the junctions j_3, j_4, and possibly also j_2 should be protected reliably from any surface breakdown.

Conclusions

Analysis of the processes in an n−p−n−p−n−p structure shows that such a structure can be used in controlled rectifiers with twice the permissible voltages in the forward and reverse directions. The voltage drop across the structure in the on state is practically the same as the voltage drop across an ordinary thyristor. The turn-on and turn-off times of six-layer structures are of the same order of magnitude as the corresponding time constants of p−n−p−n structures.

The author is grateful to I. V. Grekhov and A. I. Uvarov for discussing the problems considered in the present paper.

Literature Cited

1-2.   A. A. Lebedev, A. I. Uvarov, and V. E. Chelnokov, Radiotekhnika i Élektron., 12:677 (1967).
3.   A. A. Lebedev and A. I. Uvarov, Fiz. Tekh. Poluprov., 1:211 (1967).
4.   A. A. Lebedev, A. I. Uvarov, and V. E. Chelnokov, Radiotekhnika i Élektron., 11:1458 (1966).
5.   A. A. Lebedev and A. I. Uvarov, Radiotekhnika i Élektron., 12:686 (1967).

CURRENT–VOLTAGE CHARACTERISTICS OF p–n–p–n STRUCTURES GOVERNED BY RECOMBINATION, GENERATION, AND AVALANCHE MULTIPLICATION PROCESSES IN p–n JUNCTIONS

V. A. Kuz'min and Yu. A. Parmenov

Moscow Engineering Physics Institute

The principal parameters of the current–voltage characteristics of silicon planar thyristors are calculated taking into account the recombination, generation, and avalanche multiplication in p–n junctions. Analytic expressions are obtained for the determination of the turn-off current, critical gate current, and turn-off gain of an asymmetrical silicon planar thyristor. The dependence of the turn-off current on the gate current and the dependence of the turn-off gain on the anode current are calculated. Allowance is made for the avalanche multiplication of carriers in the central p–n junction and analytic expressions are derived for the turn-on current and the breakover voltage. It is shown that the breakover voltage of planar thyristors designed for low-voltage operation is close to the breakdown voltage of the central p–n junction, which is due to the low value of the leakage current of this junction.

The majority of modern silicon semiconductor devices (except those intended for high-power applications) are made using the planar technology.

A special feature of the planar p–n–p–n structures is the absence of leakage in p–n junctions. Consequently, the current gains of the two equivalent transistors comprising a planar thyristor depend on the current that is mainly due to recombination in the emitter p–n junctions [1]. The reverse current of the central (collector) p–n junction of such structures is governed by carrier generation and avalanche multiplication.

Analyses of the current–voltage characteristic of a thyristor, taking into account carrier recombination and generation in p–n junctions, have been given in several papers [2-4].

Muss and Goldberg's treatment [2] is limited to the graphical plotting of a current–voltage characteristic obtained using a computer.

Lebedev, Uvarov, and Chelnokov [3] analyzed a symmetrical structure in the case when recombination takes place only at one of the two emitter junctions. They assumed that avalanche multiplication is unimportant. Ryabinkin [4] calculated the current–voltage characteristic in the case when the dependence $\alpha(I)$ is linear.

330

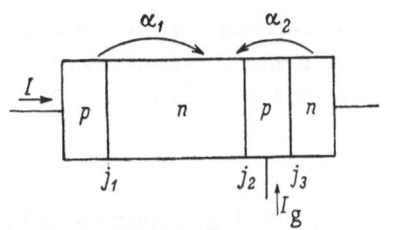

Fig. 1. Structure of a p−n−p−n thyristor.

Lebedev, Uvarov, and Chelnokov [3] obtained analytic expressions for the turn-off, turn-on, and critical gate currents. Ryabinkin [4] derived expressions for the turn-on and turn-off currents and for the breakover voltage.

We shall take into account the avalanche multiplication and calculate the static parameters of asymmetrical silicon planar thyristors: the turn-on, turn-off, and critical gate currents, the breakover voltage, the turn-off gain, and the dependence of this gain on the current.

1. Model Used in the Analysis

A typical structure of a silicon thyristor is shown in Fig. 1. We shall calculate the static current−voltage characteristic of such a thyristor, making the following assumptions.

1. The dependences of the current gains of the equivalent p−n−p and n−p−n transistors on the anode current are governed primarily by carrier recombination in the emitter junctions j_1 and j_3. According to Ryabinkin [4], these dependences can be represented by the expression

$$\alpha(I) = \alpha^0 \frac{\sqrt{1+GI}-1}{\sqrt{1+GI}+1} = \alpha^0 \frac{(\sqrt{1+GI}-1)^2}{GI}, \tag{1}$$

where

$$\left.\begin{aligned}
&G = G_1 = \frac{4I_{dp}}{\gamma_1^0 I_{r_1}^2}, \\
&\alpha^0 = \alpha_1^0 = \gamma_1^0 \varkappa_1^0, \\
&I_{dp} = \frac{qD_p p_n}{L_p} \operatorname{cth} \frac{W_n}{L_p}, \\
&I_{r_1} = \frac{qn_i W_{j_1}}{(\sqrt{\tau_n \tau_p})_{j_1}} \cdot \frac{f(b)}{\frac{q}{kT}(\varphi_{k_1} - U_1)},
\end{aligned}\right\} \tag{2}$$

for the p−n−p transistor and

$$\left.\begin{aligned}
&G = G_2 = \frac{4I_{dn}}{\gamma_3^0 I_{r_3}^2}, \\
&\alpha^0 = \alpha_2^0 = \gamma_3^0 \varkappa_2^0, \\
&I_{dn} = \frac{qD_n n_i^2}{W_p \int_0^{} N(x)\,dx}, \quad I_{r_3} = \frac{qn_i W_{j_3}}{(\sqrt{\tau_n \tau_p})_{j_3}} \cdot \frac{f(b)}{\frac{q}{kT}(\varphi_{k_3} - U_3)},
\end{aligned}\right\} \tag{3}$$

for the n−p−n transistor. Here, α^0, γ^0, and $\varkappa^0$ are, respectively, the current gain, injection efficiency, and transport factor of a transistor in the absence of recombination in the emitter p−n junction; I_{dp}, I_{dn}, I_{r_1}, I_{r_3}, are, respectively, the coefficients in front of the exponential functions in the expressions for the diffusion and recombination currents; W_{j_1}, W_{j_3}, W_n, W_p are the thicknesses of the emitter junctions j_1 and j_3 and of the n- and p-type bases, respectively; τ_n, τ_p are, respectively, the electron lifetime in heavily doped p-type silicon and the hole lifetime in heavily doped n-type silicon; φ_{k_1}, φ_{k_3}, U_1, U_2 are the contact potentials and the voltages across the junctions j_1 and j_3, respectively; N(x) is the law governing the impurity distribution in the p-type base; $f(b)$ is some function which depends on the position of the trapping level in the forbidden band and on the voltage across a p−n junction [1].

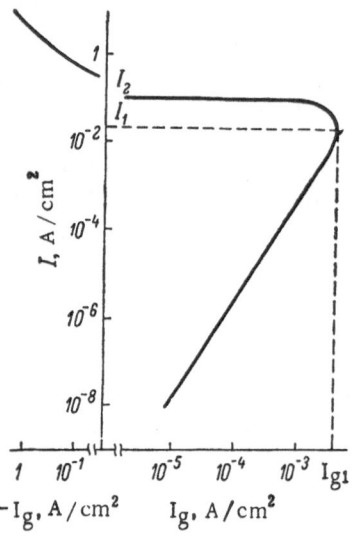

Fig. 2. Dependence $I(I_g)$.

2. The current−voltage characteristic of the central p−n junction is governed by carrier generation and avalanche multiplication, and it can be written in the form [1]

$$I_0 = MCU^{\frac{1}{3}}, \tag{4}$$

where M is the multiplication factor, which is assumed to be the same for electrons and holes,

$$M = \frac{1}{1 - \left(\frac{U}{U_0}\right)^n}, \tag{5}$$

$$C = \frac{qn_iW'_{j_2}}{2\left(\sqrt{\tau_n\tau_p}\right)_{j_2} \operatorname{ch}\left[\dfrac{E_t - E_i}{kT} + \ln\sqrt{\left(\dfrac{\tau_p}{\tau_n}\right)_{j_2}}\right]}; \tag{6}$$

E_t, E_i are, respectively, the energy level of the traps and the level of the middle of the forbidden band; $W'_{j_2} = \sqrt[3]{\dfrac{12\varepsilon\varepsilon_0}{qa_{j_2}}}$ is the thickness of the p−n junction j_2 when $\varphi_{k_2} - U_2 = 1$ V; a_{j_2} is the impurity concentration gradient in the junction j_2; U_0 is the avalanche breakdown voltage of the junction j_2.

In the $U_0 \geq 6$ V range, the breakdown voltage can be found from the formula [5]:

$$U_0 = 60\left(\frac{3 \cdot 10^{20}}{a_{j_2}}\right)^{0.4}. \tag{7}$$

3. The injection level in the p- and n-type bases is low.

4. The lifetimes in each of the regions of the p−n−p−n structure are constant.

2. Turn-off Current I_{off}, Critical Gate Current I_{gcr}, and Turn-off Gain K_I

The equation for the current flowing through the central p−n junction j_2 when $U_2 = 0$ can be represented in the form

$$I\alpha_1(I) + (I + I_g)\alpha_2(I + I_g) = I. \tag{8}$$

Substituting into Eq. (8) the values of $\alpha_1(I)$ and $\alpha_2(I)$, defined by Eq. (1), we obtain the expression

$$a_1^0G_2\left(\sqrt{1 + G_1I} - 1\right)^2 + a_2^0G_1\left(\sqrt{1 + G_2(I + I_g)} - 1\right)^2 - G_1G_2I = 0. \tag{9}$$

An attempt to solve Eq. (9) for the current I yields a quartic polynomial. However, Eq. (9) can be solved relatively simply for the gate current I_g:

$$I_g = \frac{1}{G_2}\left(1 + \sqrt{\frac{G_2}{a_2^0}I - \frac{a_1^0G_2}{a_2^0G_1}\left(\sqrt{1 + G_1I} - 1\right)^2}\right)^2 - I - \frac{1}{G_2}. \tag{10}$$

At very low and very high current densities, Eq. (10) can be simplified:

for $IG_1 \ll 1$, $IG_2 \ll 1$

$$I_g = 2\sqrt{\frac{I}{a_2^0G_2}} + \frac{1 - a_2^0}{a_2^0}I, \tag{11}$$

for $IG_1 \gg 1$, $IG_2 \gg 1$

$$I_g = \frac{I\,(1 - \alpha_1^0 - \alpha_2^0)}{\alpha_2^0}. \tag{12}$$

The inverse function $I(I_g)$, plotted using Eq. (10) for $G_1 = 10^4$ cm^2/A, $G_2 = 825$ cm^2/A, $\alpha_1^0 = 0.27$, $\alpha_2^0 = 0.93$, is shown in Fig. 2.

The dependence $I(I_g)$ can be used to determine the principal parameters of the current−voltage characteristic: the turn-off current, the critical gate current, the turn-off gain, and the dependence of this gain on the current.

We shall now consider in more detail the function $I(I_g)$. When $I_{g1} > I_g > 0$, the function $I(I_g)$ is double-valued. This means that there are two points on the current−voltage characteristic at which the condition $U_2 = 0$ is satisfied. For currents $I < I_1$, this point corresponds to the limit of the active operation of the n−p−n transistor in the common-emitter configuration. Thus, the dependence $I(I_g)$ gives the value of the collector current of the n−p−n transistor at the point separating the active operation and saturation of this transistor.

When the current is $I > I_1$, the dependence $I(I_g)$ represents the function $I_{off}(I_g)$. When $I_g = 0$, we obtain, at point $I = I_2$, the turn-off current of a p−n−p−n structure connected as a diode. In the range of negative gate currents $I_g < 0$, assuming that the load current at the turn-off point is $I = I_{off}$, we obtain the value of the turn-off gain of the thyristor $K_I = I/|I_g|$ and its dependence on the current. Finally, at the point $I = I_1$, where the function $I(I_g)$ is single-valued, we obtain the critical gate current $I_{g\,cr} = I_{g1}$. In the special case $G_1 = G_2 = G$, the turn-off current can be deduced directly from Eq. (10):

$$I_{off} = \frac{4\,(\alpha_1^0 + \alpha_2^0)}{G\,(\alpha_1^0 + \alpha_2^0 - 1)^2}. \tag{13}$$

Assuming that $\alpha_1^0 = \alpha_2^0 = \alpha$, we find that Eq. (13) becomes identical with the expression obtained in [3] for a symmetrical thyristor.

Typical Calculation. We shall consider a fast-response silicon planar thyristor and calculate the actual dependence $I(I_g)$ and the principal parameters of the static current−voltage characteristic. We shall use the following parameters of the various regions of the thyristor structure:

$$W_p = 2\ \mu, \qquad W_n = 10\ \mu,$$

$$\int_0^{W_p} N\,(x)\,dx = 2 \cdot 10^{13}\ \text{cm}^{-2}, \qquad N_d = 10^{15}\ \text{cm}^{-3}.$$

Usually the p-type base and emitter regions are produced by simultaneous diffusion and therefore we shall assume that the parameters of these regions are identical.

We shall also postulate that the lifetimes in the base regions and in the p-type emitter are governed by the concentration of the gold dopant. According to [6, 7], the minority carrier lifetime in gold-doped silicon can be deduced from the formulas

$$\tau_n = \frac{2.86 \cdot 10^7}{N_{Au}}\ \text{sec (p-type silicon)}, \tag{14}$$

$$\tau_p = \frac{7.25 \cdot 10^7}{N_{Au}}\ \text{sec (n-type silicon)}, \tag{15}$$

where N_{Au} is the concentration of gold.

TABLE 1

τ_n, sec	G_1, cm²/A	G_2, cm²/A	α_1^0	α_2^0
		$\tau_{pe} = 0.1\tau_n$		
10^{-8}	157	132	0.317	0.942
$5 \cdot 10^{-8}$	$2.5 \cdot 10^3$	$3.3 \cdot 10^3$	0.742	0.988
10^{-7}	$9.1 \cdot 10^3$	$1.3 \cdot 10^4$	0.855	0.993
		$\tau_{pe} = 0.01\tau_n$		
10^{-8}	157	13.2	0.317	0.942
$5 \cdot 10^{-8}$	$2.5 \cdot 10^3$	$3.3 \cdot 10^2$	0.742	0.988
10^{-7}	$9.1 \cdot 10^3$	$1.3 \cdot 10^3$	0.855	0.993

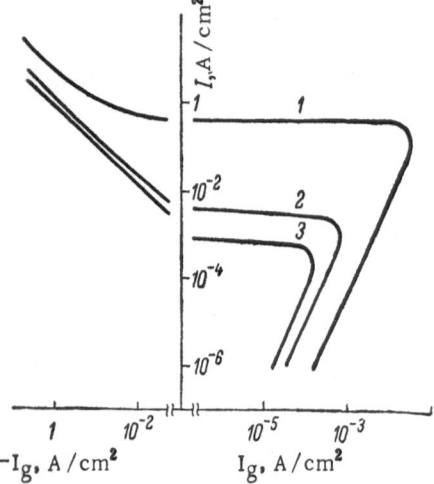

Fig. 3. Dependences $I(I_g)$ plotted in accordance with Eq. (10). τ_n, sec: 1) 10^{-8}; 2) $5 \cdot 10^{-8}$; 3) 10^{-7}.

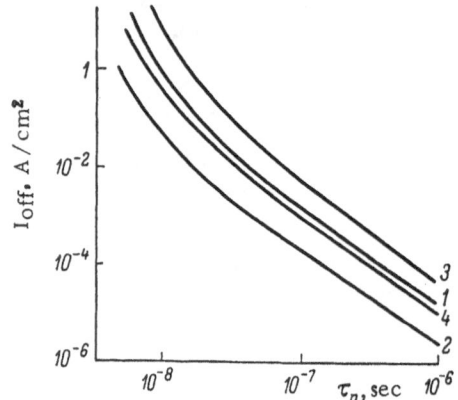

Fig. 4. Dependences of the turn-off and critical gate currents on the lifetime τ_n for two different lifetimes in the n-type emitter. 1, 3) I_{off}; 2, 4) $I_{g\,cr}$.

It follows from Eqs. (14) and (15) that when the p- and n-type regions are doped to the same degree, the lifetimes in these regions are related by the expression

$$\tau_p = 2.53\tau_n.$$

The lifetime in the n-type emitter, τ_{pe}, is usually unknown. It is shown in [2] that this lifetime may be several orders of magnitude lower than in other regions of the p−n−p−n structure. We shall assume that the lifetime in the n-type emitter is one or two orders of magnitude lower than the lifetime in the adjoining p-type base.

Thus, if we specify the lifetime in the p-type base (τ_n), we automatically set the lifetimes in all the regions of the p−n−p−n structure. The value of f (b) is assumed to be $\pi/2$, which corresponds to the position of the trapping level near the middle of the forbidden band. This is a good approximation for gold-doped silicon.

The values of $\varphi_{k_1} - U_1$, $\varphi_{k_3} - U_3$ are taken to be 0.3 V, and the injection coefficients of the junctions j_1 and j_3 are assumed to be unity. The thickness of the junction j_1 is taken to be $8 \cdot 10^{-5}$ cm and the thickness of the junction j_3 is assumed to be $1.4 \cdot 10^{-5}$ cm; these thicknesses correspond to impurity concentration gradients $a_{j_1} = 5 \cdot 10^{19}$ cm⁻⁴ and $a_{j_3} = 10^{22}$ cm⁻⁴. The values of G_1, G_2, α_1^0, α_2^0, calculated for different lifetimes τ_n and τ_{pe}, are listed in Table 1.

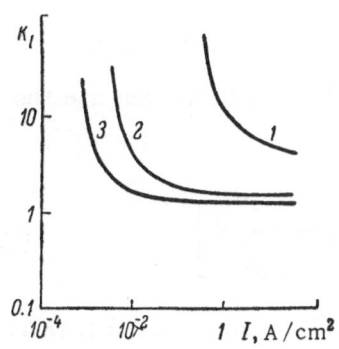

Fig. 5. Dependences of the turn-off gain on the current. τ_n, sec: 1) 10^{-8}; 2) $5 \cdot 10^{-8}$; 3) 10^{-7}.

Figure 3 shows the dependences $I(I_g)$ for three different values of the lifetime τ_n given in Table 1 and for $\tau_{pe} = 0.1\,\tau_n$. The dependences of the turn-off and critical gate currents on the lifetime τ_n are given in Fig. 4. Curves 1 and 2 correspond to the case $\tau_{pe} = 0.1\,\tau_n$, and curves 3 and 4 to the case $\tau_{pe} = 0.01\,\tau_n$. Figure 5 shows the dependence of the turn-off gain on the current for the three values of the lifetime τ_n listed in Table 1 and $\tau_{pe} = 0.1\,\tau_n$.

3. Turn-on Current I_{on} and Breakover Voltage U_{bo}

The equation which describes the dependence of the current on the voltage for $I_g = 0$ is given by

$$I = \frac{I_0\,(U)}{1 - M\,[\alpha_1\,(I) + \alpha_2\,(I)]}.$$ (16)

At low voltages, $I \approx I_0$. The breakover point, which is of greatest interest, is defined by Eq. (16) and the following equation [8]:

$$M\left[\alpha_1\,(I) + \alpha_2\,(I) + I\left(\frac{d\alpha_1}{dI} + \frac{d\alpha_2}{dI}\right)\right] = 1.$$ (17)

In general, it is difficult to find the turn-on current I_{on} and the breakover voltage U_{bo}. However, in order to determine the role of avalanche multiplication at the breakover point, it is sufficient to consider two special cases.

1. $G_1 = G_2 = G$. In this case the structure remains asymmetrical so that $\alpha_1^0 \neq \alpha_2^0$. Physically, the condition $G_1 = G_2$ is equivalent to the assumption that the lifetimes in the junctions j_1 and j_3 have close values. Substituting the values of $\alpha_1(I)$ and $\alpha_2(I)$ in Eq. (17) and solving this equation for the current, we obtain

$$I_{on} = \frac{2M\,(\alpha_1^0 + \alpha_2^0) - 1}{G\,[M\,(\alpha_1^0 + \alpha_2^0) - 1]^2}.$$ (18)

Substituting $M = 1$, $\alpha_1^0 = \alpha_2^0 = \alpha$ into Eq. (18) we obtain the value of the turn-on current deduced in [3] for a symmetrical thyristor. Using Eqs. (4) and (16) at the turn-on point, we find that

$$MCU_{bo}^{\frac{1}{3}} = I_{on}\ [1 - M\,(\alpha_1 + \alpha_2)].$$ (19)

Solving simultaneously Eqs. (18) and (19), we obtain a transcendental equation for the determination of M:

$$f\,(M) = \frac{1}{GCU_o^{\frac{1}{3}}},$$ (20)

where

$$f\,(M) = M\left(\frac{M-1}{M}\right)^{\frac{1}{3}\,n}\,[M\,(\alpha_1^0 + \alpha_2^0) - 1].$$ (21)

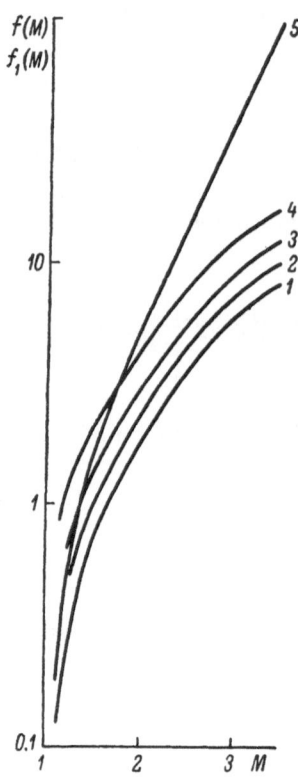

Fig. 6. Dependences of the functions $f(M)$ (curves 1-4) and $f_1(M)$ (curve 5) on M, plotted in accordance with Eqs. (20) and (28). $\alpha_1^0 + \alpha_2^0$: 1) 1.05; 2) 1.2; 3) 1.4; 4) 1.8; 5) 1.1 ($\alpha_1^0 = 0.2$).

When M > 2, the quantity $\left(\frac{M-1}{M}\right)^{\frac{1}{3} n}$ is close to unity. Consequently, in the case of large values of M we can use the simpler expression

$$M = \frac{1}{2(\alpha_1^0 + \alpha_2^0)}\left[1 + \sqrt{1 + \frac{4(\alpha_1^0 + \alpha_2^0)}{GCU_0^{\frac{1}{3}}}}\right]. \qquad (22)$$

In the case of small values of M, we must use Eq. (20). The dependences $f(M)$ on M are plotted in Fig. 6 for n = 3 and various values of $\alpha_1^0 + \alpha_2^0$.

Knowing the value of M, we can determine I_{on} using Eq. (18). If M is large, we can use Eq. (22), which yields

$$I_{on} = \frac{2\sqrt{1 + \frac{4(\alpha_1^0 + \alpha_2^0)}{GCU_0^{\frac{1}{3}}}}}{G\left[\sqrt{1 + \frac{4(\alpha_1^0 + \alpha_2^0)}{GCU_0^{\frac{1}{3}}}} - 1\right]^2}. \qquad (23)$$

The breakover voltage is found from the known values of U_0 and M:

$$U_{bo} = U_0 \sqrt[n]{\frac{M-1}{M}}. \qquad (24)$$

2. $G_1 \gg G_2$. When G_1 is large, the current gain α_1 reaches its steady-state value α_1^0 at low currents so that the dependence of α_1 on the current at the breakover point can be ignored and we can assume that $\alpha = \alpha_1^0$. In this case, Eq. (17) becomes

$$M\left(\alpha_1^0 + \alpha_2 + I\frac{d\alpha_2}{dI}\right) = 1. \qquad (25)$$

Solving Eq. (25) for the current, we obtain

$$I_{on} = \frac{(1 - \alpha_1^0 M)[M(2\alpha_2^0 + \alpha_1^0) - 1]}{G_2[M(\alpha_1^0 + \alpha_2^0) - 1]^2}. \qquad (26)$$

The value of the multiplication factor M is found from the transcendental equation

$$f_1(M) = \frac{1}{G_2 CU_0^{1/3}}, \qquad (27)$$

where

$$f_1(M) = \frac{M\left(\frac{M-1}{M}\right)^{n/3}[M(\alpha_1^0 + \alpha_2^0) - 1]}{(1 - \alpha_1^0 M)^2}. \qquad (28)$$

The dependence of the function $f_1(M)$ on M is shown in Fig. 6 for $\alpha_1^0 + \alpha_2^0 = 1.1$ and $\alpha_1^0 = 0.2$. The breakover voltage is still given by Eq. (24).

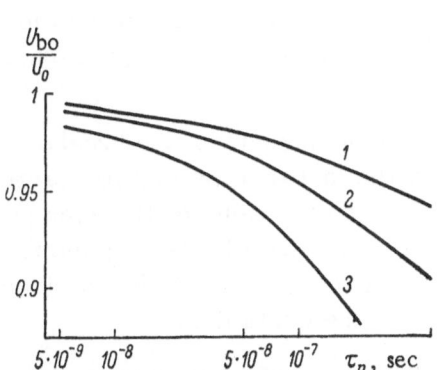

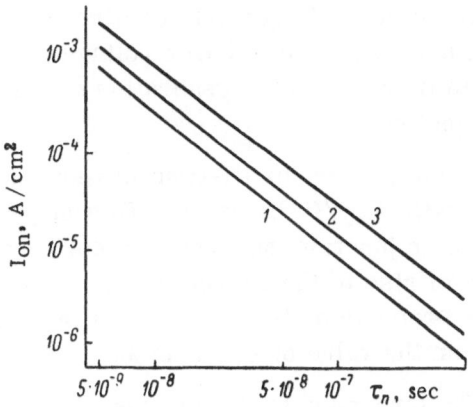

Fig. 7. Dependences of the ratio of the breakover voltage to the breakdown voltage of the central p−n junction on the lifetime τ_n, calculated using Eqs. (22) and (24) for three impurity concentration gradients in the central p−n junction. a_{j_2} (cm^{-4}): 1) 10^{21}; 2) 10^{20}; 3) 10^{19}.

Fig. 8. Dependences of the turn-on current on the lifetime τ_n, calculated using Eq. (23). a_{j_2} (cm^{-4}): 1) 10^{21}; 2) 10^{20}; 3) 10^{19}.

Typical Calculation. We shall calculate the turn-on current and the breakover voltage using the parameters of the various regions of the thyristor given in Sec. 2 and Table 1. It follows from Table 1 that G_1 and G_2 have close values when $\tau_{pe} = 0.1\tau_n$ and therefore we shall assume that $G_1 = G_2$. We shall take the value of G to be the average given by $G = (G_1 G_2)^{\frac{1}{2}}$. The value of the hyperbolic cosine in Eq. (6) will be taken to be equal to 5 [2]. Figure 7 shows the dependence of the ratio U_{bo}/U_0 on the lifetime τ_n and Fig. 8 shows the dependence of the turn-on current on the lifetime τ_n for three different values of the impurity concentration gradient in the central p−n junction a_{j_2}. The calculated dependences shown in these two figures are based on Eqs. (18), (20)-(24).

4. Discussion of Results

An analysis of the dependences presented in Figs. 3-8 yields the following conclusions about the parameters of the static current−voltage characteristics of silicon thyristors.

The turn-off current depends weakly on the gate current in the range $0 < I_g < 0.1 I_{off}$. The turn-off gain depends strongly on the current in the range $I_{off} < I < 10 I_{off}$, and when $I > 10 I_{off}$ this gain tends to its steady-state value determined by the static current gains α_1^0 and α_2^0. The steady-state value of K_I can be found easily from Eq. (12):

$$K_I = \frac{\alpha_2^0}{\alpha_1^0 + \alpha_2^0 - 1}.$$

The turn-off and the critical gate currents decrease with increasing lifetime τ_n but their ratio varies weakly and its value is ~10.

The ratio of the breakover voltage U_{bo} to the avalanche breakdown voltage of the central p−n junction U_0, calculated for the same values of τ_n, increases with increasing impurity concentration gradient in the central p−n junction a_{j_2}. Consequently, the breakover voltage of silicon planar thyristors designed for low voltages (large values of a_{j_2}) is close to the breakdown voltage of the central p−n junction, which is explained by the low reverse current of this thin junction.

In the case of high-voltage structures (small values of a_{j_2}), we should expect a larger difference between the breakover voltage and the breakdown voltage of the central p−n junction, due to an increase in the generation current with increasing thickness of the space-charge layer of the junction j_2.

When the carrier lifetime is reduced in all regions of the structure, the breakover voltage and the ratio U_{bo}/U_0 increase. This may be explained by the fact that the leakage current of the central p−n junction increases proportionally to $1/\tau_n$ whereas the value of G, which governs the rate of rise of the current gains of the equivalent transistors with increasing current, decreases proportionally to τ_n^2. The resultant slowing down of the rise of $\alpha(I)$ has a stronger influence on the value of U_{bo} than any increase in the value of the current I_0.

The turn-on current decreases with increasing lifetime τ_n and with increasing impurity concentration gradient in the central p−n junction.

Conclusions

1. An analytic expression is obtained for the turn-off and critical gate currents and for the turn-off gain of an asymmetrical silicon planar thyristor. The dependences of the turn-off current on the gate current and of the turn-off gain on the current are also determined.

2. Avalanche multiplication of carriers in the central p−n junction is taken into account and analytic expressions are obtained for the turn-on current and the breakover voltage. It is shown that the breakover voltage of planar thyristors designed for low voltages may be close to the breakdown voltage of the central p−n junction because of the low value of the leakage current of this junction.

Literature Cited

1.   C. T. Sah, R. N. Noyce, and W. Shockley, Proc. IRE, 45:1228 (1957).
2.   D. R. Muss and C. Goldberg, IEEE Trans. Electron Dev., ED-10:113 (1963).
3.   A. A. Lebedev, A. I. Uvarov, and V. E. Chelnokov, Physics of p−n Junctions, Zinatne, Riga (1966); A. A. Lebedev, Dissertation for Candidate's Degree, Physicotechnical Institute, Academy of Sciences of the USSR, Leningrad (1967).
4.   Yu. S. Ryabinkin, Radiotekhn. i Élektron., 10:2205 (1965); 11:1910 (1966).
5.   S. M. Sze and G. Gibbons, Solid State Electronics, 9:831 (1966).
6.   W. M. Bullis, Solid State Electronics, 9:143 (1966).
7.   H. F. Storm, Electro-Technology, 73:62 (1963).
8.   I. M. Mackintosh, Proc. IRE, 46:1229 (1958).

OVERLOAD ON A THYRISTOR CAUSED BY A SINGLE LARGE-AMPLITUDE CURRENT PULSE

É. F. Burtsev, I. V. Grekhov, N. N. Kryukova, É. V. Palko, and A. I. Uvarov

A. F. Ioffe Physicotechnical Institute
Academy of Sciences of the USSR, Leningrad

Overloading a thyristor with a single sinusoidal current pulse is considered. The current–voltage characteristic of the thyristor is approximated by the expression $V = V_0 + IR$. Heat conduction equations are solved making some simplifying assumptions. The temperature rise in a silicon plate is determined during and after the passage of a current pulse through the systems W–Si–W, Cu–Si–Cu, and W–Si–Cu. In particular, the temperature rise at the Si–W boundary at the end of a 10 msec sinusoidal current pulse is given by the following expression, which applies to the W–Si–W system (silicon plate 0.4 mm thick and tungsten heat sink ≥ 1 mm thick):

$$T/P_m = 0.01025 + 0.0034 V_0/V_m,$$

where $V_m = V_0 + I_m R$ and $P_m = I_m V_m$ are the peak values of the voltage and power during this pulse. The temperature rise in the Cu–Si–Cu system is approximately half that given above, other conditions being equal. The temperature in the central plane of the silicon plate during the passage of a current pulse differs considerably from the temperature at the boundaries of this plate, but at the end of the pulse (for a plate 0.4 mm thick and a pulse 10 msec long) the temperature throughout the plate is practically the same. An experimental determination was made of the temperature rise in a silicon plate at the end of a pulse. The temperature dependence of the forward voltage drop at low current densities ($\sim 30 \, \text{mA/cm}^2$) is used as the temperature-sensitive parameter. The experimental results are in satisfactory agreement with the theoretical formulas.

The considerable improvements in the parameters of high-power thyristors achieved in the last few years enable us to contemplate semiconductor rectifiers with outputs of tens and hundreds of thousands of kilowatts. The short-circuit currents in such thyristors would be very high and, therefore, the behavior of a thyristor subjected to a short-duration overload is of great interest. Usually, the protection circuit of a thyristor can disconnect the source of gate current pulses during a time interval equal to the rise time of the overload current. Therefore, in the majority of cases an overload is in the form of a single pulse. The shape of the overload current pulse in high-power rectifiers is nearly sinusoidal because of the presence of large inductances. Therefore, we shall concentrate our attention on overloading of a

339

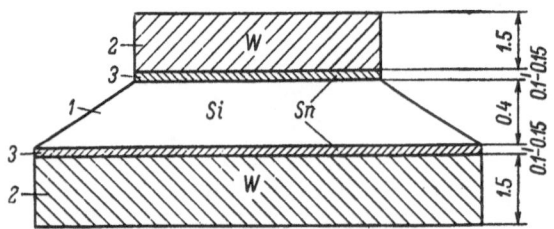

Fig. 1. Rectifying element of a thyristor.

thyristor with a single sinusoidal current pulse of 10 msec duration.

Figure 1 shows the construction and dimensions (in mm) of a cross section through a standard rectifying element of a high-power thyristor. This rectifying element consists of a silicon plate 1 (this plate is a p−n−p−n structure with metal contacts formed by chemical nickel plating) which is soldered with tin (3) to tungsten heat sinks (2). During the passage of a current pulse, practically all the heat is evolved in plate 1 (in the first approximation, the heat evolution is uniform) and this heat is dissipated by the tungsten heat sinks. The highest permissible temperatures in the silicon plate during the passage of a current pulse, $T_{1\,max}$, and after the end of the pulse, $T_{2\,max}$, are among the main factors which limit the highest permissible values of the amplitude and duration of the overload current pulse. The values of the highest permissible temperatures can be found on the basis of the following criteria:

1) if a reverse voltage is applied to a thyristor immediately after the passage of a current pulse, then $T_{2\,max} \leq 160°C$;*

2) since melting of the solder 3 is for many reasons undesirable, the maximum temperature at the silicon-tin boundary is $T_{1\,max} \leq 232°C$;

3) since the heating of the system shown in Fig. 1 by a current pulse is not uniform, mechanical stresses due to the nonuniform heating may damage the silicon plate.

Possible mechanical damage may limit the highest permissible value of the temperature rise during the passage of a pulse but this aspect will not be considered in the present paper.

Thus, in order to estimate the maximum permissible amplitude of the overload current pulse, we have to know the dependences of the temperatures T_{1max} and T_{2max} on the current, or, more exactly, on the dissipated power.

1. Calculation of the Temperature

We shall estimate the temperature of a silicon plate during the passage of an overload current pulse and determine the cooling curve after the end of the pulse by solving the equation of heat conduction for a thyristor model consisting of a thin silicon plate which is in good thermal contact with an infinite medium (tungsten) on both sides, i.e., we shall assume that the thermal resistance of the tin layers and of the contacts with these layers is negligibly small. We shall carry out our calculations making the following assumptions:

1) the problem can be solved in a one-dimensional approximation ignoring heat exchange at the lateral surfaces of the silicon plate;

2) the thermal conductivity λ, specific heat c, and density ρ of silicon and tungsten are constant and independent of temperature;

3) all the materials isotropic;

---

* The maximum temperature in a p−n−p−n structure, at which a reverse voltage can be applied to a thyristor, depends on many factors. The important factors are the rate of cooling, the resistivity of silicon, the method of protection of the p−n junctions on the surface, and the value of the applied reverse voltage. The limit $T_{2\,max} \leq 160°C$ is an approximate one: it is based on the experience gained in the use of TL-2 thyristors.

4) heat evolution during the passage of a current pulse is uniform throughout the silicon plate;

5) the forward branch of the current−voltage characteristic of the thyristor can be represented as follows: the current is zero up to a voltage V_0 after which it increases linearly with the voltage, the slope of the dependence being given by the differential resistance R.

The voltage drop across the thyristor during the passage of a sinusoidal current pulse $I = I_m \sin \omega t$ is given by

$$V = V_0 + IR = V_0 + I_m R \sin \omega t, \tag{1}$$

and the dissipated power is

$$P = IV = I_m V_0 \sin \omega t + I_m^2 R \sin^2 \omega t. \tag{2}$$

Introducing the quantity $\nu = V_0 / V_m$, where

$$V_m = V_0 + I_m R, \tag{3}$$

and expressing the square of the sine by a function with a double argument, we obtain

$$\frac{P}{P_m} = \nu \sin \omega t + \frac{(1 - \nu)}{2} (1 - \cos 2\omega t), \tag{4}$$

where

$$P_m = I_m V_m = I_m V_0 + I_m^2 R. \tag{5}$$

Thus, the time dependence of the dissipated power is

$$P = P_m \left[\nu \sin \omega t + \frac{(1 - \nu)}{2} (1 - \cos 2\omega t) \right]. \tag{6}$$

The heat conduction equations are:
in silicon

$$\frac{\partial T_1}{\partial t} = D_0 \frac{\partial^2 T_1}{\partial x^2} + \frac{q}{c_0 \rho_0}, \tag{7}$$

in tungsten

$$\frac{\partial T_{2,3}}{\partial t} = D_1 \frac{\partial^2 T_{2,3}}{\partial x^2}. \tag{8}$$

Here, $D_0 = \lambda_0 / c_0 \rho_0$ is the thermal diffusivity of silicon; $D_1 = \lambda_1 / c_1 \rho_1$ is the thermal diffusivity of tungsten; $q = (dQ/dt)/w$ is the density of the power dissipated in the silicon plate per unit volume, where

$$\frac{dQ}{dt} = \begin{cases} 0.24 P_m \left[\nu \sin \omega t + \frac{(1 - \nu)}{2} (1 - \cos 2\omega t) \right] & \text{when } 0 < \omega t < \pi \\ 0 & \text{when } \omega t > \pi, \end{cases} \tag{9}$$

and w is the thickness of the silicon plate.

Equations (7) and (8) are solved using the following initial and boundary conditions:

$$T_{t=0} = 0, \tag{10}$$

$$\left. \begin{array}{c} (T_3)_{x=0} = (T_1)_{x=0}, \quad (T_1)_{x=w} = (T_2)_{x=w}, \\ \lambda_1 \left(\frac{\partial T_3}{\partial x} \right)_{x=0} = \lambda_0 \left(\frac{\partial T_1}{\partial x} \right)_{x=0}, \quad \lambda_0 \left(\frac{\partial T_1}{\partial x} \right)_{x=w} = \lambda_1 \left(\frac{\partial T_2}{\partial x} \right)_{x=w}, \end{array} \right\} \tag{11}$$

i.e., at the boundary the temperatures of both media are equal and the heat flux is continuous. Moreover, the temperature at an infinite distance from the boundary is assumed to be finite:

$$T_{x \to \pm\infty} \to 0. \tag{12}$$

We shall calculate the temperature rise during the passage of a current pulse, and we shall measure this rise from the initial temperature of the plate, which will be taken to be zero. We shall use the operator method in our calculations. This method yields the following expressions: the temperature at the W–Si boundary (x = 0) is

$$\frac{(T_3)_{x=0}}{P_m} = \frac{0.24}{wc_0\rho_0} \left[\frac{\nu}{p} \cdot \frac{\omega}{p^2+\omega^2} \left(1 + e^{-\frac{\pi}{\omega}p} \right) + \frac{(1-\nu)}{2} \left(1 - e^{-\frac{\pi}{\omega}p} \right) \left(\frac{1}{p^2} - \frac{1}{p^2+4\omega^2} \right) \right] \frac{\operatorname{sh}\sqrt{\frac{p}{D_0}} \cdot \frac{w}{2}}{\gamma \operatorname{ch}\sqrt{\frac{p}{D_0}} \cdot \frac{w}{2} + \operatorname{sh}\sqrt{\frac{p}{D_0}} \cdot \frac{w}{2}} \tag{13}$$

and the temperature difference between the central plane and the boundaries of the silicon plate is

$$\Delta T = (T_1)_{x=-\frac{w}{2}} - (T_1)_{x=0} ,$$

$$\frac{\Delta T}{P_m} = \frac{0.24}{wc_0\rho_0} \left[\frac{\nu}{p} \cdot \frac{\omega}{p^2+\omega^2} \left(1 + e^{-\frac{\pi}{\omega}p} \right) + \frac{(1-\nu)}{2} \left(1 - e^{-\frac{\pi}{\omega}p} \right) \left(\frac{1}{p^2} - \frac{1}{p^2+4\omega^2} \right) \right] \frac{\gamma\left(\operatorname{ch}\sqrt{\frac{p}{D_0}} \cdot \frac{w}{2} - 1 \right)}{\gamma \operatorname{ch}\sqrt{\frac{p}{D_0}} \cdot \frac{w}{2} + \operatorname{sh}\sqrt{\frac{p}{D_0}} \cdot \frac{w}{2}} , \tag{14}$$

where $\gamma = \frac{\lambda_1}{\lambda_0}\sqrt{\frac{D_0}{D_1}}$. Since t = 10 msec $\gg$ w^2/2D$_0$, we can expand Eqs. (13) and (14) into two series and take only the first two terms of each series. Going back to the original functions, we find that the ratio of the temperature at the Si–W boundary to the maximum dissipated power (per unit area) can be approximated by the expression

$$\frac{(T_3)_{x=0}}{P_m} = 0.24 \left\{ \frac{\nu}{\sqrt{2\omega\lambda_1 c_1\rho_1}} F_1(z) - \frac{wc_0\rho_0}{4\lambda_1 c_1\rho_1} \sin z \left[(1-\nu)\sin z + \nu \right] + \frac{(1-\nu)\sqrt{t}}{2\sqrt{\pi\lambda_1 c_1\rho_1}} - \frac{(1-\nu)}{4\sqrt{\omega\lambda_1 c_1\rho_1}} F_2(2z) \right\} \tag{15}$$

when $0 < z < \pi$ and

$$\frac{(T_3)_{x=0}}{P_m} = 0.24 \left\{ \frac{\nu}{\sqrt{2\omega\lambda_1 c_1\rho_1}} F_1'(z) + \frac{(1-\nu)\left(\sqrt{t} - \sqrt{t-\frac{\pi}{\omega}} \right)}{2\sqrt{\pi\lambda_1 c_1\rho_1}} - \frac{(1-\nu)}{4\sqrt{\omega\lambda_1 c_1\rho_1}} F_2'(2z) \right\} \tag{16}$$

when $z > \pi$.

Similarly, the temperature drop between the central plane in the silicon plate and the Si–W boundary is given by the expression

$$\frac{\Delta T}{P_m} = 0.24 \left\{ \frac{w}{8\lambda_0} \sin z \left[\nu + (1-\nu)\sin z \right] - \frac{w^2\sqrt{\omega}}{16D_0\sqrt{\lambda_1 c_1\rho_1}} \left[\nu\sqrt{2} F_2(z) + (1-\nu) F_1(2z) \right] \right\} \tag{17}$$

when $0 < z < \pi$ and

$$\frac{\Delta T}{P_m} = -\frac{0.24w^2\sqrt{\omega}}{16D_0\sqrt{\lambda_1 c_1\rho_1}} \left[\nu\sqrt{2} F_2'(z) + (1-\nu) F_1'(2z) \right] \tag{18}$$

when $z > \pi$.

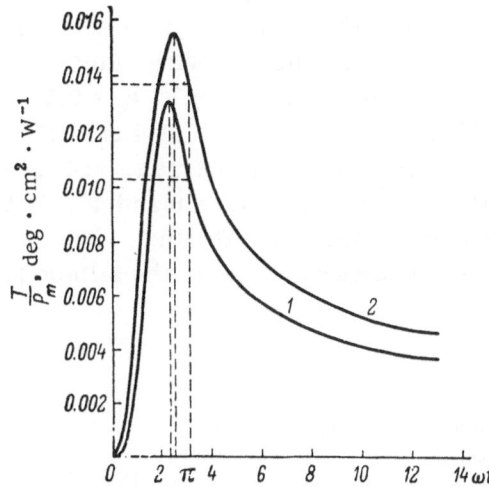

Fig. 2. Dependence of the ratio of the temperature at the Si−W boundary and the maximum power (T/P_m) on ωt: 1) $\nu = 0$; 2) $\nu = 1$.

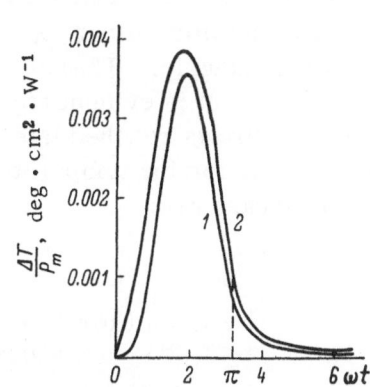

Fig. 3. Dependence of $\Delta T/P_m$ on ωt: 1) $\nu = 0$; 2) $\nu = 1$.

Here, the temperature is given in degrees; P_m is the maximum value of the power in W/cm^2; $z = \omega t$;

$$\left. \begin{aligned}
F_1(z) &= \sin z\, C(z) - \cos z\, S(z), \\
F_1'(z) &= \sin z\, [C(z) - C(z-\pi)] - \cos z\, [S(z) - S(z-\pi)], \\
F_2(2z) &= \cos 2z\, C(2z) + \sin 2z\, S(2z), \\
F_2'(2z) &= \cos 2z\, [C(2z) - C(2z-2\pi)] + \sin 2z\, [S(2z) - S(2z-2\pi)],
\end{aligned} \right\} \tag{19}$$

where

$$\left. \begin{aligned}
C(z) &= \frac{1}{\sqrt{2\pi}} \int_0^z \frac{\cos \xi}{\sqrt{\xi}}\, d\xi, \\
S(z) &= \frac{1}{\sqrt{2\pi}} \int_0^z \frac{\sin \xi}{\sqrt{\xi}}\, d\xi.
\end{aligned} \right\} \quad \text{Fresnel integrals.} \tag{20}$$

The solutions given by Eqs. (16) and (18) apply during the cooling stage. The expressions in Eqs. (15)-(18) can be simplified in the case of very small or very large currents. If the current is very small, V_m is, in practice, equal to V_0, i.e., $\nu = V_0/V_m \approx 1$. In the case of large currents, V_m is given by $I_m R$, i.e., $V_0 \ll V_m$ and $\nu = V_0/V_m \approx 0$. In these two limiting cases, the approximate solution for the temperature rise at the Si−W boundary is of the form:

for $\nu = 1$

$$\frac{(T_3)_{x=0}}{P_m} = \begin{cases} 0.01918 F_1(z) - 0.00405 \sin z & \text{when } 0 < z < \pi, \\ 0.01918 F_1'(z) & \text{when } z > \pi, \end{cases} \tag{21}$$

for $\nu = 0$

$$\frac{(T_3)_{x=0}}{P_m} = \begin{cases} 0.1356\sqrt{t} - 0.00678 F_2(2z) - 0.00405 \sin^2 z & \text{when } 0 < z < \pi, \\ 0.1356\left[\sqrt{t} - \sqrt{t - \dfrac{\pi}{\omega}}\right] - 0.00678 F_2'(2z) & \text{when } z > \pi. \end{cases} \tag{22}$$

These solutions apply when the silicon plate is 0.04 cm thick, the tungsten heat sink is thicker than 0.1 cm, and the current pulse duration is 10 msec. The following values of the parameters are used in these calculations: $\lambda_0 = 0.26$ cal. $\text{cm}^{-1} \cdot \text{sec}^{-1} \cdot \text{deg}^{-1}$, $c_0 = 0.18$ cal $\cdot \text{g}^{-1} \cdot \text{deg}^{-1}$, $\rho_0 = 2.34$ g/cm^3 for silicon; $\lambda_1 = 0.38$ cal. $\text{cm}^{-1} \cdot \text{sec}^{-1} \cdot \text{deg}^{-1}$, $c_1 = 0.034$ cal $\cdot \text{g}^{-1} \cdot \text{deg}^{-1}$, $\rho_1 = 19.3$ g/cm^3 for tungsten. The solution for very small and very large currents are presented graphically in Fig. 2. It is evident from this figure that in both cases ($\nu = 0$ and $\nu = 1$) the maximum temperature is reached later than the maximum of the current pulse. The corresponding expressions for the temperature drop between the central plane in the silicon plate and the Si−W boundary are:

for $\nu = 1$

$$\frac{\Delta T}{P_m} = \begin{cases} 0.004615 \sin z - 0.001952 F_2(z) & \text{when } 0 < z < \pi, \\ -0.001952 F_2'(z) & \text{when } z > \pi, \end{cases} \tag{23}$$

for $\nu = 0$

$$\frac{\Delta T}{P_m} = \begin{cases} 0.004615 \sin^2 z - 0.001381 F_1(2z) & \text{when } 0 < z < \pi \\ -0.001381 F_1'(2z) & \text{when } z > \pi. \end{cases} \tag{24}$$

The dependences given in Fig. 3 show that, at the end of the current pulse ($t = \pi/\omega$), the temperature is the same throughout the silicon plate. This is to be expected because the duration of the current pulse is considerably longer than the time required for the diffusion of heat throughout the silicon plate and, therefore, by the end of the pulse a quasi-stationary temperature distribution is established in the plate.

At the Si−W boundary, we have

$$\frac{(T_3)_{x=0}}{P_m} = 0.01369 \ \text{deg} \cdot \text{cm}^2 \cdot \text{W}^{-1} \text{ for } \nu = 1, \tag{25}$$

$$\frac{(T_3)_{x=0}}{P_m} = 0.01025 \ \text{deg} \cdot \text{cm}^2 \cdot \text{W}^{-1} \text{ for } \nu = 0. \tag{26}$$

For $0 < \nu < 1$ at time $t = \pi/\omega$, we obtain

$$\frac{(T_3)_{x=0}}{P_m} = [0.01025 + 0.00344\nu] \ \text{deg} \cdot \text{cm}^2 \cdot \text{W}^{-1}. \tag{27}$$

The maximum value of the temperature ($t \approx 3\pi/4\omega$) for $\nu = 1$ is reached when $\frac{(T_3)_{x=0}}{P_m} = 0.01552$ deg $\cdot \text{cm}^2$/W and the temperature drop between the central plane of the silicon plate and the Si−W boundary is given by

$$\frac{\Delta T}{P_m} = 0.00291 \ \text{deg} \cdot \text{cm}^2 \cdot \text{W}^{-1}. \tag{28}$$

Similarly, for $\nu = 0$ we find that

$$\frac{(T_3)_{x=0}}{P_m} = 0.01323 \text{deg} \cdot \text{cm}^2 \cdot \text{W}^{-1} \text{ when } \frac{\Delta T}{P_m} = 0.00293 \ \text{deg} \cdot \text{cm}^2 \cdot \text{W}^{-1}. \tag{29}$$

A similar calculation carried out for the Cu−Si−Cu system, in which the copper blocks are assumed to be infinitely large, yields the following values of the ratio of the temperature to the maximum power dissipated at the Si−Cu boundary (these expressions apply at the end of a current pulse $t = \pi/\omega$):

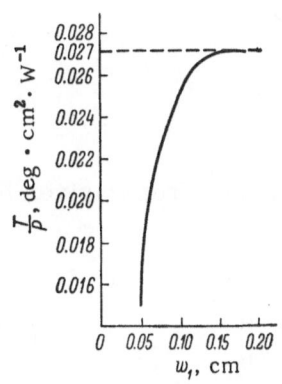

Fig. 4. Dependence of the ratio of the temperature at the Si−W boundary to the dissipated power (T/P) on the thickness of the tungsten layer w_1.

$$\frac{T}{P_m} = 0.00744 \text{ deg} \cdot \text{cm}^2 \cdot \text{W}^{-1} \quad \text{for} \quad \nu = 1 \tag{30}$$

$$\frac{T}{P_m} = 0.00557 \text{ deg} \cdot \text{cm}^2 \cdot \text{W}^{-1} \quad \text{for} \quad \nu = 0, \tag{31}$$

and hence we can see that a higher overload current is permissible if copper is used instead of tungsten in the W−Si−W structure. These calculations are based on the following values of the thermal conductivity, specific heat, and density of copper:

$$\left. \begin{array}{l} \lambda_2 = 0.96 \text{ cal} \cdot \text{cm}^{-1} \cdot \text{sec}^{-1} \cdot \text{deg}^{-1} \\ c_2 = 0.1 \text{ cal} \cdot \text{g}^{-1} \cdot \text{deg}^{-1} \\ \rho_2 = 8.8 \text{ g/cm}^3. \end{array} \right\} \tag{32}$$

The temperature at the silicon−tungsten boundary in the Cu−Si−W system (the copper and tungsten blocks are assumed to be infinitely large) is calculated in a similar manner. These calculations yield the following final expressions:

$$\frac{T}{P_m} = 0.24 \left\{ \frac{\nu \sqrt{2}}{\sqrt{\omega}\,(\sqrt{\lambda_1 c_1 \rho_1} + \sqrt{\lambda_2 c_2 \rho_2})} F_1(z) + \frac{w\,[\gamma_2(\gamma_1 + \gamma_2) - 2(1 + \gamma_1 \gamma_2)]}{2\lambda_0(\gamma_1 + \gamma_2)^2} \sin z\,[\nu + (1 - \nu)\sin z] + \right.$$
$$\left. + \frac{(1 - \nu)}{(\sqrt{\lambda_1 c_1 \rho_1} + \sqrt{\lambda_2 c_2 \rho_2})} \left[\frac{\sqrt{t}}{\sqrt{\pi}} - \frac{1}{2\sqrt{\omega}} F_2(2z) \right] \right\} \tag{33}$$
$$\text{when } 0 < z < \pi$$

and

$$\frac{T}{P_m} = 0.24 \left\{ \frac{\nu \sqrt{2}}{\sqrt{\omega}\,(\sqrt{\lambda_1 c_1 \rho_1} + \sqrt{\lambda_2 c_2 \rho_2})} F_1'(z) + \frac{(1 - \nu)}{(\sqrt{\lambda_1 c_1 \rho_1} + \sqrt{\lambda_2 c_2 \rho_2})} \left[\frac{\left(\sqrt{t} - \sqrt{t - \frac{\pi}{\omega}} \right)}{\sqrt{\pi}} - \frac{1}{2\sqrt{\omega}} F_2'(2z) \right] \right\} \tag{34}$$
$$\text{when } z > \pi.$$

Here, $\gamma_2 = \dfrac{\lambda_2}{\lambda_0} \sqrt{\dfrac{D_0}{D_2}}$.

The expression for the ratio T/P_m at the Si−Cu boundary is obtained from Eqs. (33) and (34) by interchanging γ_1 and γ_2. At the end of a current pulse ($t = \pi/\omega$), we obtain

$$\frac{T}{P_m} = 0.01019 \text{ deg} \cdot \text{cm}^2 \cdot \text{W}^{-1} \, (\nu = 1), \tag{35}$$

$$\frac{T}{P_m} = 0.00721 \text{ deg} \cdot \text{cm}^2 \cdot \text{W}^{-1} \, (\nu = 0). \tag{36}$$

i.e., the Cu−Si−W system can withstand overload currents higher than those permissible in the W−Si−W system but lower than those which can be allowed in the Cu−Si−Cu configuration.

So far, we have assumed that the tungsten block is infinitely large, while in real devices the tungsten layer has a finite thickness. In order to determine the error due to this approximation, we shall calculate the dependence of T/P at the Si−W boundary on the tungsten layer thickness w_1. We shall consider the system in which the heat flux flowing from the silicon into the tungsten through the surface x = 0 is

$$\frac{dQ}{dt} = -\lambda_1 \left(\frac{\partial T}{\partial x} \right)_0, \tag{37}$$

and the temperature at the x = w_1 surface is zero.

We shall consider a square current pulse, i.e., we shall assume that

$$\frac{dQ}{dt} = \begin{cases} \dfrac{0.24IV}{2} \text{ when } t < t_0 \ (t_0 = 10 \text{ msec }), \\ \\ 0 \quad \text{ when } \ t > t_0. \end{cases} \tag{38}$$

The ratio of the temperature at the Si−W boundary to the power per unit area is given by the following expression at the end of the pulse:

$$\frac{T}{P} = \frac{0.24\sqrt{t_0}}{\sqrt{\pi}\lambda_1 c_1 \rho_1} \left\{ 1 + \sum_{k=1}^{\infty} (-1)^k \left[2\exp\left(-\frac{k^2 w_1^2}{D_1 t_0} \right) - \frac{2kw_1\sqrt{\pi}}{\sqrt{D_1 t_0}} \left(1 - \mathrm{erf}\frac{kw_1}{\sqrt{D_1 t_0}} \right) \right] \right\}. \tag{39}$$

In the limiting case, when $w_1 \to \infty$, we obtain

$$\frac{T}{P} = \frac{0.24\sqrt{t_0}}{\sqrt{\pi}\lambda_1 c_1 \rho_1} = 0.0271 \ \mathrm{deg} \cdot \mathrm{cm^2} \cdot \mathrm{W^{-1}}.$$

The dependence of T/P on w_1 is given in Fig. 4: we can see from the figure that for $w_1 >$ 0.1 cm the value of T/P is practically equal to the value in the case of an infinitely large tungsten block.

2. Experimental Determination of the Temperature

The temperature of a p−n−p−n structure at the end of a current pulse was determined experimentally using the well-known technique in which the temperature-sensitive parameter is the forward voltage drop in the initial part of the current−voltage characteristic of a thyristor in the on state. The temperature was measured in the following way. A thyristor T, turned on by a gate signal from a current source (CS in Fig. 5a), passes a small current ($\sim$40 mA/cm^2). The dependence of the forward bias across the thyristor on the current flowing through it is determined with an oscillograph (Osc.) which measures the current through a shunt S and the voltage drop across the thyristor. A sinusoidal pulse generator (SPG) produces a single sinusoidal current pulse; the dependence of the voltage V across the thyristor before, during, and after the passage of the pulse is shown in Fig. 5b. Before the passage of the pulse $V = V_1$. During the pulse (duration t_p) the value of V is determined by the current−voltage characteristic of the thyristor. At the end of the pulse, during a short recovery time (t_r) of the blocking diodes D (Fig. 5a) the voltage across the thyristor is $V \ll V_1$ and then the voltage increases to $V_2 < V_1$. Next, during a time interval t_{cool}, required to cool the p−n−p−n structure to its initial temperature, the voltage increases back to V_1. Knowing the temperature dependence of V_1, we can use ΔV_1 (Fig. 5b) to determine the temperature rise after a time interval t_r, measured from the end of the pulse. A typical temperature dependence of V_1 is shown in Fig. 6: usually a linear dependence is observed up to 180-200°C. The slope of the linear region is within the limits 2.4-3.4 mV/deg.

Figure 7a shows the experimentally determined dependences of $T_{2\,max}$ on the power dissipated in thyristors whose p−n−p−n structures had the W−Si−W configuration shown in Fig. 1; these dependences were determined using the method just described and they were compared with calculations based on Eq. (27). In these calculations, we assumed that the temperature drop between the central plane and the boundary of the silicon plate at a time $t = \pi/\omega$ was, in agreement with the results presented in Fig. 3, negligibly small, i.e., the temperature throughout the silicon plate was the same and equal to the temperature at the W−Si boundary. The current−voltage characteristics of the investigated thyristors, determined using large-amplitude single pulses of sinusoidal current, are given in Fig. 7b. The experimental values were somewhat higher (by 10-20%) than the calculated results. This was probably due to the fact

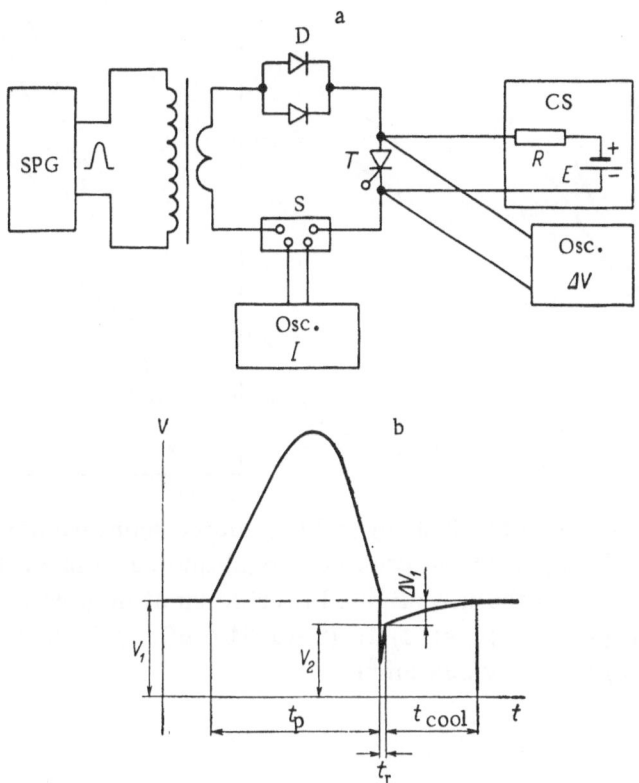

Fig. 5. Block diagram of the apparatus used to measure the temperature in a p−n−p−n structure (a) and the time dependence of the voltage across this structure (b).

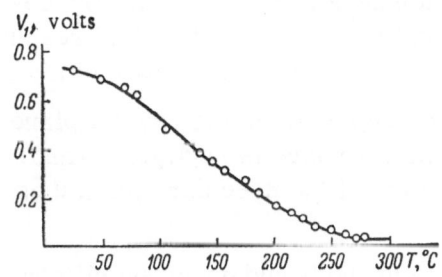

Fig. 6. Typical temperature dependence of V_1.

that the current through the p−n−p−n structure was not uniform and consequently the temperature distribution across the silicon plate was also nonuniform. The measured value of ΔV_1 was determined by the region with the highest temperature. Therefore, the experimentally determined temperature was always higher than the calculated value because the calculations were based on the assumption that the distribution of the current was uniform. However, the nonuniformity of the distribution could be very high because the power dissipated in the p−n−p−n structures of these thyristors was approximately proportional (up to a certain temperature) to their area (Fig. 8).

The restriction imposed on the amplitude of the sinusoidal current pulse by the condition $T_{2\,max} \leq 160°C$ is more stringent, for the thyristor construction shown in Fig. 1, than the restriction imposed by the requirement $T_{1\,max} \leq 232°C$. According to Eqs. (25) and (26), P_m ($\nu = 1$) = 11.687 kW/cm², P_m ($\nu = 0$) = 15.610 kW/cm² for $T_{2\,max}$ = 160°C and the temperatures $T_{1\,max}$ ($\nu = 1$) = 181°C, $T_{1\,max}$ ($\nu = 0$) = 207°C corresponding to these powers are considerably lower than the melting point of tin.

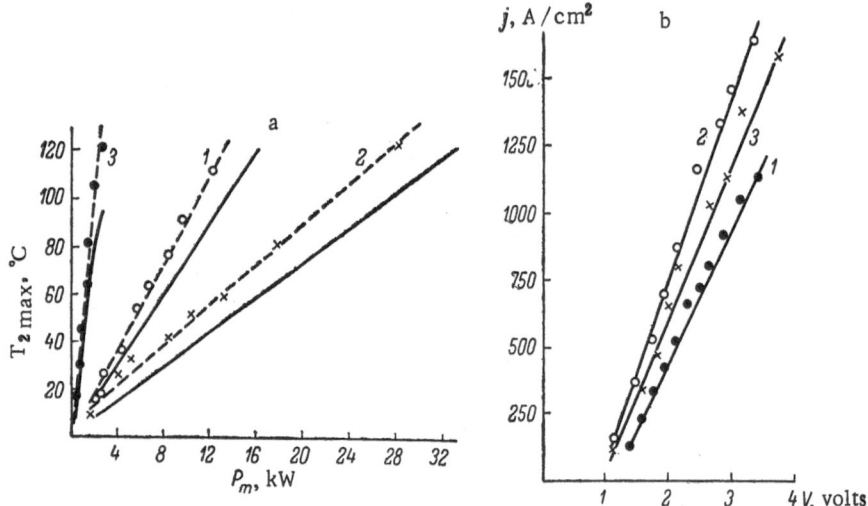

Fig. 7. Experimental (dashed) and calculated (continuous) dependences of $T_{2\,max}$ on the peak power (a) and current–voltage characteristics of the investigated thyristors at high values of the current (b). Samples: 1) 15 (S = 1.54 cm²); 2) 2242 (S = 3.05 cm²); 3) 21 (S = 0.265 cm²).

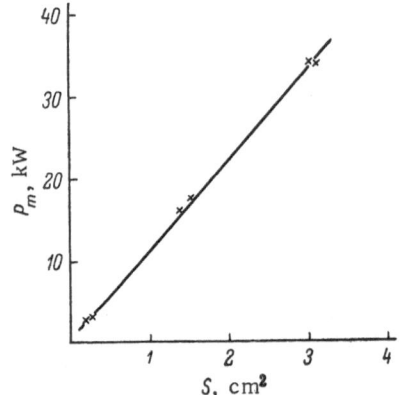

Fig. 8. Dependence of P_m on the area of a p–n–p–n structure at a fixed temperature (T = 160°C).

Conclusions

1. A calculation of the temperature of a silicon plate in a thyristor is carried out assuming that a single sinusoidal current pulse passes through this plate. It is shown that:

a) the maximum temperature at the boundary between the silicon plate and a tungsten heat sink is reached after the current pulse passes through its maximum;

b) the temperature drop between the crystal plane of the silicon plate and its boundary is negligibly small at the end of a current pulse of 10 msec duration (50 cps frequency);

c) the temperature at the boundary of the silicon plate at the end of the current pulse depends on the heat-sink material: for the Cu–Si–Cu system this temperature is about half that predicted for the W–Si–W system, other conditions being equal;

d) in the case of mass-produced thyristors the restriction on the pulse amplitude imposed by the value of the maximum permissible temperature at the end of the pulse is more stringent in the case when a reverse voltage is applied to the thyristor after this pulse ($T_{2\,max} \leq 160°C$) than the restriction imposed by the maximum permissible temperature at the silicon plate boundary ($T_{1\,max} \leq 232°C$), which is set by the melting point of tin.

2. An experimental determination of the temperature rise in a silicon plate of a thyristor at the end of a sinusoidal current pulse shows that:

a) the calculated values of the temperature rise are in good agreement with the experimental data;

b) the experimental values are somewhat higher (10-20%) than those predicted theoretically because the distribution of the current over the p−n−p−n structure area may not be sufficiently uniform;

c) this nonuniformity of the current distribution is not very strong because the temperature rise is inversely proportional (within the limits of the experimental error) to the working area of the p−n−p−n structure, other conditions being equal.

NEW OPTOELECTRONIC DEVICES MADE
OF ZINC-COMPENSATED SILICON

B. V. Kornilov

A brief description is given of the principal physical properties of p- and n-type silicon compensated with zinc. Specific suggestions are made about the possible applications of compensated silicon in new opto-electronic and electronic devices. It is shown that this material can be used in photoresistors, light amplifiers, radiation converters, memory elements, photodiodes, and other devices. The principal parameters of these devices are given. It is concluded that zinc-compensated silicon is a promising optoelectronic material.

Progress in optoelectronics is held up primarily by two factors. First, those semiconducting materials which have been investigated in detail frequently do not have the properties that are essential for the development of optoelectronic devices and for broadening the working frequency range of these devices. The second, yet equally important factor is the development of suitable planar technology and photolitographic techniques. At present, the most promising material from the point of view of applications and using the capabilities of modern planar technology is undoubtedly silicon. Using this material and the planar technology combined with photolithographic techniques, we can fabricate transistors with bases only 1 μ thick. The accidental impurity concentration in silicon is only 10^{12}-10^{13} cm^{-3}. None of the other semiconducting materials can compete with silicon in respect of these two important parameters.

A systematic investigation of the physical properties of zinc-compensated silicon, carried out in 1962-1966, convinced the present author of the great promise of this material in many optoelectronic applications. The introduction of zinc into n-type silicon, i.e., compensation with zinc, makes silicon sensitive. The effect of compensation is so great that the photoelectric properties of silicon can be made to approach the properties of such classical photoconductors as PbS and CdS.

Since the design of optoelectronic devices must be based on the knowledge of the fundamental properties of a semiconducting material, we shall start with a brief review of the principal physical properties of zinc-doped silicon.

1. Principal Physical Properties of Zinc-Compensated Silicon

Zinc dopant in silicon acts as a double acceptor with energy levels at $E_c - 0.55$ eV and $E_v + 0.31$ eV [1-3]. In addition to its acceptor properties, zinc can also act as a donor in the impurity concentration range $5 \cdot 10^{13} < N < 5 \cdot 10^{14}$ cm^{-3}. Its solubility at 1350°C is of the order of $5 \cdot 10^{16}$ cm^{-3}. Zinc has a large diffusion coefficient 10^{-6}-10^{-7} cm^2/sec, in the temperature range 900-1350°C [1]. The recombination properties of p- and n-type silicon are very different. In p-type silicon, the hole-capture cross section of a singly charged negative center is $S_p^- \sim 10^{-13}$ cm^2 according to [4] and $S_p^- = 5 \cdot 10^{-14}$ cm^2 according to [5, 6]; the electron-capture cross section of a neutral zinc atom is $S_n^0 \sim 5 \cdot 10^{-16}$ cm^2 [4, 6]. In n-type silicon the electron-capture cross section of a singly charged negative center (S_n^-), determined by the method of steady-state and transient extrinsic photoconductivity and by the method of optical charge exchange between zinc atoms, lies within the limits $10^{-20} \leq S_n^- < 10^{-18}$ cm^2 [7, 8] and the hole-capture cross section by a doubly charged negative center is 10^{-16} cm^2 [9]. The cross sections S_p^-, S_n^0, and S_n^- are practically independent of temperature in the range 100-300°K. The electron-capture cross section of a singly charged negative center increases by a factor of 3.6 when the field intensity is raised from 10^2 to 10^3 V/cm [10]. This change in the electron-capture cross section in strong electric fields is responsible for the nonlinearity in the current-voltage characteristic, which can sometimes exhibit an N-type negative resistance region. Low-frequency periodic oscillations of the current, associated with domain formation, have been observed in n-type silicon at 77°K [11]. A photoresponse delay of the order of 1 min has been reported for n-type silicon at 77°K [12]. Photoionization cross sections of zinc atoms in various charged states are given in [3]. Optical charge exchange between zinc atoms has been observed in p- and n-type silicon.

A linear-gradient diffused p−i−n junction, formed by deep zinc levels, has been reported to exhibit a strongly selective photoresponse delay in the wavelength range 1.2-1.8 μ [13].

A sudden change in the current, by 3-9 orders of magnitude, can occur in p$^+$−p−p$^+$ or n$^+$−n−n$^+$ structures with base thicknesses of the order of a few tens of microns; this effect is observed at 77°K in fields of $(1-2) \cdot 10^4$ V/cm [14]. Counter-doping or compensation of n-type silicon with zinc atoms can be used to vary the principal physical properties of this material over a wide range. For example, the resistivity of p- and n-type silicon can range, depending on the degree of compensation, from 0.3 to $2 \cdot 10^5$ $\Omega \cdot$ cm at 300°K and from 0.3 to $2 \cdot 10^{12}$ $\Omega \cdot$ cm in the temperature range 300-77°K. The hole mobility in p-type silicon can vary, depending on the degree of compensation, from 350 to 50 cm^2/V $\cdot$ sec. The hole lifetime in p-type silicon can range from $4 \cdot 10^{-11}$ to $2 \cdot 10^{-8}$ sec. The electron lifetime in p-type silicon can vary within the range $4 \cdot 10^{-9}$ to $2 \cdot 10^{-6}$ sec. The minority carrier lifetime in n-type silicon can be varied from $5 \cdot 10^{-8}$ to 10^{-4} sec and the majority lifetime from $2 \cdot 10^{-4}$ to 0.1 sec.

The new optoelectronic devices described in the following sections simply illustrate possible applications of only some properties of zinc-compensated silicon but they do not even remotely exhaust all the possibilities for exploiting other properties of this material either singly or in combination.

2. Photoresistor

A highly sensitive silicon photoresistor [15] can be built from n-type silicon in which the concentrations of zinc N_{Zn} and phosphorus N_P satisfy the inequality

$$N_{Zn} < N_P < 2N_{Zn}. \tag{1}$$

The energy levels of zinc in silicon are shown schematically in Fig. 1. Let us assume that N_1 represents the concentration of accidental recombination centers which are present either in the original crystal or which appear after high-temperature annealing used in the zinc-diffusion stage; N_2 will be used to denote the concentration of zinc atoms with an energy level at $E_c - 0.55$ eV and N_3 the concentration of zinc atoms with a level at $E_v + 0.31$ eV. Since zinc is a double acceptor, it follows that $N_2 = N_3$ [2-3]. We shall show that zinc-doped n-type silicon

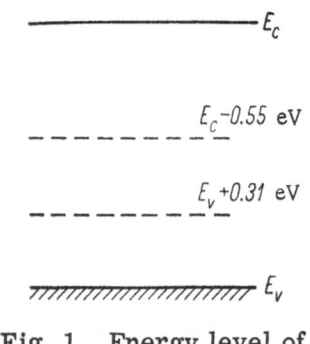

Fig. 1. Energy level of
zinc-doped silicon.

can, under certain conditions, have a high photosensitivity, which is associated with a long majority-carrier lifetime.

Carrier recombination takes place at two levels: the zinc level at $E_c - 0.55$ eV and the level of accidental centers N_1. The lower level of zinc at $E_v + 0.31$ eV does not participate in recombination because, at 300°K, it is a hole-trapping level. It is known that in order to obtain a long electron lifetime one of the impurities participating in recombination should have a small effective electron-capture cross section. Such an impurity is singly charged zinc [8]. Moreover, in order to "sensitize" zinc-doped silicon, the zinc impurity should be present in a concentration such that carrier recombination is governed by the zinc level.

The probability of electron capture by the levels N_1 and N_2 is

$$\frac{1}{\tau_n} = p_{k1} v S_{n1} + p_{k2} v S_{n2},\tag{2}$$

where p_{k1} and p_{k2} are the concentrations of "vacancies" at the levels 1 and 2; v is the thermal velocity of carriers; S_{n1} and S_{n2} are the electron-capture cross sections of the levels N_1 and N_2.

The equality of the rates of capture of holes and electrons in the steady-state case and the condition

$$S_{n2} \simeq S_{p1}\tag{3}$$

yield an approximate relationship for p_{k1}

$$p_{k1} = N_1 \frac{N_1}{N_2} \frac{S_{n2}}{S_{p2}},\tag{4}$$

where S_{p2} is the cross section for the capture of a hole by a doubly charged negative center with a level at 0.55 eV.

The lifetime τ_n is controlled by zinc atoms when the rate of capture of electrons by the N_2 levels is at least 10 times higher than the rate of capture by the N_1 levels, i.e., when

$$\frac{p_{k2} S_{n2}}{p_{k1} S_{n1}} \geqslant 10.\tag{5}$$

The ratio N_1/N_2 can be found from Eqs. (5) and (4), bearing in mind that during recombination $p_{k2} \rightarrow N_1$ and using $S_{n2} = 10^{-20}$ cm^2, $S_{p2} = 10^{-16}$ cm^2 [8, 9], $S_{n1} = 10^{-16}$ cm^2 [6]. This ratio is equal to 0.1. The concentration N_1 is usually within the limits $5 \cdot 10^{12} < N_1 \leq 5 \cdot 10^{13}$ cm^{-3}. Consequently, the concentration of zinc should be of the order of 10^{14} cm^{-3} and the value of τ_n should be 50 msec.

Crystals with the required concentration of zinc, satisfying the condition (1), were prepared by doping the melt and by diffusion annealing. The photosensitivity obtained by these two methods was practically the same.

Photoresistors obtained in this way were sensitive in the wavelength range 0.5-2 μ at T = 300°K. The sensitivity maximum was located at 0.8-0.9 μ. The photosensitivity in the impurity-absorption region of the spectrum was approximately 100 times lower than in the region of the maximum. At 293°K, the main parameters of the photoresistors were as follows: the threshold sensitivity, measured at 100 cps at the sensitivity maximum, was of the order of $5 \cdot 10^{-12}$ W $\cdot$ cps$^{1/2}$ $\cdot$ mm^{-2}; the voltage sensitivity was $\sim 10^5$ V/W; the relaxation time was 5 msec and the resistivity $\rho = 2 \cdot 10^3$ Ω $\cdot$ cm. The sensitivity rose by a factor of about 20 when the temperature was lowered to −60°C. The main parameters of these photoresistors were

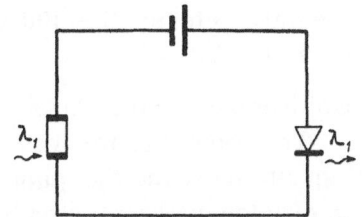

Fig. 2. Basic circuit for the amplification of light.

highly stable. The relaxation time differed by a factor of 2-3 from the majority-carrier lifetime. The voltage and threshold sensitivities of these devices were not inferior to those of better samples of PbS and their response was considerably faster (the photoresponse time constant of crystals such as PbS and CdS is usually 2-3 orders of magnitude longer than the lifetime). Thus, photoresistors made of zinc-compensated silicon were found to have a response time 2-3 orders of magnitude better than that of the more widely used photoresistors and this advantage should extend the frequency capability of modern optoelectronics.

3. Quantum Amplifier and Converter

A circuit consisting of a battery, a light-emitting diode (GaAs, SiC), and a photoresistor connected in series (Fig. 2), can have a quantum gain higher than unity.

We shall define the quantum gain k as the ratio of the number of quanta z_2 emitted by the diode to the number of quanta z_1 incident on the photoresistor. The value of z_1 is given by

$$z_1 = \frac{E}{\varepsilon_1}, \tag{6}$$

where E is the radiation power incident on the photoresistor and ε_1 is the energy of a single quantum. The number of quanta z_2 emitted by the diode is

$$z_2 = \frac{\Delta I}{e} \eta, \tag{7}$$

where ΔI is the photocurrent; η is the radiative recombination coefficient representing the number of emitted quanta per one electron; e is the electronic charge.

The change in the current flowing through the diode, caused by the illumination of the photoresistor, is

$$\Delta I = U \Delta\sigma \frac{vc}{a}, \tag{8}$$

where a, v, and c are the geometrical dimensions of the photoresistor, and

$$\Delta\sigma = e\Delta n\mu. \tag{9}$$

Under steady-state conditions, the excess electron density Δn can be written in the form

$$\Delta n = \tau g, \tag{10}$$

where

$$g = \frac{z_1 \zeta \beta}{avc}. \tag{11}$$

Here, β is the quantum efficiency and ζ is a coefficient which makes allowance for the reflected light not absorbed in the photoresistor. Using Eqs. (6)-(11), we obtain the following expression for k:

$$k = \frac{U\tau\eta\mu\zeta\beta}{a^2}. \tag{12}$$

The quantity $U\mu/a^2$ is the reciprocal of the carrier transit time. Assuming that $\tau/t_{tr} = G$, where G is the photoconductivity gain, Eq. (12) can be rewritten in the simple form

$$k = G\eta\zeta\beta. \tag{13}$$

It follows that the quantum gain is proportional to the photoconductivity gain G, the radiative recombination coefficient η, the quantum efficiency β, and the coefficient ζ. For example, the quantum gain is k = 8.4, for the following reasonable values of the parameters: U = 300 V, $\tau = 5 \cdot 10^{-3}$ sec, $\eta = 10^{-4}$, $\mu = 5 \cdot 10^3$ cm$^2 \cdot$ V$^{-1} \cdot$ sec^{-1}, $a = 0.3$ cm, $\beta = 1$, $\zeta = 1$.

For an optoelectronic pair consisting of a zinc-doped silicon photoresistor and a GaAs light-emitting diode, we obtained experimentally k = 7 for $\lambda = 0.85 \mu$. The photoresistor was kept at 77°K and the diode at 273°K. The cooling of the diode to 77°K should increase the quantum gain by about an order of magnitude because of an increase in the efficiency of emission by GaAs. The minimum converted and amplified power in such a system was 10^{-11} W $\cdot$ cps$^{1/2} \cdot$ mm^{-2}. The low-temperature range, 77-100°K, was not the optimum range from the point of view of the operation of the silicon photoresistor. At these temperatures, the photoresponse time increased strongly (up to 1 min) for intrinsic and extrinsic excitation; this would impose serious restrictions on its applications [12].

The temperature range 300-150°K was preferable for the operation of the photoresistor for two reasons: first, the electron lifetime and the photoresponse time were practically identical in this region; second, at these temperatures there was no photocurrent saturation associated with reduction of the majority-carrier lifetime in a strong field [10]. The second factor was particularly important in view of the use of strong fields (>1000 V/cm) which are required in order to obtain a high value of k.

The transmission band of a photoresistor amplifier, operating at 300°K in conjunction with a GaAs light-emitting diode, can be estimated from the formula

$$G \Delta B = \frac{1}{2\pi \tau_{\text{rel}}}, \tag{14}$$

where ΔB is the transmission band; G is the photoconductivity gain; τ_{rel} is the dielectric relaxation time, equal to the transit time of carriers in fields corresponding to the onset of flow of a space-charge-limited current. For $G = 10^2$, $\tau_{\text{rel}} = 10^{-6}$ sec ($\rho = 10^6 \Omega \cdot$ cm), we obtain $\Delta B \approx 1$ kc.

Periodic oscillations of the current, which appear in a silicon photoresistor subjected to fields exceeding 300 V/cm at 77°K [11], can be used for low-frequency modulation of infrared radiation quanta in an amplifier. Such a device not only amplifies radiation of the $\lambda = 0.86 \mu$ wavelength but it also modulates this radiation at a frequency of 1-10 sec^{-1}. The present author developed a prototype quantum amplifier with low-frequency modulation: the maximum current density was 0.6 A/cm^2, the minimum current density was 0.04 A/cm^2, and the gain was 10.

An optoelectronic pair, consisting of a silicon photoresistor and a GaAs electroluminescent diode, can be used as the basis of many devices: quantum amplifiers, infrared radiation converters, optrons, optoelectronic switches, memory elements, and optoelectronic relays.

4. p − i − n Drift Photodiode

A device with promising applications is the p−i−n photodiode. The p−i−n junction is due to formation of a space charge of ionized zinc acceptors in p- and n-type silicon. The p−n junction is obtained by the diffusion of zinc into phosphorus-doped silicon.

In contrast to ordinary silicon photodiodes, the p−i−n diode has a spectral characteristic with a nearly flat response in the wavelength range 0.45-0.80 μ (Fig. 3). The short-wavelength fall of the sensitivity is suppressed by a drift field, which appears in the p-type region because of the presence of a hole density gradient. The internal field factor, given by βW, where $\beta = \frac{1}{p(x)} \cdot \frac{dp}{dx}$, and W is the distance from the surface of the p-type region to the p−n junction, can be of the order of 10-100. Electrons generated in a surface layer about 10 μ thick

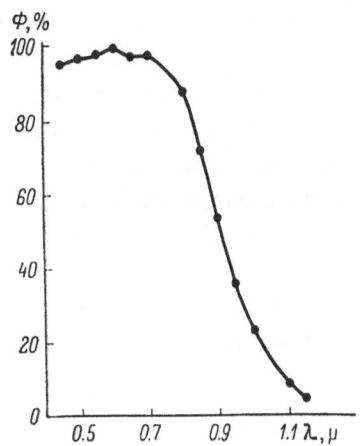

Fig. 3. Spectral characteristic of a silicon p−i−n
photodiode.

are extracted by the internal field before they have a chance to recombine. This increases the contribution of electrons to the total current and "rectifies" the spectral dependence of the sensitivity. The contribution of the hole component, which represents the absorption in the bulk of the p−i−n device, can be controlled by varying the value of W.

The integrated photosensitivity of such a photodiode is 6-8 mA/lm and it approaches the maximum sensitivity of other drift photodiodes. A silicon photodiode has a relatively low noise level due to relatively long minority carrier life-times in the p- and n-type regions. The electron (S_n^0) and hole (S_p^{2-}) capture cross sections in the relevant regions of the junction are of the order of 10^{-16} cm² [5, 6, 9], i.e., they are considerably smaller than the carrier-capture cross sections of gold-compensated silicon. The low value of the dark cur-rent (1-3 μA) of high-resistivity compensated p−i−n struc-tures is due to the presence, in the p-type layer, of an internal field which limits the electronic component of the current.

When the geometrical dimensions of the p−n junctions are reduced, a photodiode of this type can, in principle, operate at higher frequencies (100-1000 Mc). Another definite advantage of a silicon phototiode is its ability to operate at high reverse voltages (U ≥ 50 V).

A phototiode with these characteristics can find many applications. It should be par-ticularly useful in the detection of visible laser radiation. Another interesting application would be its use as a collector in an electrooptical transistor. The electroluminescent emitter of such a transistor can be made of gallium arsenide or silicon carbide. A photodiode of this type can be useful also in circuits with high-voltage power sources. Moreover, photodiodes of this type can be employed in multielement matrices for feeding numerical data into digital com-puters.

5. Other Possible Applications of Zinc-Doped Silicon

Optical pumping (charge exchange) in zinc-doped silicon can be used in memory elements with a sensitivity range extending up to 2 μ.

Another interesting application is the use of p-type silicon in new highly sensitive ther-mistors operating in the low temperature range 260-77°K [16]. The parameters and stability of these thermistors seem to be most suitable for operation in this range of temperatures. A prototype thermistor developed by the present author and Kontsevoi [16] had the following char-acteristics: temperature sensitivity coefficient $\alpha = (1/R)(dR/dT) = 7.1\%$ deg⁻¹, B = 2.303 Δ × lg R/Δ(1/T) = 5000, time constant 3-4 sec. A thermistor of this type has better characteristics and is more easily prepared than thermistors made of polycrystalline powders, which are nor-mally used at low temperatures.

A strong rise of the current by 3-9 orders of magnitude, observed at 77°K in thin p⁺−p−p⁺ structures [17], can be used in nonlinear elements.

Conclusions

Without attempting an exhaustive treatment, we have listed some silicon devices which seem to be of interest. It is quite likely that new, basically different, devices will be developed using other properties of this material. However, even the examples given here show that zinc-compensated silicon is a very promising material in the development of optoelectronics and electronic devices.

Zinc-doped silicon has advantages in many applications, including those involving changes in its properties, not only because it is such a well-known material but also because it has many advantages. Let us list some of them: 1) the large diffusion coefficient of zinc in silicon amounting to 10^{-6}-10^{-7} cm^2/sec at 1000-1200°C; 2) the relatively high solubility of electrically active zinc atoms in silicon, which is $5 \cdot 10^{16}$ cm^{-3} at 1350°C; 3) the stability of the energy levels of zinc; 4) the possibility of using silicon in the integrated circuit technology; 5) the compatibility of the spectral characteristics of silicon with those of light-emitting GaAs and SiC diodes; 6) the wide range of the majority and minority carrier lifetimes (10^{-10}-10^{-2} sec). The exceptionally promising applications of zinc-compensated silicon in optoelectronic devices are, to a large extent, due to the multiply charged behavior of the zinc dopant.

Literature Cited

1. C. S. Fuller and F. J. Morin, Phys. Rev., 105:379 (1957).
2. R. O. Carlson, Phys. Rev., 108:1390 (1957).
3. B. V. Kornilov, Fiz. Tverd. Tela, 5:3305 (1963).
4. K. D. Glinchuk and N. M. Litovchenko, Fiz. Tverd. Tela, 5:3150 (1963).
5. B. V. Kornilov, Fiz. Tverd. Tela, 7: 3458 (1965).
6. B. V. Kornilov, Fiz. Tverd. Tela, 8:201 (1966).
7. B. V. Kornilov, Fiz. Tverd. Tela, 6:3721 (1964).
8. B. V. Kornilov, Fiz. Tverd. Tela, 7:1795 (1965).
9. B. V. Kornilov and S. E. Gorskii, Fiz. Tekh. Poluprov., 2:262 (1968).
10. B. V. Kornilov and A. V. Anfimov, Fiz. Tekh. Poluprov., 1:340 (1967).
11. B. V. Kornilov and A. V. Anfimov, Fiz. Tverd. Tela, 8:3420 (1966).
12. Yu. I. Zavadskii and B. V. Kornilov, Fiz. Tekh. Poluprov., 1:1326 (1967).
13. B. V. Kornilov, Fiz. Tverd. Tela, 6:331 (1964).
14. B. V. Kornilov, Fiz. Tverd. Tela, 6:3331 (1963).
15. B. V. Kornilov and Yu. A. Kontsevoi, Author's Certificate No. 953,984, September (1963).
16. B. V. Kornilov and Yu. A. Kontsevoi, Author's Certificate No. 952,954, June (1963).
17. B. V. Kornilov and V. G. Emuranov, Author's Certificate No. 954,045, September (1963).

INVESTIGATION OF ELECTRICAL
AND LUMINESCENT PROPERTIES OF
DIFFUSED AND EPITAXIAL p–n JUNCTIONS IN
ALUMINUM-DOPED SILICON CARBIDE

V. I. Pavlichenko and I. V. Ryzhikov

An investigation was made of the current–voltage, luminescence intensity–voltage, capacitance–frequency, luminescence intensity–current, and relaxation characteristics of p–n junctions in silicon carbide. These junctions were prepared by the diffusion of aluminum from the gaseous phase or from solutions in rare-earth elements, as well as by the epitaxial growth of aluminum- or nitrogen-doped p- and n-type SiC films on SiC crystals of the opposite type of conduction (the crystals were also doped with nitrogen and aluminum). A comparison of these characteristics for high-resistivity and low-resistivity samples showed that the junctions in the low-resistivity samples were abrupt and the compensated regions in these samples were thin and had a relatively low resistivity. The fast response, high luminescence efficiency at current densities ≥ 20-50 A/cm^2, and the low noise of the low-resistivity diffused and epitaxial p–n junctions in aluminum-doped silicon carbide should be very useful in their applications as sources of nanosecond light pulses.

Investigations of the rectifying and luminescent properties of silicon carbide p–n junctions were reported in [1-4]. These junctions were prepared by the diffusion of boron either on its own or together with aluminum, as well as by the alloying of the silicon carbide with silicon–aluminum mixtures.

An analysis of the rectifying, capacitative, and luminescent characteristics of the diffused p–n junctions showed that high-resistivity compensated layers were formed near the p–n junction. These layers influenced strongly the junction characteristics and their special features: wide power-law regions in the forward branches of the current–voltage and in the luminescence intensity–voltage characteristics, a sublinear dependence of the luminescence intensity on the current flowing through the junction, a rapid rise of the voltage drop with decreasing temperature, a weak dependence of the reverse current on the bias, and a considerable temperature dependence of the reverse current. All these features were manifested less clearly or were altogether absent in the characteristics of the alloyed p–n junctions which did not have high-resistivity compensated layers.

357

The present paper describes an investigation of the rectifying, luminescent, capacitative, and time characteristics of diffused and epitaxial p−n junctions whose p-type regions were doped with aluminum (the thermal activation energy of aluminum is considerably lower than that of boron). In contrast to boron-doped junctions, the electroluminescence emitted by aluminum-doped silicon carbide was green and blue.

Samples and Measurement Method

We investigated p−n junctions in silicon carbide. These junctions were prepared by growing aluminum-doped films of p-type SiC on n-type SiC crystal substrates. These films were grown from solutions in the following rare-earth metals: dysprosium, gadolinium, thulium, or ytterbium. Other epitaxial junctions were prepared by growing crystallized n-type films on aluminum-doped p-type SiC crystals. Moreover, we studied junctions prepared by the diffusion of aluminum from the gaseous phase and from solutions in rare-earth metals.

The technology of fabrication of epitaxial and diffused junctions from aluminum-doped silicon carbide was identical with that described in [5, 6]. The diffusion of aluminum and growth of aluminum doped films were carried out in high-density pyrolytically-coated graphite crucibles under conditions preventing the penetration of boron vapor.

Crystals of n-type silicon carbide containing 10^{17}-10^{19} cm^{-3} of neutral nitrogen atoms were used as the starting material. The capacitance of the p−n junctions was measured by a bridge method.

The electroluminescence decay was investigated using an electrometer amplifier whose output voltage was proportional to the amplitude of the electrical pulses ariving at its input and, therefore, this voltage was proportional to the intensity of luminescence at a given moment.

Results of Measurements and Discussion

Figure 1 shows typical current−voltage characteristics of junctions prepared by short-duration (0.2-1 h) diffusion of Al from the gaseous phase (1) or from solutions in rare-earth metals (2), and by short-duration (0.1-0.3 h), "low-temperature" (1800-1900°C) growth of aluminum-doped p-type films on n-type SiC crystals doped fairly heavily ($N_d − N_a = 8 \cdot 10^{17}$ to 10^{19} cm^{-3}) with nitrogen atoms (3). The forward voltage drop across such junctions (type A) is 2.6-3 V for a current density of $5 \cdot 10$ A/cm^2 and the breakdown occurs at reverse voltages of 8-15 V (breakdown is accompanied by a rapid rise of the reverse current).

When lightly doped crystals (concentration of neutral nitrogen atoms $\leq 6 \cdot 10^{17}$ cm^{-3}) are used and when the duration of diffusion of aluminum is increased to 3-6 h, the resultant p−n junctions (type B) have much higher values of the forward voltage drop (up to 4-10 V) and they break down at reverse voltages up to 25-100 V (curve 4 in Fig. 1). The reverse currents through these junctions increase relatively slowly with increasing reverse voltage. Similar current−voltage characteristics are exhibited also by p−n junctions prepared by the prolonged growth (0.5-1.5 h) of aluminum-doped p-type films at high temperatures (1950-2100°C) on lightly doped crystals (curve 5 in Fig. 1).

Two types of junction, with current−voltage characteristics similar to those of type A and B junctions, are also obtained by growing recrystallized n-type films on aluminum-doped p-type SiC crystals: these junctions are similar to those of type A and B in respect of the effect on the current−voltage characteristics of the growth temperature and duration and of the resistivities of the substrates and films.

The forward branches of the current−voltage and luminescence intensity−voltage characteristics of the low-resistivity p−n junctions (type A) exhibit − over a considerable range of

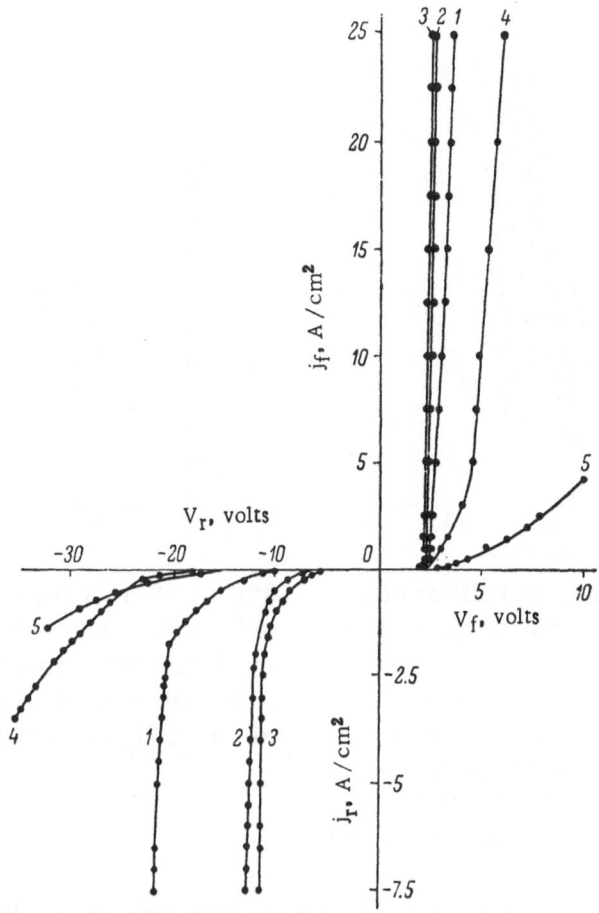

Fig. 1. Current—voltage characteristics of low-
resistivity (1-3) and high-resistivity (4, 5) diffused
and epitaxial p—n junctions in aluminum-doped
silicon carbide.

the currents (up to 3-4 orders of magnitude) — an exponential region with a slope equal to qV/bkT
(Fig. 2). The value of the factor $b \approx 2$ remains constant over the whole investigated range of
temperatures (200-500°K). The saturation current of these p—n junctions is $(2-10) \cdot 10^{-2} A/cm^2$
at room temperature.

Using these values of the saturation current and the factor b, as well as the observation
that this factor was independent of temperature, we find that the current in the exponential regions of the characteristics is controlled by carrier recombination in the space-charge layer
of the p—n junction or in the neutral part of the compensated region [7-9].

In addition to a region characterized by a factor $b \approx 2$, many samples, particularly the
lowest resistivity p—n junctions (2.3-2.5 V for I = 10 mA and $s = 0.25 \cdot 10^{-2} cm^2$), have a
steeper exponential region with a factor b = 1.4-1.8 (curve 1 in Fig. 3). These samples exhibit
a clear transition from the "Hall" exponential region to the steeper region at current densities
of $(1-5) \cdot 10^{-2} A/cm^2$. The appearance of this steep dependence in "short" p—n junctions in
silicon, prepared by ion implantation, is attributed in [10] to carrier recombination not only in
the compensated region of relatively high resistivity near the p—n junction but also in the
heavily doped region (p^+-type layer of the p—n junctions investigated in [10]). The value of the
factor b is governed by the ratio W/L_p of the compensated region and, in the limiting case of
"short" diodes ($W/L_p \ll 1$), this factor decreases to 1 [10]. Moreover, it is reported in [10]
that the value of the current at the beginning of the second region (followed by the power-law

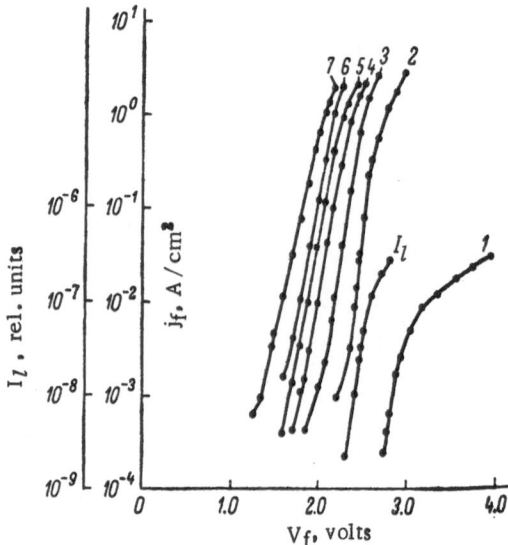

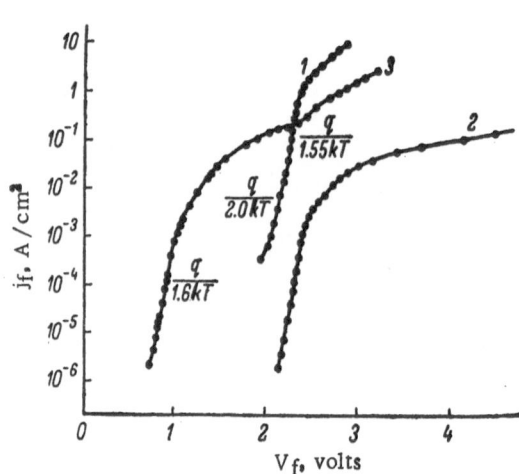

Fig. 2. Forward current−voltage characteristics of a diffused p−n junction in
aluminum-doped silicon carbide, determined in the temperature range 77-
488°K (1-7), and luminescence inten-
sity−voltage characteristic of the same
junction at 293°K (I_l). T, °K: 1) 77; 2)
213; 3) 293; 4) 368; 5) 420; 6) 463; 7) 488.

Fig. 3. Forward current−voltage characteristics of low-resistivity epitaxial
(1, 2) and diffused (3) p−n junctions in
aluminum-doped silicon carbide, recorded
at 300°K (1, 3) and 77°K (2).

dependence) of the current−voltage characteristic is proportional to the conductivity of the p^+-
type region, which depends on the implanted-ion dose.

The power-law dependence for the silicon p−n junctions [10] is $I \propto (V − V_0)^n$, and n increases from 1 to 2 when the implanted-ion dose in the p^+-type layer is increased. A similar
dependence with a slope n = 1 is observed for our p−n junctions in aluminum-doped silicon
carbide (curve 1 in Fig. 4). This slope is observed throughout the investigated temperature
range (77-500°K). When the temperature is lowered to that of liquid nitrogen, the ohmic (n ≈ 1)
region becomes more prominent (curve 2 in Fig. 4). The linear dependence of the current on
the voltage is probably due to predominance of the ohmic conduction in the contact layer [10]
because of the relatively low level of the aluminum doping in the p^+-type layer in our samples
when the conductivity of this layer does not differ greatly from that of the compensated layer.

The power-law (usually ohmic) region predominates also at room temperature in the B
p−n junctions, beginning from current densities of $(1-5) \cdot 10^{-3}$ A/cm^2 (curve 3 in Fig. 4). This
is due to the formation of transition layers of relatively high resistivity near these p−n junctions so that the beginning of the power-law (ohmic) region shifts by 2-3 orders of magnitude
in the direction of lower current densities. Such junctions exhibit, in addition to their linear
dependence of the current on the voltage, dependences characterized by values of n between 1
and 2. The limit n = 2 is usually reached when recombination of the minority carriers is dominant in the heavily doped regions because of the leakage of these carriers through the $p^+−i$
and $i−n^+$ junctions. This happens when the voltage across a diode is governed mainly by the
drop across the compensated transition layer and across the space-charge regions of these
junctions. The intermediate values, n = 1.4-1.6, are due to the simultaneous influence of the
voltage drop in the contact layer and in the high-resistivity compensated layer.

An interesting experimental observation is the presence, in the forward current−voltage
characteristics of some low-resistivity p−n junctions, of an exponential region with a factor

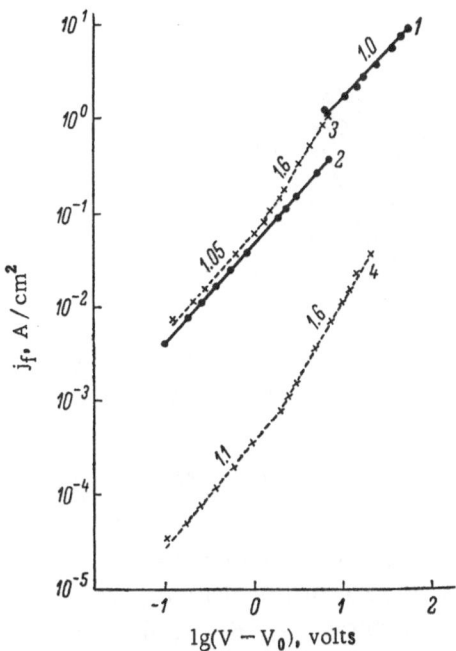

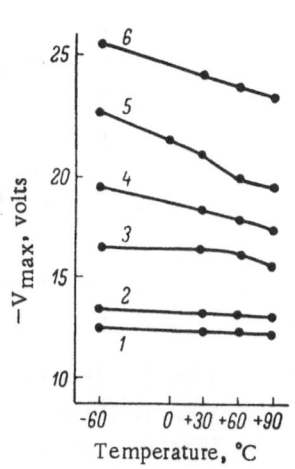

Fig. 4. Forward current−voltage characteristics in the power-law region of low-resistivity (1, 2) and high-resistivity (3, 4) p−n junctions prepared by the diffusion of aluminum. T, °K: 1, 3) 293; 2, 4) 77.

Fig. 5. Temperature dependence of the breakdown voltage of low-resistivity (1-3) and high-resistivity (4-6) p−n junctions.

b = 1.3-2.3 at voltages of 0.7-1.0 V across the junction (curve 3 in Fig. 3), i.e., at voltages considerably lower than the contact potential difference. This region may be due to the tunneling of carriers across a very thin transition layer [11].

The reverse currents through junctions of type A are low (curves 1-3 in Fig. 1) but they increase rapidly when the voltage is increased up to the breakdown value. The temperature dependence of the breakdown voltage of these junctions is relatively weak (curves 1-3 in Fig. 5). On the other hand, junctions of type B exhibit a weaker dependence of the reverse current on the voltage (curves 4 and 5 in Fig. 1) but the breakdown voltage of these junctions varies more strongly with the temperature (curves 5 and 6 in Fig. 5). The reverse current−voltage characteristics of type A junctions, plotted on a double logarithmic scale (Fig. 6), usually have two linear regions with the slopes m = 2.5-3.0 and m = 4-6 at room temperature ($I_r \propto \log V_r$). When the temperature is increased, the exponent m decreases to 2.0 and 4-4.5, respectively, and an ohmic region (m = 1.0) appears in the characteristics. The quadratic dependence of the reverse current on the applied voltage is typical of the space-charge-limited flow of current in a high-resistivity region near a p−n junction [12]. At temperatures above 450-550°K a region with a slope m = 0.6-0.8 appears in the characteristics and this region approaches m = 0.5 at still higher temperatures; this is typical of the generation−recombination mechanism of the reverse current in the space-charge region of an abrupt p−n junction, predicted by the theory of Shockley, Noyce, and Sah [7]. When the temperature is reduced to that of liquid nitrogen, only one region is observed and in this region the reverse current increases rapidly with the voltage, the slope being m = 5-7 (curve 1 in Fig. 6).

The thermal activation energy of the reverse current in the ohmic region is 0.14-0.15 eV at low temperatures and 0.46-0.52 eV at high temperatures (Fig. 7). The first of these values is in approximate agreement with the thermal activation energy of nitrogen in silicon carbide

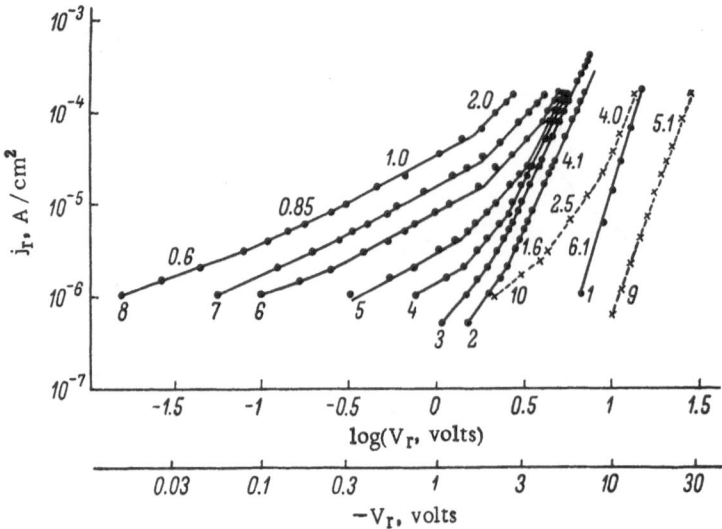

Fig. 6. Reverse current—voltage characteristics of low-resistivity (1-8) and high-resistivity (9, 10) diffused p—n junctions in aluminum-doped silicon carbide. T, °K: 1, 9) 77; 2) 213; 3, 10) 293; 4) 348; 5) 398; 6) 423; 7) 488.

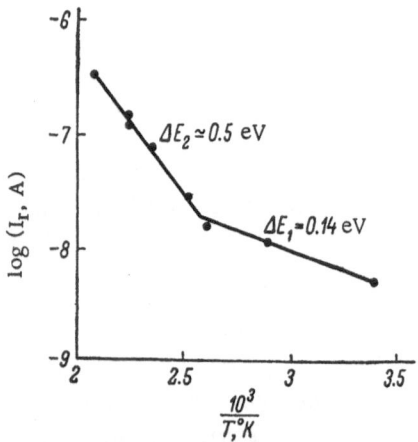

Fig. 7. Temperature dependence of the reverse current flowing through a low-resistivity diffused p—n junction in aluminum-doped silicon carbide in the linear part of the current—voltage characteristic.

crystals [13] but the nature of the impurity centers responsible for the second activation energy is not known.

The reverse current—voltage characteristics of the high-resistivity p—n junctions have a region with the slope m = 1.5-2 (curve 10 in Fig. 6) at room temperature. Such a dependence is usually observed in silicon carbide diodes with a relatively thick compensated transition region [14]. The more rapid rise of the reverse current observed for the low-resistivity diffused and epitaxial junctions at room and lower temperatures probably indicates that the thickness of the compensated layer is not great and that additional ionization of impurities by the strong field takes place in the space-charge region (these impurities are not fully ionized by the thermal mechanism at such temperatures). At higher temperatures, impurity ionization is practically complete and the slope m in the first two regions of the reverse characteristic becomes, respectively, 1 and 2. This is in agreement with the theoretically expected values for the ohmic region and for the space-charge-limited region.

A comparative investigation of the capacitance-frequency characteristics of low- and high-resistivity p—n junctions fully supports the conclusions drawn in the preceding paragraphs. The specific capacitances of the low-resistivity p—n junctions in aluminum-doped silicon carbide are relatively large, reaching values of 0.05-0.1 μF/cm^2 and 0.3-0.8 μF/cm^2 for the diffused and grown junctions. The specific capacitances of the high-resistivity junctions are 1-2 orders of magnitude lower. Moreover, the fall of the capacitance of the low-resistivity p—n junctions (curves 1 and 2 in Fig. 8) appears at much higher values of the frequency (10^6-10^7 cps) than the

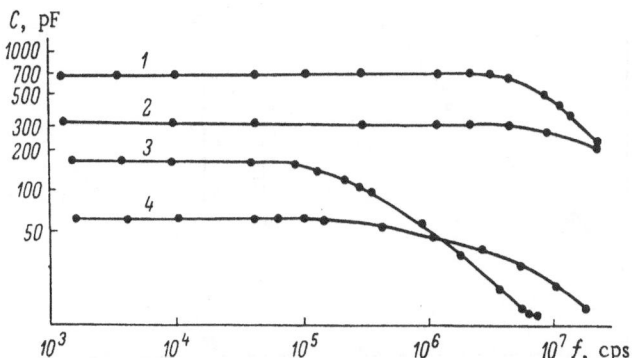

Fig. 8. Dependences of the capacitance on the frequency for low-resistivity (1, 2) and high-resistivity (3, 4) epitaxial (1, 3) and diffused (2) p−n junctions.

corresponding fall for the high-resistivity junctions, which is observed at 10^4-10^5 cps (curves 3 and 4 in Fig. 8). The capacitance−frequency characteristics of the high-resistivity junctions are similar to the corresponding characteristics of the junctions prepared by the diffusion of boron on its own or with aluminum. It is known [15] that the fall of the measured capacitance of the p−n junctions is due to the shunting influence of the capacitances of the high-resistivity compensated layers adjoining the p−n junctions.

The capacitance of the majority of the low-resistivity junctions depends on the voltage, in accordance with the formula $C \propto (\varphi_c - V)^\alpha$, where α varies from -0.42 to -0.50. On the average, the dependence is weaker for the diffused junctions and stronger for the epitaxial junctions. The voltage intercept of the capacitance−voltage characteristics $C_1^{-2} = f(V)$ of the low-resistivity p−n junctions is 2.4-2.7 V (Fig. 9a). This value, which is close to the contact potential difference of the p−n junctions, is in good agreement with the voltage intercept deduced from the forward branches of the current−voltage characteristics plotted on the linear scale. The ionized impurity concentrations near the p−n junctions, calculated from the capacitance−voltage characteristics using the well-known formula

$$N_0 = \frac{1.44 \cdot 10^8 \left(\frac{C}{S}\right)^2}{\epsilon' (\varphi_c - V)},$$

are $(0.15$-$1.5) \cdot 10^{19}$ cm^{-3}. These values are in approximate agreement (they are slightly larger) with the concentrations of neutral nitrogen atoms $N_d - N_a$, determined in preliminary measurements from the absorption spectra.

At a frequency of 10^3 cps, the capacitance of the majority of the high-resistivity junctions (Fig. 9b) varies in accordance with the law $C_1^{-3} \propto (\varphi_c - V)$, which indicates a linear distribution of the diffused aluminum with a concentration gradient of 10^{21}-$5 \cdot 10^{23}$ cm^{-4} in the p−n junction region.

On the other hand, the capacitance−voltage characteristics and the contact potential differences of the low-resistivity junctions suggest that these junctions are abrupt.

This conclusion is confirmed by the results of an investigation of oblique sections of the low-resistivity junctions, which are shallow (0.1-0.2 μ below the surface) and have an abrupt change in the distribution of the potential at the p−n junction boundary. Anodic oxidation of these junctions, revealed by oblique sections, also indicates that the structure is abrupt and not

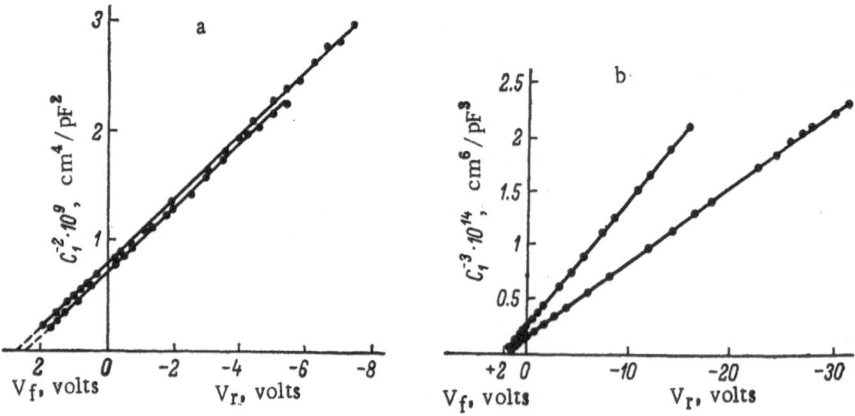

Fig. 9. Capacitance−voltage characteristics of low-resistivity (a) and high-resistivity (b) p−n junctions in aluminum-doped silicon carbide.

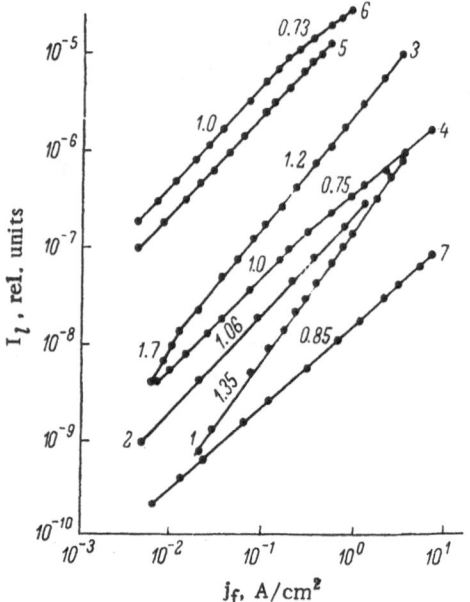

Fig. 10. Dependences of the luminescence intensity on the current density for low-resistivity (1-6) and high-resistivity (7) p−n junctions. T, °K: 1-4, 7) 293; 5-6) 77.

stepped (a stepped structure, typical of p−n junctions prepared by the diffusion of boron, is due to the presence of a thick high-resistivity transition region).

The room-temperature dependences of the luminescence intensity on the current density are linear or superlinear for the majority of the p−n junctions in aluminum-doped silicon carbide (curves 1-4 in Fig. 10): $I_l \propto j_f^n$ (n = 1.0-1.35) in the investigated current density range from 10^{-4}-10^{-3} to $5 \cdot 10^{-1}$ to $5 \cdot 10$ A/cm². At low current densities, a superlinear region with a slope n = 1.5-2 is observed quite frequently. When the temperature is lowered to that of liquid nitrogen, the linearity of the intensity−current characteristics is obeyed very closely (n = 1.00-1.05) in the current density range from $(1-5) \cdot 10^{-3}$ to $(1.5-3) \cdot 10^{-1}$ A/cm².

In a narrow range of currents, corresponding to the power-law (ohmic) region of the forward current−voltage characteristics, the intensity−current dependences of the low-resistivity p−n junctions are sublinear with a slope n = 0.75-0.8 (curves 4 and 6 in Fig. 10), in close agreement with the corresponding conductivity−current dependences $\sigma \propto I^n$ of silicon p−n junctions. It is reported in [8, 9] that, in the case of silicon, $p \propto I^{0.5}$ near a p^+−i junction and $p \propto I^{1.0}$ near an i−n^+ junction, in agreement with the theoretically expected dependence of the injected hole density on the current $p \propto I^{0.75}$. It is likely that the same mechanism is responsible for the sublinear intensity−current dependences with n = 0.75-0.85 (curve 7 in Fig. 10) observed for our high-resistivity p−n junctions throughout most of the investigated range of the currents, with the exception of low current densities (up to $5 \cdot 10^{-5}$ to $5 \cdot 10^{-4}$ A/cm²), where the current−voltage characteristics are exponential. The sublinear dependence observed for our high-resistivity junctions is due to the higher resistance of the p^+-type region in these junctions compared with the corresponding region in the low-

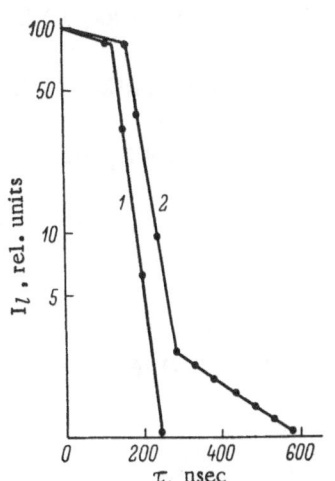

Fig. 11. Electroluminescence
relaxation of low-resistivity
(1) and high-resistivity (2)
p—n junctions in aluminum-
doped silicon carbide.

resistivity p—n junctions. The low-resistivity diffused and epitaxial junctions in aluminum-doped silicon carbide exhibit, because of their nearly linear intensity—current dependences, a high electroluminescence efficiency at high current densities [$\geq$ (2-5) $\cdot$ 10 A/cm^2]: the efficiency is 3-4 times higher than that of the best alloyed silicon carbide light sources.

The relaxation characteristics of the electroluminescence also confirm the suggested structure of the junctions. The low-resistivity p—n junctions in aluminum-doped silicon carbide have short rise (1-3 nsec) and decay (3-10 nsec) times of the light pulse. The relaxation characteristics of these junctions (curve 1 in Fig. 11) have only one steep exponential region whose duration does not exceed 0.1 μsec (the resolution limit of the apparatus) whereas the high-resistivity p—n junctions have additionally one or two slow-decay exponential regions (curve 2 in Fig. 11). The higher conductivity of silicon carbide in the compensated n-type region of the low-resistivity junction is responsible for the differences observed in the light-pulse decay times and in the structure of the relaxation characteristics of the low-resistivity and high-resistivity junctions. When a current pulse passes through a junction, the electron quasi-Fermi level is shifted upward in the compensated region near the junction. Consequently, the originally empty nitrogen levels are filled with electrons and, after the end of the current pulse, the electroluminescence decay time is governed by the probability of electron transitions from the nitrogen levels to the conduction band. The low-resistivity junctions in aluminum-doped silicon carbide are characterized by relatively high aluminum concentration gradients near the p—n junctions and therefore the compensated n-type regions near these junctions have a higher conductivity when the shallow nitrogen levels are capable of trapping electrons. On the other hand, additional exponential regions appear in the decay of the electroluminescence of the high-resistivity junctions when the impurity concentration gradient is lowered and, therefore, the conductivity of the compensated region is reduced.

The fast response and high efficiency of the electroluminescence, observed under pulsed conditions at high current densities in the low-resistivity diffused and epitaxial p—n junctions in aluminum-doped silicon carbide, make these junctions superior to the most efficient alloyed silicon carbide and gallium phosphide light sources. Moreover, the blue and green luminescence is characterized by a high value of the signal/noise ratio, which is important in pulse techniques.

Literature Cited

1.   É. E. Violin and G. F. Kholuyanov, Fiz. Tverd. Tela, 6:1696 (1964).
2.   G. F. Kholuyanov, Fiz. Tverd. Tela, 6:3336 (1964).
3.   Yu. S. Blank, Yu. A. Vodakov, and A. A. Mostovskii, Fiz. Tverd. Tela, 5:2228 (1963).
4.   V. I. Pavlichenko, I. V. Ryzhikov, T. G. Kmita, P. M. Karageorgii-Alkalaev, and A. Yu. Leiderman, Fiz. Tverd. Tela, 8:1239 (1966).
5.   Yu. S. Krasnov, T. G. Kmita, I. V. Ryzhikov, V. I. Pavlichenko, O. T. Sergeev, and Yu. M. Suleimanov, Fiz. Tverd. Tela, 10:1140 (1968).
6.   V. I. Pavlichenko and I. V. Ryzhikov, Fiz. Tekh. Poluprov., 2:1644 (1968).
7.   C. T. Sah, R. N. Noyce, and W. Shockley, Proc. IRE, 45:1228 (1957).

8.   I. V. Ryzhikov, V. I. Pavlichenko, and T. G. Kmita, Radiotekhnika i Élektronika, 12:848
     (1967).

9.   A. Yu. Leiderman and P. M. Karageorgii-Alkalaev, Radiotekhnika i Élektronika, 10:720
     (1965).

10. V. M. Gusev, V. I. Kurinnyi, I. I. Kruglov, I. V. Ryzhikov, B. V. Sestroretskii, and Yu. A.
     Sin'kov, Fiz. Tekh. Poluprov., 3:903 (1969).

11. T. N. Morgan, Phys. Rev., 148:890 (1966).

12. L. Patrick, J. Appl. Phys., 28:765 (1957).

13. G. A. Lomakina, Fiz. Tverd. Tela, 7:600 (1965).

14. V. I. Pavlichenko, I. V. Ryzhikov, and T. G. Kmita, this volume, p. 115.

15. C. T. Sah and V. G. K. Reddi, IEEE Trans. Electron Dev., ED-11(7):345 (1964).